Textbook of Developmental Biology

Textbook of Developmental Biology

Editor: Leonard Roosevelt

CALLISTO
REFERENCE
www.callistoreference.com

Callisto Reference,
118-35 Queens Blvd., Suite 400,
Forest Hills, NY 11375, USA

Visit us on the World Wide Web at:
www.callistoreference.com

ISBN: 978-1-63239-810-9 (Hardback)

Cataloging-in-publication Data

Textbook of developmental biology / edited by Leonard Roosevelt.
   p. cm.
Includes bibliographical references and index.
ISBN 978-1-63239-810-9
1. Developmental biology. 2. Biology. I. Roosevelt, Leonard.
QH491 .T48 2017
571.8--dc23

Table of Contents

Preface

Developmental biology refers to study of the growth and development of plants and animals. The main aim of developmental biology is to examine embryonic development of animals and to differentiate stem cells in organisms. Along with these, it also incorporates a detailed study of regeneration and metamorphosis. This book traces the progress of this field and highlights some of its key concepts and applications. It will also provide interesting topics for research which readers can take up. For all readers who are interested in this area of study, the case studies included in this text will serve as excellent guide to develop a comprehensive understanding. This book is a valuable compilation of topics, varying from the basic to the most complex advancement in the field of developmental biology. It aims to serve as a resource guide for students and experts alike and contribute to the growth of the discipline.

The world is advancing at a fast pace like never before. Therefore, the need is to keep up with the latest developments. This book was an idea that came to fruition when the specialists in the area realized the need to coordinate together and document essential themes in the subject. That's when I was requested to be the editor. Editing this book has been an honour as it brings together diverse authors researching on different streams of the field. The book collates essential materials contributed by veterans in the area which can be utilized by students and researchers alike.

Each chapter is a sole-standing publication that reflects each author's interpretation. Thus, the book displays a multi-facetted picture of our current understanding of applications and diverse aspects of the field. I would like to thank the contributors of this book and my family for their endless support.

Editor

There Is No Joy like Malicious Joy: Schadenfreude in Young Children

Simone G. Shamay-Tsoory[1]*, Dorin Ahronberg-Kirschenbaum[1], Nirit Bauminger-Zviely[2]

1 Department of Psychology, University of Haifa, Haifa, Israel, 2 School of Education, Bar Ilan University of Haifa, Haifa, Israel

Abstract

Human emotions are strongly shaped by the tendency to compare the relative state of oneself to others. Although social comparison based emotions such as jealousy and schadenfreude (pleasure in the other misfortune) are important social emotions, little is known about their developmental origins. To examine if schadenfreude develops as a response to inequity aversion, we assessed the reactions of children to the termination of unequal and equal triadic situations. We demonstrate that children as early as 24 months show signs of schadenfreude following the termination of an unequal situation. Although both conditions involved the same amount of gains, the children displayed greater positive expressions following the disruption of the unequal as compared to the equal condition, indicating that inequity aversion can be observed earlier than reported before. These results support an early evolutionary origin of inequity aversion and indicate that schadenfreude has evolved as a response to unfairness.

Editor: Marina Pavlova, University of Tuebingen Medical School, Germany

Funding: The authors have no support or funding to report.

Competing Interests: The authors have declared that no competing interests exist.

* Email: sshamay@psy.haifa.ac.il

Introduction

The developmental origins and proximate mechanisms behind social comparison based emotions are not well understood, despite recent progress (e.g. [1]). Social comparison based emotions involve two (or more) person situations in which one's emotions depends on the other's state [2]. The process of social comparison may trigger prosocial emotions such as empathy and compassion to the distress of others but also competitive emotions such as malicious joy or schadenfreude when facing others misfortune [3]. Schadenfreude is a relatively unstudied emotion and involves experiencing pleasure when another person faces an unfavorable event [4]. Schadenfreude is related to other competitive social comparison based emotions such as envy [5] and resentment [6] and it frequently arises in situations in which the target deserves the misfortune (e.g. [7,8]). Interestingly, while there is strong evidence for biological, evolutionary and developmental roots of prosocial empathically motivated helping behaviors (e.g. [9]) the evolutionary and developmental origins of schadenfreude are unknown.

One possibility is that schadenfreude, as well as other competitive social comparison based-emotions such as envy and jealousy, originally evolved, as a response to competition between rivals over limited resources. According to this notion, schadenfreude involves pleasure associated with gains in the context of limited resources. For example, siblings—who from conception are rivals for a parent's resources [10] may experience schadenfreude, as a response to a potential reward such as parental availability. Thus, the suffering of the sibling may be rewarding because it signals potential additional parental resources. Sibling rivalry is frequently reported in the animal kingdom, including sibling murder between baby eaglets and pelicans [11] or between

shark embryos [12], indicating that it has an evolutionary importance.

Similarly to sibling rivalry, mating rivalry may have evolved as a response to competition between same-sex individuals—who are rivals for mating partners. It has been shown that mating strategies in both men and women includes derogating other individuals as a basic mechanism for increasing self-attractiveness [13]. Based on these findings it has been proposed that schadenfreude is a psychological mechanism that responds to misfortunes that lower competitors' mate value in order to increase mating opportunities [14].

Thus, the sibling and mating rivalry accounts of schadenfreude may indicate that the distress of a rival (e.g. same-sex rival; sibling) is rewarding as it indicates a potential increase in resources such as parental attention or mating partners. The sibling and the mating rivalry accounts of schadenfreude are in line with the 'gain' hypothesis, according to which, schadenfreude is viewed as an emotion that originates from competition over limited recourses and therefore it involves a positive reaction to a potential gain during competition [15]. According to this theoretical formulation, pleasure, is a basic automatic reaction to positive rewards and malicious pleasure is the result of the potential reward rather than pleasure in the other's misfortune [15]. This suggests that schadenfreude involves a positive reaction to potential gains which may be unrelated to the suffering of the rival. Thus, if indeed, as suggested by the gain hypothesis, schadenfreude is a response to a potential gain regardless of a rivals' misfortune, than it should involve similar amounts of positive reactions in response to the termination of a competitive situation vs. a non-competitive situation if both situations involve similar amounts of gains.

Yet, an equal plausible hypothesis suggested here is that schadenfreude has evolved as a response to inequity aversion or the resistance to unfairness and inequalities. Inequity aversion predicts that individuals are sensitive to how their payoffs compare with those of others and therefore individuals may react negatively to unfair treatment [16]. According to this, schadenfreude may involve the pleasure of termination of an unpleasant unequal situation. Interestingly, it has been shown that inequity aversion develops early in children, further attesting to its evolutionary significance. Fehr, Bernhard, and Rockenbach [17] have reported that children at age 7–8 prefer resource allocations that remove advantageous or disadvantageous inequality. Other studies suggest that inequity aversion may be observed even before the age of five. It has been shown that children as young as four years old can judge situations to be undesirable based on concerns with fairness (for reviews, see [18,19,20]). In addition, Paulus, Gillis, Li, & Moore, [21] reported that preschool children involve third parties in dyadic sharing situations. Moreover, LoBue, Nishida et al., [22] have recently reported that even three years old children react negatively to disadvantageous inequality. Other reports show that even 15-month-old infants are sensitive to fairness and can engage in altruistic sharing [23].

That inequity aversion is evident early indicates that it has deep developmental roots. It has been suggested that negative reactions to an unequal reward distribution in regard to the effort invested may have been essential for the evolution of cooperation [24]. Indeed, negative reactions to inequalities have been reported not only in human adults but also in capuchin monkeys [25] and domestic dogs [26].

Considering the evolutionary significance of negative reactions to disadvantageous distribution, it is possible that schadenfreude has evolved as a positive reaction to the termination of inequity.

To test this hypothesis, in the current study we examined the emotional reactions to equal and unequal conditions in the distribution of parental attention in two and three years of children. Schadenfreude has been rarely reported in children and the only study that directly measured it reported signs of schadenfreude in 7 years old children which decreased with age [1].

It was reasoned that if schadenfreude is an emotion that originates from inequity aversion then the termination of an unequal condition should trigger more positive reactions as compared to the termination of an equal event even if the two conditions involve equal gains. Thus, we placed two years old children in a real social situation involving their mother and a peer. As opposed to previous studies which manipulated envy to provoke schadenfreude, in the current study we manipulated jealousy to elicit schadenfreude. Jealousy is the emotion children experience in a triadic situation, when there is a potentially unequal situation which raises a concern about losing exclusivity in significant relationships to a third party (e.g. [27]). In contrast, envy may involve only two-person situations, and this feeling comprises the wish to have another person's possession or success and/or the wish that the other person did not possess this desired characteristic or object [28]. Whereas envy and jealousy are somewhat different [5], these emotions are related and often co-occur [29,30] indicating that jealousy could equally be associated with schadenfreude.

The study included two main conditions each comprised of two phases. In each condition, in the first phase the mother read a book, while in the second phase the mother accidentally spilled water over the book. In the unequal condition (UNEQUAL) the mother read the book to the similar-aged peer (jealousy manipulation phase) while in the equal condition (EQUAL) the mother read the book to herself. We sought to examine if two- and three-year old children can show signs of schadenfreude following the termination of the jealousy phase (UNEQUAL) as compared to the control (EQUAL) condition.

Methods

The research has been approved by the University of Haifa Ethic committee. We contacted the parents of the children through ads, following the approval of the University of Haifa ethics committee. After obtaining written parental consent for participation, we advised the parents about the nature of the research by telephone. To reduce stress and use an ecologically valid environment, all experimental conditions were carried out in the home of the target child.

Participants

105 participants participated in the study. The participants included 35 triads including a mother (mean age = 35.486, SD = 4.461) her child (20 girls, 15 boys; mean age = 3.050, SD = 0.650) and a similar-aged peer (22 girls, 13 boys; mean age = 3.485, SD = 4.461).

Task

The EQUAL condition: Phase 1: story-reading scenario. Based on Bauminger, Chomsky-Smolkin et al. [31], the story-reading scenario included a triad comprising the target child, the mother and a peer who was a familiar preschool classmate. The session began with the mother sitting on a chair near a table on which a book and a glass of water were placed. The experimenter encouraged the two children to play with the age-appropriate toys and instructed the mother to ignore the children while completing a demographic questionnaire (2 min). As depicted in Figure 1a, upon the experimenter's signal, the mother took the book from the table and started reading the story aloud to herself (2 min).

The EQUAL condition: Phase 2: spilled water scenario. At the end of the 2 min, or if the target child showed substantial distress before that time, the mother was signaled to take the glass of water and accidently spill water over the book (Figure 1b).

The UNEQUAL condition: Phase 1: story-reading scenario. The session began with the mother sitting on a chair near a table on which a book and a glass of water were placed. Upon the experimenter's signal, the mother placed the peer on her lap and embraced the child while reading a story aloud to that child (Figure 1c).

The UNEQUAL condition: Phase 2: spilled water. At the end of the 2 min, or if the target child showed substantial distress before that time, the mother was signaled to take the glass of water and accidently spill water on the book(Figure 1d).

At the end of the experiment, the mother invited the target child to sit on her lap and hear the story.

The water spill manipulation was used to provoke schadenfreude as this emotion is frequently provoked following a misfortunate termination of a competitive situation [5]. Therefore, it was predicted that spilling water over the book following the unequal situation (reading the book to a peer) would provoke schadenfreude.

In the EQUAL situation (control condition) the mother took the book from the table and started reading the story aloud to herself, while in the UNEQUAL situation (experimental condition) the mother placed the peer on her lap and embraced the child while reading a story aloud to him/her. It should be noted that the control condition was designed to be as similar as possible to the experimental condition. It was reasoned that if the mother would

Figure 1. The EQUAL and the UNEQUAL conditions. In the EQUAL condition the mother reads a book aloud to herself while the kids are playing (Figure 1a) the mother is then signaled to take the glass of water and accidently spill water over the book (Figure 1b). In the UNEQUAL condition the mother placed the peer on her lap and embraced the child while reading a story aloud to that child (Figure 1c) and then she was signaled to accidently spill water on the book (Figure 1d). At both conditions the child were allowed to play freely.

not read the story aloud to herself than the children would not notice that she is reading a story. In both the experimental and control conditions both children could listen to the story. The only difference between the conditions was that in the experimental condition the peer was sitting on the mother's lap while she was reading the story.

Coding systems for jealousy and schadenfreude. Children's videotaped jealousy and schadenfreude-provoked behaviors, verbalizations, and affects were assessed using three coding scales: hierarchical explicitness of emotional reaction, quantity of jealousy and schadenfreurde behaviors and affect. These measures provide a comprehensive assessment of children's real-time emotional reaction and the amount of these reactions. The jealousy scales were the same scales used in previous studies [31,32,33] derived from the behaviors, verbalizations, and affects identified as jealousy indices by previous research (e.g. [34,35]). The schadenfreude ratings were novel and developed for the current study based on the validated jealousy ratings. Basically, the schadenfreude ratings were parallel and equivalent to the jealousy ratings.

Explicitness of the emotional response. The explicitness of the emotional responses in Phase 1 (story reading) and phase 2 (spilled water) of the EQUAL and UNEQUAL conditions were coded according to the following scales:

a) *Phase 1*
Hierarchical jealousy scale: This 7-point scale ranked explicitness of actions, verbalizations, and affective expressions of jealousy and in hierarchical order, from no interest at all (1) up to direct indication of the children's comparison and lack of equality, accompanied by negative affect (7) [see 31 for detailed descriptions of this scale,32]. Coders assigned the child the highest score evidenced over the 2-minute scenario. A score of 4 and above indicated explicit actions (e.g., pushing the rival aside and standing between the mother and

peer), verbalizations (e.g., "I want too"), and affects (e.g., shouting "Enough!") that reflected jealousy, whereas a score below 4 indicated only eye gaze in different degrees.

b) *Phase 2*
Hierarchical schadenfreude scale: This 7-point scale was based on the jealousy phase and ranked explicitness of actions, verbalizations, and affective expressions of schadenfreude in hierarchical order, from no interest at all (1) up to direct indication of the children's comparison and lack of equality, accompanied by positive affect (7). It is important to note that differently from the hierarchical jealousy scale here, the emotions that were coded were positive affect rather than negative emotions that were coded in the jealousy condition. Coders assigned the child the highest score evidenced over the 2-minute scenario. A score of 4 and above indicated explicit actions (e.g., jumping with happiness), verbalizations (e.g., "yes! The water spilled over the book!"), and affects (e.g., shouting "great!") that reflected schadenfreude, whereas a score below 4 indicated only eye gaze in different degrees.

Quantity of different jealousy and schadenfreude manifestations. The quantity of the emotional responses in Phase 1 (story reading) and phase 2 (spilled water) of the EQUAL and UNEQUAL conditions were coded according to the following scales:

a) *Phase 1: Behavioral jealousy coding category scale:* This scale assessed the frequency of jealousy manifestations comprising two main categories: (1) verbalizations, including attention-seeking comments (e.g., "I don't feel good") and interactive comments (e.g., repeating words from the story being read or answering questions aimed at the peer); and (2) actions including attention seeking actions (e.g., caressing mom's hair) and involvement actions (e.g., putting one's head between the book and the peer, to block the peer's view).

Scores were calculated for each category and were divided by scenario duration, with higher scores indicating a higher quantity of jealousy manifestations.

b) *Phase 2: Behavioral schadenfreude coding category scale*: As in the jealousy scale, this scale assessed the frequency of schadenfreude manifestations comprising two main categories: (1) verbalizations (e.g.,"good") and interactive comments (e.g., "can you read it to me now?"); and (2) actions including jumping, clapping hands, running, rolling on the floor. Scores were calculated for each category and were divided by scenario duration, with higher scores indicating a higher quantity of schadenfreude manifestations.

Affect scale. The affective changes in Phase 1 (story reading) and phase 2 (spilled water) of the UNEQUAL conditions were coded according to the following scales:

a) *Phase 1:* Based on Bauminger-Zvieli and Kugelmass [33], this 4-point scale was designed to assess a possible change in children's negative affect before versus during the jealousy provoking social scenario. Coding, ranging from 1 (very negative affect) to 4 (very positive affect), was executed twice: Time 1 - when the peer-rival entered the room and each child played alone with his/her toys; Time 2 – when the mother took the peer onto her lap and read him/her a story.

b) *Phase 2:* To assess a possible change in children's positive affect before versus during the schadenfreude provoking social scenario we used a similar to phase 1 affect scale. Coding, ranging from 1 (very positive affect) to 4 (very negative affect), was executed twice: Time 1 - when the mother read the story to the peer/herself (few seconds after Time 2 in phase 1); Time 2 – when the mother spilled the water over the book.

It should be noted that the scoring system was based on Bauminger-Zvieli and Kugelmass [33]. This scoring method does not allow measuring change in affect during the EQUAL situation as Time 2 is missing in this scenario (the mother does not take a peer onto her lap).

All videotapes underwent coding by two coders who separately in a counterbalancing order assigned scores to each child. The interclass correlation coefficients for the mother scenario were 0.90 for all jealousy and schadenfreude categories (verbalization, action). In the few cases of disagreement between the coders, the value used for data processing was the mean of the two coders' scores for that child.

Assessment of Children's Spontaneous Schadenfreude Expressions Reported by Mothers. To examine the relationship between the behaviors coded during the experiment and the ratings of the mothers of schadenfreude within the natural home environment, we developed for the current study a scale which asked "Has your child ever expressed jealousy/schadenfreude?" (yes/no)

Results

Explicitness: Hierarchical jealousy and schadenfreude scales

In order to assess the difference in the change in the explicitness of emotional ratings between the UNEQUAL and EQUAL conditions in phase 1 and 2, we performed an ANOVA of the rating data, testing for a significant interaction of *phase (1,2)*condition (EQUAL, UNEQUAL)*. This ANOVA showed significant main effects of the factors *phase* ($F(1,34) = 81.749$, $P<0.0001$) and

condition ($F(1,34) = 114.750$, $P<0.0001$). Significant interaction was found for *phase*condition* ($F(1,34) = 26.046$, $P<0.0001$). The main effect *phase* resulted from higher ratings for phase 1 (phase 1: $M/S.E. = 3.986/0.189$; phase $2 = 1.323/0.214$), the main effect of *condition* resulted from higher ratings in the UNEQUAL condition (UNEQUAL $= 3.986/0.204$; EQUAL $= 1.323/0.151$). As shown in Figure 2, follow-up paired t tests indicated that the difference between the UNEQUAL and the EQUAL condition was evident both in phase 1 ratings ($t (34) = 14.358$, $P = 0.0001$) and in phase 2 ratings ($t (34) = 3.353$, $P = 0.002$).

Quantity of different emotional manifestations

An ANOVA of the quantity data was carried out testing for a significant interaction of *phase (phase 1, phase 2)*condition (EQUAL, UNEQUAL) * category (verbalization/action)*. This ANOVA showed significant main effects of the factors *phase* ($F(1,34) = 24.706$, $P<0.0001$), *category* ($F(1,34) = 11.617$, $P<0.002$) and *condition* ($F(1,34) = 44.182$, $P<0.0001$). Significant interaction effects were found for *phase*condition* ($F(1,34) = 7.481$, $P<0.01$) and *phase*condition* category* ($F(1,34) = 19.001$, $P<0.0001$). As shown in Figure 3, the main effect *phase* resulted from higher frequency ratings for phase 1, the main effect of *condition* resulted from higher ratings in the UNEQUAL condition and the main effect for the *category* resulted from higher ratings of actions. Follow-up paired t tests indicated that the differences between the UNEQUAL and the EQUAL condition were evident in the action phase 1 ratings ($t (34) = 2.811$, $P<0.008$), verbalization phase 1ratings ($t (34) = 7.417$, $P<0.0001$), in the phase 2 action ratings ($t (34) = 4.964$, $P<0.0001$) and phase 2 verbalization ratings ($t (34) = 2.829$, $P<0.008$).

Affect scale

As indicated above this scale assesses the change in affect before versus during the jealousy (phase 1) and schadenfreude (phase 2) provoking social scenario. In order to assess the difference in the negative emotional manifestations between the *affect 1* and *affect 2* in the phase 1, we performed a paired t-test which indicated a significant change in negative affect ($t (34) = 16.139$, $P<0.0001$) (Fig. 4). Similarly a significant change in positive affect was found in phase 2($t (34) = 11.662$, $P<0.0001$), indicating increase positive affect following the spilled water condition (Fig. 5).

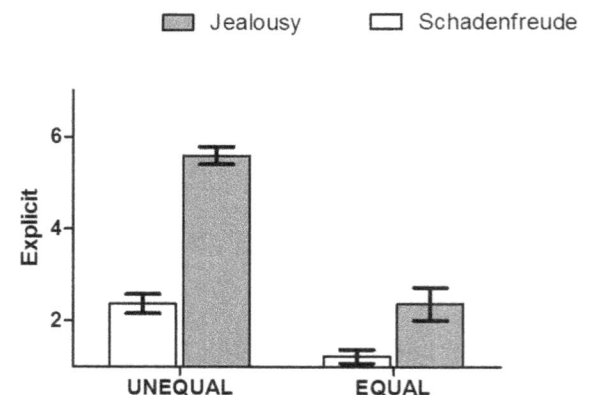

Figure 2. Explicitness: Hierarchical jealousy and schadenfreude scales. A significant main effects of the factors *phase,condition* and *phase*condition*. Follow-up paired t tests indicate that the difference between the UNEQUAL and the EQUAL condition was evident both in phase 1 ratings and in phase 2 ratings.

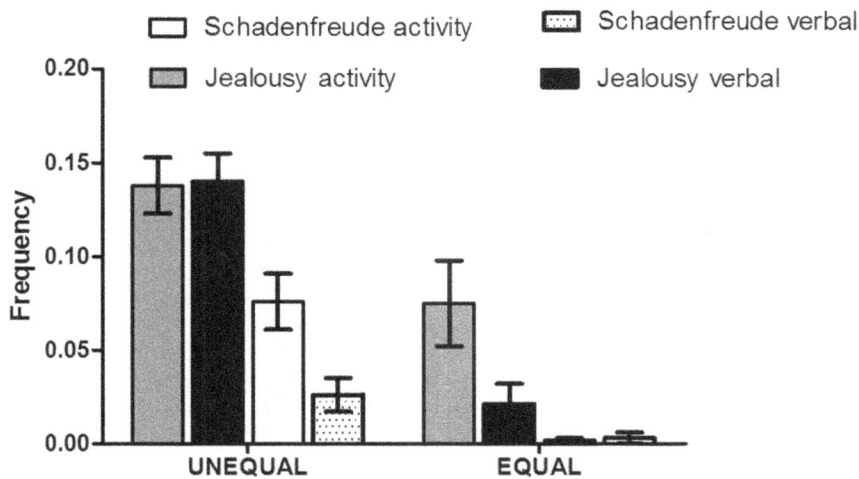

Figure 3. Quantity of different emotional manifestations. A significant interaction of *phase (phase 1, phase 2)*condition (EQUAL, UNEQUAL) * category (verbalization/action)*. Follow-up paired t tests indicated that the differences between the UNEQUAL and the EQUAL condition were evident in the action phase 1 ratings, verbalization phase 1ratings, in the phase 2 action ratings and phase 2 verbalization ratings.

Parents ratings

We carried out a MANOVA to compare the emotional responses measured in phase 2 (explicitness of emotional response, frequency of action and verbal, affect 2 phase 2) following the UNEQUAL condition and compared the reactions of children reported by their mother to have expressed schadenfreude (N = 25) to those reported not to have expressed schadenfreude in the natural home environment (N = 10). This MANOVA showed a significant effect ($F(4,30) = 5.056$, $P<0.002$), indicating an overall significant difference between the groups. Tests of between subjects effects indicated significant differences between the groups in the variables affect 2 phase 2 ($F(1,33) = 15.714$, $P<0.0001$) and activity ($F(1,33) = 15.714$, $P<0.018$) but not for the variables frequency of verbal responses ($F(1,33) = 2.11$, $P = 0.156$) and explicitness of the response ($F(1,33) = 1.911$, $P = 0.176$).

Finally, we carried out a MANOVA to compare the emotional responses measured in phase 2 (explicitness of emotional response, frequency of action and verbal, affect 2) following the UNEQUAL

condition in same sex peers (N = 21) and different sex peers (N = 14). This MANOVA showed a non-significant effect ($F(4,30) = 0.948$, $P = 0.450$), indicating no overall significant difference between the groups.

Discussion

The current study examined if jealousy towards a peer would influence schadenfreude when the peer experiences a subsequent misfortune event. In contrast, the same event occurring without jealousy was not expected to produce schadenfreude.

According to the 'gain' hypothesis, schadenfreude is viewed as a positive reaction to a potential reward. This hypothesis suggests that while in most circumstances observing others in physical or emotional pain lead to empathy (e.g. [9,36,37]), in competitive zero-sum situations we may gain from misfortunes befalling on another individual which may lead to schadenfreude [15].

In the current study, the EQUAL as well as the UNEQUAL conditions ended similarly, the water spill phase in both conditions

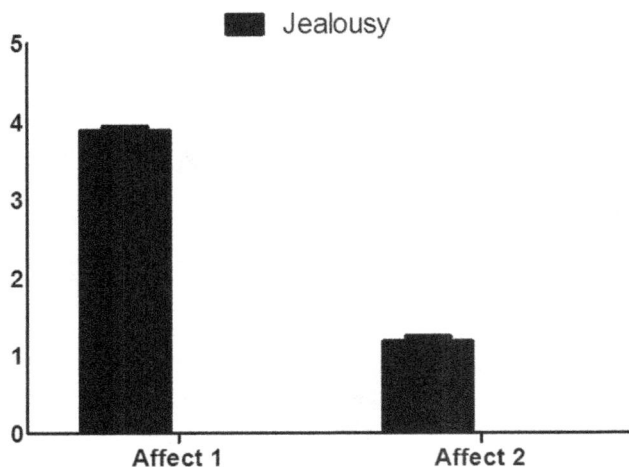

Figure 4. Affect scale. A significant change in negative affect (reduced positive affect) in the jealousy condition.

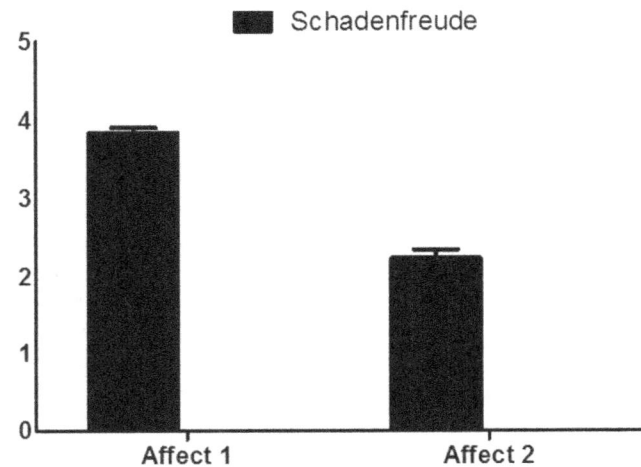

Figure 5. Affect scale. A significant change in negative affect (reduced negative affect) in the schadenfreude condition.

resulted in potentially more maternal attention. Yet, children reacted with greater emotional intensity following the UNEQUAL as compared to the EQUAL condition. Furthermore, children did not show more schadenfreude toward same sex targets as compared to opposite sex target further contradicting the model of Colyn and Gordon [38], which suggests that schadenfreude is a psychological mechanism that responds to misfortunes that lower competitors' mate value in order to increase mating opportunities.

The analyses of the hierarchical scales and quantity of different emotional manifestations scales show differences between the UNEQUAL and EQUAL conditions indicating inequality was associated with higher emotional ratings as compared to equality. While inequity produced higher jealousy ratings than equity, the termination of the inequitable situation produced higher schadenfreude ratings as compared to the termination of the equitable situation. Furthermore, the analysis of the affect scale shows a significant decrease in positive affect in the jealousy manipulation and a significant decrease in negative affect following the schadenfreude manipulation. The difference between schadenfreude ratings in the UNEQUAL and EQUAL conditions throughout the scales indicates that indeed the termination of the inequitable situation provokes schadenfreude. Furthermore, that signs of schadenfreude were observed even in the youngest children in the sample around the age of 24 months, both support the hypotheses of early evolutionary origin of inequity aversion and indicate that schadenfreude may have evolved as a response to unfair allocation of resources. Furthermore, the findings showing that children reported by their parents to have shown signs of schadefreude at home obtained higher schadenfreude ratings in the UNEQUAL condition as compared to children not reported to have shown signs of schadenfreude, further confirm that indeed the positive emotional reactions observed following the UNEQUAL condition reflect emotional reactions associated with schadenfreude.

It should be noted that while some reports on inequity aversion have found sensitivity to inequity around the age of 7 [17], others have reported that children as young as 3 years old react negatively to advantageous or disadvantageous inequality [22]. One possibility of the differences in the findings is the different methods used. While Fehr et al. [17] examined how children allocate rewards between themselves and another random partner, Lebou et al., [22] have probed the emotional reactions of children to the distribution of unequal of rewards made by others. Interestingly, although these studies used similar methods to paradigms used in the research on schadenfreude these emotions are not addressed. Thus, the current findings extend the literature on fairness and inequity aversion by putting forward the role of emotional reactions that emerges following unfair conditions.

Another interesting finding that emerged from the analysis was that jealousy ratings were higher than schadenfreude ratings, suggesting that jealousy is highly intense as compared to schadenfreude. Indeed, it has been suggested that jealousy is more intense than other social comparison based emotions such as envy [5,39] perhaps because it involves an extreme fear of loss of maternal attention. Research on jealousy shows that this emotions

appears most intensely in the majority of children between approximately 13 to 25 months [34] and can be clearly observed around the third year of life [31]. Moreover, there are even reports of forms of jealousy in babies as young as 6 months old [40], further indicating that jealousy is a powerful emotion that develops extremely early in life. Another possibility is that greater responses to negative events are related to a more basic negativity bias which refers to the psychological phenomenon by which humans pay more attention to and give more weight to negative rather than positive information [41]. Hence, the adaptive nature of negativity bias is such that jealousy in response to unfavorable comparison is likely to motivate specific behaviors for eliminating the gap between the self and the other, whereas there is little in the way of response warranted by the favorable comparison.

Finally, although the current study appears to support the hypothesis according to which schadenfreude is related to inequity aversion and not to actual potential gains, it is possible that the two hypotheses are not mutually exclusive.

Although, the termination of both the EQUAL and the UNEQUAL conditions resulted in similar potential reward, the termination of the UNEQUAL condition involved also the elimination of jealousy and therefore there was an additional emotional gain involved. Furthermore, it has been suggested that there are potential social comparison benefits behind any misfortune to the extent that it represents downward comparison and the boost to self-evaluation that might follow [5]. Jealousy, like envy represents the polar opposite of a downward comparison and therefore a misfortune befalling on someone we are jealous of, reverses the unfavorable comparison and may have an ameliorating effect on self-esteem [42].

Collectively, the current study shows for the first time that children as early as 24 months show signs of schadenfreude following the termination of an unequal situation, indicating that inequity aversion can be observed earlier than reported before. These findings imply that social comparison and sensitivity to fairness develop early in life further highlighting the evolutionary significance of positive reactions to the termination of an unfair situation. Furthermore, it has been reported that social comparison based emotions are related to different personality traits including self-esteem, neuroticism and sense of inferiority. Smith et al. [5], for example, reported that dispositional envy is negatively correlated with measures of self-esteem and positively related to depression [5]. Considering the strong relationship between envy, jealousy and schadenfreude, it is possible that individuals with low self-esteem may experience more schadenfreude. Future research may use the paradigm reported here and examine if individual differences in the tendency to feel schadenfreude among young children predicts different personality traits including low self-esteem and neuroticism.

Author Contributions

Conceived and designed the experiments: SGST. Performed the experiments: DAK. Analyzed the data: SGST. Contributed reagents/materials/analysis tools: NBZ. Wrote the paper: SGST.

References

1. Steinbeis N, Singer T (2013) The effects of social comparison on social emotions and behavior during childhood: The ontogeny of envy and Schadenfreude predicts developmental changes in equity-related decisions. Journal of Experimental Child Psychology 115: 198–209.
2. Ze'ev AB (2001) The subtlety of emotions: MIT Press.
3. Powell CA, Smith RH (2012) Schadenfreude Caused by the Exposure of Hypocrisy in Others. Self and Identity: 1–19.
4. Heider F (1958) The psychology of interpersonal relations: Psychology Press.
5. Smith RH, Kim SH (2007) Comprehending envy. Psychological bulletin 133: 46.
6. Hareli S, Weiner B (2002) Dislike and envy as antecedents of pleasure at another's misfortune. Motivation and Emotion 26: 257–277.
7. Feather NT (1999) Values, achievement, and justice: Studies in the psychology of deservingness: Plenum Publishing Corporation.

8. Feather NT (2006) Deservingness and emotions: Applying the structural model of deservingness to the analysis of affective reactions to outcomes. European review of social psychology 17: 38–73.

9. De Waal FB (2008) Putting the altruism back into altruism: the evolution of empathy. Annu Rev Psychol 59: 279–300.

10. Buss DM (2000) The dangerous passion: Why jealousy is as necessary as love and sex: Free Press.

11. Cash KJ, Evans RM (1986) Brood reduction in the American white pelican (Pelecanus erythrohynchos). Behavioral Ecology and Sociobiology 18: 413–418.

12. Joung S-J, Hsu H-H (2005) Reproduction and embryonic development of the shortfin mako, Isurus oxyrinchus Rafinesque, 1810, in the northwestern Pacific. ZOOLOGICAL STUDIES-TAIPEI- 44: 487.

13. Buss DM, Dedden LA (1990) Derogation of competitors. Journal of Social and Personal Relationships 7: 395–422.

14. Colyn LA, Gordon AK (2012) Schadenfreude as a mate-value-tracking mechanism. Personal Relationships.

15. Smith RH, Powell CA, Combs DJ, Schurtz DR (2009) Exploring the when and why of schadenfreude. Social and Personality Psychology Compass 3: 530–546.

16. Loewenstein GF, Thompson L, Bazerman MH (1989) Social utility and decision making in interpersonal contexts. Journal of Personality and Social psychology 57: 426–441.

17. Fehr E, Bernhard H, Rockenbach B (2008) Egalitarianism in young children. Nature 454: 1079–1083.

18. Smetana JG, Campione-Barr N, Metzger A (2006) Adolescent development in interpersonal and societal contexts. Annu Rev Psychol 57: 255–284.

19. Wainryb C, Brehl BA (2006) I thought she knew that would hurt my feelings: Developing psychological knowledge and moral thinking. Advances in child development and behavior 34: 131–171.

20. Turiel E (2007) The Development of Morality. Handbook of Child Psychology: John Wiley & Sons, Inc.

21. Paulus M, Gillis S, Li J, Moore C (2013) Preschool children involve a third party in a dyadic sharing situation based on fairness. Journal of Experimental Child Psychology 116: 78–85.

22. LoBue V, Nishida T, Chiong C, DeLoache JS, Haidt J (2011) When getting something good is bad: Even three-year-olds react to inequality. Social Development 20: 154–170.

23. Schmidt MF, Sommerville JA (2011) Fairness expectations and altruistic sharing in 15-month-old human infants. PloS one 6: e23223.

24. Fehr E, Fischbacher U (2004) Third-party punishment and social norms. Evolution and human behavior 25: 63–87.

25. Brosnan SF, De Waal FB (2003) Monkeys reject unequal pay. Nature 425: 297–299.

26. Range F, Horn L, Viranyi Z, Huber L (2009) The absence of reward induces inequity aversion in dogs. Proc Natl Acad Sci U S A 106: 340–345.

27. Volling BL, McElwain NL, Miller AL (2002) Emotion regulation in context: The jealousy complex between young siblings and its relations with child and family characteristics. Child development 73: 581–600.

28. Parrott WG (1991) The emotional experiences of envy and jealousy. The psychology of jealousy and envy: 3–30.

29. DeSteno DA, Salovey P (1996) Jealousy and the characteristics of one's rival: A self-evaluation maintenance perspective. Personality and Social Psychology Bulletin 22: 920–932.

30. Parrott WG, Smith RH (1993) Distinguishing the experiences of envy and jealousy. Journal of Personality and Social psychology 64: 906.

31. Bauminger N, Chomsky-Smolkin L, Orbach-Caspi E, Zachor D, Levy-Shiff R (2008) Jealousy and emotional responsiveness in young children with ASD. Cognition & Emotion 22: 595–619.

32. Bauminger N (2004) The expression and understanding of jealousy in children with autism. Development and psychopathology 16: 157–177.

33. Bauminger-Zvieli N, Kugelmass DS (2013) Mother–Stranger Comparisons of Social Attention in Jealousy Context and Attachment in HFASD and Typical Preschoolers. Journal of abnormal child psychology 41: 253–264.

34. Masciuch S, Kienapple K (1993) The emergence of jealousy in children 4 months to 7 years of age. Journal of Social and Personal Relationships 10: 421–435.

35. Miller AL, Volling BL, McElwain NL (2000) Sibling jealousy in a triadic context with mothers and fathers. Social Development 9: 433–457.

36. Batson CD (1994) Why act for the public good? Four answers. Personality and Social Psychology Bulletin 20: 603–610.

37. Decety J, Michalska KJ, Akitsuki Y (2008) Who caused the pain? An fMRI investigation of empathy and intentionality in children. Neuropsychologia 46: 2607.

38. Colyn LA, Gordon AK (2012) Schadenfreude as a mate-value-tracking mechanism. Personal Relationships 20: 524–545.

39. Salovey P, Rodin J (1986) The differentiation of social-comparison jealousy and romantic jealousy. Journal of Personality and Social psychology 50: 1100–1112.

40. Hart S, Carrington H (2002) Jealousy in 6-month-old infants. Infancy 3: 395–402.

41. Taylor SE (1991) Asymmetrical effects of positive and negative events: the mobilization-minimization hypothesis. Psychological bulletin 110: 67.

42. Van Dijk W, Ouwerkerk J, Nieweg M, Van Koningsbruggen G, Wesseling Y (2008) Why people enjoy the misfortunes of others: striving for positive self-evaluation as a motive for schadenfreude. Unpublished, VU University Amsterdam, Amsterdam.

2

Three-Dimensional Reconstructions Come to Life – Interactive 3D PDF Animations in Functional Morphology

Thomas van de Kamp[1,2]*, Tomy dos Santos Rolo[1], Patrik Vagovič[1¤], Tilo Baumbach[1], Alexander Riedel[2]

1 ANKA/Institute for Photon Science and Synchrotron Radiation, Karlsruhe Institute of Technology (KIT), Eggenstein-Leopoldshafen, Germany, 2 State Museum of Natural History (SMNK), Karlsruhe, Germany

Abstract

Digital surface mesh models based on segmented datasets have become an integral part of studies on animal anatomy and functional morphology; usually, they are published as static images, movies or as interactive PDF files. We demonstrate the use of animated 3D models embedded in PDF documents, which combine the advantages of both movie and interactivity, based on the example of preserved *Trigonopterus* weevils. The method is particularly suitable to simulate joints with largely deterministic movements due to precise form closure. We illustrate the function of an individual screw-and-nut type hip joint and proceed to the complex movements of the entire insect attaining a defence position. This posture is achieved by a specific cascade of movements: Head and legs interlock mutually and with specific features of thorax and the first abdominal ventrite, presumably to increase the mechanical stability of the beetle and to maintain the defence position with minimal muscle activity. The deterministic interaction of accurately fitting body parts follows a defined sequence, which resembles a piece of engineering.

Editor: Alistair Robert Evans, Monash University, Australia

Funding: This work was partly funded by Deutsche Forschungsgemeinschaft, DFG (www.dfg.de; RI 1817/3-1, 3-3) and the German Federal Ministry of Education and Research (www.bmbf.de; grants 05K10CKB and 05K12CK2). The authors' acknowledge support by Deutsche Forschungsgemeinschaft and Open Access Publishing Fund of Karlsruhe Institute of Technology. The funders had no role in study design, data collection and analysis, decision to publish, or preparation of the manuscript.

* Email: thomas.vandekamp@kit.edu

¤ Current address: Center for Free-Electron Laser Science, DESY, Hamburg, Germany

Introduction

Functional morphology of animals usually relies on observations of living specimens and/or the interpretation of morphological characters found in dead ones [1]. In recent years, the arrival of three-dimensional (3D) imaging techniques significantly extended the pool of available methods for morphological studies [2–5]. Digital models based on segmented datasets allow the analysis of both external and internal structures [6], and by providing a "digital copy" they facilitate a non-destructive examination of minute, brittle, and irreplaceable samples. Some animations of 3D data have been published recently as 2D movies [7–11].

Animated 3D PDF (portable document format) files, however, provide a much broader range of interactivity as opposed to movies, as the perspective can be chosen and varied and/or complex models can be masked to show only selected parts of interest, e.g. distinct muscle groups or parts of the skeleton [12,13]. Most software applications used for image stack segmentation do not offer sufficient functionality to move polygon meshes with respect to each other. Herein, we describe an approach to analyse and illustrate complex motion systems by animating 3D mesh models of static specimens with the help of 3D animation software.

We illustrate the workflow (Figure 1) based on µCT (synchrotron X-ray microtomography) data of *Trigonopterus* weevils [14]: First, the hind leg's screw-and-nut type joint [15] is animated (Figure S1); we proceed with the animation of the entire weevil, i.e., a motion system comprising 44 components (Figure S2), to clarify the functional morphology of its defensive behaviour. The latter involves death-feigning, also known as thanatosis [16]. When preparing preserved specimens we found it hard to move their rostrum and legs from thanatosis into a walking position. Movements appeared mechanically blocked and it was impossible to identify the blocking mechanism by manual examination.

Trigonopterus Fauvel is a genus of wingless weevils dwelling in primary forests of Southeast Asia and Melanesia. Its hundreds of species are spread over its range, many of them still undescribed. New Guinea appears to be a centre of its diversity with more than 300 species recorded [17,18]. Specimens are found sitting on foliage or in the litter of forest floors, but little is known of their biology. A compact thanatosis position may be a character that gained evolutionary significance with *Trigonopterus*' inability to fly. Thus, a full understanding of the passive defence mechanisms may lead to a better understanding of the genus' extraordinary diversity.

Materials and Methods

Samples

We scanned two complete specimens of *Trigonopterus vandekampi* Riedel [19] of similar body size and one specimen of *Trigonopterus oblongus* (Pascoe). One specimen of *T. vandekampi* was in walking, the others in thanatosis position. All specimens had been fixed in 100% ethanol and were critical point dried.

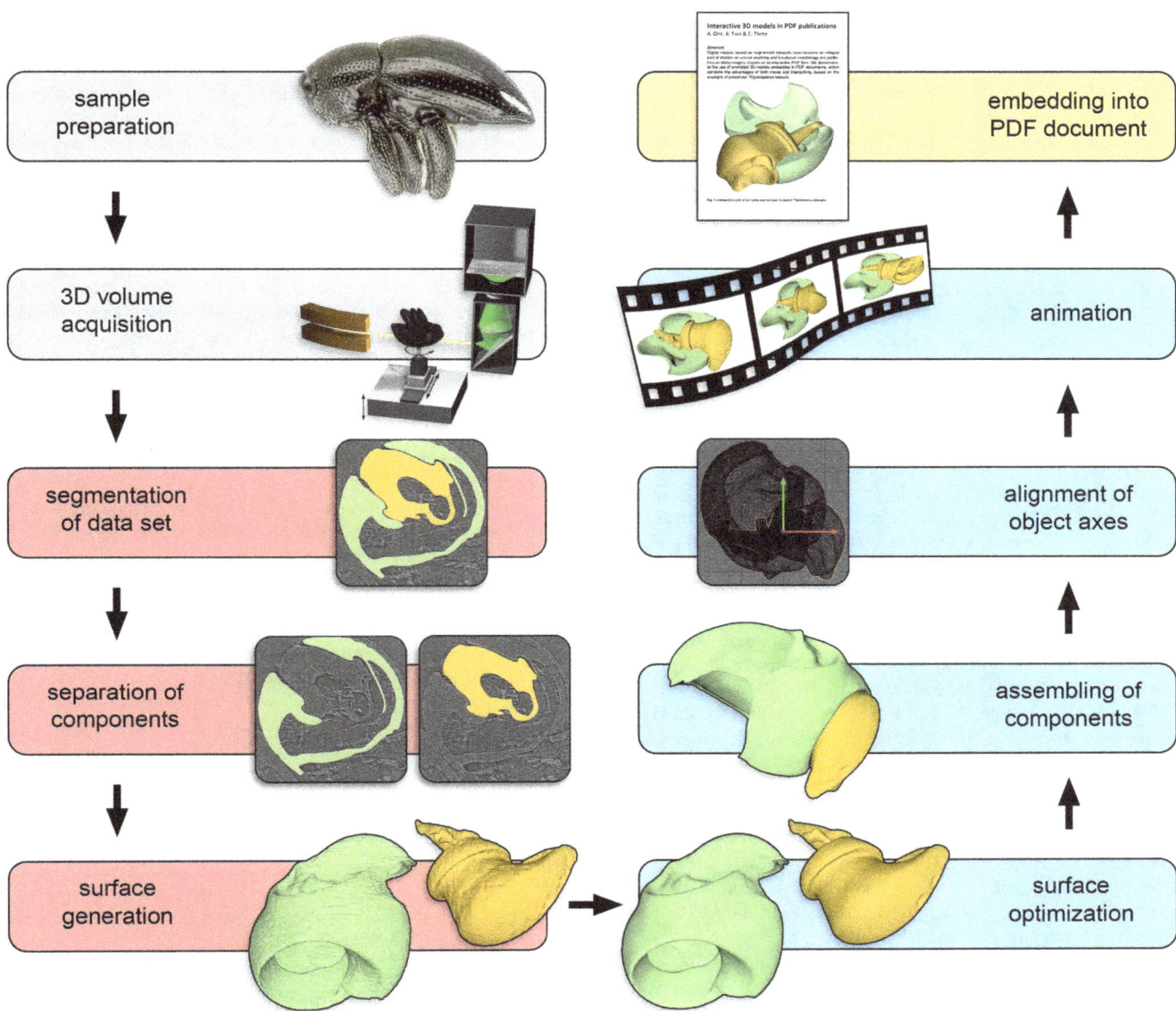

Figure 1. Flow diagram of the steps creating an interactive animated 3D model, based on the example of a screw-and-nut type hip joint of the weevil *Trigonopterus oblongus.* After acquisition of a 3D volume, scientific visualization software (e.g. Amira; red boxes) is used for creating surface models. 3D computer graphics software (here: CINEMA 4D and Deep Exploration; blue boxes) is employed for surface optimization, assembling and animation. The animated model may be embedded into a PDF document.

Synchrotron-based X-ray microtomography

Tomographic scans were performed at the microtomographic station at the TOPO-TOMO beamline of the ANKA synchrotron radiation facility located at Karlsruhe Institute of Technology, Germany. The tomographic 180° scans were taken using a filtered white beam with the spectrum peak at about 20 keV. An indirect detector system based on scintillating screen, diffraction limited optical microscope and CCD detector was used for the acquisition of the frames. For converting X-rays into visible light an LSO terbium doped scintillator was employed.

In the case of *T. oblongus*, a magnification of 18× resulted in an effective pixel size of 0.5 µm. An exposure time of 240 ms per frame was used to record 2,500 projections. Both specimens of *T. vandekampi* were scanned with a magnification of 10× and an effective pixel size of 0.9 µm. An exposure time of 2 s per frame

was used to record 1,500 projections. For all scans, a PCO 4000 14 bit CCD camera system with a resolution of 4,008×2,672 pixels served for recording the frames. Before reconstruction, the frames were processed with the phase retrieval ImageJ plugin ANKAphase [20]. Volume reconstruction was done with the PyHST software developed at the European Synchrotron Radiation Facility in Grenoble, France. Microtomographic image data are deposited in MorphDBase (accession numbers A_Riedel_20140623-M-10.1, A_Riedel_20140623-M-19.1, A_Riedel_20140623-M-15.1, A_Riedel_20140623-M-13.1, A_Riedel_20140623-M-16.1, A_Riedel_20140623-M-17.1)

Segmentation

Body sclerites were segmented and converted into individual surface components (polygon meshes), as done in other recent studies [21–23] following the procedure described in [12]. Soft tissue and connecting cuticle were not segmented unless hard to delimit from sclerites, i.e. at the attachment points of tendons. The 3D volumes were imported into Amira (version 5.4.2; FEI Visualization Sciences Group) or Avizo (version 6.2.1; FEI Visualization Sciences Group). The Image slices were segmented manually to create polygon meshes (surface models). Initially, every tenth slice was segmented with subsequent interpolation on interjacent slices. For delicate structures, smaller steps were taken to minimize interpolation errors. The interpolated labels were checked; errors and artefacts were corrected manually. After segmenting the objects each morphological structure was isolated. The *smooth labels* dialog was used for smoothing the labels (size 5; mode: 3D volume) and polygon meshes of the structures' surfaces were created with the *SurfaceGen* module at default settings.

Optimization of polygon meshes

Polygon meshes from segmented image volumes typically contain millions of polygons and numerous segmentation artefacts showing the traces of individual layers. A smooth surface facilitates reduction of the polygon count without losing too many structural details. Thus, an iterative series of surface smoothing and polygon reduction is most effective in removing segmentation artefacts and simultaneously reducing polygon count to 0.1% (Figure S3) thus greatly helping data handling in the downstream process.

For this study, the polygon count of the original meshes was reduced to 10% in Amira/Avizo. The files were subsequently saved in the Wavefront format (OBJ) to allow import into CINEMA 4D (versions 12 & 14; Maxon Computer GmbH) for subsequent smoothing and polygon reduction. The parameters were set with respect to the polygon count and the general shape of the objects.

Axis alignment, motion analysis and animation

Surface meshes may be animated using any suitable 3D program from a wide choice of software. For embedding an animated model into a 3D PDF document, the data have to be saved as Universal 3D Files (U3D) using e.g. Deep Exploration. Here, we used CINEMA 4D (Version 14) in the case of *T. oblongus* and Deep Exploration (Version 6; Right Hemisphere®; Note S2) to animate the joints of *T. vandekampi*.

Before animation, all meshes were assembled in CINEMA 4D with each component separately editable. Based on the position of the segmented sclerites in the original image stack, the individual components are automatically placed at their correct positions in the software's coordinate system. For the complex model of *T. vandekampi*, symmetric appendices (i.e. antennae and legs) were duplicated and mirrored. Object hierarchies were created and meshes of the different body parts were coloured.

Most joints of the heavily sclerotized weevil show a precise form closure of its components, so possible movements could be simulated by interactively moving one component towards its counterpart until the joint reaches the fully bent, respectively depressed position, yet avoiding any overlap of the adjacent surfaces. The joint's motion could be approximated by iterative trial and error. First, an appropriate position for the animation axes had to be found for each component of the joint. The axes were aligned by using the software's object axis tool (Figure 2). The position of an object axis was altered from three 2D perspectives (bottom, right and front view) to determine the optimal position in three-dimensional space. Positioning of the axis is highly sensitive and a tilting of only $0.1°$ from the ideal position may visibly increase artificial overlap of surfaces.

Then, one component was moved relative to the other finding its terminal positions, i.e. its fully extended and its fully depressed position, and for each a keyframe was created, thus defining the beginning and the end of the motion. Intermediate frames were interpolated automatically using linear interpolation setting. In joints with simple movements, e.g. a rotation around one stable axis, these two terminal keyframes were enough to simulate the joint's motion satisfactorily. However, in most cases the position of an animated component required realignment during the movement, and between two and six additional keyframes at intermediate positions had to ensure a precise simulation. During this process of approximating an optimal simulation, invalid arrangements could be detected by overlapping surfaces with display settings to isoparms in different 2D perspectives (e.g. bottom, right, front (Figure 2 A–C). In addition, the joints were temporarily cut to reveal any unrealistic friction of surfaces (Figure 2 E,F). Hard, guiding surfaces and soft structures, e.g. membranes or flexible tendons, which are pushed aside during movement in the living animal, had to be distinguished by the investigator.

The specimen with extended legs was segmented in part to verify the terminal position of the metacoxa. Its cavity is anteriorly open, so its movement is not strictly confined by the thorax (as is the case in pro- and mesocoxa), and thus required empiric measurement of its position with legs extended. From both positions, groups of polygon models composed of the metacoxa, metatrochanter, metafemur and parts of thorax and abdomen were loaded into the same scene and scaled to the same size. The walking position group was moved until thorax and abdomen overlapped with the ones from thanatosis. Thus assigning the final positions of the hind leg, we simulated its movement from walking position to thanatosis. Based on our field observations, the whole process of attaining thanatosis position in *Trigonopterus* takes about one second, i.e. it is faster than the eye can follow in detail. Thus, we decreased the motion speed of our animation. The precise timing of each joint's motion is considered a working hypothesis, since no video recording of the process is available. The adduction of all joints starts simultaneously as is the case in many other weevils falling into thanatosis.

Between 120 and 180 frames for the animation of each joint allowed smooth interpolation and an overall animation time of several seconds at 30 fps (Figure S2). The model of the screw joint of *T. oblongus*, which was animated in CINEMA 4D, was saved as a COLLADA 1.4 file (DAE). and imported into Deep Exploration.

For both models Deep Exploration was used to colour the mesh components and to create the final hierarchies for the meshes. Animation speed was set to 30 fps. Each model including materials and animations was subsequently saved as a Universal 3D File (U3D), containing both mesh geometry and animation sequences.It can be opened and displayed with suitable software, e.g. Deep Exploration, but for a wide dissemination the PDF format is preferable.

Embedding into PDF files

New documents were created with Adobe Acrobat (version 9 Pro Extended; Note S2) and the U3D meshes were implemented with the *3D tool*. Using default *Activation Settings* and assigning a *Poster Image* from default view, the 3D visualization parameters were set as follows: white background, CAD optimized lights, solid rendering style and default 3D conversion settings. For the reconstruction of the coxa-trochanteral joints of *T. oblongus*, the animation style was set to *Bounce*, whereas it was set to *Loop* for the animated reconstruction of *T. vandekampi*. After starting the 3D view by clicking on the poster image, several views were

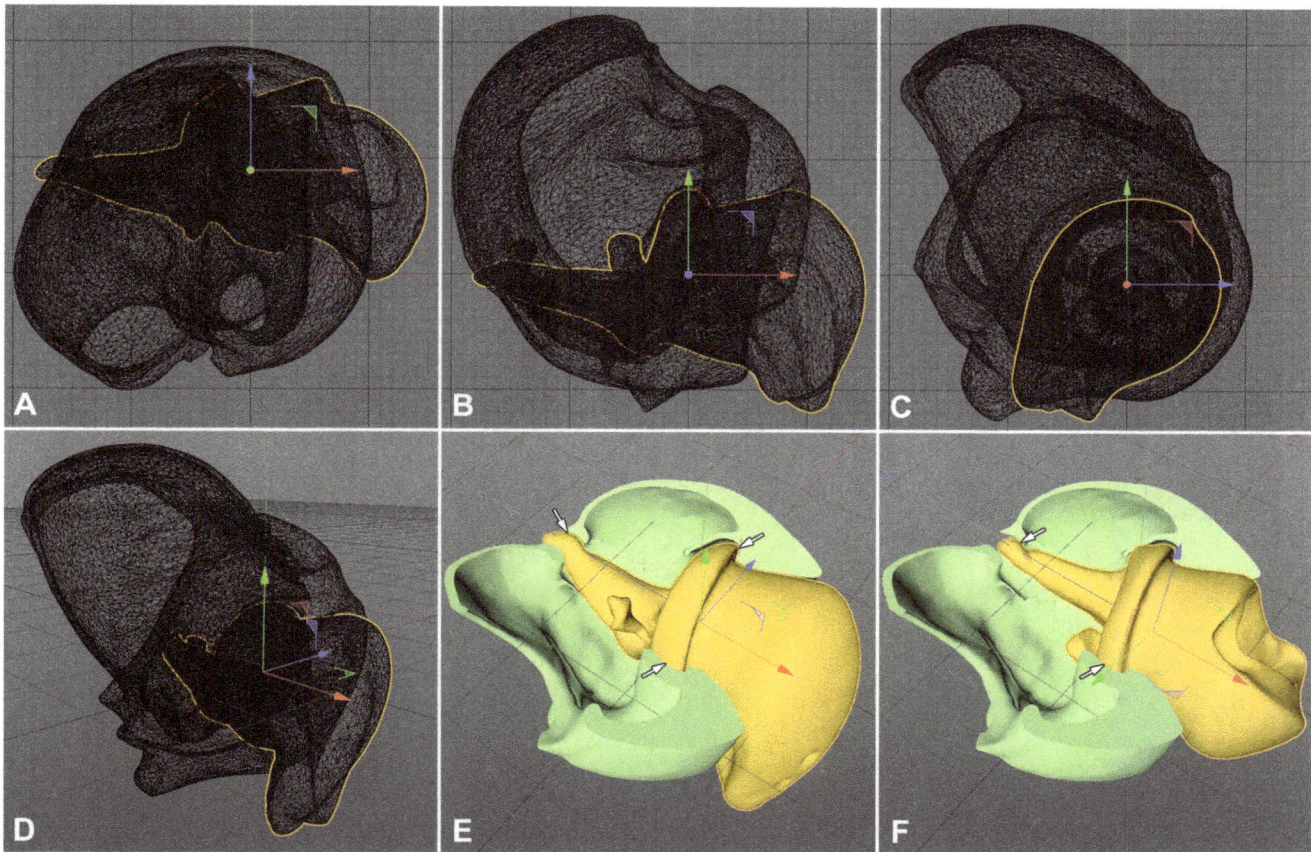

Figure 2. Axis alignment and animation of the screw joint of *Trigonopterus oblongus* **in CINEMA 4D.** (A–C) 2D views (A: bottom, B: right, C: front) displaying surface isoparms for axis alignment. The boundary of the trochanter is indicated by the yellow frame, the rotation axis by the arrows (red: X axis, green: Y axis, blue: Z axis). (**D**) Displayed surface isoparms in central perspective. (**E, F**) Same joint (Gouraud shading); coxa (green) cut by attached Boole tool, thus revealing friction surfaces of the joint parts (white arrows).

created using the *Manage Views* option from the 3D toolbar. Annotations were added to the documents, which were subsequently saved as Portable Document Format files (PDF). Animated models are deposited at Dryad (http://doi.org/10.5061/dryad.56kf4).

Results

Animation of a screw joint

For the isolated metacoxal screw joint, each coxa and trochanter were segmented separately (Figure S3). The terminal keyframes were set at 0 and 120, and four additional keyframes were needed to ensure realistic simulation for an arbitrary animation time of four seconds. The animation shows a rotation of 130 degrees with a translatory movement of 65 μm. Besides its larger size, the metacoxal joint of *T. oblongus* appears similar or identical to that of *T. vandekampi*.

Animation of a complex system - thanatosis of a Trigonopterus weevil

A digital model of *T. vandekampi* suitable to answer our questions pertaining to the functional morphology of thanatosis was created by segmenting the major body sclerites (Table S1) of the specimen in thanatosis and by animating 50 individual articulations (Table S2). The noteautomatic placement of the

individual components (i.e. the corresponding joint partners) in a consistent coordinate system as assigned by the software Amira resulted in an accurate animation of the assembled virtual beetle (Figure S2).

Trigonopterus weevil's cascade of movements to attain thanatosis

The movements of *T. vandekampi* from walking position to thanatosis and reverse follow a defined sequence (Figure 3A). Some movements of the head, thorax and the appendices may partly happen simultaneously, but there are some benchmarks (Note S1) that must be passed by one component before another component can proceed for mechanical reasons. If this sequence of motions is violated, the weevil is unable to attain a perfect thanatosis position. The functional morphology is designed in a way to maintain the thanatosis position by the interaction of multiple body parts which mechanically block an unwanted opening of appendages. The following sequence of movements and mechanisms is hypothesized based on our animated model and on extensive field observations of the defence behaviour of cryptorhynchine weevils:

1) The tarsi are lifted and nestled backwards along the posterior face of the tibial apices which causes the weevil to lose its hold and fall to the side. The bent tibiae fit into the ventrally sulcate femora, their ventral edge overlapped by the

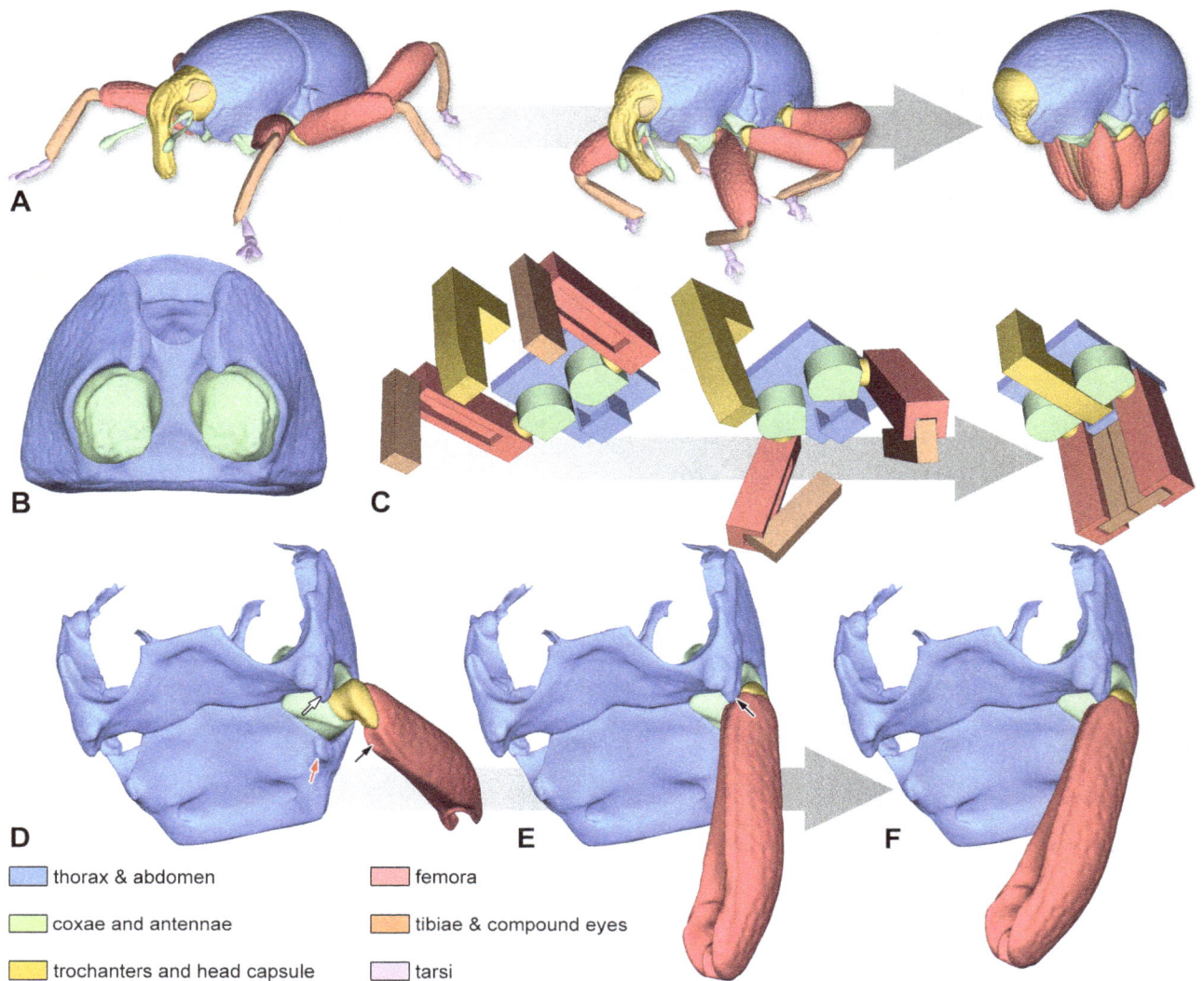

Figure 3. Blocking mechanisms of legs in _Trigonopterus vandekampi_. (A) Illustration of the movement from walking position to thanatosis. (B) Prothorax in ventral aspect; note the flattened mesial faces of the coxae and the narrow thoracic canal. (C) Simplified model of the prothoracic blocking mechanism. (D–F) Metacoxal leverage. (D) Hind leg elevated; note the depressed face of the metafemur (black arrow), the metathoracic intercoxal ridge (white arrow) and the abdominal protrusion (red arrow). (E) Inward rotation of the trochanter causes the depressed face of the femur to press against the posterior face of the intercoxal ridge (arrow). (F) The leverage effect causes the coxa to swing backwards and the joint comes to a dead stop.

anteroventral ridge of the femora. Thus, each leg forms a compact unit with potential stress relieved from the tibia-femoral joint.

2) The compact femur-tibia-units are pressed together with the left and right legs touching medially. Almost no interspaces are visible between the legs in lateral aspect. Now the tibiae cannot be unfolded since their movement is blocked: the protibiae are blocking each other mesially; the mesotibiae are blocked by the overlapping profemora, and the metatibiae are blocked by the overlapping mesofemora. To allow unfolding of tibiae and tarsi, the pro- and mesocoxae have to be rotated outwards by approx. 10°.

3) The prothoracic acetabula are mesially bordering a comparatively narrow thoracic canal; the procoxae are somewhat D-shaped in cross-section, with their flattened inner faces forming portions of the thoracic canal's lateral wall

(Figure 3B). As the rostrum fits tightly into the thoracic canal and the movement of the procoxa is confined to rotation, the latter is mechanically inhibited by the retracted rostrum. Head and prothorax now form one functional unit.

4) The head/prothorax-complex is retracted. Now, the ventral rim of the mesothoracic receptacle overlaps the retracted rostrum's tip ventrally. Thus, the head capsule must be moved forward ca. 57 µm along the beetle's body axis to allow an outward rotation of the procoxae. During thanatosis the antennae are almost fully concealed in the thoracic canal with the rostrum completely covering the opening of the thoracic canal.

Because outward rotation of the trochanteral screw joints is combined with a translatory movement (approx. 0.19 µm/°), the rotation of the protrochanters – and profemora, respectively – is inhibited by the midlegs while the prothorax is pressed against the mesothorax. The posterior surface of the profemur is concave at middle but swollen at the base. This swelling fits tightly into the concave anterior face of the mesocoxa thus blocking the rotation of the latter. The dorsal edge of the mesofemur basally forms an angulation which is posteriorly blocked by the intercoxal ridge of the metathorax tightly opposing it. Since the rotation axis of the mesotrochanter (translation: $0.19\ \mu m/°$) is almost perpendicular to the body axis, any elevating rotation is effectively blocked. Such a rotation, which is necessary to bring the leg into walking position, is only possible if the mesocoxa is turned to the side. The metacoxa differs markedly from pro- and mesocoxa as it tilts around two pivotal points. When the metatrochanter is rotating inwards and approaching its resting position (Figures 3D–F), the metafemur is pressed against the intercoxal ridge of the metathorax (Figure 3E). The leverage created causes the metacoxa to swing backwards (Figure 3F) and the metacoxa-trochanteral joint comes to a dead stop. In this position, coxa, trochanter and femur form a functional unit. The metafemur is maximally approximated to the body by the translation of the coxa-trochanteral screw joint, which is largest in the hind leg ($0.24\ \mu m/°$).

Discussion

In recent years, complex morphological 3D models based on segmented datasets have been published as PDF files [12,22,24] which allow the user to handle and examine relevant structures interactively. Other 3D models containing motion information were published as animated movies but without the option of user-interactivity other than stop-and-go. In fact, PDF files offer the opportunity to combine both motion information and interactivity. Furthermore, their file-size is only a fraction of files published in movie formats e.g. in MOV or MP4.

The recording of 3D data by e.g. CT for studies of functional morphology is ideally coupled with direct movierecording of motion, both taken simultaneously in the best case [7–10]. However, such an ideal setting is not always possible: the organisms of interest may be long extinct, too rare, or too shy to observe in a laboratory setting. High-resolution µCT recording suitable for *in vivo* imaging of small-sized specimens still pose radiation doses killing most insects within a few seconds [25]. Obviously, there remains a wide field of conditions where simultaneous recording of both motion and 3D data is impossible.

In many arthropod joints, movements are restricted by the morphology of the corresponding rigid parts, leaving very little play due to precise form closure of the components [26]. Simulations can be performed by interactively moving one component towards its counterpart until the joint reaches one endpoint. Some joints may involve uncertainty where exactly this endpoint is located, but in the described case where the limbs always reach a clearly defined and stable terminal position this was not an issue. The lack of information on the precise timing of motions may be a more serious drawback, especially when it concerns the simulation of complex and highly coordinated movements, such as the movement of two pairs of wings during flight [27,28] or six pairs of legs performing a running motion [29,30]. However, while the study of coordinative motion is out of reach without real motion data, it is still possible to investigate the qualitative movement of an isolated limb.

Although these limitations may appear quite restrictive, in the case of *Trigonopterus* weevils the attempt used in the present study proved to be highly effective for understanding the mechanisms of the weevils' defensive morphology (Note S1). The beetle's head and legs interlock mutually and with specific features of thorax and the first abdominal ventrite, presumably to increase its mechanical

stability in thanatosis. The protective posture is maintained by minimal muscle activity, and largely by the mechanical interaction of exoskeletal parts. The deterministic interaction of accurately fitting body parts follows a defined sequence, which resembles a piece of engineering and in fact a closer analysis could be of interest to the field of biomimetics. Most aspects of the complex mechanisms could be illustrated in a single PDF 3D model of relatively small data size. While being completely interactive, predefined views illustrate the different mechanisms described above. This underlines the potential of animated 3D models: preserved or extinct species can be brought to life again, at least in the digital world.

Supporting Information

Figure S1   Interactive animated 3D reconstruction of the metacoxal joint of *Trigonopterus oblongus*. Click on the figure to start interactive 3D view; switch between views by using the menu (Adobe Reader 8.1 or higher required).

Figure S2   Interactive animated 3D reconstruction of *Trigonopterus vandekampi* simulating the movements from walking position to thanatosis posture. Default views illustrating the blocking mechanisms are provided. Click on the figure to start interactive 3D view; switch between views by using the menu (Adobe Reader 8.1 or higher required).

Figure S3   Optimization of polygon meshes, exemplified with the metacoxa of *Trigonopterus oblongus*, showing surface (top) and corresponding mesh (bottom). By a consecutive series of polygon reduction and smoothing, the polygon count – and thus the file size – was reduced to ca. 1/1,000 of its original value without compromising the surface structure while simultaneously reducing labelling artefacts.

Table S1   List of separate polygon meshes created from labeled exoskeleton parts of *Trigonopterus vandekampi*.

Table S2   List of the 50 individual articulations animated to create the moving interactive model of *Trigonopterus vandekampi*. Note that femora and trochanters do not share movable articulations in the species. Joints between tarsomeres 3 and the minute tarsomeres 4 were neglected.

Acknowledgments

We thank S. Scharf for helping with the segmentation of data sets, D. Pelliccia for assistance during the tomographic scans, and R. Hofmann and R. Heine for helpful discussions. B. Ruthensteiner and A. R. Evans reviewed the manuscript and their comments lead to many improvements. The ANKA Synchrotron Radiation Facility is acknowledged for providing beamtime.

Author Contributions

Conceived and designed the experiments: TK TR PV TB AR. Performed the experiments: TK TR PV AR. Analyzed the data: TK AR. Contributed reagents/materials/analysis tools: TK TR PV TB AR. Wrote the paper: TK AR.

References

1. Homberger DG (1988) Models and tests in functional morphology: the significance of description and integration. Amer Zool 28: 217–229 doi:10.1093/icb/28.1.217.

2. Zill S, Frazier SF, Neff D, Quimby L, Carney M, et al. (2000) Three-dimensional graphic reconstruction of the insect exoskeleton through confocal imaging of endogenous fluorescence. Microsc Res Techniq 48: 367–384 doi:10.1002/(SICI)1097-0029(20000315)48:6<367::AID-JEMT7>3.0.CO;2-Y.

3. Corfield JR, Wild JM, Cowan BR, Parsons S, Kubke MF (2008) MRI of postmortem specimens of endangered species for comparative brain anatomy. Nat Protoc 3: 597–605 doi:10.1038/nprot.2008.17.

4. Westneat MW, Socha JJ, Lee W-K (2008) Advances in biological structure, function, and physiology using X-ray imaging. Ann Rev Physiol 70: 119–142 doi:10.1146/annurev.physiol.70.113006.100434.

5. Handschuh S, Baeumler N, Schwaha T, Ruthensteiner B (2013) A correlative approach for combining microCT light and transmission electron microscopy in a single 3D scenario. Front Zool 10: 44 doi:10.1186/1742-9994-10-44.

6. Betz O, Wegst U, Weide D, Heethoff M, Helfen L, et al. (2007) Imaging applications of synchrotron X-ray phase-contrast microtomography in biological morphology and biomaterials science. I. General aspects of the technique and its advantages in the analysis of millimetre-sized arthropod structure. J Microsc 227: 51–71 doi:10.1111/j.1365-2818.2007.01785.x.

7. Sahara W, Sugamoto K, Murai M, Tanaka H, Yoshikawa H (2006) 3D kinematic analysis of the acromioclavicular joint during arm abduction using vertically open MRI. J Orthop Res 24: 1823–1831 doi:10.1002/jor.20208.

8. Brainerd EL, Baier DB, Gatesy SM, Hedrick TL, Metzger KA, et al. (2010) X-ray reconstruction of moving morphology (XROMM): precision, accuracy and applications in comparative biomechanics research. J Exp Zool Part A 313: 262–279 doi:10.1002/jez.589.

9. Gatesy SM, Baier DB, Jenkins FA, Dial KP (2013) Scientific rotoscoping: a morphology-based method of 3-2D motion analysis and visualization. J Exp Zool Part A 313: 244–261 doi:10.1002/jez.588.

10. Baier DB, Gatesy SM, Dial KP (2013) Three-dimensional, high-resolution skeletal kinematics of the avian wing and shoulder during ascending flapping flight and uphill flap-running. PLOS ONE 8: e63982 doi:10.1371/journal.pone.0063982.

11. Lauridsen H, Hansen K, Wang T, Agger P, Andersen L (2011) Inside Out: Modern imaging techniques to reveal animal anatomy. PLOS ONE 6: e17879 doi:10.1371/journal.pone.0017879.

12. Ruthensteiner B, Heβ M (2008) Embedding 3D models of biological specimens in PDF publications. Microsc Res Techniq 71: 778–786 doi:10.1002/jemt.20618).

13. Murienne J, Ziegler A, Ruthensteiner B (2008) A 3D revolution in communicating science. Nature 453: 450 doi:10.1038/453450d.

14. Riedel A, Sagata K, Surbakti S, Tänzler R, Balke M (2013) One hundred and one new species of Trigonopterus weevils from New Guinea. ZooKeys 280: 1–150 doi:10.3897/zookeys.280.3906.

15. van de Kamp T, Vagovič P, Baumbach T, Riedel A (2011) A biological screw in a beetle's leg. Science 333: 52 doi:10.1126/science.1204245.

16. Bleich OE (1928) Thanatose und Hypnose bei Coleopteren. Experimentelle Untersuchungen. Z Morphol Oekol Tiere 10:1–61 doi: 10.1007/BF00419278.

17. Riedel A., Daawia D, Balke M (2010) Deep cox1 divergence and hyperdiversity of Trigonopterus weevils in a New Guinea mountain range (Coleoptera, Curculionidae). Zool. Scripta 39: 63–74 doi:10.1111/j.1463-6409.2009.00404.x.

18. Tänzler R., Sagata K, Surbakti S, Balke M, Riedel A (2012) DNA barcoding for community ecology - how to tackle a hyperdiverse, mostly undescribed Melanesian fauna. PLoS ONE 7: e28832 doi:10.1371/journal.pone.0028832.

19. Riedel A (2010) One of a thousand - a new species of Trigonopterus (Coleoptera, Curculionidae, Cryptorhynchinae) from New Guinea. Zootaxa 2403: 59–68 doi: not available.

20. Weitkamp T, Haas D, Wegrzynek D, Rack A (2011) ANKAphase: software for single-distance phase retrieval from inline X-ray phase-contrast radiographs. J Synchrotron Radiat 18: 617–629 doi:10.1107/S0909049511002895.

21. Witmer LM, Ridgely RC (2008) The paranasal air sinuses of predatory and armored dinosaurs (Archosauria: Theropoda and Ankylosauria) and their contribution to cephalic structure. Anat Rec 291: 1362–1388 doi:10.1002/ar.20794.

22. Ziegler A, Ogurreck M, Steinke T, Beckmann F, Prohaska S, et al. (2010) Opportunities and challenges for digital morphology. Biol Direct 5: 45 doi:10.1186/1745-6150-5-45.

23. Weide D, Thayer MK, Betz O (2012) Comparative morphology of the tentorium and hypopharyngeal–premental sclerites in sporophagous and non-sporophagous adult Aleocharinae (Coleoptera: Staphylinidae). Acta Zool 95: 84–110 doi:10.1111/azo.12011.

24. Baeumler N, Haszprunar G, Ruthensteiner B (2008) 3D interactive microanatomy of Omalogyra atomus (Philippi, 1841) (Gastropoda, Heterobranchia, Omalogyridae). Zoosymposia 1: 101–118 doi:10.11646/zoosymposia.1.1.9.

25. dos Santos Rolo T, Ershov A, van de Kamp T, Baumbach T (2014) In vivo X-ray cine-tomography for tracking morphological dynamics. Proc Natl Acad Sci USA 111: 3921–3926 doi:10.1073/pnas.1308650111.

26. Bögelsack G, Karner M, Schilling C (2000) On technomorphic modelling and classification of biological joints. Theory Biosc 119: 104–121 doi:10.1007/s12064-000-0007-3.

27. Willmott AP, Ellington CP (1997) The mechanics of flight in the hawkmoth Manduca sexta. I. Kinematics of hovering and forward flight. J Exp Biol 200: 2705–2722 doi:not available.

28. Willmott AP, Ellington CP (1997) The mechanics of flight in the hawkmoth Manduca sexta. II. Aerodynamic consequences of kinematic and morphological variation. J Exp Biol 200: 2723–2745 doi: not available.

29. Full RJ, Tu MS (1991) Mechanics of a rapid running insect - two-, four- and six-legged locomotion. J Exp Biol 156: 215–231 doi:not available.

30. Spence AJ, Revzen S, Seipel J, Mullens C, Full R (2010) Insects running on elastic surfaces. J Exp Biol 213: 1907–1920 doi:not available.

3

Evolution of Growth Habit, Inflorescence Architecture, Flower Size, and Fruit Type in Rubiaceae: Its Ecological and Evolutionary Implications

Sylvain G. Razafimandimbison[1]*, Stefan Ekman[2], Timothy D. McDowell[3], Birgitta Bremer[1]

1 Bergius Foundation, The Royal Swedish Academy of Sciences and Botany Department, Stockholm University, Stockholm, Sweden, **2** Museum of Evolution, Uppsala University, Uppsala, Sweden, **3** Department of Biological Sciences, East Tennessee State University, Johnson City, Tennessee, United States of America

Abstract

During angiosperm evolution, innovations in vegetative and reproductive organs have resulted in tremendous morphological diversity, which has played a crucial role in the ecological success of flowering plants. Morindeae (Rubiaceae) display considerable diversity in growth form, inflorescence architecture, flower size, and fruit type. Lianescent habit, head inflorescence, small flower, and multiple fruit are the predominant states, but arborescent habit, non-headed inflorescence, large flower, and simple fruit states occur in various genera. This makes Morindeae an ideal model for exploring the evolutionary appearances and transitions between the states of these characters. We reconstructed ancestral states for these four traits using a Bayesian approach and combined nuclear/chloroplast data for 61 Morindeae species. The aim was to test three hypotheses: 1) self-supporting habit is generally ancestral in clades comprising both lianescent and arborescent species; 2) changes from lianescent to arborescent habit are uncommon due to "a high degree of specialization and developmental burden"; 3) head inflorescences and multiple fruits in Morindeae evolved from non-headed inflorescences and simple fruits, respectively. Lianescent habit, head inflorescence, large flower, and multiple fruit are inferred for Morindeae, making arborescent habit, non-headed inflorescence, small flower, and simple fruit derived within the tribe. The rate of change from lianescent to arborescent habit is much higher than the reverse change. Therefore, evolutionary changes between lianescent and arborescent forms can be reversible, and their frequency and trends vary between groups. Moreover, these changes are partly attributed to a scarcity of host trees for climbing plants in more open habitats. Changes from large to small flowers might have been driven by shifts to pollinators with progressively shorter proboscis, which are associated with shifts in breeding systems towards dioecy. A single origin of dioecy from hermaphroditism is supported. Finally, we report evolutionary changes from headed to non-headed inflorescences and multiple to simple fruits.

Editor: Kamal Bawa, University of Massachusetts, United States of America

Funding: This study was supported by grants from the Swedish Research Council and the Knut and Alice Wallenberg Foundation to Birgitta Bremer. The funders had no role in study design, data collection and analysis, decision to publish, or preparation of the manuscript.

Competing Interests: The authors have declared that no competing interests exist.

* E-mail: sylvain.razafimandimbison@bergianska.se

Introduction

During angiosperm evolution, changes in vegetative and reproductive organs have resulted in remarkable morphological diversity, which has played an important role in the ecological success of flowering plants [1]. The fusion of clustered fruits into multiple fruits (or syncarps) has occurred repeatedly in different lineages [1–2]. Some multiple fruits are important food sources for a wide range of animals, and the evolution of this type of compound fruit has been hypothesized as a result of selection by large animals [3]. This is based on the fact that multiple fruits are generally favored and their seeds are effectively dispersed by large frugivorous dispersers [4]. Flowering plants from different groups produce edible multiple fruits that are economically important. Examples include jackfruits and breadfruits (Moraceae), pineapples (Bromeliaceae), and noni fruits (Rubiaceae). Despite their crucial roles in different ecosystems and for the human society, little is known about the evolution of multiple fruits. This is partly due to the lack of robust phylogenies for the lineages that contain species producing multiple fruits and species bearing simple fruits. Molecular-based phylogenies are essential for placing patterns of any heritable trait in an evolutionary context [5].

In the coffee family (Rubiaceae), most taxa with multiple fruits are members of the tribes Naucleeae [6–7] and Morindeae [8–10]. Fruits in Morindeae, belonging to the Psychotrieae alliance in the subfamily Rubioideae, are predominantly multiple fruits composed of two to many fully to basally fused drupaceous (fleshy) fruits (Fig. 1C–D, H), which are derived from ovaries of the adjacent flowers. This type of compound fruit is found in three (*Coelospermum* Blume, *Gynochthodes* Blume (Fig. 1H), and *Morinda* L. (Fig. 1C)) of the five genera currently recognized in the tribe, and is absent in the other two genera (*Appunia* Hook.f. and *Siphonandrium* K.Schum.) the infructescences (fruiting stage of inflorescences) of which are formed by clusters of simple, drupaceous fruits (Fig. 1B). A few members of *Coelospermum* are characterized by branched or headed infructescences bearing pedicellate (stalked), drupaceous fruits (Fig. 1F), while some *Gynochthodes* species have infructescences composed of pedicellate, drupaceous fruits grouped in umbels (flat-topped or rounded flower/fruit clusters with the pedicels arising from more or less the same point) or fascicles (tight bundles). It has been postulated by McClatchey [11] that multiple fruits of the broadly circumscribed *Morinda* (*Morinda* sensu lato), which included all lianescent and arborescent *Morinda* species with multiple fruits recently transferred to *Gynochthodes* [8], evolved from an ancestor with umbels and simple fruits by suppression of the pedicels and fusion of the ovaries of the adjacent flowers. This would imply that multiple fruits of *Morinda* and *Gynochthodes* (both sensu Razafimandimbison et al. [8–9]) are derived in Morindeae.

Besides its fruit diversity, Morindeae are also diverse in growth form, inflorescence architecture, and flower size. Lianescent (climbing) habit, headed inflorescence, and small flower are the predominant states, and occur in different genera of the tribe. This makes Morindeae an attractive model for exploring the evolutionary appearances and transitions between the major states of the four major characters (i.e., growth habit, inflorescence architecture, flower size, and fruit type) from a phylogenetic perspective. The recently published molecular phylogeny of Morindeae [9] provides a solid basis for such a study. These traits have been revealed to be evolutionarily labile in Morindeae [9], however a proper ancestral state reconstruction (hereafter called ASR) is essential in order to better understand their evolution. Moreover, a combination of these four characters has been used for circumscribing the five recognized genera of the tribe (Table 1).

Morindeae comprise ca. 100 woody climbing species (ca. 62%), with only ca. 54 arborescent (tree or tree-like) and frutescent (shrubby) species (ca. 28%), and two suffrutescent (shrubby plants having woody stems only at the base) species. Frutescent plants are common in *Appunia*. Both arborescent and frutescent plants are predominant in *Morinda* but are extremely rare in *Coelospermum* (the arborescent *C. reticulatum* (F.Muell.) Benth.) and *Gynochthodes* (frutescent *G. decipiens* (Schltr.) Razafim. & B.Bremer and the arborescent *G. trimera* (Hillebr.) Razafim. & B.Bremer). Conversely, lianescent plants (woody vines or climbers) in the tribe are found mostly in *Gynochthodes* [8–9], with only one species in *Appunia* (*A. megalantha* C.M.Taylor & Lorence, [12]), two species in *Morinda* (*M. longiflora* G.Don and *M. morindoides* (Baker) Milne-Redh., [9]), and 10 species in *Coelospermum* [13]. There are at least two different (but not mutually exclusive) hypotheses regarding the evolutionary transitions between lianescent and arborescent growth forms. The first hypothesis considers self-supporting (arborescent or frutescent) habit to be the common ancestral condition for clades containing both lianescent and arborescent/frutescent species [14–15]. The second hypothesis states that the evolutionary change from lianescent to arborescent/frutescent habit is uncommon, because "the evolution of lianescence can carry a high degree of specialization and developmental burden that might limit evolution back to self-supporting growth forms" [16].

Head inflorescences (or heads) occur when two or more flowers are borne on a common receptacle (the end of the inflorescence stalk upon which the floral organs are borne). In Morindeae, these heads are composed of two to 50 flowers that are clustered tightly on the receptacles (Fig. 1A, C–D, G). These inflorescences may contain a single head or two to several heads, which are in turn arranged into various branching forms: umbels, corymbs (flat-topped or round-topped, racemose inflorescences with the lower pedicels longer than the upper), or panicles (branched clusters of flowers). Within a head the ovaries of the adjacent flowers may be fused or free. The majority of Morindeae species with heads have fused ovaries, but in *Appunia* (Fig. 1A) and *Siphonandrium* the ovaries are free. It is important to note that the degree of the fusion of

ovaries prior to fruit development (pre-genital fusion) varies greatly among species, from only basally to completely fused. Only 27 Morindeae species (nine from *Coelospermum* and 18 from *Gynochthodes*) bear non-headed inflorescences, which are arranged in umbels, or compound umbels, or fascicles (in the latter genus) and panicles or corymbs (in the former genus). We postulated that heads evolved from non-headed inflorescences in Morindeae. In addition, inflorescences in Morindeae are mostly terminal on the shoot, sometimes leaf-opposed ("pseudo-terminal") and axillary. Axillary inflorescences are found in *Siphonandrium*, at least three species of *Appunia*, and ca. 18 species of *Gynochthodes*. Leaf-opposed inflorescences distinguish the mostly Asian, arborescent *Morinda* clade (including the Neotropical *M. royoc* L., the pantropical *M. citrifolia* L., and the African *M. chrysorhiza* (Thonn.) DC. and *M. lucida* Benth.) from the remaining *Morinda* [9]; this type of inflorescence is also known to occur in *Appunia* (e.g., *A. surinamensis* (Bremek.) Steyerm.).

In Morindeae, flowers in the same inflorescence or head appear to open successively over days or weeks [17] (Fig. 1A, C, E, G). The flowers vary greatly in size (Table 1); large flowers (corolla tube length/corolla lobe length >1) are found in *Appunia* and *Morinda* (Fig. 1A, C) and presumably pollinated by larger insects, such as long-proboscis moths [18]. Plants with small flowers (corolla tube length/corolla lobe length <1) are restricted to *Coelospermum*, *Gynochthodes*, and *Siphonandrium* (Fig. 1E, G), and are most likely to be pollinated by small insects (e.g., short-proboscis moths or small bees or flies) [19–20]. Overall, the species of *Morinda* have larger flowers than the species of *Appunia*, with the exception of the lianescent *A. megalantha* (corolla tubes of 23–24 mm long > corolla lobes of 15–17 mm long, [12]). Within *Morinda*, seven African species have much larger flowers than the remaining species of the genus. The flowers of *Coelospermum* are larger than those of the species of *Gynochthodes* [9]. Furthermore, Morindeae species vary in their breeding systems, and flowers are either bisexual or unisexual or functionally unisexual. Only the hermaphroditic condition has been reported in *Appunia* and *Morinda* [21], while the androdioecious (male and hermaphroditic individuals) [21–22], strict or functional dioecious [23–24], and hermaphroditic [21,24] conditions are all known in *Gynochthodes* and *Coelospermum* [21–24]. The New Guinean *Siphonandrium* is dioecious (Table 1).

We reconstructed the evolution of fruit type, inflorescence architecture, flower size, and growth form across a phylogeny of the tribe Morindeae. We were particularly interested in testing the following hypotheses: 1) self-supporting habit is generally plesiomorphic in clades comprising both lianescent and arborescent species [14]; 2) evolutionary changes from lianescent to arborescent/frutescent habit are less frequent than the reverse change, from arborescent/frutescent to lianescent habit [16]; 3) and Head inflorescences and multiple fruits in Morindeae evolved from non-headed inflorescences and simple fruits, respectively. The ecological and evolutionary implications of the findings of this study are discussed.

Results

The Bayesian majority rule consensus tree generated from the combined nrETS/nrITS/*trn*T-F data and shown in Figure 2 was fully resolved. Its overall topology is almost identical to that of the Bayesian majority rule consensus tree published in Razafimandimbison et al. [9].

Figure 1. Characteristics and morphological variation of the tribe Morindeae (for details see text). A–B: *Appunia debilis*; C: *Morinda citrifolia*; D: *Morinda pacifica*; E: *Coelospermum fragrans*; F: *Coelospermum balansanum*; G: *Gynochthodes kanalensis*; and H: *Gynochthodes retusa* (A–C by T. D. McDowell; D by F. Tronquet; E–G by J. T. Johansson; and H by K. Kainulainen).

Ancestral State Reconstructions of Growth form, Inflorescence Architecture, Flower Size, and Fruit Type in Morindeae

Two types of ASRs were performed (one with the outgroup *Damnacanthus indicus* C.F.Gaertn. and *Mitchella repens* L. (tribe Mitchelleae), hereafter called ASR with outgroup, and the other without the outgroup, hereafter called ASR without outgroup), to infer the ancestral states of growth form, inflorescence architecture, flower size, and fruit type (characters 1–4, respectively) at seven important nodes of Morindeae (nodes A–G, Fig. 2). The results of these ASRs are summarized in Tables 2,3. For growth form (character 1) and flower size (character 3) the ratios q_{01}/q_{10} from the ASRs with outgroup, respectively, were 4.820 and 2.691 for Morindeae (Table 2). This indicates that the rates of changes from lianescent to arborescent habit and from large to small flower were higher than the rates of the reverse directions, from arborescent to lianescent habit and from small to large flower. In contrast, for inflorescence architecture (character 2) and fruit type (character 4), the ratios q_{01}/q_{10}, respectively, were 0.676 and 0.677 for Morindeae (Table 2); this means that the rates of changes from non-headed to headed inflorescence and from multiple to simple fruit are higher than the rates of the reverse changes, from headed to non-headed inflorescence and from simple to multiple fruit.

The node of the Morindeae-Mitchelleae clade (= the Mitchelleae-Morindeae common ancestor) (Fig. 2) was inferred with strong and moderate support, respectively, as large flower and multiple fruit in the ASR with outgroup; however, the results of this analysis were inconclusive for growth habit and inflorescence architecture (Table 3). The outcomes of the ASRs with and without outgroup were very similar for the seven important nodes of Morindeae (nodes A–G, Fig.2) (Tables 2,3). At node A (the

Morindeae common ancestor) the lianescent habit, head inflorescence, large flower, and multiple fruit states were inferred; however the support was weak for head inflorescence, moderate for lianescent habit and multiple fruit, and strong for large flower. Within Morindeae lianescent habit was strongly inferred at node B (the *Morinda-Coelospermum-Gynochthodes* clade), node C (*Coelospermum-Gynochthodes* clade), node E (*Morinda*), node F (*Coelospermum*), and node G (*Gynochthodes*), while arborescent habit was resolved at node D (*Appunia*). For inflorescence architecture (character 2) head inflorescence was inferred at nodes B-G; the support was moderate for nodes B–C and F–G but strong for nodes D–E. For flower size (character 3) large flower was strongly inferred at nodes B and D–E, whereas nodes C and F–G were unambiguously resolved as small flower. Finally, for fruit type (character 4) fused fruit (multiple fruit) was strongly inferred at nodes B–C and nodes E–G, however simple fruit was highly resolved at node D (Tables 2,3).

Discussion

We performed ASRs with and without outgroup in order to assess the influence of the outgroup taxa on the outcomes. The fact that the results of these two ASRs are very similar for the seven nodes of Morindeae (nodes A–G, Fig. 2, Tables 2,3) suggests that the inclusion of the outgroup taxa (*Damnacanthus indicus* and *Mitchella repens*) has almost no effect on the analyses. The ASR with outgroup infers large flower with strong support and multiple fruit with moderate support at the node of the Morindeae-Mitchelleae clade. Lianescent habit, head inflorescence, large flower, and multiple fruit are inferred at node A (Morindeae) (Tables 2,3). If these inferences are correct, these states are interpreted as plesiomorphic for Morindeae and are plesiomorphic within Morindeae, with respect to the later changes (i.e., arborescent

Table 1. Morphological characteristics and other important information of the five recognized genera of the tribe Morindeae.

	Appunia Hook.f.	*Coelospermum* Blume	*Gynochthodes* sensu lato Blume	*Morinda* sensu stricto L.[1]	*Siphonandrium* K.Schum.[2]
Geographic distribution	Neotropics	Tropical Asia and Australasia	Tropical Asia, Australasia, and Madagascar	Pantropical	New Guinea
Number of species	Ca. 12	Ca. 11	Ca. 95	Ca. 40	1
Growth habit	Mostly frutescent	Mostly woody lianescent	Mostly woody lianescent,	Mostly arborescent and frutescent	Lianescent
Inflorescence architecture	Head inflorescences	Mostly non-headed inflorescences	Mostly head inflorescences	Head inflorescences	Head inflorescences
Flower size	Large (corolla tubes 5–10 (23–24) mm long > corolla lobes 0.5–7 (15–17) mm) long	Small (corolla tubes 3–7 (11) mm long < corolla lobes 4.5–16 mm) long	Small (corolla tubes 0.7–5.5 mm long < corolla lobes 1.5–11 mm) long	Large (corolla tubes 5–40 (80) mm long > corolla lobes 1–14 (22) mm) long	Small (corolla tubes ca. 3 mm long > corolla lobes ca. 5 mm long)
Breeding systems	Hermaphroditic	Androdioecious or dioecious or functionally dioecious	Androdioecious or dioecious or functionally dioecious	Hermaphroditic	Dioecious
Fruit type	Simple, drupaceous fruits	Mostly simple, drupaceous fruits	Mostly multiple fruits	Multiple fruits	Simple, drupaceous fruits

[1]All lianescent, dioecious species of *Morinda* with small flowers have recently been transferred to *Gynochthodes* [8].
[2]Filaments of *Siphonandrium* are tightly fused and its anthers are glued together, all forming a staminal tube. This feature is unique within Morindeae.

Evolution of Growth Habit, Inflorescence Architecture, Flower Size, and Fruit Type in Rubiaceae...

19

Figure 2. Bayesian majority rule consensus tree from the combined nrETS/nrITS/_trn_T-F data of 61 Morindeae taxa. Values above nodes are the posterior probabilities. Capital letters A–G denote selected nodes whose state probabilities were estimated for the states of the four characters (1–4). Data shown across the tips are growth habit (character 1: 0= lianescent, 1= arborescent, 2= herbaceous), inflorescence architecture (character 2: 0= headed inflorescences (heads); 1= non-headed inflorescences), flower size (character 3: 0= large, 1= small), and fruit type (character 4: 0= simple fruits, 1= fused or multiple fruits). SF and LF stand for small and large fruits, respectively.

habit, non-headed inflorescence, small flower, and simple fruit) in the group.

Evolution of Growth form in Morindeae and its Ecological and Evolutionary Implications

Two-thirds of the species in Morindeae are represented by the lianescent species of _Gynochthodes_, while only two _Morinda_ and one _Appunia_ species are lianas. Conversely, a single species of _Coelospermum_ (_C. reticulatum_) and two _Gynochthodes_ (_G. decipiens_ and _G. trimera_, not investigated in this study) species are arborescent. The fact that lianescent habit is inferred at node A (Morindeae) means that this state is plesiomorphic at node B (the _Morinda-Coelospermum-Gynochthodes_ clade), node C (the _Coelospermum-Gy-nochthodes_ clade), and nodes E–G (_Morinda_, _Coelospermum_, and _Gynochthodes_, respectively). Arborescent habit is inferred as apomorphic at node D (_Appunia_), and seems to have arisen at least three times within Morindeae: _Appunia_, _Morinda_ (node E: _M. butchii_ Urb. to _M. citrifolia_ L., Fig. 2), and _Coelospermum_ (node F: _C. reticulatum_, Fig. 2). Our findings provide no support for the reported prevalence of a plesiomorphic arborescent habit in lineages containing both lianescent and arborescent plants [14]. In fact, a plesiomorphic lianescent habit and multiple independent origins of arborescent from lianescent habit have recently been inferred for the primarily lianescent subfamily Secamonoideae in the family Apocynaceae [16] and the family Menispermaceae [25]. Moreover, the rate of change from lianescent to arborescent habit in Morindeae is significantly higher than the reverse change, from arborescent to lianescent habit ($q_{01}/q_{10} = 4.820>1$, Table 2); this is inconsistent with Lahaye et al.'s [16] claim that "the evolution of lianescence can carry a high degree of specialization

and developmental burden that might limit evolution back to self-supporting growth forms". Based on the evidence presented above we argue that evolutionary changes between arborescent and lianescent habits can be reversible, and that their frequency and trends seem to vary between groups. In addition, the weak-stem condition of shrubs and treelets in _Appunia_ (observations by T. D. McDowell) and the scandent- or vining-branch condition of shrubs or treelets in the Neotropical _Morinda royoc_ [26] may be viewed as a reflection of their origins from lianescent forms. We find no evidence of any reversal from arborescent to lianescent habit in the 61 Morindeae species included in this study. On the other hand, the sole lianescent species (_A. megalantha_) in the otherwise arborescent _Appunia_, not investigated in this study due to lack of material for sequencing, may represent a unique case of an evolutionary reversal from arborescent to lianescent habit in the tribe.

Furthermore, the acquisition of arborescent habit in _Appunia_ seems to have coincided with the diversification of the genus in the Neotropics. Consequently, the evolutionary changes from lianescent to arborescent habit within Morindeae may in part be attributed to reduced competition for open ground and a scarcity of host trees for climbing plants in more open habitats [15–16]. This could explain the abundance of some species of the Asian, arborescent _Morinda_ in sparse forests on hill slopes or open disturbed forests and the common occurrence of many Asian, lianescent _Gynochthodes_ in forests or thickets on mountains [27]. The pantropical, arborescent _Morinda citrifolia_ L. is also commonly found on seashores and sparse forests throughout its geographic ranges [26–27]. Similarly, the Neotropical _M. royoc_ is common in pine savannas and coastal strands [26]. Finally, five _Appunia_ species of the Neotropical Guianas region are shrubs, which frequently

Table 2. Bayesian reconstruction of ancestral states in the four characters (1–4) at seven nodes (A–G) across a posterior sample of trees including Morindeae but no outgroup.

Coded character states	Character*			
	1 (growth form)	2 (inflorescence architecture)	3 (flower size)	4 (fruit type)
	lianescent =0 and arborescent =1	headed =0 and non-headed =1	large =0 and small =1	simple =0 and fused =1
q_{01}/q_{10}	4.820	0.676	2.691	0.677
κ	0.961 (0.363–1.508)	0.711 (0.111–1.315)	1.124 (0.239–1.950)	0.787 (0.142–1.332)
Node A (Morindeae)	0.861	0.584	0.980	0.370
Node B (the _Morinda-Coelospermum-Gynochthodes_ clade)	0.992	0.586	0.878	0.181
Node C (the _Coelospermum-Gynochthodes_)	0.998	0.584	0.000	0.185
Node D (_Appunia_)	0.007	0.909	1.000	0.989
Node E (_Morinda_)	0.860	0.910	1.000	0.025
Node F (_Coelospermum_)	0.998	0.565	0.000	0.214
Node G (_Gynochthodes_)	1.000	0.583	0.000	0.186

*For each character 1–4, the following information is provided: the ratio of the average rate q_{01} to the average rate q_{10}, the average and 95% highest posterior density (HPD) of κ, and the marginal posterior probabilities of having state 0 in each of the seven nodes (A–G). As all characters are binary, the marginal posterior probability of having state 1 is one minus the probability of state 0. The 95% HPD of κ excludes 0 in all cases, which is a strong indication that branch lengths carry information about the amount of change in the morphological characters.

Table 3. Bayesian reconstruction of ancestral states in the four characters (1–4) at seven nodes (A–G) across a posterior sample of trees including Morindeae as well as the outgroup taxa *Damnacanthus indicus* and *Mitchella repens* (Mitchelleae).

	Character			
	1 (growth form)	2 (inflorescence architecture)	3 (flower size)	4 (fruit type)
Coded character states	lianescent (0), arborescent (1), and herbaceous (2)	headed (0) and non-headed (1)	large (0) and small (1)	simple (0) and fused (1)
κ	0.981 (0.394–1.516)	0.671 (0.061–1.259)	1.144 (0.221–1.996)	0.784 (0.128–1.312)
Node of Morindeae-Mitchelleae clade	0.258, 0.216, 0.524	0.513	0.996	0.359
Node A (Morindeae)	0.733, 0.199, 0.067	0.566	0.994	0.371
Node B (the *Morinda-Coelospermum-Gynochthodes* clade)	0.970, 0.011, 0.019	0.568	0.886	0.166
Node C (the *Coelospermum-Gynochthodes*)	0.991, 0.001, 0.008	0.565	0.000	0.171
Node D (*Appunia*)	0.007, 0.980, 0.013	0.916	1.000	0.987
Node E (*Morinda*)	0.733, 0.202, 0.065	0.917	1.000	0.023
Node F (*Coelospermum*)	0.990, 0.001, 0.009	0.540	0.000	0.207
Node G (*Gynochthodes*)	0.998, 0.000, 0.002	0.564	0.000	0.172

*For each character 1–4, the average and 95% highest posterior density of κ is provided. For character 1, we also provide the marginal posterior probabilities of having state 0, 1, and 2, respectively, in each of the eight selected nodes (the Mitchelleae-Morindeae root node and nodes A–G). For character 2–4, we provide the marginal posterior probabilities of having state 0 for the same eight nodes. The 95% HPD of κ excludes 0 in all cases, which is a strong indication that branch lengths carry information about the amount of change in the morphological characters.

occur at forest edges, in clearings along riverbanks, and in disturbed, opened sites (observations by T. D. McDowell).

Evolution of Inflorescence Architecture and its Evolutionary Implications

The majority of species in Morindeae with head inflorescence belong to *Gynochthodes*, although they occur in all five recognized genera (Table 1). Conversely, nine of the 11 *Coelospermum* species and 18 *Gynochthodes* species have non-headed inflorescences. If the weakly inferred head inflorescence at node A (Morindeae) is correct, this state is interpreted as plesiomorphic within Morindeae (for nodes B–G); this is inconsistent with our hypothesis of a derived head inflorescence within the tribe. The inferred plesiomorphic head inflorescence for nodes B–G, although weakly supported for nodes B–C and F–G, is consistent with highly to moderately supported plesiomorphic multiple fruits in nodes B–C and F–G. Multiple fruits can only be produced by taxa with head inflorescences, although plants with headed inflorescences can also produce simple fruits (e.g., *Appunia* (node D), Fig. 2). The evolutionary changes from headed to non-headed inflorescence occurred at least four times within Morindeae: twice each in *Coelospermum* (the Australian *C. reticulatum* and the New Caledonian *C. balansanum* group, Fig. 2) and *Gynochthodes* (the *G. coriacea* group and the Australian *G. retropila* (Halford & A.J.Ford) Razafim. & B.Bremer, Fig. 2). This is, to our current knowledge, the first report of evolutionary changes from headed to non-headed inflorescences in Rubiaceae.

The findings of this study raise new interesting questions. We do not know if the formation of non-headed from head inflorescences passes through the development of pedicels (umbels) followed by the formation of inflorescence branches in the umbellate forms to produce elongated, branched inflorescences. Alternatively, the non-headed inflorescence could be derived from a branched inflorescence of many heads if flower number was reduced to leave only one flower per receptacle. Unfortunately, discrete state ASR cannot tell us anything about the intermediate evolutionary

changes leading to the formation of non-headed inflorescences in *Coelospermum* and *Gynochthodes*. Detailed comparative morphological and developmental studies combined with phylogeny are essential in order to elucidate the underlying developmental basis between the states of inflorescence architecture in Morindeae [28–29].

Evolution of Flower Size in Morindeae and its Ecological and Evolutionary Implications

Almost all species of Morindeae with large flowers belong to the arborescent *Appunia* and *Morinda*, with the exception of the sole lianescent *Appunia* species, *A. megalantha* [12], and the two lianescent *Morinda* species, *M. longiflora* and *M. morindoides*. Conversely, Morindeae plants with small flowers are mostly the lianescent species of *Coelospermum*, *Gynochthodes*, and *Siphonandrium*, except the two arborescent *Gynochthodes* species (*G. decipiens* and *G. trimera*) and the lianescent *G. sublanceolata* Miq. Our ASRs strongly infer large flowers at the Morindeae-Mitchellae root node as well as node A (Morindeae), meaning that this state is plesiomorphic for the tribe Morindeae, the *Morinda-Coelospermum-Gynochthodes* clade (node B), *Appunia* (node D), and *Morinda* (node E). Small flowers are derived for the *Coelospemum-Gynochthodes* clade (node C), *Coelospermum* (node F), and *Gynochthodes* (node G). In other words, small flowers seem to have evolved only once from the large flowers within Morindeae (Fig. 2). It is worth noting that *G. sublanceolata* and *G. decipiens*, with large flowers but not included in this study, may represent one or two cases of reversals from small to large flowers.

The *Coelospemum-Gynochthodes* clade (node C) contains over 60% of the species in the tribe, and produce small flowers with inconspicuous colors that are most likely to be pollinated by small insects (e.g., short-proboscis moths or small bees or flies). Pollinators with progressively shorter proboscis may have been driving the transition from large to small flowers and an accompanying increase in speciation rate. Furthermore, change from large to small flowers in the *Coelospemum-Gynochthodes* clade appears to have been associated with a gender dimorphism transition. Androgynoecious (male and hermaphroditic) and dioecious conditions are only known from the lianescent species of *Gynochthodes* and *Coelospermum* [19–24] (Table 1). Thus, the high incidence of dioecy in the *Coelospermum-Gynochthodes* clade is correlated with woody, climbing growth habit, small flowers pollinated probably by unspecialized pollinators, and fleshy fruits. This pattern is consistent with those that have been reported from island habitats and various tropical forests [30–37]. Therefore, this study presents further support for the importance of these traits in the evolution of dioecy. On the other hand, it is important to note that all hermaphroditic members of *Appunia* and *Morinda* with large flowers also have the woody (but arborescent or frutescent) habit and fleshy fruits. This suggests that woodiness and fruit fleshiness alone cannot fully predict dioecy in the tribe Morindeae. In sum, the members of the *Coelospermum-Gynochthodes* clade display island syndrome characteristics, which are consistent with the fact that many of their species are indeed island endemics [13,24].

In contrast to the *Coelospermum-Gynochthodes* clade, the large, mostly white flowers of *Appunia* and *Morinda* may be pollinated by larger insects, such as long-proboscis moths. This is consistent with the report on the Asian, arborescent *Morinda coreia* Buch.-Ham. being pollinated by hawkmoths in India [18]. The fact that the species of *Appunia* and *Morinda* are hermaphroditic suggests a single origin of dioecy in the *Coelospermum-Gynochthodes* clade from hermaphroditism. Members of *Appunia* and *Morinda* are predominantly distributed in continental areas (Africa mainland, continental Asia, and South and Central America), and show characteristics of the mainland pollinations and floral traits [30,33–34,36–38].

Evolution of Multiple Fruits in Morindeae and its Ecological and Evolutionary Implications

Most Morindeae, about 90% of the species, bear multiple fruits. The majority of these species belong to *Gynochthodes* and *Morinda*, with only three species (two investigated in this study, Fig. 2) in *Coelospermum*. The infructescences of *Appunia* and *Siphonandrium* are composed of simple, drupaceous fruits. Our ASRs with moderate certainty infer multiple fruits at node Morindeae-Mitchelleae and node A (Morindeae). If correct, this state is plesiomorphic for Morindeae, the *Coelospermum-Gynochthodes* clade (node C), *Morinda* (node E), *Coelospermum* (node F), and *Gynochthodes* (node G) (Tables 2,3). This is inconsistent with the hypothesis of a derived multiple fruit for the broadly delimited *Morinda* (including the lianescent *Morinda* species transferred to *Gynochthodes* sensu Razafimandimbison et al. [8–9]), as postulated by McClatchey [11]. Simple, drupaceous fruits are derived for *Appunia* (node D) and seem to have arisen at least five times within Morindeae: once in *Appunia*, twice each in *Coelospermum* (the Australian *C. reticulatum* and the New Caledonian *C. balansanum* group), and *Gynochthodes* (the *G. coriacea* group and the Australian *G. retrophila*) (Fig. 2). This is, to our knowledge, the first report of an evolutionary transition from multiple to simple fruits in Rubiaceae. Within the *Coelospermum-Gynochthodes* clade (node C) the evolutionary change from multiple to simple fruits coincides with that of from headed to non-headed inflorescences. However, it is interesting that *Appunia* (node D) seems to have retained the plesiomorphic headed inflorescences but acquire simple fruits.

Like the acquisition of arborescent habit, the derivation of simple fruits in *Appunia* seems to have coincided with the divergence of the *Appunia* lineage in the Neotropics. The change from multiple to simple fruits in this genus is in part attributed to shifts in seed dispersal vectors. Seeds of the simple, drupaceous fruits of *Appunia* species are presumably dispersed by birds, whereas seeds of the larger multiple fruits are dispersed effectively by large frugivorous animals [3]. The same mechanism seems to underlie the evolutionary change from multiple to simple fruits within the *Coelospermum-Gynochthodes* clade.

The degree of ovary fusion prior to fruit development in head inflorescences varies greatly from only a basal, partial fusion to completely fused ovaries among *Morinda* and *Gynochthodes*. This variation, which is rarely mentioned by Rubiaceae systematists [26], merits consideration for its ecological implications. Clusters of simple fruits of *Appunia* are likely to be dispersed individually by frugivorous birds. Multiple fruits composed of partly to fully fused ovaries are presumably dispersed as single units, while those formed by basally fused ovaries could well be dispersed individually by frugivorous birds or as single units by larger frugivorous dispersers. Furthermore, we suspect that in many members of *Gynochthodes* and *Morinda* ovaries of the adjacent flowers are basally fused prior to and during maturation of the anthers, and that ovary fusion extends midway during fructification. This type of ovary fusion was reported for *Breonia richardsonii* Razafim. in the tribe Naucleeae of the subfamily Cinchonoideae (Rubiaceae) by Razafimandimbison [39].

Future Perspectives

The Bayesian phylogenetic approach used here provides a sound framework for examining the evolution of distinctive vegetative (growth habit) and reproductive traits (flower, inflorescence, and fruit structures), which have broad ecological importance and potential impact on our understanding of speciation and diversity. Methods, which rely upon mapping discrete character states across a phylogeny, inevitably reduce the complexity of character variation among a diverse group of species. Thus, the arborescent habit includes all non-liana woody shrubs and trees (large trunked trees (e.g., noni, *Morinda citrifolia*), suffrutescent plants (e.g., *M. buchii* Urb.), shrubs or treelets with scandent- or vining-branches (e.g., *Morinda royoc*), and weakly branching treelets (e.g., *Appunia debilis* Sandwith)). Similarly, the character states "large flower" and "small flower" and their diagnosis based upon corolla tube/lobe ratio summarize diverse flower sizes. The presence or absence of head inflorescences involves the complication of comparing much-branched inflorescences with unbranched inflorescences: either may have flowers in heads or not. Fruit fusion, though variable in degree, is summarized in the character states as simple or multiple. Despite the simplification of diverse characteristics into discrete character states, the essential outcomes of these analyses are clearly evident across the phylogenetic span of this inquiry: repeated shifts have occurred in the evolution of the growth habit, inflorescence architecture, flower size, and fruit across the species of the Morindeae. Moreover, the direction of these evolutionary changes has often been unexpected and at odds with currently accepted hypotheses. Finally, the findings of this study provide a new context for viewing patterns of character evolution and examining their ecological and developmental basis.

Materials and Methods

Taxon Sampling and Data Collection

The sampling used for this study coincided with the molecular phylogenetic study of Morindeae by Razafimandimbison et al. [9], on the basis of which new generic limits of the tribe were established. This latter study resulted in the transfer of all lianescent, dioecious *Morinda* species to *Gynochthodes* and all species of *Sarcopygme* Setch & Christoph. to *Morinda* [8–9]. Accordingly, the newly combined names of *Morinda* and *Gynochthodes*, respectively, were utilized in this study to replace the names of the sampled

Sarcopygme and lianescent *Morinda* used in Razafimandimbison et al. [9]. Five Morindeae taxa (*Appunia tenuiflora* (Benth.) Jacks & Hook.f., *Morinda royoc* L. 2, and *Gynochthodes candollei* Montrouz. 2, 4, and 5) with incomplete sequences were excluded from this study to decrease the percentage of missing information in the combined nrETS/nrITS/*trn*T-F matrix and obtain a well-resolved phylogeny of Morindeae for basing our ASRs. We investigated a total of 66 taxa, and all information about the voucher specimens and sequences used in the study is published in Razafimandimbison et al. [9].

All morphological characteristics of the five genera of Morindeae summarized in Table 1 were based on data from field notes made by SGR (for *Coelospermum, Gynochthodes,* and the paleotropical *Morinda*) and by TDM (for *Appunia* and the Neotropical *Morinda*). This was coupled with data compiled by SGR from herbarium specimens on loan from many herbaria (BR, K, L, MO, P, S, TAN, TEF, UPS, [40]) and the literature [8–9,12–13,21–22,24,26–27].

Laboratory Work and Phylogenetic Analyses

The protocols used for DNA extraction, amplification, and sequencing are outlined in Razafimandimbison et al. [9]. The alignment of the combined nrETS/nrITS/*trn*T-F data was re-adjusted after the removal of *A. tenuiflora, M. royoc* 2, and *G. candollei* 2, 4, and 5. We treated each of the three gene regions as a separate partition and selected likelihood models following Razafimandimbison et al. [9]. As a consequence, we applied separately parameterized GTR+ G models to the *trn*T-F and nrITS partitions and a separately parameterized HKY+ G model to the nrETS partition. The gamma distributed rate heterogeneity across sites was approximated with four discrete categories. Flat Dirichlet priors were applied to the state frequencies and to the substitution rates of the GTR model, whereas a flat beta distribution was used as prior for the transition-to-transversion rate. A uniform prior on the interval (0.1, 50) was applied to the gamma curve shape parameter α. The prior on branch lengths was an exponential distribution with mean 0.1. Rate heterogeneity across partitions was modeled according to a proportional model with a flat Dirichlet prior. Tree topologies were treated a priori equally likely. Three runs of Metropolis-coupled MCMC was run for 25×10^6 generations, each run starting from a random tree with initial branch lengths set to 0.1. Each run included four chains, three of which were incrementally heated to a temperature of 0.15 to ensure swap rates between adjacent chains between 10 and 70%. Every 1000th generation of the cold MCMC chain was sampled. Stationarity and convergence of runs, as well as the correlation of split frequencies between the runs were checked using the program AWTY [41]. We checked the effective sample size (ESS) of parameters using the program Tracer v.1.5.0 [42]. Trees sampled from the first 12.5×10^6 generations were discarded as burn-in. All saved trees (after excluding burn-ins) from the three independent runs were pooled for a consensus tree.

Reconstruction of Ancestral States

A variety of comparative phylogenetic methods have been used for reconstructing ancestral states of characters and mapping character changes across lineages: maximum parsimony [43–44], maximum likelihood [45], Bayesian inference [46–48], and stochastic character mapping [49]. The influences of method choice in reconstructing ancestral states of characters are well documented [2,50]; it has recently been demonstrated that homoplasious characters are sensitive to choice of method [2].

The Bayesian approach implemented in the computer program BayesTraits v. 1.0 [48] appears to preserve the highest amount of uncertainty in ASR of discrete characters [2,50]. It takes into account both phylogenetic uncertainty and branch length, and also permits one to explore a variety of models for character transition and to investigate nodes of interest [2,50]. We performed ASRs of the four characters of the tribe Morindeae (growth habit, inflorescence architecture, flower size, and fruit type) using the software BayesTraits as described by [48] and on two posterior tree samples, one in which all outgroup taxa (Fig. 2) had been pruned and one in which we kept the outgroup taxa *Mitchella repens* and *Damnacanthus indicus* of the tribe Mitchelleae, known from previous studies to be the closest relatives of the Morindeae [51]. *Pagamea guianensis* Aubl. and *Gaertnera phyllostachya* Baker were pruned from the analyses, because they represent the poorly sampled tribe Gaertnereae and appear on long branches in the phylogeny. Before proceeding with the ASR, we checked trees for the node-density artifact [52] using the on-line implementation at http://www.evolution.reading.ac.uk/pe/index.html. The following four discrete characters were reconstructed for the ingroup taxa: growth form lianescent (0), arborescent (including frutescent and suffrutescent plants of *Morinda*, the weak-stemmed shrubs or treelets of *Appunia*, and the scandent- or vining branched shrubs or treelets of *Morinda royoc* L., i.e., all non-liana woody shrubs and trees) (1), and herbaceous (only relevant for the outgroup taxon *M. repens*) (2); inflorescence headed (0) or non-headed (1); flowers large (corolla tube length/corolla lobe length >1) (0) or small (corolla tube length/corolla lobe length <1) (1); and fruits simple (0) or fused (1). State probabilities were estimated for the following seven selected nodes in the Bayesian majority rule consensus tree (Fig. 2): Morindeae (node A), the *Morinda-Coelospermum-Gynochthodes* clade (node B), the *Coelospermum-Gynochthodes* clade (node C), *Appunia* (node D), *Morinda* (node E), *Coelospermum* (node F), and *Gynochthodes* (node G). Node A corresponds to the root of the tree when the outgroup taxa had been excluded. In addition, we reconstructed the root node (joining the Mitchelleae and Morindeae) in the analyses involving the two outgroup taxa of Mitchelleae. Reversible-jump MCMC was used to integrate over models. For single binary characters, there are four possible models, one two-rate model in which forward (q_{01}) and backward (q_{10}) rates are free, one single-rate model in which q_{01} and q_{10} are constrained to be equal, two single-rate models in which either q_{01} or q_{10} is estimated, and the reverse rate is fixed to zero. Ratios of q_{01} to q_{10} deviating from 1 indicate that the rate of change in one direction is higher than in the opposite direction.

We used a uniform prior on the models and an exponential prior on rates, the mean of which was seeded by a uniform hyperprior on the interval (0, 10). By applying an exponential prior on rates we say that moderate rates are a priori more likely than high rates and that strong evidence from the data is required to accept high-rate estimates. We also included the branch-length transformation parameter κ in the model [53]. This parameter raises original branch lengths to the κ power. If $\kappa = 0$, all branches are equally long, i.e., change is independent of branch lengths. If $\kappa = 1$, branch lengths are not modified and change is perfectly proportional to the original branch lengths. $\kappa > 1$ indicates that change accelerates with increasing branch length and $0 < \kappa < 1$ indicates that change decelerates with decreasing branch lengths. The prior on κ is a uniform distribution on the interval (0, 5) (A. Meade, pers. com.). The MCMC was run for 220×10^6 generations, the first 20×10^6 of which were discarded as burnin. A sample was saved from the posterior every 1000th generation. The rate deviation of the normal distribution was set to obtain an MCMC acceptance rate between 20% and 40%. Each analysis was conducted three times to check that similar harmonic mean likelihoods were obtained across runs.

Acknowledgments

We thank the following herbaria for allowing access to their collections: BR, CAY, K, L, MO, P, S, TAN, TEF, UPS, and US; the DGF (Direction Générale des Forêts) and MNP (Madagascar National Parks) in Madagascar for issuing collecting permits for SGR; Missouri Botanical Garden, Madagascar Program for logistical support; Parc Botanique et Zoologique de Tsimbazaza and Missouri Botanical Garden, Madagascar Program (Lalao Andriamahefarivo and Faranirina Lantoarisoa) for arranging collecting permits for SGR; Kent Kainulainen for technical assistance with Figure 1; and Andrew J. Ford, Johan T. Johansson, Frédéric Tronquet, and Kent Kainulainen for kindly providing photos of Morindeae taxa.

Author Contributions

Conceived and designed the experiments: SGR. Performed the experiments: SGR SE. Analyzed the data: SGR SE. Contributed reagents/materials/analysis tools: BB. Wrote the paper: SGR SE TDM BB.

References

1. Endress PK (1994) Diversity and evolutionary biology of tropical flowers. Cambridge: University Press. 511p.
2. Xiang QY, Thomas DT (2008) Tracking character evolution and biogeographic history through time in Cornaceae - Does choice of methods matter? J Syst Evol 46: 349–374.
3. Eyde RH (1985) The case for monkey-mediated evolution in big-bracted dogwoods. Arnodia 45: 2–9.
4. Corlett RT (1998) Frugivory and seed dispersal by vertebrates in the Oriental (Indomalayan) Region. Biol Rev Cambridge Phil Soc 73: 413–448.
5. Rowe N, Speck T (2005) Plant growth forms: an ecological and evolutionary perspective. New Phytol 166: 61–72.
6. Razafimandimbison SG, Bremer B (2001) Tribal delimitation of Naucleeae (Rubiaceae): inference from molecular and morphological data. Syst Geogr Pl 71: 515–538.
7. Razafimandimbison SG, Bremer B (2002) Phylogeny and classification of Naucleeae (Rubiaceae) inferred from molecular (nrITS, *rbc*L, and *trn*T-F) and morphological data. Am J Bot 89: 1027–1041.
8. Razafimandimbison SG, Bremer B (2011) Nomenclatural changes and taxonomic notes in the tribe Morindeae (Rubiaceae). Adansonia 33: 281–307.
9. Razafimandimbison SG, McDowell TD, Halford DA, Bremer B (2009) Molecular phylogenetics and generic assessment in the tribe Morindeae (Rubiaceae-Rubioideae): how to circumscribe *Morinda* L. to be monophyletic? Mol Phylogenet Evol 52: 879–886.
10. Razafimandimbison SG, McDowell TD, Halford DA, Bremer B (2010) Origin of the pantropical and nutriceutical *Morinda citrifolia* L. (Rubiaceae): comments on its distribution range and circumscription. J. Biogeogr 37: 520–529.
11. McClatchey WC (2003) Diversity of growth forms, and uses in the *Morinda citrifolia* L. complex. In: Nelson SC, editor. Proceeding of the 2002 Hawai'i Noni Conference. Honolulu: University of Hawaii at Manoa. 5–10.
12. Taylor CM, Lorence D (2010) Rubiacearum Americanum Magna Hama Pars XXII: Notable new species of South American *Coutarea*, *Morinda*, *Patima*, and *Rosenbergiodebndron*. Novon 95–105.
13. Johansson JT (1988) Revision of *Caelospermum* Blume (Rubiaceae, Rubioideae, Morindeae). Blumea 33: 265–297.
14. Speck T, Rowe NP, Civeyrel L, Classen-Bockhoff R, Neinhuis C, et al. (2003) The potential of plant biomechanics in functional biology and systematics. In: Stuessy TF, Mayer V, Hörandl E, editors. Deep morphology: toward a renaissance of morphology in plant systematics. Lichtenstein: ARG Ganter Verlag. 241–271.
15. Whitlock BA, Hale AM (2011) The phylogeny of *Ayenia*, *Byttneria*, and *Rayleya* (Malvaceae s.l.) and its implication for the evolution of growth forms. Syst Bot 36: 129–136.
16. Lahaye R, Giveyrel L, Speck T, Rowe NP (2005) Evolution of shrub-like growth forms in the lianoid subfamily Secamonoideae (Apocynaceae s.l.) of Madagascar: phylogeny, biomechanics, and development. Amer J Bot 92: 1381–1396.
17. Robbrecht E (1988) Tropical woody Rubiaceae. Opera Bot Belg 1:1–271.
18. Raju AJS, Rao SP, Ezradaman V, Zafar R, Kalpana PR, et al. (2004) The hawkmoth *Macroglossum gyrans* and its interaction with some plant species at Visakhapatnam. Zoos' Print J 19: 1595–1598.
19. Halford DA, Ford AJ (2009) Two species of *Morinda* L. (Rubiaceae) from northeast Queensland. Austrobaileya 8: 81–90.
20. Halford DA, Ford AJ (2009) *Coelospermum purpureum* Halford & A.J. Ford (Rubiaceae), a new species from north-east Queensland. Austrobaileya 8: 69–76.
21. Burck MW (1883) Sur l'organisation florale chez quelques Rubiacées. Suite. Ann Jard Bot Buitenzorg 3: 109.
22. Johansson JT (1994) The genus *Morinda* (Morindeae, Rubioideae, Rubiaceae) in New Caledonia. Taxonomy and phylogeny. Opera Bot 122: 5–67.
23. Liu Y, Luo Z, Wu X, Bai X, Zhang D (2012) Pollinators with progressively shorter proboscis may have been driving the transition from large to small flowers and an accompanying increase in speciation rate. Pl Syst Evol 298: 775–785.
24. Wong KM (1984) A synopsis of *Morinda* (Rubiaceae) in the Malay Peninsula, with two new species. Malayan Nat J 38: 89–98.
25. Ortiz RD, Kellogg CEA, Werff HVD (2007) Molecular phylogeny of the mooseed family (Menispermaceae): implications for morphological diversification. Am J Bot 94: 1425–1438.
26. Burger W, Taylor CM (1993) Family # 202 Rubiaceae. In: Burger W, editor. Flora Costaricensis. Fieldiana Bot. 33: 1–333.
27. Tao C, Taylor CM (2011) Rubiaceae. Fl China 19: 220–229.
28. Endress PK (2010) Disentangling confusions in inflorescence morphology: patterns and diversity of reproductive shoot ramification in angiosperms. J Syst Evol 48: 225–239.
29. Feng CM, Xiang QY, Franks RG (2011) Phylogeny-based developmental analyses illuminate evolution of inflorescence architectures in dogwoods (*Cornus* s.l., Cornaceae). New Phytol 191: 850–869.
30. Bawa KS (1980) Evolution of dioecy in flowering plants. Ann Rev Ecol Syst 11: 15–39.
31. Bawa KS (1982) Outcrossing and the incidence of dioecism in island floras. Am Nat 119: 866–871.
32. Bawa KS (1994) Pollination of tropical dioecious angiosperms: a reassessment? No, not yet. Am J Bot 81: 456–460.
33. Bawa KS, Bullock SH, Perry DR, Coville RE, Grayum MH (1985) Reproductive biology of tropical lowland rain forest trees. II. Pollination systems. Am J Bot 72: 346–356.
34. Bawa KS, Opler PA (1975) Dioecism in tropical forest trees. Evolution 29: 167–179.
35. Muenchov GE (1987) Is dioecy associated with fleshy fruit? Am J Bot 74: 287–293.
36. Renner S, Ricklefs RE (1995) Dioecy and its correlates in the flowering plants. Am J Bot 82: 596–606.
37. Weller SG, Sakai AK (1999) Using phylogenetic approaches for the analysis of plant breeding system evolution. Ann Rev Ecol Syst 30: 167–199.
38. Tetsuto A (2006) Threatened pollination systems in native floras of the Ogasawara (Bonin) Islands. Ann Bot 98: 317–334.
39. Razafimandimbison SG (2002) A systematic revision of *Breonia* (Rubiaceae-Naucleeae). Ann Missouri Bot Gard 89: 1–37.
40. Holmgren PK, Holmgren NH, Barnett LC (1990) Index herbarium. Part I: the herbaria of the world, 8th edition. New York: New York Botanical Garden. 693 p.
41. Nylander JAA, Wilgenbusch JC, Warren DL, Swofford DL (2008) AWTY (Are We There Yet?): A system for graphical exploration of MCMC convergence in Bayesian phylogenetics. Bioinformatics 24: 581–583.
42. Rambaut A, Drummond AJ (2009) Tracer version 1.5. Edinburgh: University of Edinburgh. Available from http://tree.bio.ed.ac.uk/software/tracer/.
43. Maddison DR, Maddison WP (1992) MacClade: Analysis of phylogeny and character evolution. Version 3.0. Sunderland: Sinauer Associates.
44. Maddison D, Maddison WP (2007) Mesquite: a modular system for evolutionary analysis (online). Version 2.01. Available from http://mesquiteproject.org.
45. Pagel M (1999) The maximum likelihood approach to reconstructing ancestral character states of discrete on phylogenies. Syst Biol 48: 612–622.
46. Huelsenbeck JP, Ronquist F, Nielsen R, Bollback JP (2001) Bayesian inference of phylogeny and its impact on evolutionary biology. Science 294: 2310–2314.
47. Ronquist F (2004) Bayesian inference of character evolution. Trends Ecol Evol 9: 475–481.
48. Pagel M, Meade A, Barker D (2004) Bayesian estimation of ancestral states on phylogenies. Syst Biol 53: 673–684.
49. Huelsenbeck J, Bollback JP (2001) Empirical and hierarchical Bayesian estimation of ancestral states. Syst Biol 50: 351–366.
50. Ekman S, Andersen HL Wedin M (2008) The limitations of ancestral state reconstruction and the evolution of the ascus in the Lecanorales (Lichenized Ascomycota). Syst Biol 57: 141–156.
51. Razafimandimbison SG, Rydin C, Bremer B (2008) Evolution and trends in the Psychotrieae alliance (Rubiaceae): A rarely reported evolutionary change from one-seeded carpels to many-seeded carpels. Mol Phylogenet Evol 48: 207–223.
52. Venditti C, Meade A, Pagel M (2006) Detecting the node-density artifact in phylogeny reconstruction. Syst Biol 55: 637–643.
53. Pagel M (1994) Detecting correlated evolution on phylogenies: a general method for the comparative analysis of discrete characters. Proc Royal Soc London, ser B 255: 37–45.

Phenotypic Variation in Infants, Not Adults, Reflects Genotypic Variation among Chimpanzees and Bonobos

Naoki Morimoto[1]*, **Marcia S. Ponce de León**[2], **Christoph P. E. Zollikofer**[2]*

1 Laboratory of Physical Anthropology, Graduate School of Science, Kyoto University, Kyoto, Japan, 2 Anthropological Institute, University of Zurich, Zurich, Switzerland

Abstract

Studies comparing phenotypic variation with neutral genetic variation in modern humans have shown that genetic drift is a main factor of evolutionary diversification among populations. The genetic population history of our closest living relatives, the chimpanzees and bonobos, is now equally well documented, but phenotypic variation among these taxa remains relatively unexplored, and phenotype-genotype correlations are not yet documented. Also, while the adult phenotype is typically used as a reference, it remains to be investigated how phenotype-genotye correlations change during development. Here we address these questions by analyzing phenotypic evolutionary and developmental diversification in the species and subspecies of the genus *Pan*. Our analyses focus on the morphology of the femoral diaphysis, which represents a functionally constrained element of the locomotor system. Results show that during infancy phenotypic distances between taxa are largely congruent with non-coding (neutral) genotypic distances. Later during ontogeny, however, phenotypic distances deviate from genotypic distances, mainly as an effect of heterochronic shifts between taxon-specific developmental programs. Early phenotypic differences between *Pan* taxa are thus likely brought about by genetic drift while late differences reflect taxon-specific adaptations.

Editor: David Caramelli, University of Florence, Italy

Funding: This work was supported by the Swiss National Science Foundation (no. 3100A0-109344/1) and Japan Society for Promotion of Science Research Fellowship for Young Scientists (no. 251133). The funders had no role in study design, data collection and analysis, decision to publish, or preparation of the manuscript.

Competing Interests: The authors have declared that no competing interests exist.

* Email: morimoto@anthro.zool.kyoto-u.ac.jp (NM); zolli@aim.uzh.ch (CPEZ)

Introduction

The ready accessibility of population-wide genotypic and phenotypic data from humans and our closest relatives, the great apes, has spurred a large number of studies investigating the relationship between patterns of genotypic and phenotypic evolution. One central issue is the relative role of neutral versus adaptive evolutionary processes in shaping genotypic and phenotypic variation. A steadily growing number of studies indicates that variation of cranial morphology among modern human populations, and between modern humans and fossil hominins (species related more closely to modern humans than to great apes) largely reflects the effects of genetic drift, while only a small proportion of variation can be attributed to selection [1,2,3,4,5,6,7,8,9,10]. Fossil hominin aDNA now also permits insights into earlier phases of human population and evolutionary history at an unprecedented level of detail [11,12,13,14,15]. These analyses are limited, however, by the "aDNA preservation horizon", which is currently around 50,000 years BP for fossil hominin nDNA, and around 400,000 years BP for mtDNA from temperate zones [16].

One possible solution to investigate genotype-phenotype evolution beyond this horizon is to study living great ape species as a model system. The genus *Pan* represents the best model for this purpose, since it is our closest living relative, its species, subspecies and population structure is now genetically well-documented [17,18,19,20], and population history and genetic diversification

are well understood [18,19,21,22,23]. To date, two *Pan* species, *P. troglodytes* (common chimpanzee) and *P. paniscus* (bonobo) are recognized, and *P. troglodytes* is subdivided into four subspecies (*P. t. troglodytes*, *P. t. schweinfurthii*, *P. t. verus* and *P. t. ellioti*) [19]. Also, these *Pan* taxa have been the subject of detailed anatomical [24,25,26,27,28], morphological [29,30,31,32,33], phylogeographic [17,19,23,34], and behavioral [32,35,36,37,38,39,40] studies.

The extant *Pan* taxa are closely related to each other, which represents several advantages for comparative analyses. First, genotypic differences between taxa are small compared to variation within each taxon, such that the number of genes associated with phenotypic differentiation during (sub-) speciation is expected to be comparatively small [41]. Second, diversity among *Pan troglodytes* taxa represents patterns of incipient speciation, which are not yet blurred by long-term processes of taxon-specific specialization and/or convergence [42,43]. Also, we may note that the estimated time frame of *Pan* speciation [19,23] is comparable to that of our own genus *Homo* (ca. 2 million years).

Despite the increasing knowledge about *Pan* taxa, it still remains to be explored how changes at the level of the genotype are linked to changes at the level of the phenotype during speciation. The first aim of this study is thus to provide new phenotypic data documenting the evolutionary divergence of *Pan* taxa, and to relate this new evidence to the well-established body of genotypic evidence. While evolutionary studies traditionally focus on variation in craniodental features e.g. [44,45], we study here morphological variation of the femoral shaft (= diaphysis). The femur is a functionally highly constrained element of the postcranial skeleton, and can thus be expected to be under strong stabilizing selection.

Most studies exploring genotype-phenotype relationships in great apes and humans have naturally focused on adult morphologies. This is because taxon-specific morphological features are thought to be more clearly expressed in adults than in juveniles. However, there is clear evidence that the phenotypes of early ontogenetic stages, and patterns of developmental change, are highly informative about patterns of evolutionary divergence at the levels of skeletal structure e.g. [46,47,48,49,50,51,52,53], of locomotor behaviors [35,37], and of social interactions [54]. The second aim of this study is thus to expand the scope of genotype-phenotype comparisons by taking into account the perspective of ontogeny. Here we explore how genotype-phenotype relationships change during the development of the femoral diaphysis in the different *Pan* taxa, and relate this information to evolutionary change at the level of the genotype and phenotype. Specifically, we explore when during ontogeny the effects of drift versus selection become evident in taxon-specific phenotypes.

Measuring genotype-phenotype relationships is a complex endeavor, both theoretically and practically, and requires several model assumptions. In the standard model of quantitative population genetics, phenotypic variance V_P is the combination of genetic variance V_G and environmental variance V_E: $V_P = V_E + V_G$. Empirical data and theoretical considerations indicate that, for complex traits, phenotypic variance can be approximated by $V_P = V_E + V_A$, where V_A represents additive genetic variation (the portion of phenotypic variation that can be explained by the cumulative effects of allelic variation) [55]. The question of interest here is how V_P and V_A evolve in segregating populations. In a constant environment (V_E = const.), $V_P = V_A$, such that phenotypic variation reflects additive genotypic variation. Under these basic model assumptions, effects of drift and selection are typically estimated by comparing neutral genotypic distances with non-neutral distances [56,57,58,59,60]. The former distances (F_{ST}: genetic variation within subpopulation relative to total genetic variation [61,62]) are estimated from non-coding genetic markers thought to evolve under no selection such as STRs (short tandem repeats) and non-coding SNPs (single nucleotide polymorphisms) [63]. The latter distances are typically estimated from continuous quantitative genetic traits (Q_{ST}: evaluated in analogy to F_{ST} [64]) assuming additive genetic effects [64]. The question is whether Q_{ST} is equal to, smaller than, or larger than F_{ST}, which indicates neutral evolution, uniform or stabilizing selection, and diversifying selection, respectively [65].

Q_{ST} can be estimated from phenotypic distance P_{ST} [66] using a measure of heritability (h^2, proportion of additive genetic variance to phenotypic variance, V_A/V_P) [66,67,68,69,70]. In wild populations, heritability h^2 is often unknown and needs to be estimated from largely comparable lab studies. Furthermore, h^2 tends to change due to *in-vivo* environmental effects that accumulate during an individual's lifetime, and due to developmental changes in gene activation patterns [71,72,73]. In any case, estimates of h^2 affect the distance measures expressed by Q_{ST}, such that estimating the relative contribution of additive genetic and *in-vivo* environmental effects to P_{ST} remains a challenge [74].

A further challenge of $F_{ST} - Q_{ST}$ comparisons is the practical difficulty in measuring genotypic and phenotypic distances. Genotypic distances have been typically calculated using population-specific allele frequencies [75] (e.g., in Nei's standard distance D_a [76] and Cavalli-Sforza and Edwards chord distance D_{CH}

[77]). One problem is that sample sizes of wild populations are often limited, which makes it difficult to estimate population-specific allele frequencies and within-population variation. Complementary methods have thus been proposed, e.g. Principal Components Analysis (PCA) of genetic data [78,79]. While phenotypic distances have traditionally been evaluated from arrays of linear and angular measurements, geometric morphometrics (GM) offers elegant methods to quantify complex patterns of phenotypic variation [80,81,82]. In GM, biological form is typically measured by the spatial configuration (3D geometry) of anatomical points of reference, so-called landmarks [83,84]. Alternatively, various methods of GM have been developed to quantify the shape of landmark-free biological structures such as outlines [85], endocranial cavities [86] and longbone shafts [46,87]. One key feature of all GM methods is that phenotypic variation can simultaneously be represented in physical (three-dimensional) space by means of graphical interpolation and in multivariate space by means of PCA. PCA thus provides an ideal means to compare multivariate genotypic and phenotypic data independent of underlying population models.

Materials and Methods

Volumetric data of the femora of $N = 146$ *Pan* specimens were acquired with computed tomography (CT) ($N = 50$ *Pan troglodytes troglodytes*, $N = 39$ *P.t. schweinfurthii*, $N = 26$ *P. t. verus*, $N = 31$ *P. paniscus*; see Figs. S1 and S2, Table S1, and Text S1 and S2 for details on sample structure). *P. t. troglodytes* and *P. t. verus* specimens were obtained from the collections of the Anthropological Institute and Museum of the University of Zurich (AIMUZH), *P. t. schweinfurthii* specimens were obtained from the collections of the Royal Africa Museum, Tervuren, Belgium (MRA), and *P. paniscus* specimens were obtained from AIMUZH and MRA (Table S1). Each taxon is represented by four consecutive ontogenetic stages from infancy to adulthood. These were defined according to dental eruption: m2 (second deciduous molar erupted), M1, M2, M3 (first, second, third permanent molars erupted). In *Pan*, m2, M1, M2 and M3 erupt approximately at 0.5–0.83, 3, 7 and 11 years after birth, respectively [88].

Because femoral epiphyses are not yet ossified during the early stages of ontogeny, we focus on diaphyseal morphology. Effects of *in-vivo* bone modification in the femur have been studied in various *Pan* taxa, and it has been shown that ontogenetic changes in femoral morphology reflect an underlying developmental program that is fairly independent of environmental influences [87]. In other words, environmental variance V_E remains approximately constant throughout ontogeny [31,87,89] (see Text S3), which is an important prerequisite to estimate Q_{ST} from P_{ST} [74].

To quantify a specimen's diaphyseal surface morphology the transverse radius of curvature was evaluated for each point of the external (subperiosteal) surface, as specified in ref. [87]. The data of all specimens were then analyzed by means of morphometric mapping (MM) methods [87,90] (Fig. S3 and Text S1). MM is a landmark-free geometric morphometric method that permits dense sampling of data from smooth surfaces. It is thus well suited to quantify even subtle morphological differences in femoral shaft form between different taxa and/or developmental stages [87,91,92,93]. To correct for size differences between specimens, size is normalized by diaphyseal length and the median value of the radius of curvature. Shape variation is then decomposed into statistically independent shape components, which represent multivariate descriptors of the total femoral diaphyseal morphology. Since MM establishes a direct link between femoral geometry and its multivariate representation, patterns of inter- and intra-group variation can be visualized in multivariate shape space ("morphospace"; Fig. 1) as well as in real (physical) space (Fig. 2). To infer the femoral diaphyseal morphology and its developmental pattern in the last common ancestor (LCA) of *Pan* taxa, the phylogenetic tree of *Pan* taxa was projected onto the morphospace using a model of squared-change parsimony under a Brownian motion model [94] for each ontogenetic stage (Fig. S4) using the software package MorphoJ [95]. Also, MM was used to infer the infant and adult femoral diaphyseal morphology of the LCA (Fig. 2).

Mean femoral diaphyseal shape was calculated for each taxon at each ontogenetic stage i, and inter-taxon phenotypic (*i.e.*, morphometric) distance matrices M_i were calculated for each stage. As a phenotypic distance metric, the Euclidean distance in morphospace was used. Between-taxon quantitative genetic differentiation (Q_{ST}) was also estimated for each ontogenetic stage. To this end, pairwise Q_{ST}s were evaluated from P_{ST}s with the software RMET 5.0 [96,97], using PC scores (PC1–3) and a standard estimation of heritability $h^2 = 0.55$. This procedure resulted in stage-specific distance matrices Q_i.

Genotypic distances between *Pan* taxa (matrices F) were calculated from sequence datasets. The sequence data of 150,000 bp on 15 non-coding autosomal regions in $N = 74$ *Pan* specimens were obtained from GenBank (accession number: JF725992–727161 [22]). Inter-taxon genotypic distances were evaluated with various methods; Nei's standard distance D_a [76], Cavalli-Sforza and Edwards chord distance D_{CH} [77], and Euclidean distances in Patterson's PC space D_{PPC} [78,79]. Further, F_{ST} and R_{ST} from published sources were also used to construct genotypic distance matrices ([18,19,21,22]; refs. [18] and [19] use the same marker set) (Table S2).

Overall, three kinds of between-taxon distance matrices F (genotypic), M (phenotypic) and Q (quantitative genetic) were evaluated, and these matrices were used for $F-M$ and $F-Q$ ($F_{ST} - Q_{ST}$ [P_{ST}]) comparisons. The similarity between these distance matrices was evaluated with principal coordinate analysis (PCO), and assessed statistically with the Mantel test and resampling statistics (see Text S1 and Fig. S3 for details on PCO and resampling statistics). In brief, PCO transforms a between-taxon distance matrix into a "taxon constellation" (i.e., locations of taxa relative to each other in multivariate space). To assess the coincidence between genotypic and phenotypic taxon constellations, we used Procrustes analysis. This method superimposes two or more different constellations using a least-squares criterion. The Mantel test was performed using Relethford's MANTEL 3.1 (software programs RMET and MANTEL are available at http://employees.oneonta.edu/relethjh/programs/).

The fact that more than two *Pan* taxa are studied here facilitates rather than complicates F_{ST}–Q_{ST} comparisons. For $K = 2$ groups (populations or taxa), one F_{ST} distance is compared with one Q_{ST} distance. These need to be scaled appropriately with an estimate of h^2 to permit significant implications on neutral versus adaptive evolution, but h^2 is typically unknown. For $K > 2$ groups (this study: $K = 4$), the structures of two $K \times K$ distance matrices (F and Q) are compared, and scaling issues can be addressed with methods of matrix-matrix correlation and multidimensional scaling (MDS) such as the PCO method used here e.g. [2,7,98,99,100]. Assuming that $h^2(i) = $ const. for all groups at a given ontogenetic stage i, MDS will thus scale P_{ST} and Q_{ST} relative to F_{ST} even without explicit estimates of $h^2(i)$ (refs. [10,101]).

These matrix-matrix comparisons permit to assess whether the structure of a phenotypic (M) or quantitative-genetic (Q) distance matrix is similar to, or deviates from, a putatively neutral genotypic distance matrix F. Similarity would imply that M and Q are scaled versions of F (scaling factor h^2). An important assumption is that the genetic markers to estimate F_{ST} follow neutral evolution. This is critical to evaluate the relative role of neutral and adaptive processes from phenotypic data. The genetic markers used here to estimate F_{ST} represent non-coding regions

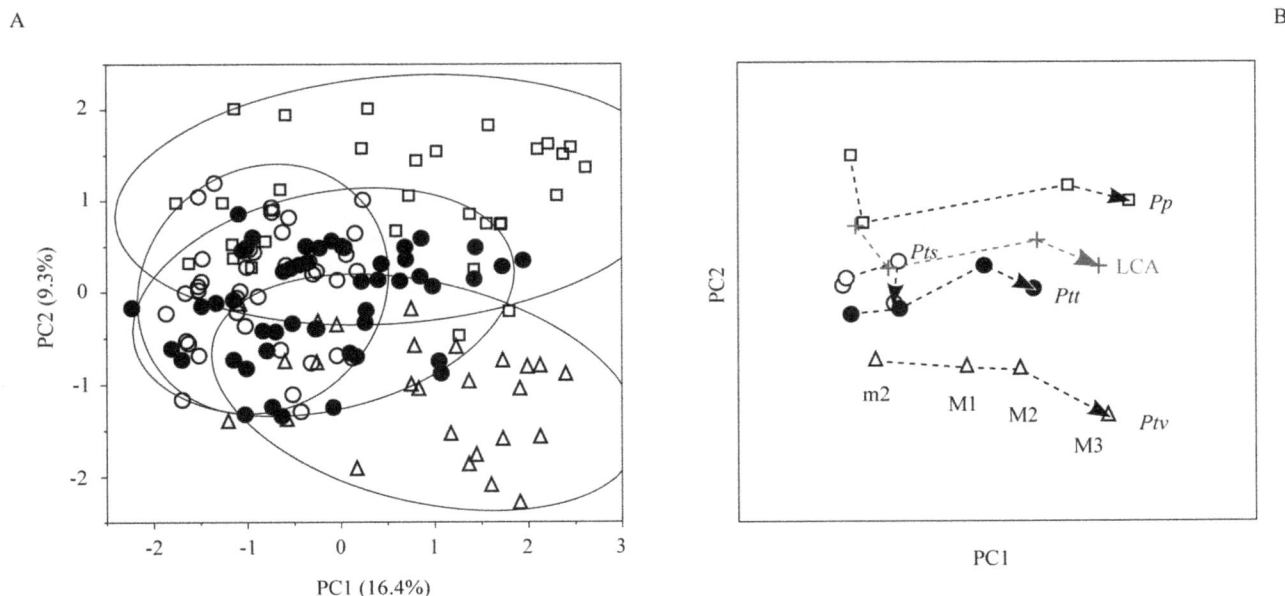

A

B

Figure 1. Femoral diaphyseal shape variation in an ontogenetic sample of *Pan* taxa. A: Variation along the first two principal components of shape, PC1 and PC2 (filled circles: *P.t. troglodytes*, open circles: *P.t. schweinfurthii*, open triangles: *P.t. verus*, open squares: *P. paniscus*). Solid outlines show 95%-density ellipses for each taxon. B: plot of mean shapes at consecutive ontogenetic stages. m2: second deciduous molar erupted; M1/M2/M3: permanent molars 1/2/3 erupted. Gray symbols and dashed line indicate the inferred shape at each ontogenetic stage and ontogenetic trajectory of the last common ancestor.

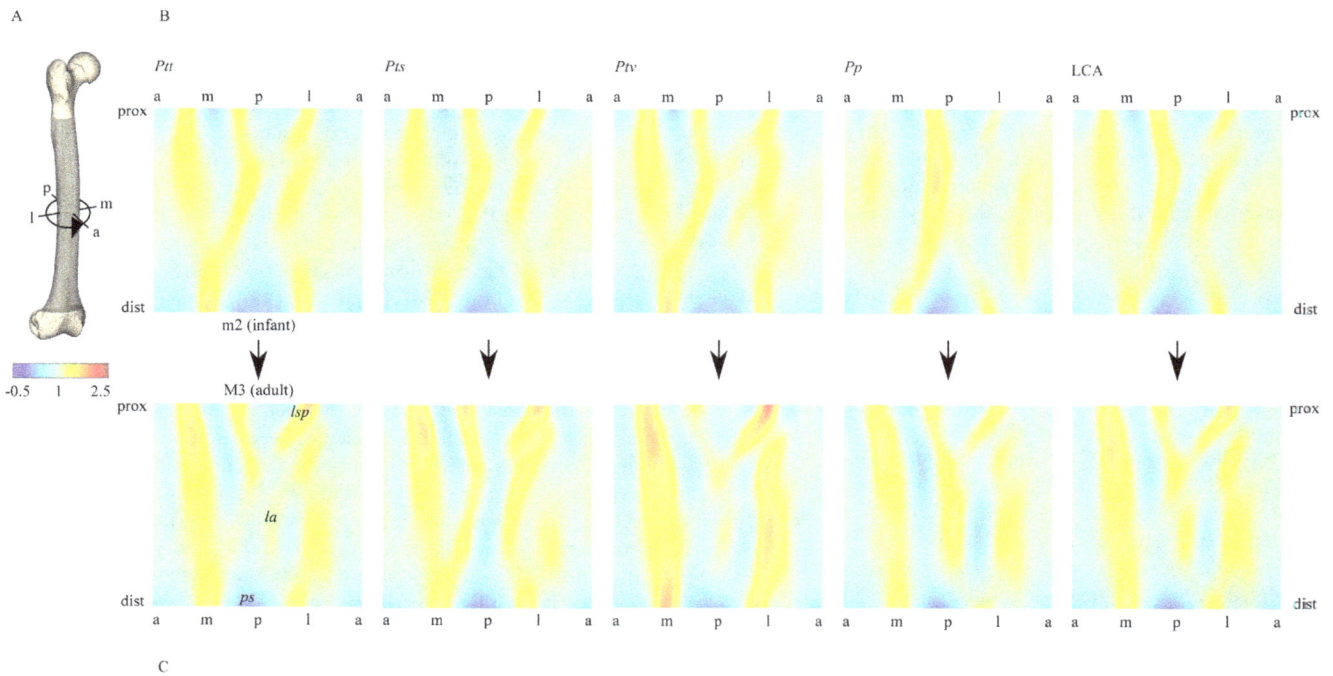

Figure 2. Taxon-specific femoral diaphyseal shapes. A: principle of morphometric map generation (anterior [0°] → medial [90°] → posterior [180°] → lateral [270°] → anterior [360°]). B, C: morphometric maps of taxon-specific morphologies at ontogenetic stages m2 (B, infant) and M3 (C, adult) (false-color images of external surface curvature [relative units]). *la*: linea aspera, *lsp*: lateral spiral pilaster, *ps*: popliteal surface.

[18,19,21,22], so it is reasonable to assume that variation reflects neutral processes.

Results

Fig. 1 shows commonalities and differences in femoral diaphyseal shape and shape variation between *Pan* taxa. The first two principal components represented here (PC1 and PC2) account for 25.7% of the total shape variation in the sample. There is substantial overlap between taxon-specific distributions of *P. t. troglodytes* and *P. t. schweinfurthii*, but almost no overlap between *P. paniscus* and *P. t. verus* (Fig. 1A). At each ontogenetic stage, taxon-specific mean shapes are statistically different from each other (Fig. 1B, Table S3). Furthermore, taxon-specific ontogenetic trajectories (see SI and refs. [102,103]) have statistically similar directions through morphospace (Fig. 1B and Table S4). Trajectories differ from each other, however, in their length (mostly along PC1), and in their location in morphospace (mostly along PC2) (Fig. 1B). Trajectories of *P. t. troglodytes* and *P. t. schweinfurthii* are in close vicinity, but the trajectory of the latter taxon is significantly shorter than that of the former. Compared to these taxa, the trajectory of *P. paniscus* is significantly longer (Fig. 1B, Table S5).

Differences between trajectories are already present at the m2 (infant) stage, indicating that taxon-specific femoral shape is established early during ontogeny. The differences in trajectory length indicate that the shape differences between *Pan* taxa increase toward adulthood. Longer trajectories indicate a larger total amount of femoral shape change during ontogeny, and possibly higher rates of shape change. Fig. 2 visualizes the corresponding real-space patterns of femoral diaphyseal shape change from infant to adult for each taxon. Each stage- and taxon-specific diaphyseal shape is represented here with a morphometric map (MM), which represents surface structures around (x-axis) and

along (y-axis) the femoral diaphysis. MMs visually confirm that taxon-specific femoral shape is present already at the m2 (infant) stage, and that taxon-specific features become more pronounced toward the M3 (adult) stage.

Using methods of squared-change parsimony [94], it is possible to infer the ontogenetic trajectory of the LCA of *Pan* taxa. The LCA trajectory lies between the trajectory of *P. paniscus* and the average trajectory of *P. troglodytes* taxa (Figs. 1, 2, S4). The length of the LCA trajectory is comparable to that of *P. t. troglodytes*, *P. t. verus*, and *P. paniscus*, but is longer than that of *P. t. schweinfurthii*.

All measures of genotypic distances (F_{ST}, D_a, D_{CH}, D_{PPC}) are highly correlated with each other (Table S6; Mantel test). Genotypic distances (F_{ST} and R_{ST}) evaluated from different marker sets [18,19,21,22] (Table S2) are also concordant with each other (Fig. S5), indicating that potential noise due to the small sample sizes of these studies does not greatly affect the results [104]. In all further comparative analyses we use D_{PPC} because evaluation of this distance measure does not presuppose estimation of within-group variance.

To assess the congruence between genotypic and phenotypic distance matrices, we projected the genotypic and phenotypic PCO data into the same multidimensional space and aligned them with Procrustes Analysis. Patterns of phenotypic similarity among *Pan* taxa (P_{ST}) are overall congruent with patterns of genetic similarity (D_{PPC}, F_{ST}) (Figs. 3A, S5, Tables 1, S6, S7). Figs. 3A and S5 show that the match between genotypic and phenotypic data is closest at the m2 (infant) stage (Table 1; $p < 0.05$, Mantel test). While taxa advance along their ontogenetic trajectories, patterns of phenotypic variation tend to deviate from the pattern of genetic variation (Fig. 3A, S5). These results are statistically supported by a resampling test (Fig. 3B). **F–M** correlation is highest at the m2 (infant) stage ($R^2 = 0.80$, $p = 0.02$), and is lowest at the M3 (adult) stage ($R^2 = 0.20$, $p = 0.37$). Likewise, the **F–M** correlation between genotypic and phenotypic distances evaluated by a Mantel test is

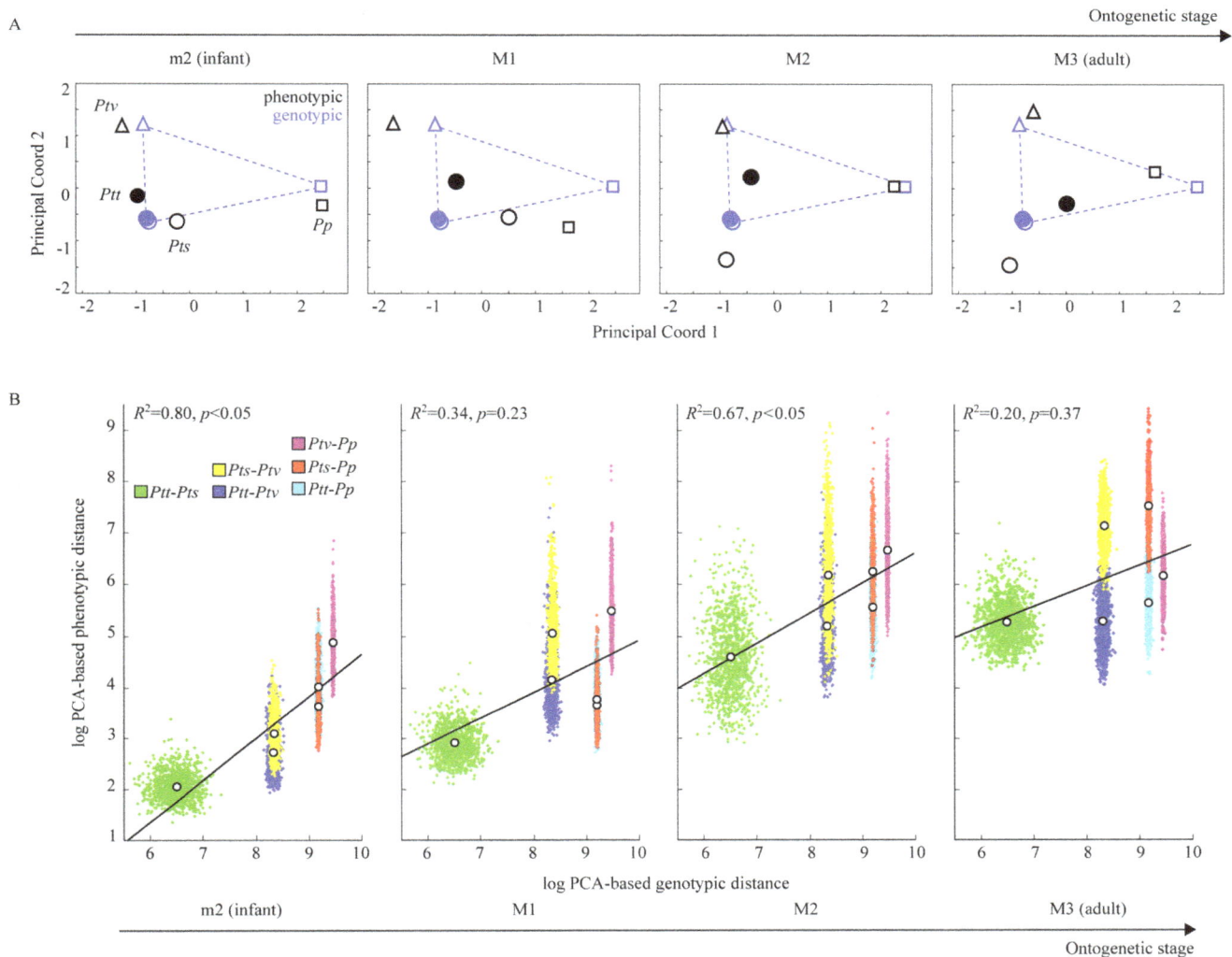

Figure 3. Comparison of genotypic and phenotypic distances between *Pan* taxa. A: Principal Coordinates Analysis (PCO) permits representation of genotypic and phenotypic distance data in the same multivariate space. The four subgraphs show phenotypic data (black dots) for consecutive ontogenetic stages m2, M1, M2, and M3, and genotypic data (same blue dots for all stages). For graphical clarity genotypic data points, which are independent of ontogenetic stage, are connected with dashed lines. Note that during ontogeny the phenotypic distance configuration departs from the neutral genetic distance configuration (see also Fig. S5). B: Correlation between phenotypic and neutral genetic distances between taxa. Each point cloud consists of 1000 randomly sampled phenotypic and genotypic distances between individuals belonging to different *Pan* taxa (resampling procedures are explained in Text S1). Correlation of phenotypic and neutral genetic distances is highest at the m2 (infant) stage and declines towards adulthood (M3). Genetic and phenotypic distances are normalized by their respective median values. Note overall increase of phenotypic distance between taxa toward adulthood.

highest at the m2 stage (Tables 1 and S7). **F–M** correlation is also significant at the M2 stage, but to a lesser extent than at the m2 stage. The decline in **F–M** correlation from infancy to adulthood thus follows a non-monotonous pattern.

The results of **F–Q** comparisons (i.e., standard $F_{ST} - Q_{ST}$ tests) are similar to the results obtained with PCA/PCO analyses (Table 1). The correlation between F_{ST} and Q_{ST} [P_{ST}] is highest at the m2 (infant) stage ($R^2 = 0.72$, $p < 0.01$), and lowest at the M3 (adult) stage ($R^2 = 0.10$, $p = 0.35$). The finding that correlation between genotypic and phenotypic markers decreases during ontogeny is thus independent of the method of genotypic and phenotypic distance measurement.

Discussion

Investigating the evolutionary divergence between populations and/or closely related taxa at the level of genes and phenes, and inferring underlying processes of selection and drift, has become an important research topic in primatology and anthropology [1,2,3,4,5,6,7,8,9,10]. Progress in this field is fostered by the availability of ever-increasing volumes of genomic and phenomic data, and sophisticated analytical tools to compare patterns of genotypic and phenotypic variation. While DNA sequence data provide *static* structural information about the genome, data at any level above the DNA (from the transcriptome to morphology) provide *dynamic* structural information about the phenotype, which changes during ontogeny. Interestingly, the effect of ontogenetic time on correlations between genotypic and phenotypic variation is still relatively unexplored. For example, ontogenetic time does

Table 1. Correlation between genotypic and genotypic distance matrices.

		m2 (infant)	M1	M2	M3 (adult)
genotypic distance[1]−phenotypic distance[2] (Mantel[3])	R^2	**0.84**	0.15	**0.64**	0.18
	p	**<0.01**	0.2609	**<0.01**	0.3478
genotypic distance−phenotypic distance (resampling[4])	R^2	**0.80**	0.34	**0.67**	0.20
	p	**0.015**	0.23	**0.045**	0.37
F_{ST}−Q_{ST} test[5] (Mantel)	R^2	**0.72**	0.40	**0.67**	0.10
	p	**<0.01**	0.087	**<0.01**	0. 3478

[1]Euclidean distance in Patterson's PC space.
[2]Euclidean distance in morphospace (shape PCs).
[3]correlation (R^2) and significance levels (p) evaluated with Mantel test (1000 permutations).
[4]evaluated with resampling statistics (see methods; Fig. S3C).
[5]estimate of heritability h^2: 0.55.

not appear as an explicit variable in the standard equations relating V_P to V_A, nor is it typically considered explicitly in F_{ST}−Q_{ST} comparisons.

To fill this gap, we studied femoral diaphyseal shape change in the genus *Pan* and compared patterns of phenotypic divergence (both during development and evolution) with patterns of genotypic divergence. The results presented here yield several new insights into evolutionary and developmental links between genotypic and phenotypic diversification in *Pan*. Before any general inferences can be drawn, it should be reminded, however, that the genotypic and phenotypic data sets studied here represent subsets of the total genotypic/phenotypic evidence that is potentially available for such studies.

The close correspondence between genotypic and phenotypic distances at the earliest ontogenetic stage analyzed here (the m2 stage) gives rise to two alternative hypotheses; H0: if the molecular markers of refs. [18,19,21,22] track neutral evolution then the observed pattern of phenotypic evolution is "neutral-like" within the constraints imposed by stabilizing selection (often described as "wandering around an adaptive optimum" [105,106,107]); H1: if the pattern of phenotypic distances between taxa is the result of selection and adaptation, then the molecular markers are non-neutral and carry an adaptive signal. Given the good evidence for neutrality in the molecular markers [108] used here, hypothesis H1 is less likely. Also, the congruence of the genotypic distance patterns evaluated from different marker types (Fig. S5) suggests that H1 is less likely, since one would expect that selection acts differently on different marker types. Our data thus support hypothesis H0, which implies that morphological variation of the femoral diaphysis in infant *Pan* reflects neutral evolutionary diversification between taxa rather than taxon-specific adaptation.

While phenotypic distances between *Pan* taxa at the m2 stage are in good concordance with genotypic distances ($R^2 = 0.8$; Fig. 3B), correlations are lower at later ontogenetic stages, and reach a value of $R^2 = 0.2$ at adulthood (Figs. 3, S5; Table 1). As already reported in earlier studies [74,109,110], correlations between molecular and phenotypic markers are typically low, and this has been interpreted in two ways: (1) that (non-coding) molecular marker variation does not adequately represent the quantitative genetic variation of coding genes that becomes manifest in the phenotype, and (2) that environmental variation has a significant influence on V_P, and hence on Q_{ST}.

The ontogenetic data presented in this study provide an empirical basis to test these hypotheses. The high correlation ($R^2 = 0.80$) between inter-taxon molecular and phenotypic varia-

tion at the m2 stage (Fig. 3B) indicates that, during early ontogeny, molecular marker variation indeed represents quantitative genetic variation. Departure from genotypic-phenotypic correspondence during later ontogenetic stages might indicate *in-vivo* modification of the femoral shaft morphology, indicating an increasing contribution of V_E to V_P over ontogenetic time. Given the evidence from earlier studies investigating *in-vivo* effects on femoral shaft morphology [31,87,89,111], however, this interpretation is unlikely, and V_E remains fairly constant from infancy to adulthood [87]. Another possible explanation is size allometry, implying that the observed pattern of phenotypic divergence reflects differences in adult body mass among *Pan* taxa. Since direct data on body mass are available for only few specimens in this study, we use the taxon-specific body masses reported in the literature [112] to test this hypothesis. Taxon-specific means of PC scores at adulthood are not correlated with adult body masses of *Pan* taxa (Fig. S6, Table S8). It is thus unlikely that the observed pattern of divergence is due to allometry.

After excluding major environmental and allometric effects, it appears most likely that phenotypic divergence is caused by genetically determined taxon-specific developmental programs. This implies that the genetic variance V_G changes during ontogenetic time t: $V_P(t) = V_E + V_G(t)$. In the present case, it is not known whether $V_G(t)$ can be approximated by additive genetic variance $V_A(t)$ alone, or whether non-additive effects have to be taken into account. Several alternative hypotheses must thus be considered to explain the observed pattern of phenotypic divergence. Under the additive genetic variance model [$V_P(t) = V_E + V_A(t)$], our hypothesis is that the genes mediating early ontogeny (up to the m2 stage) evolved by neutral processes ($Q_{ST} \sim F_{ST}$), whereas the genes mediating late ontogeny (from m2 to adulthood) evolved under selection ($Q_{ST} > F_{ST}$), probably as an adaptation to taxon-specific locomotor regimes. An alternative hypothesis is that non-additive effects V_N are a function of developmental time: $V_P(t) = V_E + V_A(t) + V_N(t)$. With the currently available empirical evidence, we cannot decide between these hypotheses. In any case, the molecular markers used here to estimate V_G are unlikely to represent variation in the actual coding genes that cause V_P to increase over ontogenetic time [110].

In spite of these uncertainties, our data permit inferences on the developmental mechanisms that cause taxon-specific differences in femoral diaphyseal shape, and to speculate on their genetic basis. As shown in Fig. 1B, taxon-specific ontogenetic trajectories set out at similar locations along PC1, but differ in their length. This pattern indicates differences in taxon-specific *rates* of development from the m2 stage onward, resulting in significant differences between adult morphologies. Evolutionary divergence via differential developmental rates is well-known as heterochrony. It thus appears that heterochronic shifts played a major role in the development of the adult femoral morphologies of *Pan* taxa. Such shifts might be effected by changes in a small number of developmental genes [113,114], which are difficult to trace with standard molecular markers, but might be further investigated with whole-genome comparisons [23].

It has been shown that a marked paedomorphic pattern is expressed in the skull relative to the postcranial skeleton in bonobos (*P. paniscus*) compared to common chimpanzees (*P. troglodytes*) [33,115,116]. The present study shows that the femur also exhibits heterochronic variation among *Pan* taxa. It is interesting to note that the femoral diaphysis of bonobos exhibits peramorphic development compared to common chimpanzees. This mosaic structure of evolutionary developmental modification is in concordance with the observation made earlier that *P. paniscus* is not just a paedomorphic chimpanzee [116,117]. It remains to be elucidated whether cranial and postcranial ontogenies are governed by the same set of "heterochrony genes", which have different local effects, or whether different sets of heterochrony genes are expressed locally [113,118].

Currently, we can only speculate about the adaptive significance of taxon-specific heterochronic modifications of femoral development, since more comparative field data are necessary to specify the diversity of locomotor behaviors and their ontogeny in all *Pan* taxa. The inferred femoral diaphyseal morphology and developmental trajectory of the *Pan* LCA indicates that the peramorphic pattern as in *P. paniscus*, *P. t. troglodytes* and *P. t. verus* represents the primitive state whereas the paedomorphic (rate hypomorphic) pattern as in *P. t. schweinfurthii* represents the derived state. The inferred femoral diaphyseal morphology of the LCA at the adult stage is relatively close to the morphology of adult *P. paniscus* and *P. t. troglodytes*. The locomotor repertoire of the LCA might thus have been close to that of adult *P. paniscus* and *P. t. troglodytes*.

The data presented here provide empirical insights into the role of neutral and adaptive evolutionary mechanisms at the level of genes and phenes. In the system studied here, it appears that – among the closely related *Pan* taxa – early developmental genes evolve mostly neutrally and produce neutral taxon-specific phenotypes, while selection acts on late developmental genes (most likely on those involved in the regulation of developmental rates) and produces adaptive phenotypes.

Evidence for this pattern of evolution has also been found in the hominin clade. For example, the pattern of genotypic and phenotypic divergence between *Homo sapiens* and *H. neanderthalensis* is concordant with a model of neutral evolution by mutation and drift [6,8]. Also, parallel ontogenetic trajectories and heterochronic divergence during late ontogeny are reported for *Homo sapiens* and *H. neanderthalensis* [51]. Likewise, it appears that genetic and phenotypic divergence in early *Homo* and between modern human populations is governed to a large extent by neutral processes [1,3,5,10,119,120]. Our data indicate that this pattern of evolution might be more general than currently thought and characteristic not only for *Homo* but also for the taxa descending from the last common ancestor of humans and chimpanzees. It remains to be tested whether the observed patterns of developmental diversification in *Pan* also characterize the developmental diversification in other great ape taxa.

As a general outcome of this study, we may state that the phenotype of early developmental stages conveys a better neutral phylogenetic signal than the adult phenotype. This finding is in contrast with the traditional notion that the fully-developed adult phenotype is most significant for taxonomy and phyletic inference. The close match between patterns of neutral molecular and phenotypic variation during early ontogeny, however, indicates that immature individuals are of special relevance to infer phylogenetic relationships, although taxon-specific features are less expressed in early stages of ontogeny (Fig. 2B) compared to late stages (Fig. 2C). Femoral diaphyseal morphology of hominoids provides a good example. While adult-based studies often show similarities of femoral diaphyseal morphology among great apes to the exclusion of humans e.g. [121,122,123], at an early developmental stage humans and chimpanzees are grouped together to the exclusion of gorillas [46]. Furthermore, our data may explain why previous meta-analyses showed a generally low correlation of F_{ST} and Q_{ST} in adult phenotypes [74,110,124]. Generalizing our findings to hominoid (and hominin) evolution, the comparison of immature and adult phenotypes will permit a better discrimination between phyletic and adaptive signals in the phenotype.

Supporting Information

Figure S1 Geographical distribution and taxonomy of *Pan* (modified from ref. [22]).

Figure S2 Sample structure by taxon and age class. A, distribution of femoral diaphyseal length (measured as the linear distance between proximal and distal epiphyseal lines). B: distribution of femoral diaphyseal cross-sectional area (measured as the median of cross-sectional areas between proximal and distal epiphyses). Filled circles: *P.t. troglodytes*, open circles: *P.t. schweinfurthii*, open triangles: *P.t. verus*, open squares: *P. paniscus*. Age classes: m2: second deciduous molar erupted; M1/M2/M3: permanent molars 1/2/3 erupted. Each symbol represents a specimen; black lines/whiskers indicate mean and range; red boxes and whiskers indicate first/third quartiles and median.

Figure S3 Principle of morphometric mapping. A, 3D representation of the right femur. B, principle of cylindrical projection (anterior [0°] → medial [90°] → posterior [180°] → lateral [270°] → anterior [0°]).

Figure S4 Phylogenetic tree in morphospace. The phylogenetic tree (blue lines; diamonds indicate the inferred state of last common ancestor at each ontogenetic stage) of the genus *Pan* is projected onto the shape space using a model of squared-change parsimony. A: m2 (infant), B: M1, C: M2, D: M3 (adult) stage. Gray symbols and line indicate the inferred ontogenetic trajectory of the last common ancestor.

Figure S5 Phenetic and genetic similarity between *Pan* taxa. Principal Coordinates Analysis (PCO) of phenetic and genetic distance data. Phenetic data (black) are given for consecutive ontogenetic stages (connected with dashed lines). Genetic data (color) are from ref. [18] (blue), ref. [19] (green), ref. [21] (red), and ref. [22] (magenta). Note that during ontogeny the phenetic distance configuration departs from the genetic distance configuration.

Figure S6 Correlation of taxon-specific means of adult body weight and PC scores. Taxon-specific means of adult body weight was calculated as a mean of male and female body weight taken from the literature [112].

Table S1 Specimen list. The following specimens are used in this study. AIMUZH: Anthropological Institute and Museum of University of Zurich. MRA: Royal Africa Museum, Tervuren, Belgium.

Table S2 Genetic distances between *Pan* taxa (F_{ST} and R_{ST}).

Table S3 Phenetic distances between taxon-specific mean shapes.

Table S4 Divergence of ontogenetic vector.

Table S5 F-test on taxon-specific variance along PC1.

Table S6 Correlation of genetic and phenetic distances.

Table S7 Correlation between phenetic and genetic

Table S8 Correlation between PC scores and taxon-specific adult body masses.

Text S1 Materials and methods.

Text S2 Habitats of *Pan* taxa.

Text S3 *In-vivo* bone modification in the femur of *Pan* taxa.

Acknowledgments

We thank P. Jans, E Gillissen and W Coudyzer for help with sample preparation and CT scanning. The comments of H. Bagheri, M. Kobayashi, and T. Marques-Bonet are greatly acknowledged. We are also grateful to the anonymous reviewers for their valuable comments and suggestions.

Author Contributions

Conceived and designed the experiments: NM MSPDL CPEZ. Performed the experiments: NM. Analyzed the data: NM. Contributed reagents/materials/analysis tools: NM MSPDL CPEZ. Contributed to the writing of the manuscript: NM MSPDL CPEZ.

References

1. Ackermann RR, Cheverud JM (2004) Detecting genetic drift versus selection in human evolution. Proc Natl Acad Sci U S A 101: 17946–17951.
2. Roseman CC (2004) Detecting interregionally diversifying natural selection on modern human cranial form by using matched molecular and morphometric data. Proceedings of the National Academy of Sciences of the United States of America 101: 12824–12829.
3. Roseman CC, Weaver TD (2004) Multivariate apportionment of global human craniometric diversity. American Journal of Physical Anthropology 125: 257–263.
4. Harvati K, Weaver TD (2006) Human cranial anatomy and the differential preservation of population history and climate signatures. Anat Rec A Discov Mol Cell Evol Biol 288: 1225–1233.
5. Roseman CC, Weaver TD (2007) Molecules versus morphology? Not for the human cranium. Bioessays 29: 1185–1188.
6. Weaver TD, Roseman CC, Stringer CB (2007) Were neandertal and modern human cranial differences produced by natural selection or genetic drift? Journal of Human Evolution 53: 135–145.
7. Smith HF, Terhune CE, Lockwood CA (2007) Genetic, geographic, and environmental correlates of human temporal bone variation. American Journal of Physical Anthropology 134: 312–322.
8. Weaver TD, Roseman CC, Stringer CB (2008) Close correspondence between quantitative- and molecular-genetic divergence times for Neandertals and modern humans. Proc Natl Acad Sci U S A 105: 4645–4649.
9. von Cramon-Taubadel N, Weaver TD (2009) Insights from a quantitative genetic approach to human morphological evolution. Evolutionary Anthropology 18: 237–240.
10. Betti L, Balloux F, Hanihara T, Manica A (2010) The relative role of drift and selection in shaping the human skull. American Journal of Physical Anthropology 141: 76–82.
11. Reich D, Green RE, Kircher M, Krause J, Patterson N, et al. (2010) Genetic history of an archaic hominin group from Denisova Cave in Siberia. Nature 468: 1053–1060.
12. Krause J, Fu Q, Good JM, Viola B, Shunkov MV, et al. (2010) The complete mitochondrial DNA genome of an unknown hominin from southern Siberia. Nature 464: 894–897.
13. Green RE, Krause J, Briggs AW, Maricic T, Stenzel U, et al. (2010) A draft sequence of the Neandertal genome. Science 328: 710–722.
14. Hawks J (2013) Significance of Neandertal and Denisovan genomes in human evolution. Annual Review of Anthropology 42: 433–449.
15. Sankararaman S, Patterson N, Li H, Pääbo S, Reich D (2012) The date of interbreeding between Neandertals and modern humans. PLoS Genetics 8: e1002947.
16. Meyer M, Fu Q, Aximu-Petri A, Glocke I, Nickel B, et al. (2014) A mitochondrial genome sequence of a hominin from Sima de los Huesos. Nature 505: 403–406.
17. Gonder MK, Disotell TR, Oates JF (2006) New genetic evidence on the evolution of chimpanzee populations and implications for taxonomy. International Journal of Primatology 27: 1103–1127.
18. Becquet C, Patterson N, Stone AC, Przeworski M, Reich D (2007) Genetic structure of chimpanzee populations. PLoS Genetics 3: e66.
19. Gonder MK, Locatelli S, Ghobrial L, Mitchell MW, Kujawski JT, et al. (2011) Evidence from Cameroon reveals differences in the genetic structure and histories of chimpanzee populations. Proceedings of the National Academy of Sciences of the United States of America 108: 4766–4771.
20. Auton A, Fledel-Alon A, Pfeifer S, Venn O, Ségurel L, et al. (2012) A fine-scale chimpanzee genetic map from population sequencing. Science 336: 193–198.
21. Fischer A, Pollack J, Thalmann O, Nickel B, Pääbo S (2006) Demographic history and genetic differentiation in apes. Current Biology 16: 1133–1138.
22. Fischer A, Prufer K, Good JM, Halbwax M, Wiebe V, et al. (2011) Bonobos fall within the genomic variation of chimpanzees. PLoS ONE 6: e21605.
23. Prado-Martinez J, Sudmant PH, Kidd JM, Li H, Kelley JL, et al. (2013) Great ape genetic diversity and population history. Nature 499: 471–475.
24. Champneys F (1871) On the muscles and nerves of a chimpanzee (*Trogloáytes niger*) and a *Cynocepalus anubis*. J Anat Lond 6: 176–211.
25. Crass E (1952) Musculature of the hip and thigh of the chimpanzee: a comparison to man and other primates. PhD thesis, Univ Wisconsin.
26. Sigmon BA (1974) A functional analysis of pongid hip and thigh musculature. Journal of Human Evolution 3: 161–185.
27. Stern JT (1972) Anatomical and functional specializations of human gluteus maximus. American Journal of Physical Anthropology 36: 315–338.
28. Morimoto N, Zollikofer CPE, Ponce de León MS (2011) Femoral morphology and femoropelvic musculoskeletal anatomy of humans and great apes: a comparative virtopsy study. Anatomical Record 294: 1433–1445.
29. Bourne GH, editor (1969) The Chimpanzee, Vol. 1: Anatomy, Behavior, and Diseases of Chimpanzees. Basel/New York: Karger.
30. Bourne GH, editor (1971) The Chimpanzee, Vol. 4: Behavior, Growth, and Pathology of Chimpanzees. Basel/New York: Karger.
31. Carlson KJ, Doran-Sheehy DM, Hunt KD, Nishida T, Yamanaka A, et al. (2006) Locomotor behavior and limb bone morphology in individual free-ranging chimpanzees. Journal of Human Evolution 50: 394–404.
32. Doran DM (1993) Comparative locomotor behavior of chimpanzees and bonobos - the influence of morphology on locomotion. American Journal of Physical Anthropology 91: 83–98.
33. Lieberman DE, Carlo J, Ponce de León MS, Zollikofer CP (2007) A geometric morphometric analysis of heterochrony in the cranium of chimpanzees and bonobos. J Hum Evol 52: 647–662.
34. Morin PA, Moore JJ, Chakraborty R, Jin L, Goodall J, et al. (1994) Kin selection, social-structure, gene flow, and the evolution of chimpanzees. Science 265: 1193–1201.
35. Doran DM (1992) The ontogeny of chimpanzee and pygmy chimpanzee locomotor behavior: a case-study of paedomorphism and its behavioral-correlates. Journal of Human Evolution 23: 139–157.
36. Doran DM (1996) Comparative positional behabior of the African apes. In: McGrew MC, Marchant LF, Nishida T, editors. Great Ape Societies. Cambridge: Cambridge University Press.

37. Doran DM (1997) Ontogeny of locomotion in mountain gorillas and chimpanzees. J Hum Evol 32: 323–344.
38. Doran DM, Jungers WL, Sugiyama Y, Fleagle J, Heesy C (2002) Multivariate and phylogenetic approaches to understanding chimpanzee and bonobo behavioral diversity. In: Boesch C, Hohmann G, Marchant LF, editors. Behavioural diversity in Chimpanzees and Bonobos. Cambridge: Cambridge University Press. 14–34.
39. Goodall J (1986) The Chimpanzees of Gombe. Cambridge: Harvard University Press.
40. McGrew MC, Marchant LF, Nishida T, editors (1996) Great Ape Societies. Cambridge: Cambridge University Press.
41. Nei M (2007) The new mutation theory of phenotypic evolution. Proc Natl Acad Sci U S A 104: 12235–12242.
42. West-Eberhard MJ (2005) Developmental plasticity and the origin of species differences. Proc Natl Acad Sci U S A 102 Suppl 1: 6543–6549.
43. Shaw KL, Mullen SP (2011) Genes versus phenotypes in the study of speciation. Genetica 139: 649–661.
44. Collard M, Wood B (2000) How reliable are human phylogenetic hypotheses? Proceedings of the National Academy of Sciences 97: 5003–5006.
45. Strait DS, Grine FE (2004) Inferring hominoid and early hominid phylogeny using craniodental characters: the role of fossil taxa. Journal of Human Evolution 47: 399–452.
46. Morimoto N, Zollikofer CPE, Ponce de León MS (2012) Shared human-chimpanzee pattern of perinatal femoral shaft morphology and its implications for the evolution of hominin locomotor adaptations. PLoS ONE 7: e41980.
47. Geiger M, Forasiepi AM, Koyabu D, Sanchez-Villagra MR (2013) Heterochrony and post-natal growth in mammals - an examination of growth plates in limbs. Journal of Evolutionary Biology 20: 12279.
48. Koyabu D, Endo H, Mitgutsch C, Suwa G, Catania KC, et al. (2011) Heterochrony and developmental modularity of cranial osteogenesis in lipotyphlan mammals. Evodevo 2: 21.
49. Wilson LAB, Sánchez-Villagra MR (2011) Evolution and phylogenetic signal of growth trajectories: the case of chelid turtles. Journal of Experimental Zoology Part B: Molecular and Developmental Evolution 316B: 50–60.
50. Sánchez M (2012) Embryos in Deep Time: The Rock Record of Biological Development: University of California Press. 265 p.
51. Ponce de León MS, Zollikofer CPE (2001) Neanderthal cranial ontogeny and its implications for late hominid diversity. Nature 412: 534–538.
52. Ackermann RR (2005) Ontogenetic integration of the hominoid face. Journal of Human Evolution 48: 175–197.
53. Gunz P, Neubauer S, Golovanova L, Doronichev V, Maureille B, et al. (2012) A uniquely modern human pattern of endocranial development. Insights from a new cranial reconstruction of the Neandertal newborn from Mezmaiskaya. Journal of Human Evolution 62: 300–313.
54. Palagi E, Cordoni G (2012) The right time to happen: play developmental divergence in the two Pan species. PLoS ONE 7: e52767.
55. Hill WG, Goddard ME, Visscher PM (2008) Data and theory point to mainly additive genetic variance for complex traits. PLoS Genetics 4: e1000008.
56. Morgan TJ, Evans MA, Garland T Jr, Swallow JG, Carter PA (2005) Molecular and quantitative genetic divergence among populations of house mice with known evolutionary histories. Heredity 94: 518–525.
57. Whitlock MC (2008) Evolutionary inference from Q_{ST}. Molecular Ecology 17: 1885–1896.
58. Smith HF (2009) Which cranial regions reflect molecular distances reliably in humans? Evidence from three-dimensional morphology. American Journal of Human Biology 21: 36–47.
59. Brommer JE (2011) Whither P_{ST}? The approximation of Q_{ST} by P_{ST} in evolutionary and conservation biology. Journal of Evolutionary Biology 24: 1160–1168.
60. Edelaar PIM, Burraco P, Gomez-Mestre I (2011) Comparisons between Q_{ST} and F_{ST}–how wrong have we been? Molecular Ecology 20: 4830–4839.
61. Wright S (1969) Evolution and the Genetics of Populations, Vol. II. The Theory of Gene Frequencies. Chicago: University of Chicago Press.
62. Wright S (1978) Evolution and the Genetics of Populations, Vol. IV. Variability Within and Among Natural Populations. Chicago: University of Chicago Press.
63. Holsinger KE, Weir BS (2009) Genetics in geographically structured populations: defining, estimating and interpreting FST. Nature Reviews Genetics 10: 639–650.
64. Spitze K (1993) Population-structure in Daphnia obtusa: quantitative genetic and allozymic variation. Genetics 135: 367–374.
65. Leinonen T, McCairns RJS, O'Hara RB, Merila J (2013) Q_{ST}–F_{ST} comparisons: evolutionary and ecological insights from genomic heterogeneity. Nature Reviews Genetics 14: 179–190.
66. Leinonen T, Cano JM, MÄKinen H, MerilÄ J (2006) Contrasting patterns of body shape and neutral genetic divergence in marine and lake populations of threespine sticklebacks. Journal of Evolutionary Biology 19: 1803–1812.
67. Merilä J, Björklund M, Baker AJ (1997) Historical demography and present day population structure of the greenfinch, carduelis chloris-an analysis of mtDNA control-region sequences. Evolution 51: 946–956.
68. Storz JF (2002) Contrasting patterns of divergence in quantitative traits and neutral DNA markers: analysis of clinal variation. Molecular Ecology 11: 2537–2551.
69. Saint-Laurent R, Legault M, Bernatchez L (2003) Divergent selection maintains adaptive differentiation despite high gene flow between sympatric rainbow smelt ecotypes (Osmerus mordax Mitchill). Molecular Ecology 12: 315–330.
70. Slate J (2013) From beavis to beak color: a simulation study to examine how much qtl mapping can reveal about the genetic architecture of quantitative traits. Evolution 67: 1251–1262.
71. Atchley WR (1984) Ontogeny, timing of development, and genetic variance-covariance structure. American Naturalist 123: 519–540.
72. Charmantier A, Perrins C, McCleery RH, Sheldon BC (2006) Age-dependent genetic variance in a life-history trait in the mute swan. Proceedings of the Royal Society B-Biological Sciences 273: 225–232.
73. Lesser KJ, Paiusi IC, Leips J (2006) Naturally occurring genetic variation in the age-specific immune response of Drosophila melanogaster. Aging Cell 5: 293–295.
74. Pujol B, Wilson AJ, Ross RIC, Pannell JR (2008) Are Q_{ST}–F_{ST} comparisons for natural populations meaningful? Molecular Ecology 17: 4782–4785.
75. Kalinowski ST (2002) Evolutionary and statistical properties of three genetic distances. Molecular Ecology 11: 1263–1273.
76. Nei M, Tajima F, Tateno Y (1983) Accuracy of estimated phylogenetic trees from molecular data. II. Gene frequency data. Journal of Molecular Evolution 19: 153–170.
77. Cavalli-Sforza LL, Edwards AW (1967) Phylogenetic analysis. Models and estimation procedures. Am J Hum Genet 19: 233–257.
78. Patterson N, Price AL, Reich D (2006) Population structure and eigenanalysis. PLoS Genetics 2: e190.
79. Price AL, Patterson NJ, Plenge RM, Weinblatt ME, Shadick NA, et al. (2006) Principal components analysis corrects for stratification in genome-wide association studies. Nat Genet 38: 904–909.
80. Slice DE (2007) Geometric morphometrics. Annual Review of Anthropology 36: 261–281.
81. Mitteroecker P, Gunz P (2009) Advances in geometric morphometrics. Evolutionary Biology 36: 235–247.
82. Zollikofer CPE, Ponce de León MS (2005) Virtual Reconstruction. A Primer in Computer-Assisted Paleontology and Biomechanics. Hoboken: NJ: John Wiley & Sons.
83. Bookstein F (1991) Morphometric Tools for Landmark Data: Geometry and Biology. Cambridge: Camnridge University Press.
84. Gunz P, Mitteroecker P, Bookstein FL (2005) Semilandmarks in Three Dimensions. In: Slice DE, editor. Developments in Primatology: Progress and Prospects. New York: Springer.
85. Kuhl F, Giardina C (1982) Elliptic Fourier features of a closed contour. Computer graphics and image processing 18: 236–258.
86. Specht M, Lebrun R, Zollikofer CPE (2007) Visualizing shape transformation between chimpanzee and human braincases. Visual Computer 23: 743–751.
87. Morimoto N, Zollikofer CPE, Ponce de León MS (2011) Exploring femoral diaphyseal shape variation in wild and captive chimpanzees by means of morphometric mapping: a test of Wolff's Law. Anatomical Record 294: 589–609.
88. Bolter DR, Zihlman AL (2011) Brief communication: dental development timing in captive Pan paniscus with comparisons to Pan troglodytes. American Journal of Physical Anthropology 145: 647–652.
89. Carlson K, Sumner D, Morbeck M, Nishida T, Yamanaka A, et al. (2008) Role of nonbehavioral factors in adjusting long bone siaphyseal atructure in free-ranging Pan troglodytes. International Journal of Primatology 29: 1401–1420.
90. Zollikofer CPE, Ponce de León MS (2001) Computer-assisted morphometry of hominoid fossils: the role of morphometric maps. In: De Bonis L, Koufos G, Andrews P, editors. Phylogeny of the Neogene Hominoid Primates of Eurasia. Cambridge: Cambridge University Press. 50–59.
91. Bondioli L, Bayle P, Dean C, Mazurier A, Puymerail L, et al. (2010) Technical note: Morphometric maps of long bone shafts and dental roots for imaging topographic thickness variation. American Journal of Physical Anthropology 142: 328–334.
92. Puymerail L (2013) The functionally-related signatures characterizing the endostructural organisation of the femoral shaft in modern humans and chimpanzees. Comptes Rendus Palevol.
93. Puymerail L, Ruff CB, Bondioli L, Widianto H, Trinkaus E, et al. (2012) Structural analysis of the Kresna 11 Homo erectus femoral shaft (Sangiran, Java). Journal of Human Evolution 63: 741–749.
94. Maddison WP (1991) Squared-change parsimony reconstructions of ancestral states for continuous-valued characters on a phylogenetic tree. Systematic Biology 40: 304–314.
95. Klingenberg CP (2011) MorphoJ: an integrated software package for geometric morphometrics. Molecular Ecology Resources 11: 353–357.
96. Relethford JH, Blangero J (1990) Detection of differential gene flow from patterns of quantitative variation. Human Biology 62: 5–25.
97. Relethford JH, Crawford MH, Blangero J (1997) Genetic drift and gene flow in post-famine Ireland. Human Biology 69: 443–465.
98. Sæther SA, Fiske P, Kålås JA, Kuresoo A, Luigujõe L, et al. (2007) Inferring local adaptation from Q_{ST}–F_{ST} comparisons: neutral genetic and quantitative trait variation in European populations of great snipe. Journal of Evolutionary Biology 20: 1563–1576.
99. Chapuis E, Martin G, Goudet J (2008) Effects of selection and drift on G matrix evolution in a heterogeneous environment: a multivariate Qst–Fst test with the freshwater snail Galba truncatula. Genetics 180: 2151–2161.

100. Martin G, Chapuis E, Goudet J (2008) Multivariate Q_{st}–F_{st} comparisons: a neutrality test for the evolution of the G matrix in structured populations. Genetics 180: 2135–2149.

101. Relethford JH (2004) Boas and beyond: Migration and craniometric variation. American Journal of Human Biology 16: 379–386.

102. Zollikofer CPE, Ponce de León MS (2006) Neanderthals and modern humans - chimps and bonobos: similarities and differences in development and evolution. In: Harvati K, Harrison T, editors. Neanderthals Revisited: New Approaches and Perspectives. New York: Springer. 71–88.

103. Penin X, Berge C, Baylac M (2002) Ontogenetic study of the skull in modern humans and the common chimpanzees: neotenic hypothesis reconsidered with a tridimensional Procrustes analysis. Am J Phys Anthropol 118: 50–62.

104. Willing E-M, Dreyer C, van Oosterhout C (2012) Estimates of genetic differentiation measured by Fst do not necessarily require large sample sizes when using many SNP markers. PLoS ONE 7: e42649.

105. Hunt G (2007) The relative importance of directional change, random walks, and stasis in the evolution of fossil lineages. Proceedings of the National Academy of Sciences of the United States of America 104: 18404–18408.

106. Haller BC, Hendry AP (2014) Solving the paradox of stasis: squashed stabilizing selection and the limits of detection. Evolution 68: 483–500.

107. Wagner GP (1996) Apparent stabilizing selection and the maintenance of neutral genetic variation. Genetics 143: 617–619.

108. Kirk H, Freeland JR (2011) Applications and implications of neutral versus non-neutral markers in molecular ecology. Int J Mol Sci 12: 3966–3988.

109. McKay JK, Latta RG (2002) Adaptive population divergence: markers, QTL and traits. Trends in Ecology & Evolution 17: 285–291.

110. Reed DH, Frankham R (2001) How closely correlated are molecular and quantitative measures of genetic variation? A meta-analysis. Evolution 55: 1095–1103.

111. Carlson KJ, Lublinsky S, Judex S (2008) Do different locomotor modes during growth modulate trabecular architecture in the murine hind limb? Integrative and Comparative Biology 48: 385–393.

112. Smith RJ, Jungers WL (1997) Body mass in comparative primatology. Journal of Human Evolution 32: 523–559.

113. Somel M, Franz H, Yan Z, Lorenc A, Guo S, et al. (2009) Transcriptional neoteny in the human brain. Proceedings of the National Academy of Sciences of the United States of America 106: 5743–5748.

114. Somel M, Liu X, Tang L, Yan Z, Hu H, et al. (2011) MicroRNA-driven developmental remodeling in the brain distinguishes humans from other primates. PLoS Biology 9: e1001214.

115. Mitteroecker P, Gunz P, Bookstein FL (2005) Heterochrony and geometric morphometrics: a comparison of cranial growth in *Pan paniscus* versus *Pan troglodytes*. Evolution & Development 7: 244–258.

116. Shea BT (1983) Pedomorphosis and neoteny in the pygmy chimpanzee. Science 222: 521–522.

117. Shea BT (1983) Allometry and heterochrony in the African apes. American Journal of Physical Anthropology 62: 275–289.

118. Khaitovich P, Enard W, Lachmann M, Pääbo S (2006) Evolution of primate gene expression. Nature Reviews Genetics 7: 693–702.

119. Lynch M (1989) Phylogenetic hypotheses under the assumption of neutral quantitative-genetic variation. Evolution 43: 1–17.

120. Roseman CC (2004) Detecting interregionally diversifying natural selection on modern human cranial form by using matched molecular and morphometric data. Proc Natl Acad Sci U S A 101: 12824–12829.

121. Lovejoy CO, Meindl RS, Ohman JC, Heiple KG, White TD (2002) The Maka femur and its bearing on the antiquity of human walking: Applying contemporary concepts of morphogenesis to the human fossil record. American Journal of Physical Anthropology 119: 97–133.

122. Ruff CB (2002) Long bone articular and diaphyseal structure in old world monkeys and apes. I: Locomotor effects. American Journal of Physical Anthropology 119: 305–342.

123. Carlson KJ (2005) Investigating the form-function interface in African apes: Relationships between principal moments of area and positional behaviors in femoral and humeral diaphyses. Am J Phys Anthropol 127: 312–334.

124. Leinonen T, O'Hara RB, Cano JM, Merilä J (2008) Comparative studies of quantitative trait and neutral marker divergence: a meta-analysis. Journal of Evolutionary Biology 21: 1–17.

5

Sex Allocation in a Polyembryonic Parasitoid with Female Soldiers: An Evolutionary Simulation and an Experimental Test

Max Bügler[1], Polychronis Rempoulakis[2], Roei Shacham[3], Tamar Keasar[2]*, Frank Thuijsman[4]

1 Chair of Computational Modeling and Simulation, Technische Universität München, Munich, Germany, 2 Biology and Environment, University of Haifa, Tivon, Israel, 3 Evolutionary and Environmental Biology, University of Haifa, Haifa, Israel, 4 Knowledge Engineering, Maastricht University, Maastricht, The Netherlands

Abstract

Parasitoid wasps are convenient subjects for testing sex allocation theory. However, their intricate life histories are often insufficiently captured in simple analytical models. In the polyembryonic wasp *Copidosoma koehleri*, a clone of genetically identical offspring develops from each egg. Male clones contain fewer individuals than female clones. Some female larvae develop into soldiers that kill within-host competitors, while males do not form soldiers. These features complicate the prediction of *Copidosoma*'s sex allocation. We developed an individual-based simulation model, where numerous random starting strategies compete and recombine until a single stable sex allocation evolves. Life-history parameter values (e.g., fecundity, clone-sizes, larval survival) are estimated from experimental data. The model predicts a male-biased sex allocation, which becomes more extreme as the probability of superparasitism (hosts parasitized more than once) increases. To test this prediction, we reared adult parasitoids at either low or high density, mated them, and presented them with unlimited hosts. As predicted, wasps produced more sons than daughters in all treatments. Males reared at high density (a potential cue for superparasitism) produced a higher male bias in their offspring than low-density males. Unexpectedly, female density did not affect offspring sex ratios. We discuss possible mechanisms for paternal control over offspring sex.

Editor: Gabriele Sorci, CNRS, Université de Bourgogne, France

Funding: The study was supported by the Israel Science Foundation (http://www.isf.org.il/), grant number 414/10. The funders had no role in study design, data collection and analysis, decision to publish, or preparation of the manuscript.

Competing Interests: The authors have declared that no competing interests exist.

* E-mail: tkeasar@research.haifa.ac.il

Introduction

Sex allocation theory exemplifies the importance of frequency-dependent selection in population ecology. The theory successfully explains the 1:1 sex ratio observed in many species, by positing a fitness advantage to the rarer sex. At equal reproductive value of male function and female function in the population, both sexes have identical fitness and equilibrium is reached (reviewed in [1], [2]). Extensions of the basic model predict the ecological conditions that select for "extraordinary sex ratios" [3]. Such circumstances include mating before dispersal among the offspring of a small number of females (Local Mate Competition), which favors the evolution of a female-biased sex allocation [4]. Deviations from equal sex allocation also occur under unequal competitive ability of males and females, which favors over-production of the weaker competitors [5]. Models that predict the effects of life-history features on sex allocation are often analytical, and focus on predicting the steady-state sex ratios rather than the dynamics leading to them [1].

Parasitoid wasps, hymenopterans whose larvae feed on the tissues of an arthropod host and eventually kill it, have a haplo-diploid sex determination system. That is, diploid females develop from fertilized eggs, while haploid males develop from unfertilized ones. This provides parasitoids with much flexibility in sex allocation, which is reflected in sex ratio shifts in response to various environmental conditions. For example, parasitoid sex

ratios are affected by the presence of mated conspecifics [6], the sex of eggs laid by conspecifics [7], host size [8,9], host age [10] and previous parasitism of the host [11]. In addition, maternal investment in parasitoids is restricted to the production of eggs and choice of a host for oviposition, thus resource allocation towards producing sons and daughters is similar and minimal. The proportion of sons at egg-laying (the primary sex ratio) therefore approximates the allotment of resources to male production (sex allocation). These characteristics make parasitoids popular models for testing predictions of sex allocation models. However, to make testable predictions, models need to incorporate the often complex life-history features of the parasitoids. These features include details on the mating systems, body sizes, developmental rates and competitive abilities of males and females, spatial distribution of hosts and dispersal abilities of adult parasitoids. It is often difficult to introduce this biological realism into standard analytical models of sex allocation. It is no less challenging to generate testable predictions from the models, and to subject them to experimental examination.

In the present study we aim to harness individual-based evolutionary simulation modeling to address these difficulties. We model sex allocation in the parasitoid *Copidosoma koehleri* (Encyrtidae: Hymenoptera) as a test case, because the unique life-history of this parasitoid challenges analytical sex allocation modeling. *C. koehleri* is a koinobiont egg-larval parasitoid of the potato tuber moth, *Phthorimaea operculella* (Gelechiidae: Lepidoptera). One egg is generally laid per ovipositor insertion into a host egg, but hosts often receive an additional egg during a later oviposition, thus becoming super-parasitized [12]. Eggs that develop within super-parasitized hosts may be of one sex or of different sexes. The eggs develop polyembryonically within the host, i.e. cleave repeatedly after oviposition to form clones of same-sex genetically identical wasps. Female clones contain more individuals, on average, than male clones (mean±SD clone sizes: 45.7±10.9 for females, 32.4±10.4 for males, [13]). Body size correlates negatively with the number of individuals in a brood (the total number of wasps emerging from a host, [14]). Female clones contain a morphologically distinct soldier larva that develops precociously, attacks competitors and dies before emerging from the host [15]. A high proportion of hosts contain two or more clones [16], which are usually mothered by different females [12,17]. In hosts that contain both a male and a female clone, the female soldier eliminates some of the males, reducing the proportion of emerging males in the brood [16]. The duration of development is similar for males and females, thus both sexes emerge together in mixed-sex broods. Females are sexually receptive immediately after emergence, and often mate with male brood-mates in mixed-sex broods in the laboratory (pers. obs.). The extent of local mating in natural populations is unknown, but was shown to be high in the related species *C. floridanum* [18]. These intricacies complicate the prediction of ESS sex ratios in *Copidosoma* and other polyembryonic parasitoids, to the extent that "the development of specific theory is required" [2].

Qualitative considerations suggest that sex allocation in *C. koehleri* should balance between conflicting selective pressures: Equal reproductive value of male function and female function in large populations should result in equal sex allocation. However, males form smaller clones than females, and are outcompeted by them in mixed-sex broods. This generates a potential for Local Resource Competition between male and female larvae that develop within the same host. Such competition may favor an increased parental allocation towards production of the weaker competitor, that is, a male-biased primary sex ratio [5]. On the other hand, mating before dispersal may select for the production of excess females (a female-biased primary sex ratio) because of Local Mate Competition [3]. The evolved parental strategy regarding sex allocation should reflect the relative importance of these selective factors.

We predicted *C. koehleri*'s sex allocation using a simulation model, which allowed us to capture the details of the wasps' life history and to follow the evolutionary dynamics of sex ratios in the simulated populations. The general simulation approach follows Lewis et al. [19], who used an individual-based model, evolving according to a Genetic Algorithm, to study factors that constrain optimal sex allocation in parasitoids. Lewis et al. [19] assumed that parasitoid females select the sex of each offspring according to host size, and modeled the assessment of host sizes using Artificial Neural Networks. In contrast, our model does not include an Aritificial Neural Network component. Instead, sex allocation is a genetically determined trait that is not affected by the parasitoids' host encounter experience. We tested the main prediction generated by the model in a laboratory experiment.

Methods

Modeling

We set up an individual-based model, implemented in Java, to simulate the population dynamics of hosts and wasps for hundreds of generations, and to track the evolution of sex allocation in the

parasitoid population. The model description follows the ODD (Overview, Design concepts, Details) protocol [20,21]:

Purpose. JWasp is an agent-based discrete time simulation model designed to explore the behavior of *Copidosoma* parasitoid wasps, having the capability of choosing the sex of their offspring. Since it is unclear on what basis the wasps perform the sex choice, the model has been designed to evolve strategies according to different criteria, based on a random set of starting strategies.

State variables and scales. The model consists of three intuitive components. These are the environment, host and wasp models. The environment model holds three sets, which contain all entities in the simulation (Table 1). The Wasp model represents the individual wasps (the agents, Table 2). A Strategy is described using the variables of Table 3.

Process overview and scheduling. The model is based on the following list of rules:

- Female wasps lay eggs into hosts. The hatched larvae feed on the host and eventually kill it. After maturing into adults, they disperse.
- Generations do not overlap (i.e., parents are dead before the offspring disperse).
- Unfertilized virgin females can only produce male eggs.
- Fertilized females can freely choose the sex of their eggs.
- Sons of virgins have lower virility than sons of fertilized females.
- A fraction of the wasps can mate on the host, before dispersal.
- Some "soldier larvae" develop from female eggs. They kill unrelated competing larvae within the same host.
- An excess of eggs within a host will prematurely kill it and all other contained eggs.

Since generations do not overlap, each generation is initialized with a new set of hosts. The simulation allows each wasp to lay an egg at each time step, where the number of steps is the life span of the wasp. Furthermore, depending on its virility, each male can mate with one or several virgin females. After all time steps are completed the eggs proliferate, and the wasps' offspring develop, and possibly mate, inside the host according to their respective survival distribution. Their daughters receive a strategy that is a linear combination of the parents' strategies according to the "Strategy inheritance" parameter. Sons inherit their mothers' strategies.

Design concepts - Basic principles. The simulation model is individual-based, and does not have a spatial component. It tracks large populations of parasitoids and their hosts. The initial composition of these populations can include individuals with several life-history strategies. The frequencies of the strategies may change over the course of the simulation due to selection and recombination, until a single stable strategy evolves. The modl's

Table 1. The environment component of the simulation model.

Variable name	Brief description
Female wasps	Set of female wasps in simulation
Hosts	Set of hosts in simulation
Male wasps	Set of male wasps in simulation

Variables are listed in alphabetical order.

Table 2. The environment component of the simulation model.

Variable name	Brief description
Eggs laid	Counts the number of eggs produced by a female
Female genes	ID of mother
Females fertilized	The number of females fertilized by a male
Generations	Number of generations to simulate
ID	Unique ID
Lifespan	The lifespan of a wasp in number of eggs it can lay
Male genes	ID of male that fertilized a female
Sex	Boolean indicating the sex
Strategy	Strategy description
Virgin	Boolean indicating virginity of a female
Virgin's son	Boolean indicating whether a male is a virgin's son

Variables are listed in alphabetical order.

parameter values can either be fixed throughout the simulation, or be allowed to evolve. In the present work we allowed the sex allocation random variable to evolve, and kept the remaining components of host and wasp strategies constant. Fixed parameter values were estimated from previous laboratory studies of *P. operculella*'s and *C. koehleri*'s life-histories [14–17], and are listed in Table S1.

Design concepts – Emergence. The emergent outputs of the model are the primary and secondary sex ratios of the wasp population, that is, the proportion of male eggs laid by a female and the proportion of adult males in the population. We examined how these sex ratios are affected by the wasps' mating system and by the risk of intra-specific competition due to superparasitism.

Design concepts – Adaptation. The wasps base their sex choice on the state of the host encountered. The larvae developing inside the host interact with other larvae by competing for resources (exploitation competition) and through possible aggres-

sion of "soldiers" (interference competition). Females outcompete males, hence the adaptive benefit of producing females should depend on the extent of competition for host resources. As in other sex allocation models, the adaptive benefit from each sex is also negatively frequency-dependent, i.e. production of the rare sex should be advantageous. In addition, the wasps' degree of mating before dispersal may affect the adaptive benefit of male vs. female production according to Local Mate Competition theory. The model predicts the combined effect of these, possibly conflicting, selective forces on the evolution of sex ratios.

Design concepts – Objectives. The male wasps' main goal is to fertilize a maximal number of females. The females' objective is to distribute their available eggs in a manner that would maximize their offspring survival. For the evolutionary development, the long term average size of the population is used as a fitness measure for comparing between strategies.

Design concepts – Learning. The model incorporates no learning. The wasps' behavior is genetically determined.

Design concepts – Prediction. Agents do not predict the future, they just act based on the rules matching the current situation. The success of their behavior is evaluated through their reproductive performance, using a natural-selection-like procedure.

Design concepts – Sensing. Female wasps can sense the content of a host to some degree when choosing the sex of their next egg. There are three different possible information levels: (a) No information: the female only distinguishes empty hosts from parasitized ones. (b) Information on relatedness: the wasp also senses whether a parasitized host contains genetically related competitors. (c) Information on sex: the wasp senses the relatedness and sex of the competitors.

Design concepts – Interaction. Males and females interact only by mating and by competing inside a host. Females compete with each other through their egg-laying behavior. When encountering a host they react to the history of that host.

Design concepts – Stochasticity. Hosts are presented to wasps randomly. Furthermore, the wasps' sex choices and survival rates are random variables, sampled from the respective distributions.

Table 3. The strategy component of the model.

Variable name	Brief description
Egg count influence	Influence of the number of eggs in a host on the probability to lay an additional egg
Host count	Number of hosts in simulation
Host limit	Carrying capacity of the host
Initial sex distribution	Initial proportion of males
Initial virgin ratio	Initial ratio of virgins
Initial wasp count	Initial number of wasps introduced into the simulation
In-host mating ratio	Proportion of wasps mating on the host, before dispersal
Sex choice distributions	The probability to lay a male egg. There is a value for empty hosts, and for hosts containing each combination of sex (male, female, mixed) and relatedness (related, unrelated, mixed)
Strategy inheritance	Proportion of male genes inherited by daughters
Survival distributions	For each possible host allocation, the number of wasps developing from a given egg is sampled from a normal distribution. We defined distribution parameters for male and female single-sex broods, and for male and female wasps within related and unrelated mixed-sex broods
Virgin son virility	Number of females that a virgin's son can fertilize
Virility	Number of females that a mated female's son can fertilize

Variables are listed in alphabetical order.

Design concepts – Collectives. The agents are not grouped and related wasps do not work together directly. Related embryos inside a host do form a collective, though, as they do not attack each other.

Design concepts – Observation. Data are extracted from the model by observing the number of male and female wasps, the population size, the occupation of the hosts, the number of eggs and the number of premature host deaths. If multiple strategies are simulated, we observe the number of distinct strategies in the simulation, which will change in frequency due to recombination and selection, and will eventually converge to a single strategy.

Initialization. The model is initialized with a number of empty hosts and a randomly sampled wasp population according to the initialization parameters. Different strategies can be provided through the user interface. The initial number of wasps using each strategy is defined by the user.

Input data. After initialization the model does not take any additional input.

Submodels - Wasp model. Each female wasp is presented with a host at each time step. If the host passes the egg-laying criterion of the wasp it receives an egg. If a wasp is fertilized, it chooses the sex of the egg based on information about the host. An unfertilized female can only lay male eggs. The only action a male wasp can perform is to mate with a female.

Submodels - Host model. Hosts whose attractiveness exceeds the threshold are accepted for parasitism. An egg count influence parameter describes the hosts' decrease in attractiveness after having been parasitized by a wasp egg. Parasitized hosts that are above threshold may potentially be superparasitized, either by the same female or by a different one. If the host gets too crowded it dies prematurely. Freshly-emerged wasps may mate on the host. Eventually the host will die and the wasps will disperse as the next generation.

To search for evolutionarily stable sex allocations, we started with 125 random viable strategies (sex allocation values that ranged 0–1) for each of three simulated scenarios, which are described in the next paragraph. "Viable" means that a population consisting of individuals using such allocation exclusively would not go extinct. We split this set into 25 subsets of five strategies each. Each strategy was initially represented by 100 wasp females. In a first round, each subset was used to simulate the dynamics of the competing sex allocation strategies. Successful strategies eventually took over the population by producing more offspring, which inherited their parents' strategies, compared to less successful sex allocation strategies. In a second round, the 25 winners of the first round were again randomly split into five subsets, and were allowed to compete to produce five winning strategies. Finally, the most successful strategy among these five was determined in a third round of simulations. This evolution and selection protocol is summarized schematically in Fig. 1. It was replicated five times for each set of conditions. We report on the mean±SD winning sex allocation in each set of replicates.

We first simulated a "symmetric scenario", which included equal embryonic proliferation and survival of male and female larvae, and mating that occurs before dispersal, i.e. a mating structure that promotes Local Mate Competition. Under this scenario, males and females have equal developmental prospects, therefore a female-biased sex allocation is expected to evolve. Next, we simulated two "asymmetric scenarios", where females proliferate more than males and outcompete them in super-parasitized hosts. In one set of simulations, the wasps mate after dispersal, while in a second set they mate with individuals from the same host (if available) immediately after emerging from the host. The comparison between the "symmetric" and "asymmetric"

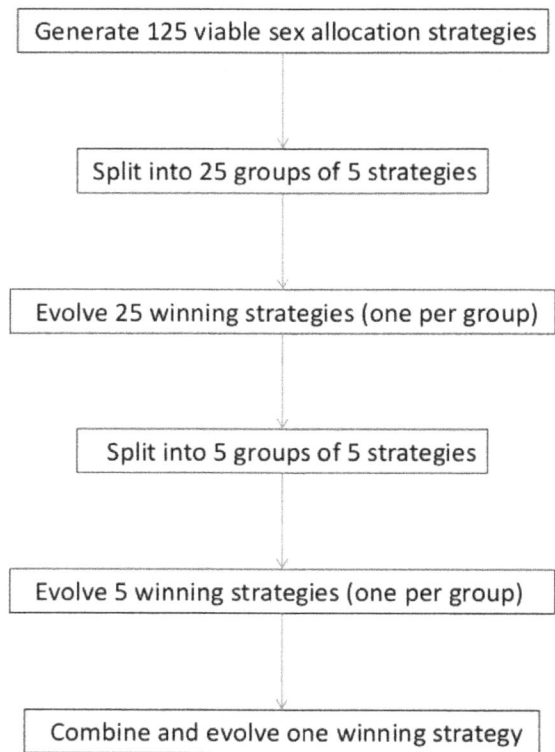

```
┌─────────────────────────────────────────────┐
│ Generate 125 viable sex allocation strategies │
└─────────────────────────────────────────────┘
                      │
                      ▼
┌─────────────────────────────────────────────┐
│      Split into 25 groups of 5 strategies     │
└─────────────────────────────────────────────┘
                      │
                      ▼
┌─────────────────────────────────────────────┐
│   Evolve 25 winning strategies (one per group) │
└─────────────────────────────────────────────┘
                      │
                      ▼
┌─────────────────────────────────────────────┐
│       Split into 5 groups of 5 strategies     │
└─────────────────────────────────────────────┘
                      │
                      ▼
┌─────────────────────────────────────────────┐
│    Evolve 5 winning strategies (one per group) │
└─────────────────────────────────────────────┘
                      │
                      ▼
┌─────────────────────────────────────────────┐
│    Combine and evolve one winning strategy    │
└─────────────────────────────────────────────┘
```

Figure 1. Steps in the evolution and selection of the wasps' sex allocation strategy. This protocol was applied to each of the modeled scenarios and host acceptance thresholds.

scenarios should reveal the selective effect of the females' competitive advantage on sex allocation. We further expected the comparison between the two "asymmetric" scenarios to reflect the effect of mating structure on the predicted sex ratio. Each of the three scenarios was simulated for nine host acceptance thresholds. Low host acceptance thresholds in our simulation model are equivalent to high levels of superparasitism, as already-parasitized hosts are accepted by wasps and receive an additional parasitoid egg. Therefore, by varying host acceptance thresholds we were able to predict the evolved sex allocations under rising risk of superparasitism, leading to increasing levels of between-clone competition.

Experiment

Insects and laboratory conditions. Laboratory-reared hosts and parasitoids were used for the experiments. Their rearing followed a standard protocol [22]. The temperature during rearing and throughout the experiment was $27 \pm 2°C$, and relative humidity was ~60%. Hosts used for the experiment were 0- to 24-h-old moth eggs. Wasps were 0- to 24-h old males and females from single-sex broods.

Experimental design. The experiment was designed to test the simulations' main predictions (see Results below). Specifically, we aimed to examine whether sex allocation in *C. koehleri* deviates from equality, and whether it is affected by the risk for superparasitism. We reared the wasps at either high or low density before exposing them to hosts, to generate an expectation of either high or low risk of superparasitism. This manipulation follows Shuker et al. [23], who found that the rearing density of adult wasps affected sex allocation in *Nasonia* parasitoids. 30

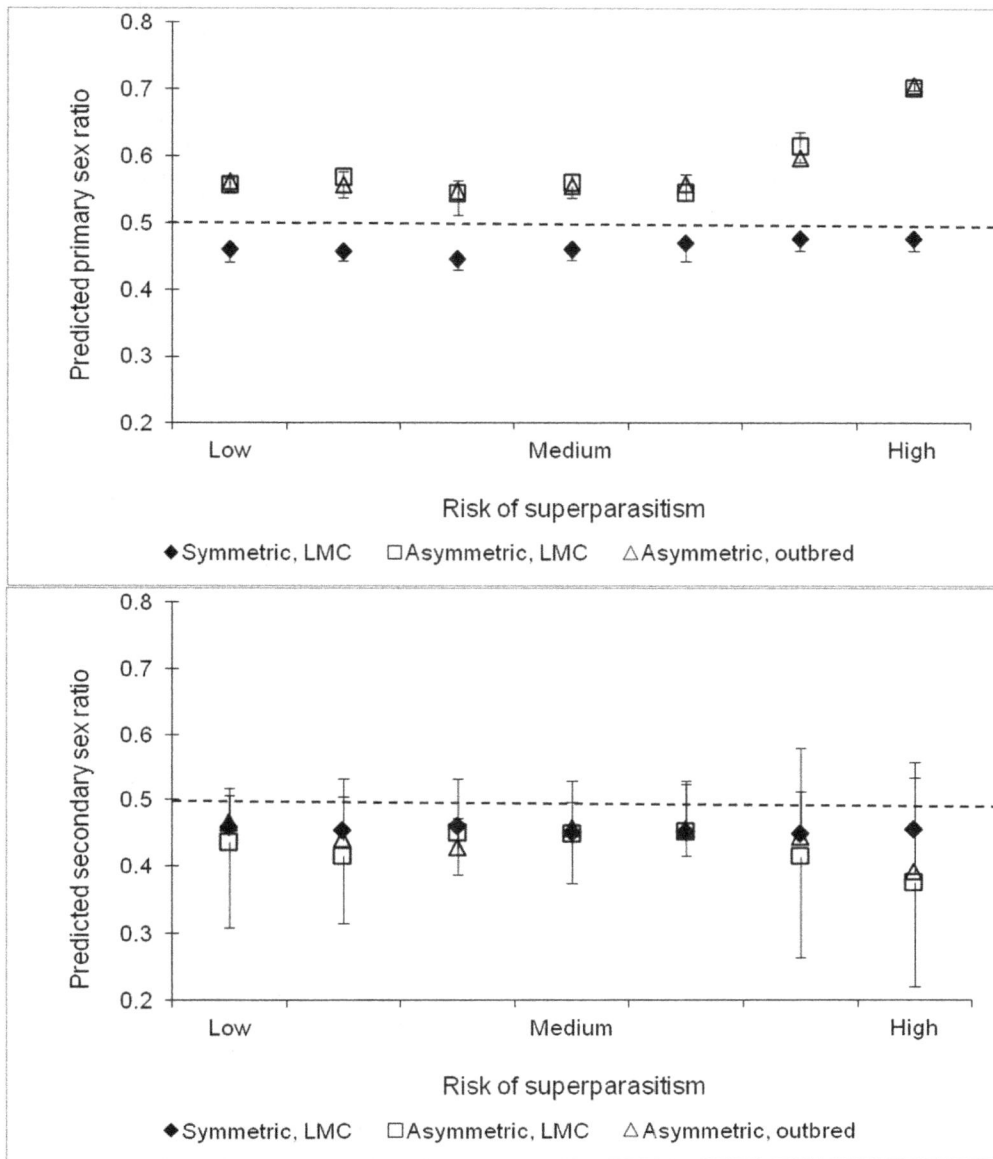

Figure 2. Evolved primary (top) and secondary (bottom) sex ratios as a function of superparasitism level (modeled by varying host acceptance thresholds). ◆ – assuming equal numbers of emerging males and females per clone, and mating before dispersal. □ – assuming developmental advantage to females and mating before dispersal. Δ – assuming developmental advantage to females and dispersal before mating. The simulations crashed at the lowest simulated value of risk of superparasitism because of extinction of the parasitoid population. They also collapsed at the highest superparasitism risk level, because the heavily superparasitized hosts died prematurely. The graph therefore shows the seven superparasitism values that allowed stable population dynamics. The dashed line marks the 1:1 sex allocation.

female-only and 30 male-only broods were used for the experiment. Upon emergence from the host, 20 wasps from each brood were moved to a 13×100 mm (diameter×length) test tube for 48 hours. These wasps formed the high-density (HD) treatment. Eight additional wasps from the same brood were housed in pairs in four additional test tubes of the same dimensions, and formed the low-density (LD) treatment. The wasps were supplied with honey as a food source, but not with hosts, during this time. We then allowed wasps from all four density combinations to mate (LD♂×LD♀, LD♂×HD♀, HD♂×LD♀, HD♂×HD♀). For each mating combination, we placed two males and two females in a petri dish for 6 hours. This was done in two replicates, i.e., 16 individuals (2/dish×2

replicates×4 treatments) from each brood. Thus, total sample size was 2 replicates×4 mating combinations×30 parental clones = 240. Mating normally occurs within 5 minutes of encounter between males and females in *C. koehleri*, and lasts for a few seconds. Females copulate repeatedly when housed with males (pers. obs.). We therefore expected the females to be mated after 6 hours of cohabitation with males. >50 hosts (moth eggs), haphazardly selected from the insectary culture, were then introduced into each dish, and the wasps were allowed to parasitize them for 24 hours. As *C. koehleri* females parasitize ca. 20 hosts during their first day after emergence [24], the number of hosts offered in the experiment did not limit the wasps' fecundity. The distribution of host sizes is not expected to differ among the four experimental treatments. Thus, any potential effects of host size on sex allocation should be similar across treatments. Hosts were reared on potatoes until parasitoid pupation. Each host mummy, containing a brood of wasp pupae, was then transferred into an individual test tube. Parasitoids were sexed after emerging from the hosts. Parasitoid brood sizes were determined in a subsample of 150 broods, collected from 15 replicates from each treatment.

Experimental data analysis. Two-way ANOVAs were used to analyze the effect of maternal and paternal rearing density on the proportions of male, female and mixed-sex broods. Proportions were arcsine-transformed prior to analysis. We used the proportion of mixed-sex broods (0.35) as an estimate for the frequency of superparasitized hosts in the whole dataset. Thus, we assumed that 0.35 of the all-male and of the all-female broods resulted from two eggs of the same sex, and that all mixed-sex broods developed from one male and one female egg. Based on these assumptions, we calculated the estimated proportion of male eggs (primary sex ratio) produced in each of the treatments.

Males comprise ca. 1/3 of mixed-sex broods [14]. Therefore we estimated the proportion of males in each treatment (the secondary sex ratio) as (total # wasps in all-male broods+1/3 of the # wasps in mixed-sex broods)/total # wasps.

Results

Simulation Model

The simulation model predicts the evolution of a female-biased primary sex ratio when male and female wasps have similar developmental prospects, and mating occurs before dispersal (the "symmetric scenario"). This prediction is not affected by the risk of superparasitism (Fig. 2, top). Secondary sex ratios are also female-biased, at all levels of superparasitism (Fig. 2, bottom). In the two scenarios with competitive asymmetry between the sexes, an excess of male-egg production is predicted. The male bias in the primary sex ratio is predicted to increase with higher risk of superparasitism, but is unaffected by the parasitoids' mating structure (Fig. 2, top). However, the predicted proportion of males in the population (secondary sex ratio) is lower than 0.5, and decreases further at high levels of superparasitism (Fig. 2, bottom).

Laboratory Experiment

The experiment manipulated the effect of wasp rearing density, a possible cue for the risk of superparasitism. We expected wasps reared at high density to anticipate more superparasitism in their hosts than wasps reared at low density. According to the model's prediction, we expected that wasps in all treatments would produce a male-biased primary sex ratio, but that the bias would be more extreme in the high-density treatments than in the low-density treatments. We further expected primary sex ratios to be influenced by maternal rearing densities if sex allocation is controlled by females, and by paternal rearing densities if sex allocation is under male control. If both males and females influence sex allocation in *C. koehleri*, then primary sex ratios are expected to be highest when both parents are maintained at high density, lowest when both parents are kept at high density, and intermediate for the remaining two treatments.

The frequencies of male, female and mixed-sex (both male and female) broods are reported in Fig. 3. The proportions of male-only broods were significantly affected by paternal rearing density, but not by maternal density (two-way ANOVA on arcsine-transformed data: $F_{3, 189} = 2.077$, $P = 0.10$ for the complete model; $F_{1, 189} = 1.94$, $P = 0.018$ for paternal density; $F_{1, 189} = 0.04$, $P = 0.841$ for maternal density). Similarly, the proportions of

female-only broods were significantly influenced by paternal densities only ($F_{3, 189} = 3.345$, $P = 0.02$ for the complete model; $F_{1, 189} = 9.932$, $P = 0.002$ for paternal density; $F_{1, 189} = 0.044$, $P = 0.834$ for maternal density). Interactions between maternal and paternal densities were non-significant. The proportions of mixed-sex broods were not significantly affected by parental rearing conditions. In conclusion, all-male broods increased in frequency, at the expense of all-female broods, in the two treatments with high paternal density.

The occurrence of mixed-sex broods in ca. 35% of the hosts indicates that they were parasitized more than once. In *C. koehleri*, superparasitism occurs in all sex combinations: male-male, female-female and female-male [16]. We estimated the proportions of male eggs laid in the whole dataset, by conservatively assuming that 35% of the single-sex hosts were parasitized by two wasp eggs as well. This provides an estimate of the primary sex ratio in the four density treatments (Fig. 4, black bars). In line with our working hypothesis, this ratio is male-biased in all treatments. The bias is especially pronounced in the treatments of high male density, consistent with a paternal influence on sex allocation.

Wasps from all treatments were provided with hosts without restriction. This allowed us to study the effects of parental sex allocation on secondary sex ratios without possible confounding effects from varying competition levels among developing offspring. Indeed, the proportion of mixed-sex broods (an estimate for the level of superparasitism) did not vary significantly across treatments. This suggests that manipulation of wasp rearing density affected the parent's anticipation of superparasitism, but not the degree of superparasitism experienced by their offspring. The secondary sex ratios (proportion of adult male wasps) in all treatments were close to 0.5 (Fig. 4, empty bars), in spite of the much higher estimated proportion of male eggs laid.

Discussion

Polyembryonic parasitoids provide prime examples of within-family conflicts over sex allocation, because the evolutionarily stable sex ratios may differ from the perspectives of different family members [2]. Sons and daughters are more related to their clone-mates (which share their sex) than to other siblings, while mothers are equally related to their sons and daughters. Thus, daughters benefit from lower sex ratios than sons in many situations, while mothers favor intermediate sex ratios. Gardner et al. [25] suggested that this conflict is resolved in favor of the daughters, through the evolution of the female soldiers that kill male sibs and thereby reduce the secondary sex ratio. Our model and experimental data suggest that, at least in *C. koehleri*, the soldiers' aggression is counteracted by a male-biased primary sex ratio determined by the parents. Thus, our experiment points at the potential of the parents to influence offspring sex allocation in *C. koehleri*, consistent with their putative role in this evolutionary conflict.

In line with standard sex allocation theory, our simulation predicts a female-biased primary sex allocation under LMC and equal survival/proliferation of both sexes (Fig. 2, top, "symmetric scenario"), at all levels of superparasitism. The predicted secondary sex ratio remains female-biased (Fig. 2, bottom, "symmetric scenario"). The introduction of advantages in proliferation and competition to female larvae selects for a male-biased primary sex ratio, whether mating occurs before or after dispersal. This implies that developmental asymmetry drives the evolution of sex allocation more strongly than mating structure in our system. In *Nasonia* parasitoids, on the other hand, LMC is predicted to affect sex allocation much more strongly than the

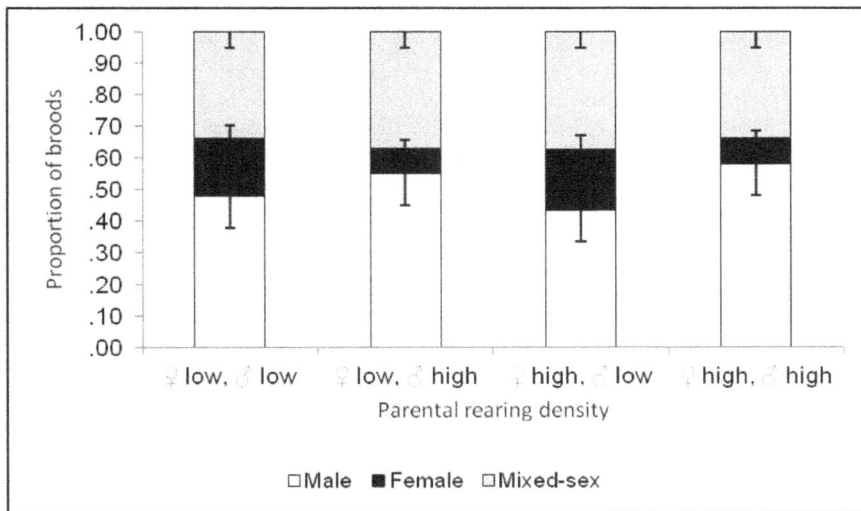

Figure 3. Mean (and associated SD) proportions of male-only, female-only and mixed-sex broods, produced by parents that were housed at high vs. low density for 48 hours after emergence.

competitive asymmetries between male and female larvae [5]. A possible reason for these different findings is that the female soldier caste of *C. koehleri* generates a powerful mechanism of sex ratio regulation through larval competition, which is absent from monoembryonic species. Increasing the risk of superparasitism selects for even more male-biased primary sex ratios in our model. This may compensate for the greater proportion of male larvae killed by female soldiers. In both "asymmetrical scenarios", secondary sex ratios are lower than 0.5 in spite of the male-biased primary sex allocation (Fig. 2, bottom), due to the female competitive advantage. The lowest secondary sex ratios are predicted at the highest levels of superparasitism, when female soldiers most frequently encounter and kill male competitors within the host.

The experimental results are consistent with the main predictions of the model, in that primary sex ratios were male biased in all treatments. A similar excess of male broods was also found in a field-collected sample of hosts parasitized by *C. koehleri* [16]. Also in line with the model, an environmental signal for increased risk of superparasitism (high rearing density) increased the male bias even further. By setting up all four parental density combinations, we learned that this increase in sex ratio was in response to high rearing density of fathers, but not of mothers. An alternative interpretation for the elevated sex ratio is that high parasitoid density serves as a signal for reduced future local mate competition (because it predicts more founding wasps per host), rather than increased future competition for host resources [5]. An additional interpretation is that females use increased wasp densities as a

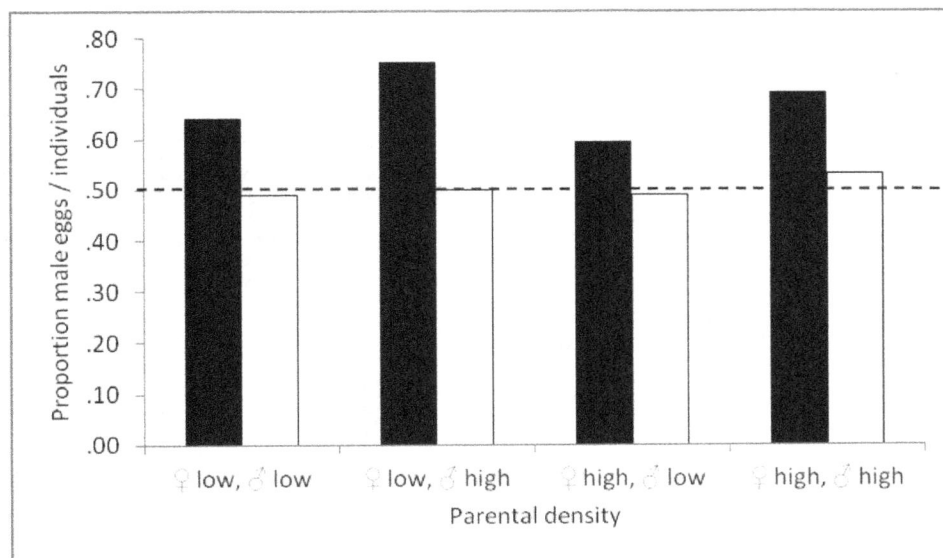

Figure 4. Estimated proportions of male eggs (primary sex ratio, black bars), and proportions of male individuals (empty bars) produced by parents that were housed at high vs. low density for 48 hours after emergence. The dashed line indicates the 1:1 sex ratio.

signal of better prospects of finding mates, and thus a higher potential for producing daughters in the next generation. Because of negative frequency dependence, this reduces the fitness benefit from producing daughters, favoring a higher proportion of males among the offspring of mated females [26]. These two alternative interpretations assume mating before dispersal, while our model predicts little effect of mating structure on sex allocation in *C. koehleri*.

Past studies have traditionally stressed the role of mothers in mediating offspring sex allocation, because of their greater involvement in parental care. Such control can involve fertilization (e.g., not using sperm of some mates [27,28]), developmental conditions that benefit embryos of one sex [29] or food provisioning to young that favors one of the sexes [30]. However, evidence for the ability of males to also affect their offspring's sex is gradually accumulating [31], as exemplified in a few studies on hymenopterans: in the parasitoid *Encarsia*, females develop on 'normal' (whitefly) hosts, whereas males can hyperparasitize and kill the female larvae of their own species, thereby reducing the proportion of females [32]. In *Nasonia*, males from different strains sire offspring that vary in sex ratios [28]. Mate-guarding after copulation by male *Urolepis rufipes* parasitoids increases the proportion of daughters among the offspring [33]. Similarly, nest-guarding by males in the mud-daubing wasp *Trypoxylon politum* allows their mates more time for nest provisioning, a possible strategy for enhancing the production of daughters [34].

By what mechanism might male rearing density have affected offspring sex ratios in the present experiment? A possible proximate mechanism involves parental epigenetic control over sex allocation. Such control has been demonstrated in *Nasonia vitripennis*, where maternal input of transformer (Nvtra) messenger RNA is required for female development from fertilized eggs [35]. A major pathway of epigenetic control involves regulation of DNA expression levels through gene methylations. Methylations, as well as homologues to vertebrate genes that control them, are documented in a wide range of hymenopterans. This provides a potential molecular mechanism whereby parental environment could affect offspring sex ratios and resolve kin conflicts [36]. Thus, males may potentially affect offspring sex allocation through substances transmitted in their ejaculates during copulation, possibly counteracting female control over fertilization.

Hymenopteran males transmit their genes through daughters only. Thus, if males can indeed influence offspring sex, they would

be expected to favor female production under all circumstances. This prediction does not agree with our finding that offspring primary sex ratios varied with paternal rearing densities. A possible interpretation is that males kept at high density were less successful in mating or fertilizing their mates than low-density males, leading to a higher proportion of haploid, male offspring. This interpretation is non-adaptive and considers high-density rearing as a constraint on male fitness. Another possible non-adaptive pathway could involve paternally-transmitted selfish genetic elements that act as sex-ratio modulators. In the parasitoids *Trichogramma* and *Nasonia*, a nuclear extra chromosome called PSR is carried by males only. PSR turns eggs destined to develop as females into males, enhancing its own transmission into the offspring generation [37]. Such modulation of sex ratios could account for our experimental results, if the extent of modulation depends on male rearing density.

To conclude, our study demonstrates the power of individual-based evolutionary modeling in exploring frequency-dependent traits in organisms, and in generating testable hypotheses about them. The simulation approach is particularly useful for species with complex life-histories, which commonly challenge analytical modeling. While we applied the model to simulate the evolution of sex allocation in polyembryonic parasitoids, the individual-based approach may be fruitfully applied to additional life-history traits and organisms as well. Male rearing densities affected offspring sex allocation in our experiment. The mechanism and evolutionary significance underlying this finding still need to be elucidated.

Acknowledgments

We thank Nina Dinov, Yael Keinan, Miriam Kishinevsky and Lihie Ohayon for technical assistance.

Author Contributions

Conceived and designed the experiments: TK FT. Performed the experiments: MB PR RS. Analyzed the data: MB TK. Wrote the paper: MB PR RS TK FT.

References

1. Seger J, Stubblefield JW (2002) Models of sex ratio evolution. Pages 2–25 in Hardy ICW (Editor) Sex ratios: concepts and research methods. Cambridge University Press.
2. West SA (2009) Sex Allocation. Princeton University Press (Monographs in Population Biology Series).
3. Hamilton WD (1967) Extraordinary sex ratios. Science 156: 477–488.
4. Macke E, Magalhães S, Bach F, Olivieri I (2011) Experimental evolution of reduced sex ratio adjustment under local mate competition. Science 334: 1127–1129.
5. Sykes EM, Innocent TM, Pen I, Shuker DM, West SA (2007) Asymmetric larval competition in the parasitoid wasp *Nasonia vitripennis*: a role in sex allocation? Behav Ecol Sociobiol 61: 1751–1758.
6. King BH (2002) Sex ratio response to conspecifics in a parasitoid wasp: test of a prediction of local mate competition theory and alternative hypotheses. Behav Ecol Sociobiol 52: 17–24.
7. Lebreton S, Chevrier C, Darrouzet E (2010) Sex allocation strategies in response to conspecifics' offspring sex ratio in solitary parasitoids. Behav Ecol 21: 107–112.
8. Godfray HCJ (1994) Parasitoids: Behavioral and evolutionary ecology. Princeton University Press (Princeton, N.J.).
9. Ode PJ, Heinz KM (2002) Host-size-dependent sex ratio theory and improving mass-reared parasitoid sex ratios. Biol Cont 24: 31–41.
10. Colinet H, Salin C, Boivin G, Hance TH (2005). Host age and fitness-related traits in a koinobiont aphid parasitoid. Ecol Entomol 30: 473–479.

11. Shuker DM, Pen I, Duncan AB, Reece SE, West SA (2005) Sex ratios under asymmetrical local mate competition: theory and a test with parasitoid wasps. Am Nat 166: 301–316.
12. Keinan Y, Kishinevsky M, Segoli M, Keasar T (2012) Repeated probing of hosts: an important component of superparasitism. Behav Ecol 23: 1263–1268.
13. Morag N, Bouskila A, Segoli M, Rapp O, Keasar T, et al. (2011) The mating status of mothers, and offspring sex, affect clutch size in a polyembryonic parasitoid wasp. Anim Behav 81: 865–870.
14. Keasar T, Segoli M, Barak R, Steinberg S, Giron D, et al. (2006) Costs and consequences of superparasitism in the polyembryonic parasitoid *Copidosoma koehleri*. Ecol Entomol 31: 277–283.
15. Segoli M, Bouskila A, Harari A, Keasar T (2009a) Developmental patterns in the polyembryonic wasp *Copidosoma koehleri*. Arthropod Struct Devel 38: 84–90.
16. Segoli M, Bouskila A, Harari A, Keasar T (2009b) Brood size in a polyembryonic parasitoid wasp is affected by the relatedness among competing larvae. Behav Ecol 20: 761–767.
17. Segoli M, Keasar T, Bouskila A, Harari A (2010) Host choice in a polyembryonic wasp depends on the relatedness to a previously parasitizing female. Physiol Entomol 35: 40–45.
18. Lindsay CB (2009) Factors affecting caste development, brood type, and sex ratios of a polyembryonic wasp. M.Sc. thesis, university of Georgia.
19. Lewis HM, Tosh CR, O'Keefe S, Shuker DM, West SA, et al. (2010) Constraints on adaptation: explaining deviation from optimal sex ratio using artificial neural networks. J Evol Biol 23: 1708–1719.

Sex Allocation in a Polyembryonic Parasitoid with Female Soldiers: An Evolutionary Simulation...

43

20. Grimm V, Berger U, Bastiansen F, Eliassen S, Ginot V, et al. (2006) A standard protocol for describing individual-based and agent-based models. Ecol Model 198: 115–126.

21. Grimm V, Berger U, DeAngelis DL, Polhill JG, Giske J, et al. (2010) The ODD protocol: a review and first update. Ecol Model 221: 2760–2768.

22. Berlinger MJ, Lebiush-Mordechi S (1997) The potato tubermoth in Israel: a review of its phenology, behavior, methodology, parasites and control. Trends Entomol 1: 137–155.

23. Shuker DM, Reece SE, Lee A, Graham A, Duncan AB, et al. (2007) Information use in space and time: sex allocation behaviour in the parasitoid wasp *Nasonia vitripennis*. Anim Behav 73: 971–977.

24. Kfir R (1981) Fertility of the polyembryonic parasite *Copidosoma koehleri*, effect of humidities on life length and relative abundance as compared with that of *Apanteles subandinus* in potato tuber moth. Ann Appl Biol 99: 225–230.

25. Gardner A, Hardy ICW, Taylor PD, West SA (2007) Spiteful soldiers and sex ratio conflict in polyembryonic parasitoid wasps. Am Nat 169: 519–533.

26. Crowley PH, Saeki Y, Switzer PV (2009) Evolutionarily stable oviposition and sex ratio in parasitoid wasps with single-sex broods. Ecol Entomol 34: 163–175.

27. Henter HJ (2004) Constrained sex allocation in a parasitoid due to variation in male quality. J Evol Biol 17: 886–896.

28. Shuker DM, Sykes EM, Browning LE, Beukeboom LW, West SA (2006) Male influence on sex allocation in the parasitoid wasp *Nasonia vitripennis*. Behav Ecol Sociobiol 59: 829–835.

29. Grant VJ, Chamley LW (2010) Can mammalian mothers influence the sex of their offspring peri-conceptually? Reproduction 140: 425–433.

30. House CM, Simmons LW, Kotiaho JS, Tomkins JL, Hunt J (2010) Sex ratio bias in the dung beetle *Onthophagus taurus*: adaptive allocation or sex-specific offspring mortality? Evol Ecol 25: 363–372.

31. Shuker DM, Moynihan AM, Ross L (2009) Sexual conflict, sex allocation and the genetic system. Biol Lett 5: 682–685.

32. Hunter MS (1993) Sex allocation in a field population of an autoparasitoid. Oecologia 93: 421–428.

33. King BH, Kuban KA (2012) Should he stay or should he go: male influence on offspring sex ratio via postcopulatory attendance. Behav Ecol Sociobiol 66: 1165–1173.

34. Brockmann HJ, Grafen A (1989) Mate conflict and male-behavior in a solitary wasp, *Trypoxylon (Trypargilum) politum* (Hymenoptera, Sphecidae). Anim Behav 37: 232–255.

35. Verhulst EC, Beukeboom LW, van de Zande L (2010). Maternal control of haplodiploid sex determination in the wasp *Nasonia*. Science 328: 620–623.

36. Drewell RA, Lo N, Oxley PR, Oldroyd BP (2012) Kin conflict in insect societies: a new epigenetic perspective. Trends Ecol Evol 27: 367–373.

37. van Vugt JJFA, de Jong H, Stouthamer R (2009) The origin of a selfish B chromosome triggering paternal sex ratio in the parasitoid wasp *Trichogramma kaykai*. Proc Roy Soc Lond B 276: 4149–4154.

6

Evolutionary and Biological Implications of Dental Mesial Drift in Rodents: The Case of the Ctenodactylidae (Rodentia, Mammalia)

Helder Gomes Rodrigues[1]*, **Floréal Solé**[1]¤, **Cyril Charles**[1], **Paul Tafforeau**[2], **Monique Vianey-Liaud**[3], **Laurent Viriot**[1]*

1 Team "Evo-Devo of Vertebrate Dentition", Institut de Génomique Fonctionnelle de Lyon, Unité Mixte de Recherche 5242 Centre National de la Recherche Scientifique, Ecole Normale Supérieure de Lyon, Université Claude Bernard Lyon 1, Lyon, France, 2 European Synchrotron Radiation Facility, Grenoble, France, 3 Laboratoire de Paléontologie, Institut des Sciences de l'Évolution de Montpellier, Unité Mixte de Recherche 5554 Centre National de la Recherche Scientifique, Université Montpellier 2, Montpellier, France

Abstract

Dental characters are importantly used for reconstructing the evolutionary history of mammals, because teeth represent the most abundant material available for the fossil species. However, the characteristics of dental renewal are presently poorly used, probably because dental formulae are frequently not properly established, whereas they could be of high interest for evolutionary and developmental issues. One of the oldest rodent families, the Ctenodactylidae, is intriguing in having longstanding disputed dental formulae. Here, we investigated 70 skulls among all extant ctenodactylid genera (*Ctenodactylus*, *Felovia*, *Massoutiera* and *Pectinator*) by using X-ray conventional and synchrotron microtomography in order to solve and discuss these dental issues. Our study clearly indicates that *Massoutiera*, *Felovia* and *Ctenodactylus* differ from *Pectinator* not only by a more derived dentition, but also by a more derived eruptive sequence. In addition to molars, their dentition only includes the fourth deciduous premolars, and no longer bears permanent premolars, conversely to *Pectinator*. Moreover, we found that these premolars are lost during adulthood, because of mesial drift of molars. Mesial drift is a striking mechanism involving migration of teeth allowed by both bone remodeling and dental resorption. This dental innovation is to date poorly known in rodents, since it is only the second report described. Interestingly, we noted that dental drift in rodents is always associated with high-crowned teeth favoring molar size enlargement. It can thus represent another adaptation to withstand high wear, inasmuch as these rodents inhabit desert environments where dust is abundant. A more accurate study of mesial drift in rodents would be very promising from evolutionary, biological and orthodontic points of view.

Editor: Alistair Robert Evans, Monash University, Australia

Funding: The ANR "Bouillabaisse" Program, a Centre National de la Recherche Scientifique postdoctoral grant, and ESRF. The funders had no role in study design, data collection and analysis, decision to publish, or preparation of the manuscript.

* E-mail: helder.gomes.rodrigues@ens-lyon.fr (HGR); Laurent.Viriot@ens-lyon.fr (LV)

¤ Current address: Royal Belgian Institute of Natural Sciences, Direction Earth and History of Life, Brussels, Belgium

Introduction

The interest of studying rodents among all mammals is stressed by their ecological ubiquity, coupled to their flourishing diversity (about 2300 species [1]). The evolutionary success of rodents is probably due to their small size, their high reproductive rates, their short breeding cycle, and their extensive range of dental characteristics. These dental particularities notably rely on the very high number of crown morphologies [2,3,4], enamel microstructure patterns [5,6], and masticatory functions [7,8,9]. These variations have been extensively described in both extant and extinct forms. However, the mechanisms involved in the formation and maintenance of the dentition (i.e. development, eruption, replacement) remain to be accurately documented in rodents. Studies concerning this topic have mainly dealt with the mouse [10,11,12,13], the usual model for mammalian biology. To date, we lack a global view regarding the diversity of mechanisms associated with rodent dentitions, and only rare discoveries showed very innovative dental systems in rodents [14]. In this context, this study aims at better understanding the underlying mechanisms of the establishment and replacement of the dentition of gundis (Ctenodactylidae), whose extant species present peculiar dental formulae. Their study might permit the opening of a new window on the knowledge of these dental issues.

The Ctenodactylidae encompass four endemic African genera: *Ctenodactylus*, *Felovia*, *Massoutiera*, and *Pectinator*. Originally, it was a highly diversified Asian group, notably during the Oligocene period (33.9–23 Ma; [15,16,17]). Then, this group dispersed into Europe, Arabia and Africa during the Miocene (23–5.3 Ma). Their evolutionary framework has been recently discussed in two main phylogenetical studies, involving all the ctenodactylid genera on one hand [18], and the crown group Ctenodactylinae on the other hand [19]. The dentition of extant ctenodactylids is characterized by high-crowned teeth covered by an important layer of cementum. They generally present one or two premolars and three molars in each jaw quadrant. One of the most striking issues is the potential presence of a third lower premolar (P_3 or dP_3, if deciduous) in *Pectinator* [18,20,21]. Indeed, a P_3 was never observed in Rodentia, while this tooth is present in their ancestors and in their closest relatives, the Lagomorpha (i.e. rabbits and hares). However, detailed studies of early dental development in mice and squirrels demonstrated the occurrence of rudimentary dental buds developing in the diastemal area of the mandible, in front of presumptive functional teeth [22,23,24]. As the mineralization of these rudimentary buds is disrupted, they were assigned to aborted germs of premolars lost over evolution [25,26]. Inasmuch as the complete development of a P_3 might be still possible (e.g. beginning of mineralization in squirrels, which have one of the most primitive dentitions among rodents), the confirmation of the occurrence of a P_3 in *Pectinator* could be of high interest for a better understanding of the underlying developmental processes involved in the reduction of mammalian dentitions.

Questions also arise concerning the actual occurrence of permanent premolars replacing the fourth upper and lower deciduous premolars (dP^4 and dP_4), and concerning their loss. In fact, only molars are present in adult specimens of *Ctenodactylus*, *Felovia* and *Massoutiera* [2,18,21,27,28]. The loss of premolars during the beginning of the adulthood is not rare in mammals. It was mentioned for instance in elephants, sirenians, kangaroos and wallabies [29,30,31,32,33]. These losses are induced by the forward pushing action of the erupting molars at the rear of the jaw, which leads to the mesial drift of all the cheek teeth, coupled to the remodeling of the surrounding alveolar bone. Then, the most anterior teeth, which are premolars, no longer fit within the jaw and are pushed out of the dental row. Since mesial drift has been recently found in rodents, and more precisely in African mole-rats (Bathyergidae [14]), we can hypothesize that a comparable mechanism is involved in the early loss of premolars in Ctenodactylidae.

The aim of this study is to more precisely describe the unusual dental characteristics of each extant ctenodactylid, and try to identify the mechanisms involved (e.g. mesial drift). These analyses notably benefit from high resolution microtomographic data of the dentitions at various ages which allow to precisely investigate the dental development, replacement, loss and possible drift.

Materials and Methods

In this study, 22 skulls of *Ctenodactylus gundi*, 10 skulls of *Felovia vae*, 21 skulls of *Massoutiera mzabi* and 17 skulls of *Pectinator spekei* were investigated. These investigated specimens are housed in the Museum National d'Histoire Naturelle (MNHN) of Paris (France), and in the Naturhistorisches Museum of Basel (Switzerland).

High quality images of one or two skulls of each species were obtained using propagation phase contrast X-ray synchrotron microtomography at the European Synchrotron Radiation Facility (ESRF, Grenoble, France). Experiments were performed on the beamline BM5. One skull (*C. gundi* MNHN CG 1986-255) was scanned in 2008 using a monochromatic beam set at 25 keV using a double crystal Si111 Bragg monochromator. We used an indirect detector based on a 10 μm thick gadolinium oxide scintillator coupled with lenses based optic to a FReLoN CCD camera (Fast Readout Low Noise Charge Coupled Device). This system provides an isotropic voxel size of 7.39 μm. In order to have moderate phase contrast effect, we used a propagation distance of 500 mm. These data were reconstructed using filtered-

backprojection algorithm, without phase retrieval, hence in edge detection mode. All the other specimens scanned at the ESRF for this study were imaged also on BM5 beamline in 2011, by using a pink beam configuration obtained by combination of a 125 μm thick LuAG scintillator and a lead glass based filter (equivalent to 0.7 mm of lead). Low energies of the spectrum were removed with 3 mm of aluminum and 2 mm of copper. Thanks to respective Kedge of the scintillator (63.31 keV) and filter (88 keV) and to the BM5 spectral properties, this configuration delivers a beam in which most of the detected photons are in the energy range between these two Kedges. It provides a quite narrow bandwidth (pink beam), allowing rapid high quality scans in propagation phase contrast mode (900 mm of propagation), without any effect of beam hardening due to the low absorption by the sample. The relatively high energy used for these scans is not problematic as the use of phase contrast brings a very high level of information. In order to make segmentation of the data more efficient, a single distance phase retrieval process was used [34–35] to reconstruct data linked to mineral density without the edge detection effect. Synchrotron microtomography has been proven to be very useful for very precise imaging of small elements, such as teeth [14]. The use of high quality pink beam coupled with single distance phase retrieval allows, for this type of sample, quality of data comparable to monochromatic beam, but with acquisition times 5 to 10 times shorter due to the higher flux of photons. One skull (*P. spekei*, MNHN CG 1995-19) was imaged using a GE phoenix nanotom 180 at energy of 100 keV with a cubic voxel of 5.64 μm. 3D renderings and virtual slices were then performed using VGStudio Max 2.0 software. Non-invasive virtual extractions of entire dentition (i.e. crown and roots) were realized for a more accurate analysis.

X^n and X_n respectively refer to the nth upper and n^{th} lower cheek tooth, and Xn for both. Dental measurements were taken from the right upper dentition (U1–U5, Fig. 1) and right lower dentition (L1–L4, Fig. 1) to test the hypothesis of mesial drift. The mesial-most point of the lower dentition and the posterior part of the zygomatic arch for the upper dentition represent the starting points (i.e. references) and the mesial or distal base of molars the final points for each measurement, which were calculated with LAS Core (Leica®) software. Premolars were not included in such measurements because of their loss or replacement, contrary to molars which are not affected. Variations of U5 correspond to the maxillary growth. Skull lengths were also measured with a caliper to characterize the developmental stages. Variations of each distance were examined by linear regressions. Occurrence of drift was assessed by checking if the linear regression slopes are equal (null hypothesis: measurements are constant) or different from zero. Significant differences observed from Student's t-test indicated that slopes could be used to characterize the presence and the relative importance of mesial drift.

Results

Overall Dental Characteristics of Extant Ctenodactylid Rodents

Here, we strictly focused our observations on the number and replacement of cheek teeth for each genus of extant ctenodactylids (Fig. 2). Dental occlusal morphologies have not been described here since they have already been accurately studied in previous works [18,19,21].

– Dental formula of *Pectinator*: dP^{3-4}/dP_4, P^4/P_4, M^{1-3}/M_{1-3} (Fig. 2A–B)

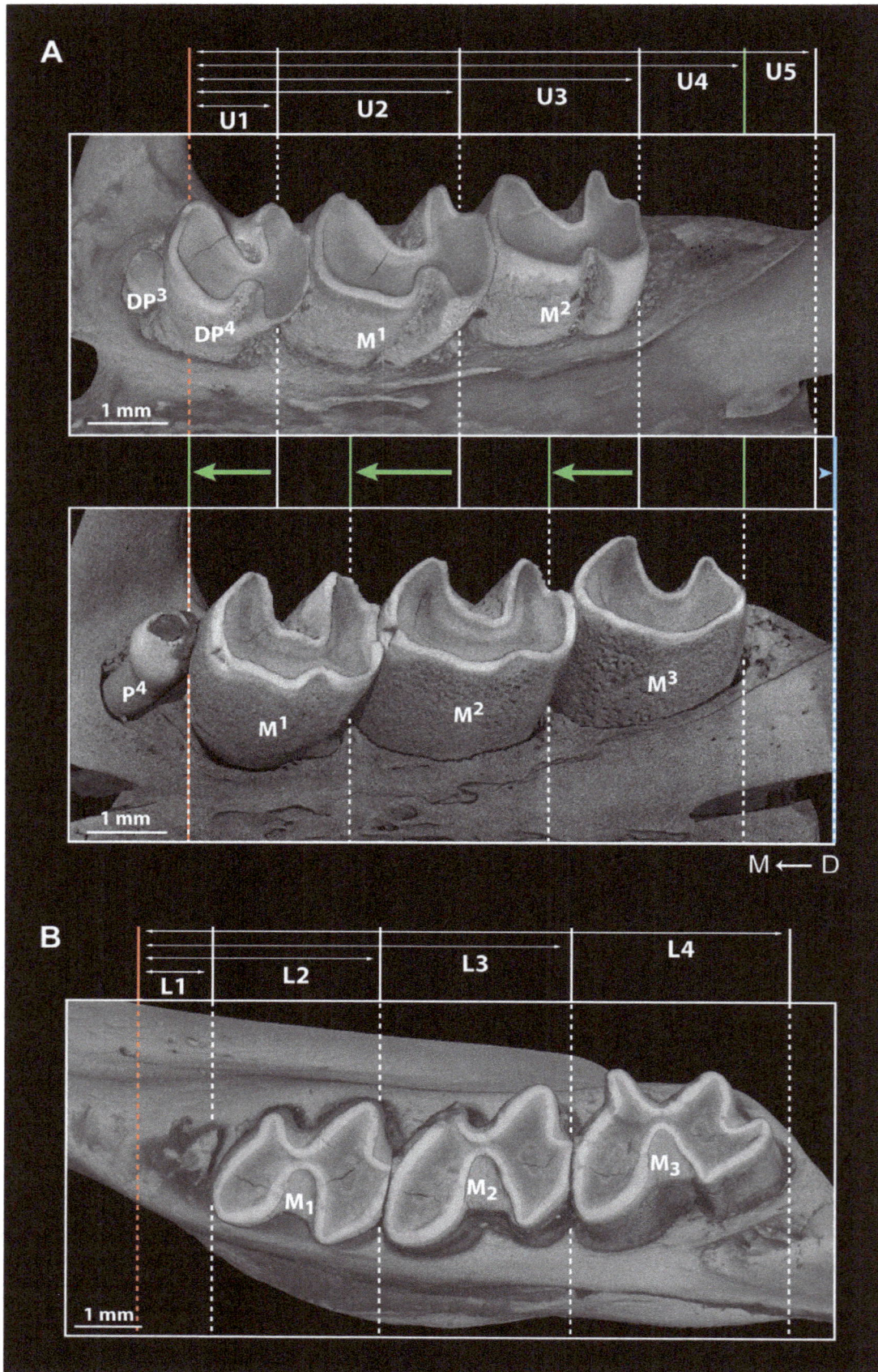

A

U1 U2 U3 U4 U5

DP³ DP⁴ M¹ M²

1 mm

P⁴ M¹ M² M³

1 mm

M ← D

B

L1 L2 L3 L4

M₁ M₂ M₃

1 mm

Figure 1. Characterization of dental mesial drift by using skull measurements (see material and methods for details). A, X-ray synchrotron and conventional microtomographic 3D rendering of left upper cheek tooth rows of *Pectinator spekei* (MNHN-CG1895-461 and 1995-19); green arrows indicate dental drift, blue arrow indicates maxillary growth. B, X-ray synchrotron microtomographic 3D rendering of left lower cheek tooth row of *Felovia vae* (MNHN-CG 1989-22). D → M stands for distal to mesial direction.

The investigated neonate skull (misidentified specimen: "*C. gundi*" MNHN-CG2006-198, (e.g. [18])) includes alveoli of dP^3, in addition to dP4 and M1, but it does not display any evidence for the presence of dP_3. One juvenile specimen (MNHN-CG 1895-461, Fig. 1A) indicates that dP^3 is brachydont (i.e. low crowned) and single-rooted, while dP^3 is three-rooted and includes a wide lingual root. Upper molars also have three roots which tend to merge and are highly reduced on M^1 and M^2 because of their strong hypsodonty (i.e. high crown). However, they are not euhypsodont (i.e. without root, [36,37]), as roots are still present. With regard to the lower teeth, dP_4 possesses one mesial and one distal root, and molars have two reduced roots, and the same morphological trends as seen on upper molars are observed. Slightly older specimens (MNHN-CG1981-503 and 1995-19) show that dP4 are replaced by smaller and single-rooted P4 after the full eruption of hypsodont M3, and that dP^3 is shed. The tooth present in front of P_4 is definitively not a P_3, but it rather corresponds to the mesial fragment of dP_4 which is not totally resorbed. The permanent premolar is indeed smaller that the deciduous one, and when erupting, it does not contribute to the complete shedding of the latter. Cementum is present on enamel of molars and dP4, notably in crown folds; a very thin layer partially surrounds the teeth. The dental formula of both adult and old specimens includes a P4, and three molars (M1-3) in each jaw quadrant.

– Dental formula of *Ctenodactylus*: dP^4/dP_4, M^{1-3}/M_{1-3} (Fig. 2C–D)

Lataste [27] was the first to pay attention to the sequence of dental eruption in *Ctenodactylus*. He established that the genus possesses dP4, P4, and three molars (M1-3). He also defined seven chronological stages concerning the sequence of dental eruption. The youngest specimen (1st stage) presents erupting dP4 and M1. According to him, P^4 is present from the 4th stage, erupts after M^3, and is lost at the 6th stage. Our analysis of Lataste's skulls permitted to establish that the "tiny tuberculous" P^4 observed by Lataste is in fact a worn dP^4 (e.g. MNHN-CG1963-921). Among the whole sample of investigated *Ctenodactylus gundi*, no specimen possesses either a P^4, or a P_4. Similarly, Vianey-Liaud et al. [18] did not find any specimen with P4. During the adulthood, the deciduous premolars are lost without being replaced; the older dentition thus comprises only molars in addition to the incisor. More generally, all the cheek teeth are euhypsodont, their seemingly single root remaining open during their whole life. They are homogeneously covered by a very thin layer of cementum.

– Dental formula of *Felovia*: dP^4/dP_4, M^{1-3}/M_{1-3} (Fig. 2E–F)

Several authors noted the presence of P^4 and possibly P_4 in this genus. The studied specimens indicate that only dP^4 and dP_4 are present. The presence of mesial alveoli is due to the loss of dP^4, which occurs late as in *Ctenodactylus*. No skull of a neonate or of a young specimen could be studied. All the premolars are single-rooted, and molars are euhypsodont. Cementum strongly fills enamel crown folds, while the covering of the whole tooth is thinner and more heterogeneous.

– Dental formula of *Massoutiera*: dP^4/dP_4, M^{1-3}/M_{1-3} (Fig. 2G–H)

As for *Ctenodactylus* and *Felovia*, the presence of P^4 and possibly P_4 has been proposed for *Massoutiera*. The investigation of specimens indicated that, as for the two former genera, only dP^4 and dP_4 are present in addition to molars, and they are then lost as well. No dP^3 was observed in the youngest specimen, which nonetheless possesses erupting dP4 and M1. All the premolars are single-rooted, and molars are euhypsodont. Although cementum is slightly thicker in enamel crown folds, the covering of tooth is relatively more homogeneous than in *Felovia*.

Evidence of Mesial Drift in Ctenodactylidae

Measurements. Measurements on lower and upper dental rows showed that mesial drift occurs in all species (Fig. 3, Table 1, Table S1). The displacement of teeth is obvious for first molars, since there is a significant diminution of measurements involving both mesial (L1, U1) and distal sides of M1 (L2, U2; Table 1), which is emphasized by negative values for upper dentitions (Fig. 3B, D, H), except for *Felovia* (Fig. 3F). Such observations are less marked in *Felovia* (L2, U2, Fig. 3E–F) and in lower molars of *Massoutiera* (L2, Fig. 3G), perhaps because we lack juvenile specimens for these genera. A drift of second molars occurs as well (L2, U2). However, the decrease of the measurements is only significant for the distal part of M_2 (L3) of *Pectinator* and *Ctenodactylus* (Fig. 3A, C, Table 1). The other measurements of M2 (L3, U3) are nearly similar from juvenile to adult forms, since the slopes of linear regression are not significantly different from 0 in *Felovia* (Fig. 3E–F) and for M^2 of *Pectinator* and *Ctenodactylus* (U3, Fig. 3B, D), while there is a slight significant increase of values in *Massoutiera* (Fig. 3G–H, Table 1). This result should be linked to the extended growth of teeth, which induces a slight enlargement of molars, and this might weaken the observable effect of mesial drift. The important growth of M3 can be noticed with a significant increase of measurements (L4, U4), and it is concomitant with both the maxillary growth (U5) and the dentary growth (proportional to L4).

Histological results. Evidence of mesial drift can be observed in both bone and dental tissues (Fig. 4). *Pectinator* is the most striking case due to the loss of dP^3 and the replacement of dP4 (Fig. 1). In virtual cross-section of upper cheek teeth (Fig. 4A), bone resorption is shown on mesial side of M^1 and M^2 by the serrated aspect of the distal part of interalveolar septa, and the entire alveolar wall shows an etched surface (Fig. 4C–D). Distally to teeth, bone apposition (or formation) is conversely illustrated by the presence of numerous openings of vascular channels in bone septa and alveolar surface (Fig. 4B–C). In this area, dental resorption occurs because the enamel layer is reduced to absent (see M^1 and M^2) and the outline is irregular due to the indirect compressive force of erupting distal molars. Dental resorption is strongly efficient at the mesial-most side of the cheek tooth row. Intense resorption notably affects deciduous premolars on various sides. In juveniles, mesial roots of dP^4 tend to be resorbed by compressive forces induced by mesial drift (Fig. 4G). Resorption of dP^3 is mesially obvious at the collar level, because of dental progression impeded by the physical constraints applied by the cortical bone delimiting the diastema, which is denser than the

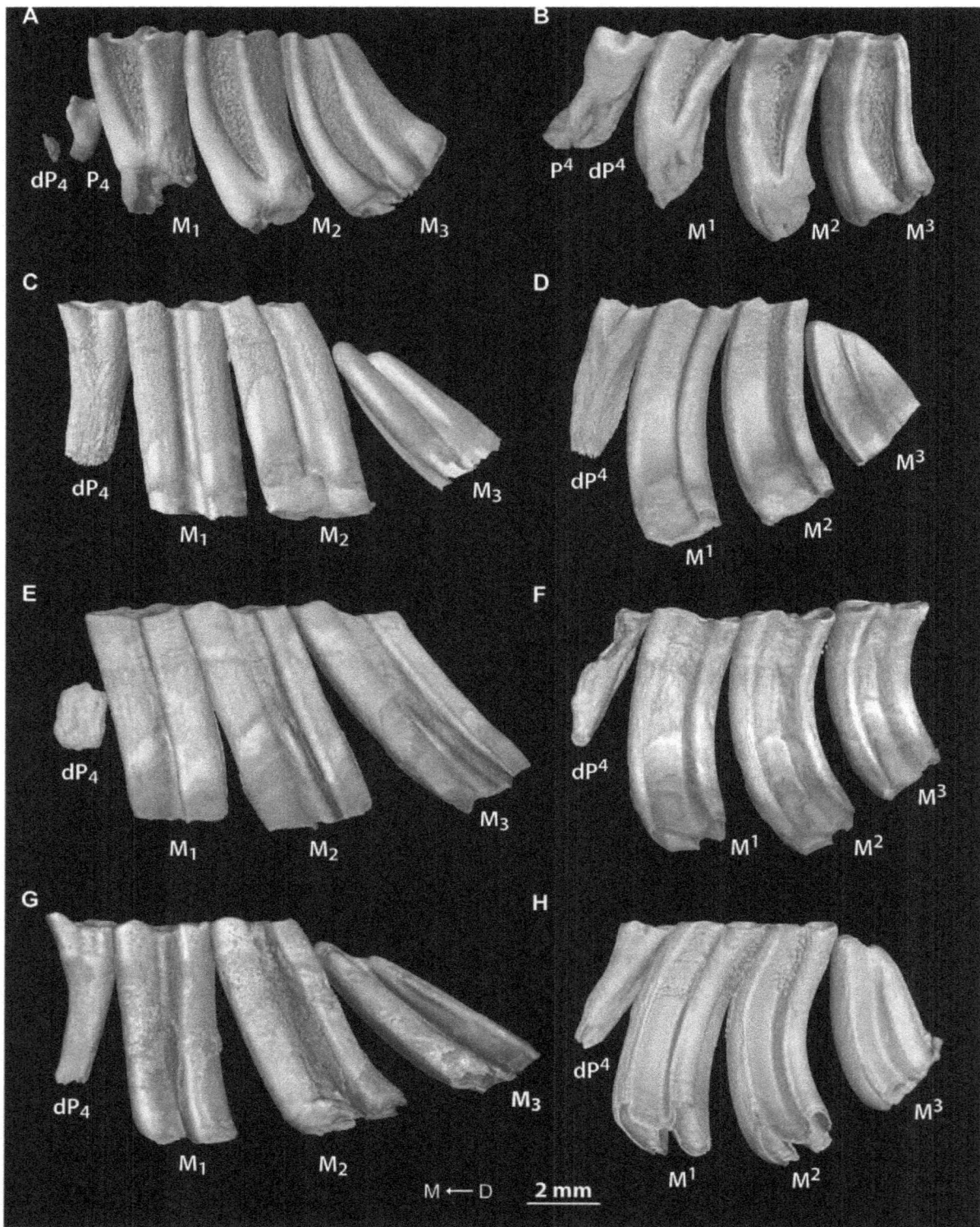

Figure 2. Lower and upper cheek tooth rows of each extant ctenodactylid in lateral view. X-ray synchrotron microtomographic 3D renderings of A–B, *Pectinator spekei* (MNHN-CG1893-226), C–D, *Ctenodactylus gundi* (MNHN-CG1986-255), E–F, *Felovia vae* (MNHN-CG1989-22) and G–H, *Massoutiera mzabi* (MNHN-CG1955-2). D → M stands for distal to mesial direction.

alveolar bone. In older specimens, dP^3 is shed and P^4 starts to protrude into the bone and leads to the mesial resorption of dP^4 in addition to the mesial one resulting from mesial drift (Fig. 4E–F,

H–I). Similar physiological mechanisms are observed on lower dentition (Fig. 4J–K). The remains of dP$_4$, which stand in front of

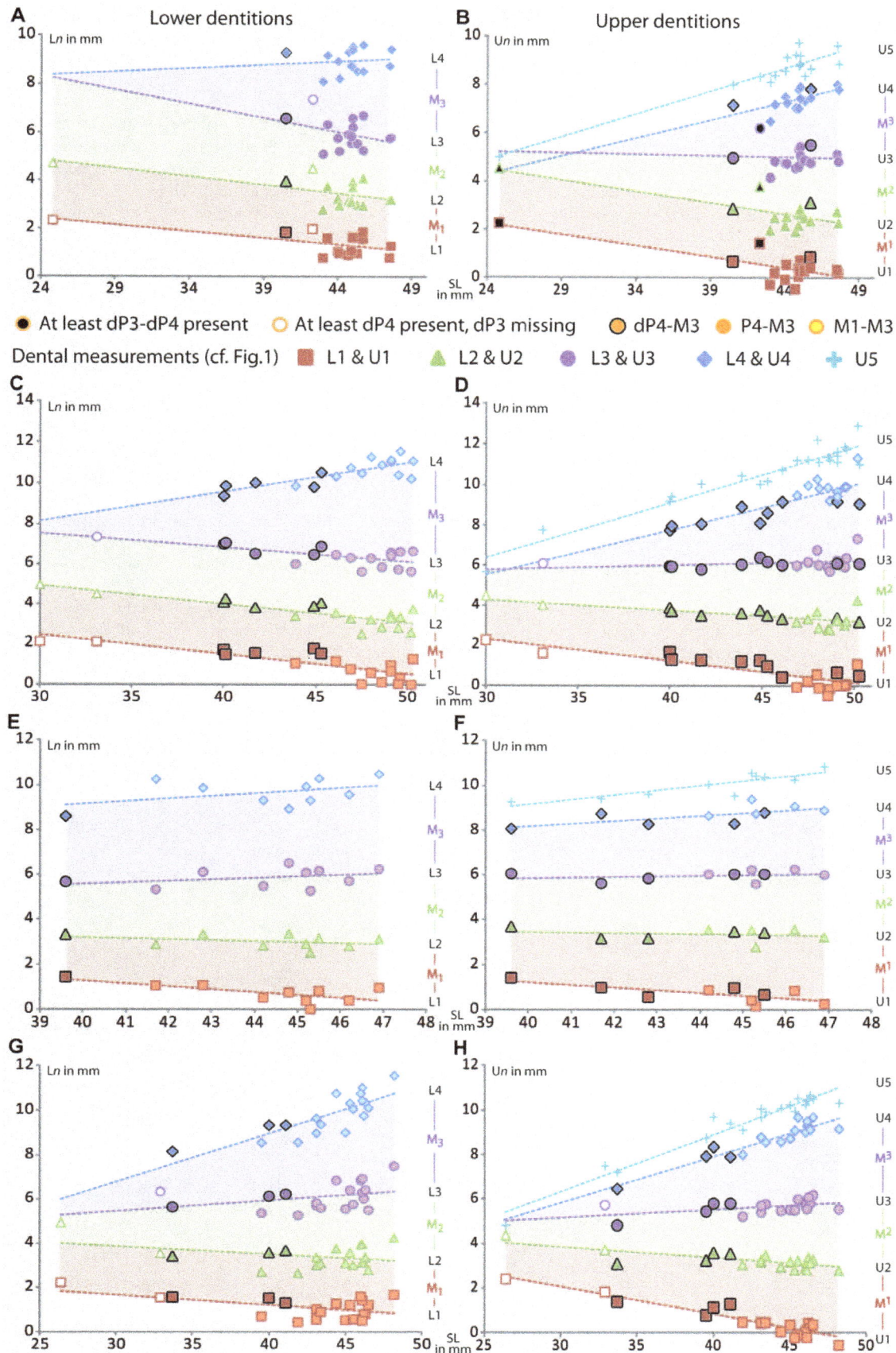

A Lower dentitions

B Upper dentitions

● At least dP3-dP4 present   ○ At least dP4 present, dP3 missing   ◉ dP4-M3   ● P4-M3   ○ M1-M3

Dental measurements (cf. Fig.1)   ■ L1 & U1   ▲ L2 & U2   ● L3 & U3   ◆ L4 & U4   ✚ U5

Figure 3. Graphs including skull measurements highlighting mesial drift for each extant ctenodactylid. A–B, *Pectinator spekei*, C–D, *Ctenodactylus gundi*, E–F, *Felovia vae* and G–H, *Massoutiera mzabi*. L*n*: measurements for right lower dentition; U*n*: measurements for right upper dentition, SL: Skull length. Red, green and violet areas show the evolution of M1, M2 and M3 length respectively, compared to skull length.

P4, are resorbed inside the bone by both distal and mesial compressions, and the remaining alveoli are filled with bone.

Similar impacts of mesial drift are noticed in *Ctenodactylus*, *Felovia* and *Massoutiera*. The pushing action of growing mesial molars is solely responsible for the loss of deciduous premolars resulting from an intense resorption (Fig. S1). In *Felovia* and *Massoutiera*, the layer of cementum is generally thicker on the mesial side than on the distal side of teeth, which is another consequence of dental migration.

Discussion

Importance of Premolar Development to Reassess Ctenodactyline Evolution

We found that an important layer of cementum is present in crown folds of all extant ctenodactylines, including *Pectinator*, contrary to the results of previous studies [18,19,21]. Our study of extant ctenodactylids again demonstrated that *Pectinator* differs from other genera in being less hypsodont and in having a more primitive dental formula involving the presence of dP^3, P^4, and P$_4$, which is consistent with its plesiomorphic dental morphology [18,19,21]. The tooth previously proposed as a dP$_3$ actually corresponds to the anterior root of dP$_4$, which remains in front of the erupted P$_4$. Schrenk [38] studied the ontogenetic development of the skull of *Ctenodactylus gundi*. Interestingly, he noted the presence of a dP^3 on several sections of the skulls, as in *Pectinator*.

He indicated the presence of dP^3 in embryos at stage 2 (Skull length: 12 mm) and stage 4 (Skull length: 22 mm). However, it does not seem that the incipiently mineralized tooth is indeed a dP^3 at the stage 4, inasmuch as it is located behind the dP^4. It might rather correspond to the mesial part of M^1. The occurrence of a dP^3 bud in *Ctenodactylus* means that this tooth starts to develop, but later regresses, contrary to *Pectinator*. Such abortion of tooth development prior to its mineralization has already been evidenced in the mouse and in a few squirrels [22,23,24]. This aborted tooth has also been hypothetically assigned to a premolar lost during evolution [26,39]. Observation of unmineralized dP^3 bud in these species is consistent with the fact this tooth was recently lost (i.e.; Pleistocene, 2.6 Ma-11,700 years) in *Ctenodactylus*, as in *Massoutiera* and *Felovia* [18] whose skull development has never been investigated. The developmental follow-up of such teeth is relevant for a better understanding of the evolutionary and developmental dynamics of the ctenodactyline dentition. In other words, it could inform about the real place of *Ctenodactylus* in the evolution of ctenodactylines, since it seems morphologically closer to the more primitive genus *Pectinator* [40], but it is phylogenetically closer to the more derived *Felovia-Massoutiera* clade ([19], Fig. 5).

Conversely to what has been proposed by numerous authors [19,21,27], P4 are absent in *Massoutiera*, *Felovia* and *Ctenodactylus*. It is worth mentioning that dP4 are more substantially worn in *Felovia*, *Massoutiera* and *Ctenodactylus* than they are in *Pectinator*,

Table 1. Data of linear regressions and Student's t-test calculated for on each variable for each species.

		L1	L2	L3	L4	U1	U2	U3	U4	U5
Pectinator	a	−0.055	−0.069	−0.115	0.030	−0.091	−0.094	−0.012	0.152	0.190
	t	2.852	2.995	10.893	1.313	4.077	4.123	0.491	4.567	9.773
	df	15	15	15	15	15	15	15	15	15
	p	**0.010**	**0.007**	**<0.001**	0.204	**0.001**	**0.001**	0.629	**<0.001**	**<0.001**
	b	3.753	6.496	11.079	7.596	4.417	6.772	5.521	0.642	0.319
Ctenodactylus	a	−0.095	−0.096	−0.071	0.142	−0.105	−0.057	0.023	0.223	0.275
	t	5.114	5.775	5.701	7.357	6.199	4.168	1.663	6.528	14.291
	df	20	20	20	20	20	20	20	20	20
	p	**<0.001**	**<0.001**	**<0.001**	**<0.001**	**<0.001**	**<0.001**	0.112	**<0.001**	**<0.001**
	b	5.287	7.859	9.652	3.903	5.434	6.009	5.123	−1.123	−1.854
Felovia	a	−0.126	−0.042	0.067	0.121	−0.120	−0.022	0.030	0.120	0.212
	t	2.552	1.008	1.058	1.384	2.868	0.514	0.900	2.558	4.813
	df	8	8	8	8	8	8	8	8	8
	p	**0.034**	0.343	0.321	0.204	**0.021**	0.621	0.394	**0.034**	**0.001**
	b	6.318	4.886	2.912	4.294	6.008	4.309	4.621	3.381	0.681
Massoutiera	a	−0.046	−0.036	0.050	0.219	−0.123	−0.050	0.036	0.212	0.256
	t	2.594	1.671	2.223	6.958	11.343	4.924	3.478	5.909	18.479
	df	19	19	19	19	19	19	19	19	19
	p	**0.018**	0.111	**0.039**	**<0.001**	**<0.001**	**<0.001**	**0.003**	**<0.001**	**<0.001**
	b	3.049	4.949	3.924	0.217	5.752	5.351	4.081	−0.563	−1.356

L*n* refers to measurements on right lower dentition; U*n* refers to measurements on right upper dentition; a: slope of linear regression, df: degree of freedom (significance at α<0.05 is indicated in bold), b: intercept.

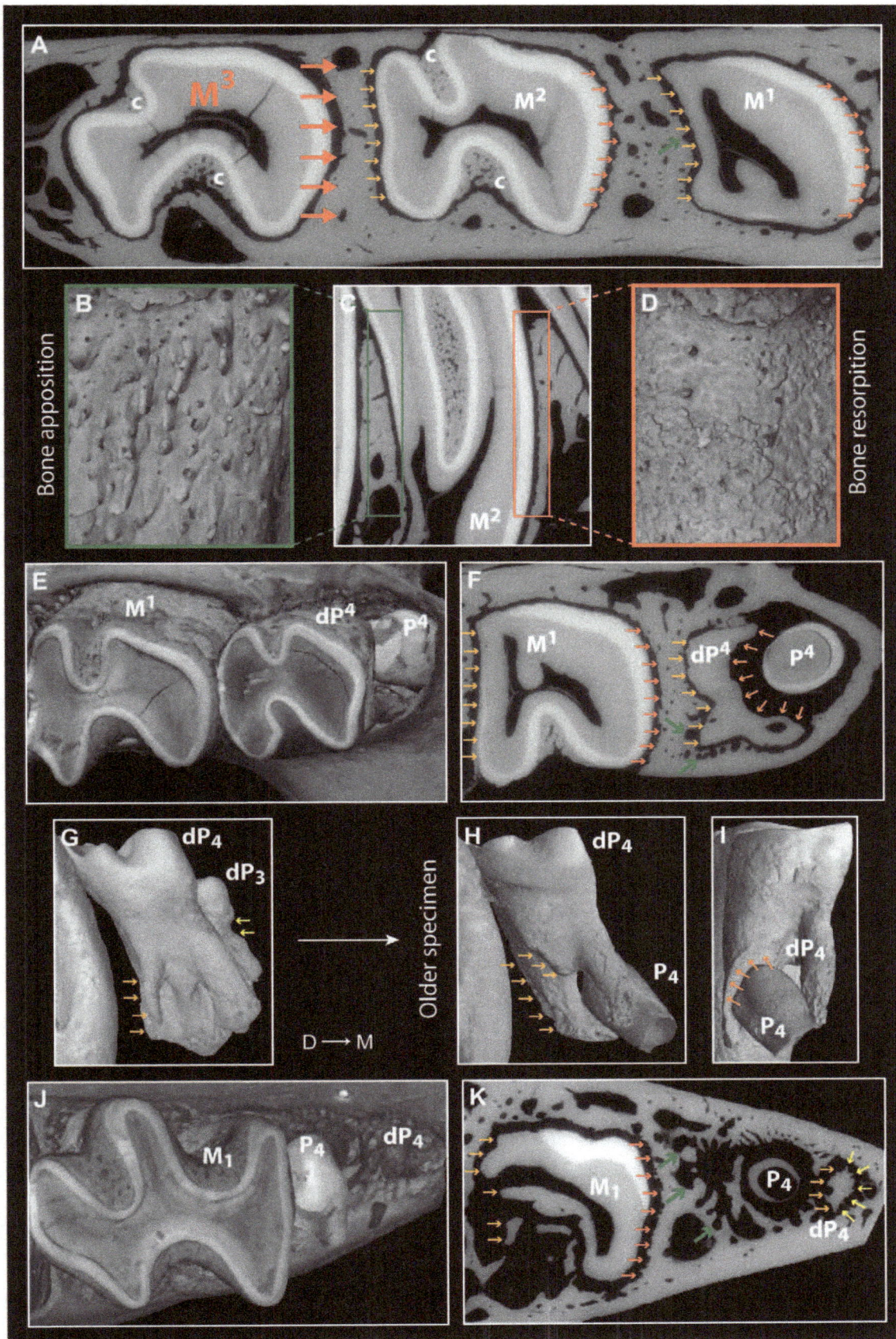

Figure 4. Bone and dental evidences of mesial drift in *Pectinator spekei*. A, F and K, X-ray synchrotron microtomographic virtual cross-section. C, X-ray synchrotron microtomographic virtual longitudinal section. B, D-E, G-J, X-ray synchrotron microtomographic 3D renderings. A-I, maxillary and upper cheek teeth (MNHN-CG1893-226 and 1895-461 for G); J-K, lower cheek teeth (MNHN-CG1893-226). Large red arrows display the orientation of compressive force; small red arrows indicate bone resorption by pointing the scalloped outline of alveolar wall; green arrows indicate bone apposition by pointing openings of vascular channels; small yellow to orange arrows indicate the different sites of dental resorption; c stands for cementum; D → M stands for distal to mesial direction.

because of their retention in these genera. Thus, the teeth generally considered to be permanent premolars, due to their smaller size, actually correspond to worn deciduous premolars. As far as we know, one cannot say whether a beginning of development occurs for permanent premolars, which could be then stopped before mineralization.

Interest of Dental Drift in Rodent Evolution with a Special Focus on the Ctenodactylid Radiation

Dental mesial drift occurs in all extant ctenodactylids. This mechanism involves premolars and the first two molars, while M3 likely represent the main forward pressures by both growing and erupting. The hypothesis of a pressure originating from distal molars can be supported by the fact that M2 could also exert pressure. For instance, the growth of these teeth leads to the dP^3 loss in *Pectinator*, while M3 is just starting to develop. The mode of pressure could be notably illustrated by the mesial bending of M3 (Fig. 2), while erupting toward more mesial teeth. It can be considered that the bone locus dedicated to the dentition in ctenodactylids cannot include all the cheek teeth because of their greater length. Consequently, the eruption of distal molars can drive mesial drift because of their size. Similarly, Humans have an analogous problem because of their reduced jaw inducing a lack of

places for wisdom teeth. More generally, we can assume that when mesial drift occurs in mammals, the length of the dentition (per jaw quadrant) is greater than the length of the corresponding bone locus due to either too large teeth (e.g. some primates and macropodids, and elephants) or extrateeth (e.g. manatees, silvery mole-rats, and pygmy rock wallabies). Such mechanism in ctenodactylids corresponds to a moderate mesial drift (mMD, Fig. 5) since only one tooth per jaw quadrant is lost. In a different way, the silvery mole-rat (*Heliophobius*) displays a strong mesial drift (stMD; Fig. 5), because many teeth are shed as the consequence of new teeth constantly erupting. At the opposite, the Cape mole-rat (*Georychus*) displays a slight mesial drift (sMD, Fig. 5) involving a weak displacement of teeth without dental loss [14].

To date, the above-mentioned species constitute the only known rodents having mesial drift. It is worth noticing that they all possess hypsodont teeth. This dental character probably favors the presence of mesial drift in rodents. Hypsodonty is the major evolutionary trend in rodents to withstand the effect of intensive wear. Therefore, mesial drift represent a superimposed mechanism partly originating from the significant enlargement of the occlusal surface of high-crowned molars during wear in ctenodactylids, whereas it is related to supernumerary teeth in the silvery mole-rat. It is also linked to the reduction of premolar size, affecting first the

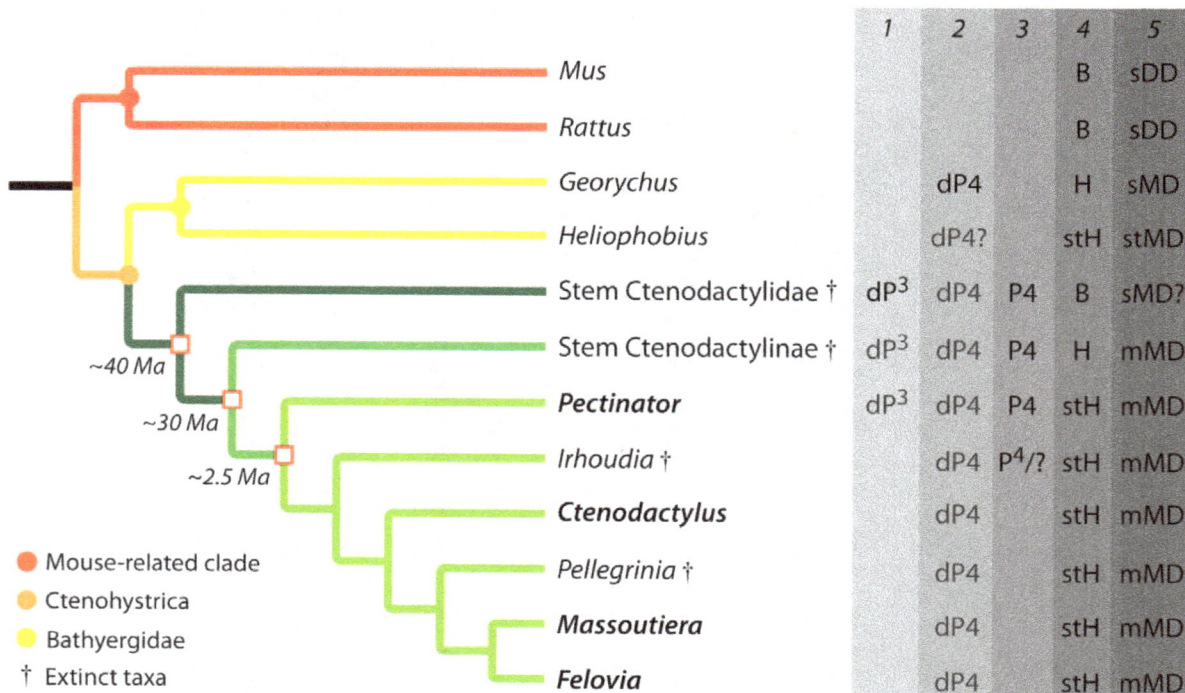

	1	2	3	4	5
Mus				B	sDD
Rattus				B	sDD
Georychus		dP4		H	sMD
Heliophobius		dP4?		stH	stMD
Stem Ctenodactylidae †	dP^3	dP4	P4	B	sMD?
Stem Ctenodactylinae †	dP^3	dP4	P4	H	mMD
Pectinator	dP^3	dP4	P4	stH	mMD
Irhoudia †		dP4	P^4/?	stH	mMD
Ctenodactylus		dP4		stH	mMD
Pellegrinia †		dP4		stH	mMD
Massoutiera		dP4		stH	mMD
Felovia		dP4		stH	mMD

~40 Ma
~30 Ma
~2.5 Ma

● Mouse-related clade
● Ctenohystrica
● Bathyergidae
† Extinct taxa

Figure 5. Simplified phylogeny of rodents showing main taxa bearing dental drift. Relationships between the main rodent clades were defined via a molecular analysis [61], while relationships within ctenodactylids were determined according to morphological analyses [18,19]. 1, Presence of third upper premolar. 2, Presence of fourth deciduous premolars; 3, Presence of fourth definitive premolars. 4, Molar crown height: B Brachydont, H Hypsodont, stH strongly hypsodont. 5, Presence of dental drift: sDD slight Distal Drift, sMD slight Mesial Drift, mMD moderate Mesial Drift, stMD strong Mesial Drift. Green branches represent the Ctenodactylidae; in grey: tooth lost during lifetime.

permanent ones, during the course of ctenodactylid evolution (i.e. since the Oligocene, [18]).

Such innovation also represents another means to withstand an important component of abrasive matter found in both plants and exogenous particles. Indeed, extant ctenodactylids generally usually live in rocky slopes and crevices in desert or semi-desert areas [41], where silica phytoliths present in grasses, and dust on the herbaceous layer are the most abundant sources of abrasive matter. This adaptation could appear early in the ctenodactylid evolution, since the Miocene and the radiation of first ctenodacty-lines [19], whose dentition is marked by the reduction of premolar size and the spread of semi-hypsodont forms (i.e equal height between lingual and vestibular sides, [18,42,43]). The first apparent evidence of mesial drift can be observed during the Upper Miocene, in *Metasayimys jebeli*, which shows strongly reduced P4 compared to dP4, and loss of dP^3 during growth [21], as in *Pectinator*. This evolutionary trend could be parallel with the concomitant and massive expansion of grasses in open environments during the Miocene [44,45].

According to some authors, a few rodents, such as laboratory rats and mice, display slight distal drift (sDD, Fig. 5) of molars [46,47,48]. It was suggested that this migration corresponds to the posterior lengthening of the jaw during development [49]. One can consider that without any assumption of effective pressure, this dental displacement is the result of a distal shift (i.e., virtual displacement) involving only bone growth [33], rather than a true drift. Nonetheless, bone remodeling of alveolar sockets was clearly demonstrated in such cases. It is difficult to assume that masticatory movements are mainly involved in dental migration because of their propalinal direction of mastication (from posterior to anterior side; [9]). Consequently, distal shift is probably the main component of displacement, which induces slight losses of approximal contact between molars leading to drift to recover contact, as demonstrated when a tooth is experimentally moved [46]. However, further studies are needed to understand the actual origin of physiological distal drift in these rodents, to be then compared to the characteristics of mesial drift from evolutionary and biological points of views.

Concluding Remarks on Dental Drift Involving Biological and Biomedical Prospects

Rats and mice, in addition to rabbits, monkeys, cats and dogs, are frequently used for addressing orthodontics issues [48,50,51]. A closed coil spring is generally applied on animal's dentition to study *in vivo* the impact of tensile forces. Such studies permit the evaluation of the overall consequences of orthodontic drift on the dentition, as the rates of bone remodeling and dental resorption for biomedical perspectives. Various orthodontic appliances are used in humans to withstand physiological mesial drift of teeth frequently leading to dental misalignments. In order to be as efficient as possible, their characteristics are previously defined by means of such experimental drift. More generally, the effects of orthodontic drift (artificial or experimental forces) are more accurately studied than those of physiological drift (natural forces) [52], while a better knowledge of this last component is useful for both biological and biomedical studies. In addition, orthodontic drift only involves tensile forces, whereas physiological drift is induced by compressive forces as observed in rodents. As a result, it is necessary to draw comparisons between these two mechanisms whose consequences could be slightly dissimilar at least at the histological level. In this way, a more precise investigation of mesial drift in rodents is needed.

The last noteworthy point concerns dental resorption resulting from the activity of odontoclasts, which is far less studied than bone remodeling driven by both osteoclasts and osteoblasts [53]. In most cases, the analysis of dental resorption refers to the eruption of permanent teeth replacing deciduous teeth by resorbing their roots [54,55], but it can also be linked to orthodontic drift [56]. This last mechanism, when mesially oriented, frequently leads to mesial root resorption, whereas physiological mesial drift leads to distal root resorption. In addition to dentine and cementum, enamel can be affected by physiological mesial drift. Moreover, three different phases of dental resorption are noticed, when driven by compressive forces: (1) distal compression (mesial drift), (2) apical compression (dental eruption), and (3) mesial compression (the tooth reaches the end of the bone locus dedicated to the dentition, and it is an indirect consequence of mesial drift). The rodents showing mesial drift represent thus a rare opportunity to accurately study the resorption in the different point of tooth and on the various tissues constituting the tooth. That could also permit the proper assessment of the putative range of odontoclast activities which affect teeth, compared to osteoclast activities affecting bone [51,57,58]. Such prospects involving this new way of investigations in rodents could also be promising regarding other aspects of dental drift such as cementum repairs [59], the roles and status of periodontal ligament (i.e. responsible for teeth anchorage; [47,60]) and transseptal fibers (i.e. linking adjacent teeth; [46]).

Supporting Information

Figure S1 Bone remodeling and dental resorption in upper dentition of Ctenodactylidae. Synchrotron micro-tomographic virtual cross-sections of A, *Ctenodactylus gundi* (MNHN-CG1986-255), B, *Felovia vae* (MNHN-CG1989-22), and C, *Massoutiera mzabi* (MNHN-CG1960-3741). D → M stands for distal to mesial direction.

Table S1 List of investigated specimens including data on dentitions for each Ctenodactylidae. Abbreviations: L*n* refers to measurements on right lower dentition; U*n* refers to measurements on right upper dentition; MNHN: Museum National d'Histoire Naturelle; BSL: Basel (Naturhistorisches Museum). Symbols: [X*n*] signifies the tooth is shed but the alveolus is still present; (X*n*) means the tooth is erupting.

Acknowledgments

We are grateful to the curators V. Nicolas and J. Cuisin from the MNHN of Paris, and to L. Costeur from the Naturhistorisches Museum of Basel who allowed the access to the collections of Ctenodactylidae, and permitted the loan of specimens. We acknowledge the ESRF (Grenoble) and the team of the beamline BM5 to have supported that research project by giving access to their experimental stations. We want to thank Cyrielle Charles for her technical help in Basel, and to F. Gomes Rodrigues and M. Bigot for living accommodation in Paris. We also thank H. Magloire for his fruitful comments on the manuscript. Thank you very much to A. Evans and the two anonymous reviewers for the improvement of the manuscript.

Author Contributions

Conceived and designed the experiments: HGR FS CC LV. Performed the experiments: HGR FS CC PT. Analyzed the data: HGR FS. Contributed reagents/materials/analysis tools: HGR FS CC PT. Wrote the paper: HGR FS CC PT MVL LV.

References

1. Wilson DE, Reeder DME (2005) Mammal species of the world: a taxonomic and geographic reference (3rd ed.): Johns Hopkins University Press. 2142 p.
2. Stehlin HG, Schaub S (1951) Die Trigonodontie der simplicidentaten Nager. Schweizerische Paläontologische Abhandlungen 67: 1–385.
3. Marivaux L, Vianey-Liaud M, Jaeger J-J (2004) High-level phylogeny of early Tertiary rodents: dental evidence. Zoological Journal of the Linnean Society 142: 105–134.
4. Misonne X (1969) African and Indo-Australian Muridae evolutionary trends. Annales du Musée Royal d'Afrique Centrale Tervuren 172: 1–219.
5. Martin T (1997) Incisor enamel microstructure and systematics in rodents. In: von Koenigswald W, Sander PM, editors. Tooth Enamel Microstructure. Rotterdam: Balkema. 163–175.
6. von Koenigswald W (2004) The three basic types of schmelzmuster in fossil and extant rodent molars and their distribution among rodent clades. Palaeontographica Abt A 270: 95–132.
7. Butler PM (1985) Homologies of molars cusps and crests and their bearing on assessments of rodent phylogeny. In: Luckett P, Hartenberger J-L, editors. Plenum. 381–402.
8. Charles C, Jaeger J-J, Michaux J, Viriot L (2007) Dental microwear in relation to changes in the direction of mastication during the evolution of Myodonta (Rodentia, Mammalia). Naturwissenschaften 94: 71–75.
9. Lazzari V, Charles C, Tafforeau P, Vianey-liaud M, Aguilar J-P, et al. (2008) Mosaic convergence of rodent dentitions. Plos One 3: 1–13.
10. Jarvinen E, Salazar-Ciudad I, Birchmeier W, Taketo MM, Jernvall J, et al. (2006) Continuous tooth generation in mouse is induced by activated epithelial Wnt/β-catenin signaling. Proceedings of the National Academy of Science 103: 18627–18632.
11. Charles C, Hovorakova M, Youngwook A, Lyons DB, Marangoni P, et al. (2011) Regulation of tooth number by fine-tuning levels of receptor-tyrosine kinase signaling. Development 138: 4063–4073.
12. Chlastakova I, Lungova V, Wells K, Tucker AS, Radlanski RJ, et al. (2011) Morphogenesis and bone integration of the mouse mandibular third molar. European Journal of Oral Sciences 119: 265–274.
13. O'Connell DJ, Ho JWK, Mammoto T, Turbe-Doan A, O'Connell JT, et al. (2012) A Wnt-Bmp feedback circuit controls intertissue signaling dynamics in tooth Organogenesis. Science signaling 5: 1–10.
14. Gomes Rodrigues H, Marangoni P, Šumbera R, Tafforeau P, Wendelen W, et al. (2011) Continuous dental replacement in a hyper-chisel tooth digging rodent. Proceedings of the National Academy of Science 108: 17355–17359.
15. Wang B (1997) The mid-Tertiary Ctenodactylidae (Rodentia, Mammalia) of eastern and central Asia. Bulletin of the American Museum of Natural History 234: 1–88.
16. Schmid-Kittler N, Vianey-Liaud M, Marivaux L (2006) The Ctenodactylidae (Rodentia, Mammalia). In: Daxner-Höck G, editor. Oligocene-Miocene Vertebrates from the Valley of Lakes (Central Mongolia): Morphology, phylogenetic and stratigraphic implications: Annalen des Naturhistorischen Museums in Wien. 173–215.
17. Vianey-Liaud M, Schmidt-Kittler N, Marivaux L (2006) The Ctenodactylidae (Rodentia) from the Oligocene of Ulantatal (Inner Mongolia, China). Palaeovertebrata 34: 111–205.
18. Vianey-Liaud M, Gomes Rodrigues H, Marivaux L (2010) A New Oligocene Ctenodactylinae (Rodentia, Mammalia) from Ulantatal (Nei Mongol): new insight on the phylogenetic origins of the modern Ctenodactylidae. Zoological Journal of the Linnean Society 160: 531–550.
19. López-Antoñanzas R, Knoll F (2011) A comprehensive phylogeny of the gundis (Ctenodactylinae, Ctenodactylidae, Rodentia). Journal of Systematic Palaeontology 9: 379–398.
20. Ellerman JR (1940) The families and Genera of living Rodents; History) BMN, editor.
21. Jaeger JJ (1971) Un Cténodactylidé (Mammalia, Rodentia) nouveau, Irhoudia bohlini n. g., n. sp., du Pleistocène inférieur du Maroc. Rapports avec les formes actuelles et fossiles. Notes du Service Géologique du Maroc 31: 113–140.
22. Luckett WP (1985) Superordinal and intraordinal affinities of rodents: Developmental evidence from the dentition and placentation. In: Luckett WP, Hartenberger, J.-L., editors. Evolutionary relationships among rodents: A multidisciplinary analysis. NATO ASI Ser. Life Sciences. 227–276.
23. Tureckova J, Lesot H, Vonesch J-L, Peterkova R, Peterka M, et al. (1996) Apopotosis is involved in disappearance of the diastemal dental primordial in mouse embryo. International Journal of Developmental Biology 40: 483–489.
24. Viriot L, Lesot H, Vonesh J-L, Peterka M, Peterkova R (2000) The presence of rudimentary odontogenic structures in the mouse embryonic mandible requires reinterpretation of developmental control of first lower molar histomorphogenesis. International Journal of Developmental Biology 44: 233–240.
25. Viriot L, Peterkova R, Peterka M, Lesot H (2002) Evolutionary implications of the occurrence of two vestigial tooth germs during early odontogenesis in the mouse lower jaw. Connective Tissue Research 43: 129–133.
26. Prochazka J, Pantalacci S, Churava S, Rothova M, Lambert A, et al. (2010) Patterning by heritage in mouse molar row development. Proceedings of the National Academy of Science 107: 15497–15502.
27. Lataste F (1885) Sur le système dentaire du genre Ctenodactylus Gray. Le Naturaliste 3: 21–22.
28. Lavocat R (1961) Le gisement de vertébrés Miocènes de Beni-Mellal (Maroc) : Etude systématique de la faune de mammifères et conclusions générales. Notes et Mémoires du Service Géologique du Maroc 155: 29–94; 109–144.
29. Sikes SK (1971) The natural history of the African elephant. London: Weidenfeld & Nicolson.
30. Moss JP, Picton DCA (1982) Short-term changes in the mesiodistal position of teeth following removal of approximal contacts in the monkey Macaca fascicularis. Archives or oral Biology 27: 273–278.
31. Domning DP, Hayek L-AC (1984) Horizontal tooth replacement in the Amazonian manatee (Trichechus inunguis). Mammalia 48: 105–127.
32. Lanyon JM, Sanson GD (2006) Degenerate dentition of the dugong (Dugong dugon), or why a grazer does not need teeth: morphology, occlusion and wear of mouthparts. Journal of Zoology 268: 133–152.
33. Lentle RG, Hume I (2010) Mesial drift and mesial shift in the molars of four species of wallaby: the influence of chewing mechanics on tooth movement in a group of species with an unusual mode of jaw action. In: Coulson G, Eldridge M, editors. Macropods: The Biology of Kangaroos, Wallabies and Rat-Kangaroos: CSIRO Publishing. 127–137.
34. Paganin D, Mayo SC, Gureyev TE, Miller PR, Wilkins SW (2002) Simultaneous phase and amplitude extraction from a single defocused image of a homogeneous object. Journal of Microscopy 206: 33–40.
35. Sanchez S, Ahlberg PE, Trinajstic K, Mirone A, Tafforeau P (in press) Three dimensional synchrotron virtual paleohistology: A new insight into the world of fossil bone microstructures. Microscopy and Microanalysis.
36. Mones A (1982) An equivocal nomenclature: What means hypsodonty? Paläontologische Zeitschrift 56: 107–111.
37. von Koenigswald W (2011) Diversity of hypsodont teeth in mammalian dentitions – construction and classification. Palaeontographica, Abt A: Palaeozoology – Stratigraphy 294: 63–94.
38. Schrenk F (1989) Zur Schädelentwicklung von Ctenodactylus gundi (Rothmann 1776) (Mammalia, Rodentia). Courier Forschungsinstitut Senckenberg 108: 1–241.
39. Gomes Rodrigues H, Charles C, Marivaux L, Vianey-Liaud M, Viriot L (2011) Evolutionary and developmental dynamics of the dentition in Muroidea and Dipodoidea (Rodentia, Mammalia). Evolution and Development 13: 260–268.
40. George W (1979) The chromosomes of the Hystricomorphous Family Ctenodactylidae (Rodentia, Sciuromorpha) and the four living genera. Zoological Journal of the Linnean Society 65: 261–280.
41. Nowak RM (1999) Walker's mammals of the world, Vol. II, 6th ed.; Hopkins J, editor: Johns Hopkins University Press.
42. Vianey-Liaud M (1976) Les Issiodoromyinae (Rodentia, Theridomyidae) de l'Eocène supérieur à l'Oligocène inférieur en Europe Occidentale. Palaeovertebrata, Montpellier, 7: 1–115.
43. Patterson B, Wood AE (1982) Rodents from the Deseaden Oligocene of Bolivia and the relationships of the Caviomorph. Bulletin of the Museum of Comparative Zoology, Harvard 149: 371–543.
44. Edwards EJ, Osborne CP, Strömberg CAE, Smith SA, Consortium CG (2010) The origins of C4 grasslands: integrating evolutionary and ecosystem science. Science 328: 587–591.2.
45. Strömberg CAE (2011) Evolution of grasses and grassland ecosystems. Annual Review of Earth and Planetary Sciences 39: 517–544.
46. Roux D, Woda A (1994) Biometric analysis of tooth migration after approximal contact removal in the rat. Archives of Oral Biology 39: 1023–1027.
47. Johnson RB, Martinez RH (1998) Synthesis of Sharpey's fiber proteins within rodent alveolar bone. Scanning Microscopy 12: 317–327.
48. Ren Y, Maltha JC, Kuijpers-Jagtman AM (2004) The rat as a model for orthodontic tooth movement-a critical review and a proposed solution. European Journal of Orthodontics 26: 483–490.
49. Sicher H, Weimann JP (1944) Bone growth and physiologic tooth movement. American Journal of Orthodontics and Oral Surgery 30: 109–132.
50. Wise GE, King GJ (2008) Mechanisms of tooth eruption and orthodontic tooth movement. Journal of Dental Research 87: 414–434.
51. Kiliç N, Oktay H, Ersöz M (2010) Effects of force magnitude on tooth movement: an experimental study in rabbits. European Journal of Orthodontics 32: 154–158.
52. Lasfargues JJ, Saffar JL (1992) Effects of prostaglandin inhibition on the bone activities associated with the spontaneous drift of molar teeth in the rat. The Anatomical Record 234: 310–316.
53. Wang Z, McCauley LK (2011) Osteoclasts and odontoclasts: signaling pathways to development and disease. Oral Disease 17: 129–142.
54. Marks SCJ, Schroeder HE (1996) Tooth eruption: theories and facts. The Anatomical Record 245: 374–393.
55. Harokopakis-Hajishengallis E (2007) Physiologic root resorption in primary teeth: molecular and histological events. Journal of Oral Science 49: 1–12.
56. Sringkarnboriboon S, Matsumoto Y, Soma K (2003) Root resorption related to hypofunctional periodontium in experimental tooth movement. Journal of Dental Research 82: 486–490.
57. Lasfargues JJ, Saffar JL (1993) Inhibition of prostanoid synthesis depresses alveolar bone resorption but enhances root resorption in the rat. The Anatomical Record 237: 458–465.

58. Sasaki T (2003) Differentiation and Functions of Osteoclasts and Odontoclasts in Mineralized Tissue Resorption. Microscopy research and technique 61: 483–495.

59. Jäger A, Kunert D, Friesen T, Zhang D, Lossdörfer S, et al. (2008) Cellular and extracellular factors in early root resorption repair in the rat. European Journal of Orthodontics 30: 336–345.

60. Saffar JL, Lasfargues JJ, Cherruau M (1997) Alveolar bone and the alveolar process: the socket that is never stable. Periodontology 2000 13: 76–90.

61. Fabre P-H, Hautier L, Dimitrov D, Douzery E (2012) A glimpse on the pattern of rodent diversication: a phylogenetic approach. BMC Evolutionary Biology 12: 88.

7

Fiat or *Bona Fide* Boundary—A Matter of Granular Perspective

Lars Vogt[1]*, Peter Grobe[2], Björn Quast[1], Thomas Bartolomaeus[1]

1 Institut für Evolutionsbiologie und Ökologie, Universität Bonn, Bonn, Germany, **2** Forschungsmuseum Alexander Koenig Bonn, Bonn, Germany

Abstract

Background: Distinguishing *bona fide* (i.e. natural) and *fiat* (i.e. artificial) physical boundaries plays a key role for distinguishing natural from artificial material entities and is thus relevant to any scientific formal foundational top-level ontology, as for instance the Basic Formal Ontology (BFO). In BFO, the distinction is essential for demarcating two foundational categories of material entity: object and *fiat* object part. The commonly used basis for demarcating *bona fide* from *fiat* boundary refers to two criteria: (i) intrinsic qualities of the boundary bearers (i.e. spatial/physical discontinuity, qualitative heterogeneity) and (ii) mind-independent existence of the boundary. The resulting distinction of *bona fide* and *fiat* boundaries is considered to be categorial and exhaustive.

Methodology/Principal Findings: By referring to various examples from biology, we demonstrate that the hitherto used distinction of boundaries is not categorial: (i) spatial/physical discontinuity is a matter of scale and the differentiation of *bona fide* and *fiat* boundaries is thus granularity-dependent, and (ii) this differentiation is not absolute, but comes in degrees. By reducing the demarcation criteria to mind-independence and by also considering dispositions and historical relations of the bearers of boundaries, instead of only considering their spatio-structural properties, we demonstrate with various examples that spatio-structurally *fiat* boundaries can nevertheless be mind-independent and in this sense *bona fide*.

Conclusions/Significance: We argue that the ontological status of a given boundary is perspective-dependent and that the strictly spatio-structural demarcation criteria follow a static perspective that is ignorant of causality and the dynamics of reality. Based on a distinction of several ontologically independent perspectives, we suggest different types of boundaries and corresponding material entities, including boundaries based on function (locomotion, physiology, ecology, development, reproduction) and common history (development, heredity, evolution). We argue that for each perspective one can differentiate respective *bona fide* from *fiat* boundaries.

Editor: Vladimir N. Uversky, University of South Florida College of Medicine, United States of America

Funding: The authors have no support or funding to report.

Competing Interests: The authors have declared that no competing interests exist.

* E-mail: lars.m.vogt@gmail.com

Introduction

Data integration, data comparability, and the development of data and metadata standards are becoming more and more important in times of increased communication via the World Wide Web and an increasing importance of online databases in academia. Ontologies and other techniques of the Semantic Web thereby play a key role for reliably communicating and managing data within and between databases. This also applies to the life sciences, for which different ontologies for different domains and different purposes already exist (cf., BioPortal; http://bioportal.bioontology.org). Unfortunately, these ontologies often differ considerably [1–3], resulting in incompatibilities and inconsistencies between the contents of the databases that use them and in how these contents are being represented in them. Therefore, in order to achieve common data and metadata standards, ontologies must be standardized as well. Formal top-level ontologies [1,4], as for instance the Descriptive Ontology for Linguistic and Cognitive Engineering (DOLCE) or the Basic Formal Ontology (BFO), play a key role in this respect. They are intended to provide domain- and purpose-independent theories within a formal framework of axioms and definitions for most general terms and concepts, which can be used as a top-level template and formal framework for developing domain reference ontologies and terminology-based application ontologies [1–3,5].

Among many other things, formal top-level ontologies must provide explicit and unambiguous definitions for top-level categories of foundational types of material entity, which scientists from all domains and research interests can agree upon. Smith [6,7] introduced the distinction of two foundational types of boundaries of physical entities, on which BFO's top-level distinction between *fiat* and *bona fide* material entities is based on:

1. ***Bona fide* boundaries:** natural or *mind-independent* boundaries [7,8], which are physical boundaries in the things themselves that exist independently from human perception [6–10].
2. ***Fiat* boundaries:** artificial (i.e. artifact of cognition) or *mind-dependent* boundaries, which are non-physical boundaries that depend on human decision and thus are the products of mental activities [6–9].

The BFO calls the two corresponding top-level categories of material entity 'object' and 'fiat object part', respectively (http://www.ifomis.org/bfo; [11]).

In their very general meaning of *mind-dependent* and *mind-independent*, however, the two attributes '*fiat*' and '*bona fide*' can be applied in various contexts and are not restricted to boundaries of physical entities. Thus, one can even talk about *fiat* concepts in a conceptual sense, or about perceptual, ecological, geometrical, legal, administrative, political and linguistic fiats (e.g., [7,8,12]). Obviously, the distinction of *fiat* and *bona fide* goes beyond the physical realm and is very general. The categories of *bona fide* and *fiat* boundary, when they are used in the context of physical boundaries, however, are very specific and depend on specific spatio-structural properties and thus properties that are intrinsic to the physical entity the boundary bounds (e.g. [6–9]).

Material entities, however, possess additional natural properties, besides their spatio-structural properties, as for instance functional dispositions or historical relations. Strictly confining the application of the attributes '*bona fide*' and '*fiat*' for boundaries to the presence of specific spatio-structural properties does not necessarily follow from the general notion of *fiatness* and *bona fideness*, since not only spatio-structural properties but also dispositions or historical relations can exist independently of human cognitive acts. As a consequence, dispositions and historical relations can be differentiated into mind-dependent and mind-independent ones as well. What is the reason for not referring to dispositions and historical relations for distinguishing *fiat* and *bona fide* boundaries and instead restricting them to spatio-structural properties? Moreover, considering the importance of boundaries for the demarcation of top-level categories of material entity, why do we rest the decision of whether a material entity is an object or a *fiat* object part exclusively on spatio-structural grounds? Is the distinction between *bona fide* and *fiat* boundary only categorial and absolute within this purely spatio-structural context?

We start with discussing the distinction of *bona fide* and *fiat* boundaries and their relation to the distinction of *bona fide* objects and *fiat* object parts within the spatio-structural context. By focusing on some borderline cases we discuss whether the distinction is categorial or rather granularity-dependent. Then we take a closer look at biological entities and aspects of continuity and connectedness characteristic of them. We argue that in the biological domain the granularity-dependence of boundaries can be found in all levels of granularity. We point out the role of *bona fide* landmarks for recognizing *fiat* boundaries, concluding that both *fiatness* of boundaries and *fiatness* of material entities comes in degrees. Since reality is dynamic and biological entities actively participate in many biological processes (i.e. evolution, embryogenesis, physiology, etc.), we argue that the criteria used for distinguishing *fiat* from *bona fide* boundaries and *fiat* from *bona fide* entities, respectively, must not only consider intrinsic spatio-structural qualities but also the dispositions and historical relations of material entities. By providing adequate examples from biology, we demonstrate that in many cases material entities that are demarcated by spatio-structurally *fiat* boundaries are nevertheless *bona fide* units in the sense that their existence as natural units is in fact mind-independent. Thus, it can be demonstrated that the differentiation into mind-dependent and mind-independent boundaries is not only granularity-dependent in a spatio-structural sense, but also perspective-dependent. We distinguish a structural from a functional and a historical perspective for the biological domain, thereby contrasting structural anatomy (*form*) with functional anatomy and historical/evolutionary anatomy. Finally, we draw the consequences for ontology design and formal top-level ontologies in terms of the requirement for integrating several perspective-dependent taxonomies of top-level categories of material entity and their relations to each other.

Results

2.1 The Spatio-Structural Notion of Top-Level Categories of Boundary and Material Entity

The ontological relation between types of boundaries and types of material entities is strong, because the existence of boundaries depends on the higher-dimensional entities they bound, i.e. their hosts [7,9,13]. Traditionally, a *bona fide* material object is characterized as an entity that extents in space and that can be demarcated clearly and unambiguously from its respective environment (i.e. its complement—the universe without this particular material entity). It possesses a single continuous outer boundary, usually referred to as its surface, which symmetrically demarcates the object from its complement and vice versa [9]. Because the boundary belongs to its material host and not to the complement, the respective material object is considered to be closed and its complement to be open. These outer boundaries are called *bona fide* boundaries [6] (also called *natural boundary*, [14]) and can be demarcated on grounds of "*some interior physical discontinuity or some qualitative heterogeneity among the parts of the object (some sharp gradient of material constitution, color, texture, electric charge, etc.)*" [8]. *Bona fide* boundaries are physical boundaries in the things themselves and "*exist independently of all human cognitive acts – they are a matter of qualitative differentiations or discontinuities in the underlying reality*" [7]. The surface of your skin and the surface of an apple in a fruit basket represent examples of *bona fide* boundaries of material objects.

Because every material entity consists of divisible matter, it can be divided spatially along inner boundaries into its constitutive parts. Inner boundaries are not necessarily *bona fide* boundaries, because they do not necessarily have to follow any physical discontinuity or qualitative heterogeneity. A boundary that is not *bona fide* is called a *fiat boundary* (also called *artificial boundary*, [14]):

"*the demarcations induced by fiat boundaries are not grounded in any intrinsic features of the underlying reality, and correspond only to cognitive phenomena such as those induced by our use and understanding of political maps and cadastral surveys*" [9].

According to Smith and Varzi [9], "*the categorial distinction between fiat and bona fide boundaries is absolute*", meaning that no instance of the type '*fiat* boundary' instantiates the type '*bona fide* boundary'—their extensions do not overlap. *Fiat* boundaries are considered to be non-physical boundaries that exclusively depend on acts of human decision:

"*we cannot directly see fiat boundaries*" [15] and they "*are the products of our mental and linguistic activity, and of associated conventional laws, norms and habits*" [8].

In other words, *fiat* boundaries are arbitrarily imposed [16], not grounded in the autonomous mind-independent world [14], but are "*human-demarcation-induced*" [7,9] boundaries. They "*are in a sense potential in that they do not actually separate anything from anything— they do not mark any actual discontinuity*" [9]. Instead, they represent boundaries, which owe their existence to conventional laws, agreements, political decrees and habits, and they do not separate anything in reality [7–10]. The moment one cuts an object along one of its *fiat* boundaries, one divides it into two new objects and the formerly *fiat* boundary of the original object would be gone and, so to speak, replaced by two newly created *bona fide* boundaries.

The Equator, but also the inner boundary demarcating your right thumb from the rest of your body, are good examples for *fiat* boundaries. A *fiat* inner boundary of a *bona fide* object constitutes the *fiat* outer boundary of one of the object's *fiat* parts. Contrary to a *bona fide* boundary, a *fiat* boundary is shared by the two *fiat* parts it demarcates (i.e. each part possesses its own *fiat* boundary, but the two boundaries are considered to *coincide*; see [9,10]).

The differentiation between *bona fide* and *fiat* boundaries is important if one wants to distinguish between the two basic top-level categories of material entity, *bona fide objects* and *fiat objects parts*. *Bona fide* objects are bound completely by a continuous *bona fide* outer boundary, whereas *fiat* object parts are limited by at least one *fiat* boundary [6–10]. Accordingly, *bona fide* objects are assumed to exist independent of human cognitive activities. *Fiat* object parts, on the other hand, owe their existence to the recognition and the establishment of *fiat* boundaries through partitioning activities. These are based on decisions or conceptually guided demarcations that demand a symbolic, reflective or linguistic capacity on the part of a human being [7]. The resulting ontology of *fiat* entities is thus concept-dependent and accessible only to linguistic human beings. Therefore, following this notion of boundaries, one has to conclude that whereas your body and an apple in a fruit basket exist independently of any human cognitive acts, boundaries delimiting an apple on a tree, the thumb of your right hand, your right upper arm, mesodermal tissue and the active center of an enzyme would not. Instead, their existence would depend on the cognitive activity of a human agent.

2.2 Borderline Cases

2.2.1 Vagueness and Indeterminate Boundaries—Are Boundaries always Crisp?.
When thinking of physical boundaries (as contrasted with e.g. political, legal or linguistic boundaries) we usually think of clear-cut lines separating one entity from its environment. In case of three-dimensional entities the boundaries are surfaces and thus entities of two dimensionality—they are not three-dimensional bodies themselves. And in case of surfaces the physical boundaries are lines and not regions. One could conclude that, regardless of whether boundaries are *fiat* or *bona fide*, they are necessarily *crisp* in this respect that they are always of a dimensionality one less than their hosts' dimensionality.

When dealing with real entities and their actual boundaries, however, a supposedly different picture emerges. In some cases, like for instance the boundaries of many geographical objects such as deserts, dunes, or the Caribbean Sea, it seems that a given entity cannot be delineated by crisp boundaries—we have troubles to identify a single surface that demarcates a desert, dune, or the Caribbean Sea. Instead, they are delineated by *border zones*, i.e. by boundary-like *regions*, which are indeterminate to some degree [6–10,17]. Consequently, we would have to distinguish between *crisp* (i.e. sharp) and *indeterminate* (i.e. fuzzy, vague) boundaries [18,19], with crisp boundaries always possessing a dimensionality one lower than their host. Thus, one could ask whether an indeterminate boundary, on the other hand, always shares the dimensionality with its host.

An indeterminate boundary could be interpreted as the region in which an ontologically crisp boundary must be located, but we simply cannot narrow down its actual location. In this case, the indeterminacy would represent a conceptual issue that is owed to linguistic or epistemological problems instead of ontological ones (cf. [14]). Thus, we must distinguish an *ontological* from an *epistemic-conceptual* interpretation of the indeterminacy of some physical

boundaries. The epistemic-conceptual interpretation of the indeterminacy problem, which is favored by various authors (e.g. [7,8,10,17,20,21]), shifts the indeterminacy problem to a supposed vagueness of the respective concepts, instead of looking for an ontological reason (e.g. quantum mechanics). They argue that indeterminate boundaries can be defined in principle, but they cannot be determined precisely. Interestingly, the examples that are commonly used in this context usually refer to *fiat* entities. Therefore, one could also argue that the vagueness of the respective boundaries is owed to their *fiat* nature and thus to the *fiatness* of the entities involved. It is thus the vagueness of the corresponding concepts themselves, resulting from their mind-dependent conceptual nature, that is responsible for the indeterminacy (e.g. [8,17]). Examples are for instance entities that are heterogeneously composed, with two clearly distinguishable poles that are not sharply delimited from each other by a crisp boundary. Instead, they are bound by a region in which they merge seamlessly, as it is often the case with different and overlapping chemical gradients in biological objects (e.g. the polarity of a blastula during embryogenesis, with the animal pole and the vegetale pole).

The indeterminacy of time dependent boundaries, like for instance coastlines or river banks, is another example for epistemic-conceptually indeterminate boundaries. Coastlines are shifting borders, and since they are crisp and *bona fide* at any given moment in time, their indeterminacy can be attributed to the problem of the time dependence of their actual location [9]. Their indeterminacy does not pose fundamental problems to ontology design, since the vagueness is just an epiphenomenon of the dynamics of reality.

Thus, as long as indeterminate boundaries can be restricted to *fiat* entities or attributed to the dynamics of reality, they do not pose fundamental problems to ontology design, as they nicely match with the basic categorial distinction of *bona fide* and *fiat* boundaries and the accompanied basic categorial distinction of *bona fide* objects and *fiat* entities respectively.

But are *bona fide* boundaries really always crisp at a given moment in time? The distinction of what counts as physical discontinuity or qualitative heterogeneity on the one hand and physical continuity and qualitative homogeneity on the other hand, is not crisp. There are cases of ontological indeterminacy—*bona fide* boundaries that are less *bona fide* than others, so to speak.

It cannot be denied that discontinuities come in various degrees of abruptness. The cutting edge of a sharp knife comes more abruptly than the edge of the white cliffs of Dover. From a purely perceptual point of view, the edges of the letters that appear on your screen while typing an email are crisper than the color changes in a rainbow. Obviously, **discontinuity is a matter of scale and therefore of granularity**. With increasing resolution spatial boundaries of physical entities become increasingly fuzzy. This results in what has been called the *Problem of the Many* [22,23], which Lewis characterizes as follows:

"Think of a cloud — just one cloud, and around it a clear blue sky. Seen from the ground, the cloud may seem to have a sharp boundary. Not so. The cloud is a swarm of water droplets. At the outskirts of the cloud, the density of the droplets falls off. Eventually they are so few and far between that we may hesitate to say that the outlying droplets are still part of the cloud at all; perhaps we might better say only that they are near the cloud. But the transition is gradual. Many surfaces are equally good candidates to be the boundary of the cloud. Therefore many aggregates of droplets (…) are equally good candidates to be the cloud. Since they have equal claim, how can we say that the cloud is one of

these aggregates rather than another? But if all of them count as clouds, then we have many clouds rather than one. And if none of them count, each one being ruled out because of the competition from the others, then we have no cloud. How is it, then, that we have just one cloud? And yet we do." [23]

Whereas one could argue that the *Problem of the Many* represents a purely linguistic and epistemological problem of referencing and the use of language rather than an ontological problem, it nonetheless results from the vagueness or fuzzyness of boundaries. The relevance of this problem becomes apparent when considering that at the subatomic level, any physical entity resolves into a swarm of subatomic particles, of which no clearly determinable boundary can be specified—the location and shape of the outer surface of a material entity involves some degree of arbitrariness at these fine levels of granularity [9,14,17]. Thus, at least at very fine levels of granularity, the idea of abrupt physical discontinuities seems to be questionable [8].

At coarser levels of granularity, however, the dichotomy between *fiat* and *bona fide* boundaries and their respective physical entities can be maintained and seems to be a reasonable distinction [8,9,17]. From which granularity level onwards this dichotomy applies depends on the entity and cannot be determined universally. In other words, when increasing the resolution some entities turn fuzzy in coarser levels of granularity than others.

Whether the indeterminacy at finer levels of granularity and the *granularity-dependence* of the *bona fideness* of boundaries as such can be solely ascribed to conceptual and epistemological issues, or whether it is evidence for an underlying ontological vagueness, represents an open question.

2.2.2 Continua in Biological Objects and the Distinction of *Fiat* and *Bona Fide* Boundaries—Is this Distinction really Absolute?.

According to the above discussed spatio-structural notion of boundaries, the question whether a given boundary is a *bona fide* boundary or a *fiat* boundary only depends on the question of what counts as physical discontinuity or qualitative heterogeneity on the one hand and what as physical continuity and qualitative homogeneity on the other hand. It is the question of where to draw the line between these two conditions. This decision immediately concerns the distinction of objects and *fiat* object parts.

According to its definition in BFO, an object is a material entity that is maximally self-connected and self-contained [11]. This definition draws on a *principle of connectedness* that allows only 'true' or 'false' as possible values, while at the same time functioning as a *principle of unity* for the entity to be delineated (e.g. [24]). This is insofar problematic, as *connectedness* comes in degrees, which at its turn results from an underlying continuity problem [12,17] that frequently impedes consistently distinguishing *bona fide* from *fiat* boundaries.

When considering *physical* connectedness, one has to deal with various types of connectedness, ranging from gravitational forces to all kinds of electro-chemical bonds (covalent bonds, ionic bonds, metallic bonds, hydrogen bonds, etc.) and physical connections and junctions (screws, bolts, staples, nails, etc.). Part of the problem is also the fact that what counts as maximally self-connected is granularity-dependent and deals with the problem of how aggregations of *bona fide* objects of finer levels of granularity constitute a single *bona fide* object of a coarser level of granularity—how does an aggregate of individual cells, with each cell having its own *bona fide* boundary, constitute a single multicellular organism at a coarser level of granularity that also possesses its own boundary (cf. [25,26])?

Continuity: Although continuity problems affect all kinds of material entities, they are especially serious in biology. Biological objects are the product of evolution and exhibit a high degree of variability that constitutes a complex network of relations of similarities and differences between all objects of the enlivened nature. Accordingly, biologists have to deal with a continuum of forms and functions that spans a complex morphological property space, in which usually no two objects occupy exactly the same place [12]. This alone poses considerable conceptual problems [27], and many of the concepts used for referring to specific types of entities within this continuum are delineated by *fiat* (cf. *fiat concepts*; [8]). But this problem of delineating different types of entities is not exclusively of conceptual nature: if, due to the high degree of variability, no two cells are identical, any aggregation of cells exhibits qualitative heterogeneity between any two neighboring cells. When distinguishing biological objects above the cellular level one is thus confronted with the question where to draw the line between relevant and irrelevant heterogeneity, since heterogeneity is always present but not always relevant. As a consequence, at supracellular levels of granularity the distinction of *bona fide* and *fiat* concepts is not crisp anymore and involves some fuzziness that cannot be explained by referring to conceptual problems alone. Instead, this fuzziness is the result of an underlying ontological continuum.

Connectedness: Another problem in biology is the fact that, except for whole organisms, all anatomical objects of the cellular and supracellular levels of granularity are connected to neighboring anatomical objects via conduits, tunnels, vessels, ducts, nerve cords, intercellular spaces, pores, channels, and junctions (cf. [12,28]). These connections are products of evolution, are functionally necessary, and are characteristic to complex systems of interacting subsystems, such as multicellular organisms and their parts. As a consequence, anatomical objects possess regions within their otherwise *bona fide* outer boundary that are *fiat* [12]. If the dichotomy of *fiat* and *bona fide* boundaries is absolute, as it has been claimed [9,10], the respective entities would have to be treated as *fiat* entities. Consequently, no *bona fide* biological objects would exist at levels of granularity coarser than the molecular level, since even organelles exhibit such connections.

Some authors argue, however, that in many cases the fraction of *fiat* boundaries is very small compared to the total outer boundary of anatomical objects and, therefore, can be ignored in these cases (e.g. [28]). The question, then, is what are the criteria on which to decide whether a given portion of *fiat* boundaries can be ignored? How much *fiatness* can be tolerated? And how could this fit with the claimed categorial and thus absolute distinction of *fiat* and *bona fide* boundary?

Granularity-dependence: Apparently, regarding continuity and connectedness we are, again, dealing with a granularity-dependence of the *bona fideness* of boundaries. Unlike the *general granularity-dependence* of physical *bona fide* boundaries discussed above, however, in which *all* physical boundaries lose their overall *bona fideness* at very fine levels of granularity and become fuzzy, the *specific granularity-dependence* of biological objects is gradual. For example the liver of a cow: The liver is surrounded by a compact layer of extracellular matrix and the peritoneum, which at its turn surrounds most organs of the trunk. The liver is thus located in a large basal swell of the peritoneum, which almost completely surrounds it. As a consequence, one can easily demarcate the liver from the rest of the cow at first glance by using traditional preparation techniques and inspection by eye. After closer inspection, however, one will realize that the liver is connected to the rest of the body by various blood vessels, bile ducts and nerve cords. Moreover, when using a light microscope, one will realize after preparation that the number of vessels, bile canaliculi and nerve cords is much higher than expected. However, on the light microscopic level of resolution one would still think that the combination of cell membranes of the outer most liver cells provides the liver a continuous *bona fide* boundary that is only interrupted by the lumina and axons of the afore noticed vessels, capillaries and nerve cords. After increasing resolution by using electron microscopy, however, one will see that even the cell membranes themselves do not provide *bona fide* boundaries for their cells, because each membrane is interrupted by gap junctions, which connect the cytoplasm of the liver cells with each other and with the surrounding tissue. As a consequence, the total outer boundary of each liver cell is interrupted by a small fraction of *fiat* boundaries and the fraction of *fiat* boundaries of the total outer boundary of the cow liver is larger than expected by light microscopy. Therefore, at some level of granularity, the *bona fide* total boundary of a supramolecular biological object will turn into a *bona fide* boundary interrupted by very few portions of *fiat* boundaries. When further increasing the resolution, the proportion of *fiat* to *bona fide* boundary often increases as well. This is independent of and qualitatively different from the *general granularity-dependence* of *bona fide* boundaries discussed further above, which only comes into effect at very fine levels of granularity and results in a switch from a *bona fide* total boundary to a *fiat* total boundary.

The *specific granularity-dependence* poses fundamental problems to ontology design in the biomedical domain. Since the connections (i.e. conduits, tunnels, vessels, ducts, nerve cords, intercellular spaces, pores, channels, and junctions) play a fundamental causal role in those biological systems to which they belong, they are of genuine interest to the biomedical domain and cannot be ignored in ontology design. Whereas for most domains the *general granularity-dependence* of the *bona fideness* of boundaries might be unproblematic, since it is usually restricted to levels of granularity that are outside their scope and focus, this is not the case for the *specific granularity-dependence* of biological objects, which is not restricted to the finer levels of granularity. As a consequence, at least the *specific granularity-dependence* of the *bona fideness* of physical boundaries and therefore also the granularity-dependence of the distinction of objects and *fiat* object parts must be accounted for in ontology design in the biomedical domain. For an approach how this can be achieved in principle via using different representations for the same type of entity for different levels of granularity see Vogt *et al.* [26].

2.2.3 *Fiat* Boundaries and *Bona Fide* Landmarks—Spatio-Structural *Fiatness* comes in Degrees.

The discussion above demonstrates that any given *bona fide* boundary can be *bona fide* at some coarser level of granularity and *fiat* at finer ones. It seems as if *fiatness*, or *bona fideness* respectively, comes in degrees. This impression is reinforced when considering that the possibility to reliably specify and re-locate any given *fiat* boundary requires some *bona fide* landmarks and coordinates (i.e. *bona fide* parts of the same or lower dimensionality that are used to locate a *fiat* boundary, as for instance the juncture of two blood vessels as a *bona fide* landmark for locating a *fiat* boundary between the two parts of the vessel, the part before and the part after the juncture) or other pragmatic or even scientifically justified criteria [7,8,29]. In other words, *fiat* entities are to varying degrees supervenient on *bona fide* objects on finer levels of granularity [7], or some other unambiguously identifiable landmarks. From this follows that, because their specification and re-location involves real properties of the underlying factual materials, *fiat* entities usually owe their existence not exclusively to human *fiat* [10]. These real properties

also constrain the range of locations of *fiat* boundaries that are relevant in the scientific discourse.

In its pure and strictest meaning, *fiatness* implies *mind-dependence*, and thus a *fiat* boundary is a boundary that is determined by human *fiat*, lacking any natural indication. *Fiat* boundaries in this sense, however, would be inapplicable in any practical context. Instead, *fiat* boundaries of interest usually rest on:

a) threshold values as *fiat* landmarks within a continuous heterogeneous field, the values being based on some legal or otherwise specified convention and agreement, as for instance isobars, the International Date Line, or meters over mean sea level;

b) *bona fide* mathematical and topological landmarks, for example the center of mass of a material entity, the upper and lower hemisphere of a rotating sphere, or the saddle point of a curve (cf. [9,10]);

c) *bona fide* landmarks within a homogeneous field that are based on natural units, as for instance the classification of chemical elements based on their characteristic number of units of mass;

d) spatio-structural *bona fide* landmarks, as for instance the demarcation of an apple hanging from a tree, which is supervenient on the branching point of the stalk from the branch it is connected to, or the *fiat* boundaries of your right upper arm, which are supervenient on your armpit and elbow that relate to the position and function of your humerus and its associated muscles, all of which are *bona fide* objects themselves;

e) the identification of causal subsystems, i.e. spatio-structural parts that actively participate in causal processes, which are characteristic to the subsystem and that play a causal role within the system as a whole, as for instance the apple on a tree as a unit of reproduction, your thumb or your right upper arm as units of locomotion, the mesodermal germ layer as a unit of embryogenesis, or the active center of an enzyme as a physiological (biochemical) unit.

These examples of different types of *fiat* boundaries demonstrate the broad range of degrees of *fiatness* that can be involved when dealing with biological entities, ranging from (a), full-blown *fiatness* that is exclusively based on convention, to (e), for which we argue that it is actually a *bona fideness*, since it is exclusively based on real properties of a causally dynamic reality. To the latter case belong all those entities that are delimited on grounds of their causal properties and dispositions—functional units, as for instance your right upper arm. Whereas one can argue that the boundaries of such functional units are to some degree fuzzy and indeterminate, this indeterminacy is granularity-dependent like all other *bona fide* boundaries, and one could argue that they can be assigned to conceptual rather than ontological reasons.

A biological object can be looked upon from very different perspectives (e.g. spatio-structural, developmental, physiological, evolutionary), with each perspective putting a different focus on the real properties of the object. As a consequence, when partitioning an object, the resulting partition will differ from perspective to perspective [30]. Whether a given partition will yield *fiat* parts or *bona fide* objects will not only depend on the real properties of the entity and the level of granularity of focus, but also on the perspective applied and thus, to a certain degree, also on the interests of the person conducting the partition.

Unfortunately, many aspects of the ontological theory of *fiatness* have been developed in the context of geographical use cases and applications (e.g. [6–10,17]). Anatomical structures and biological objects in general have been touched upon only briefly and await closer examination in terms of consequences for the criteria of distinguishing *fiat* from *bona fide* boundaries and their status as a categorial ontological distinction (for exceptions see [12,28,31]). Since biological entities actively participate in many different types of causal processes as causal agents in a dynamic reality, we will take a closer look at the different types of *dispositions* and *historical relations* that can be used for partitioning and demarcating biological material entities. Biological processes range from evolution to individual development and embryogenesis, from physiological processes within an organism to all kinds of ecological and social interaction. We think that the discussion about what criteria must be used for distinguishing *fiat* from *bona fide* boundaries, and *fiat* from *bona fide* entities, respectively, must not only consider intrinsic spatio-structural properties but also the causal roles and historical relations of material entities and the ontological nature of the respective systems in which they constitute subsystems.

2.3 Evidence for the Mind-Independence of some Non-Structural Boundaries: I. Bearers of Function
(Dispositions—Future-Oriented 'Universal' Causality)

In the following we will provide examples of biological entities that, when following BFO's definition, would have to be treated as *fiat* object parts although their boundaries do not rest on any acts of human decision and thus exist independent of any mental or linguistic activities. However, they differ from the usually used examples for *fiat* entities in that the properties used for delineating the entities are not exclusively restricted to intrinsic spatio-structural properties. Instead, they rely on spatio-structurally delimitable *bona fide* landmarks in combination with dispositions (i.e. potential for causal interaction) or historical relations for delineating the corresponding entities. Although the specification of the actual location of the respective boundaries often involves fuzziness, they nevertheless delineate entities that exist independently of all human cognitive acts. In this sense, these entities are therefore truly *bona fide*.

2.3.1 Dispositions Independent of Morphogenesis. Three different basic types of functional units that refer to dispositions that do not involve morphogenesis can be distinguished for biological material entities.

1) **Locomotory dispositions & functional units of locomotion:** Mobile organisms usually possess various parts that are bearers of the disposition to move or to be moved relative to the position of the organism as a whole. Your right upper arm, for instance, is spatio-structurally bounded by a portion of *bona fide* boundary, i.e. the surface of your skin, but also by a portion of *fiat* boundary, i.e. the demarcation from your right forearm and your trunk. According to Smith's [6–10] notion of *fiatness*, your right upper arm would therefore be a *fiat* body part. Moreover, following this notion of *fiatness*, its demarcation would exclusively rest on grounds of mental and linguistic activities—your right upper arm would represent an artificially delimited part of your body. The recognition of your right upper arm as a functional unit of locomotion, however, is not exclusively the product of mental and linguistic activities, and its delimitation from the rest of your body is not arbitrary and does not merely rest on acts of human decision. When leaving aside BFO's strictly spatio-structural framework and, instead, employing a framework of functional systems that focuses on the locomotory musculo-skeletal system, your upper arm is a *bona fide* entity—a genuine natural (i.e. mind-independent) unit of locomotion

that is delimited by its locomotory dispositions. Your right upper arm is a functional unit or element of locomotion that can move or be moved independent from the rest of your body. Granted, purely spatio-structurally, your upper arm is delimited from the rest of your body involving *fiat* boundaries. However, these *fiat* boundaries rest on *bona fide* landmarks, as for instance the proximal and distal limits of your right humerus, which is a *bona fide* object, and are thus also spatio-structurally not completely *fiat* in a strict sense. In fact, they are very close to *bona fideness*. This *bona fideness* is additionally affirmed by the locomotory function of the respective entity.

In the same way one can argue that your right forearm, your hand and each of your fingers are *bona fide* functional units of locomotion, although they are *fiat* entities from a strictly spatio-structural point of view. The important point here is that their delimitation is not the product of mental or linguistic activities, but reflects reality, only from a locomotory-functional perspective instead of a purely spatio-structural perspective. This can result in spatio-structurally fuzzy delimited *bona fide* entities, as the example of your right upper arm indicates.

2) **Physiological dispositions & functional units of physiology:** Organisms usually possess various parts that are bearers of the disposition to actively participate in physiological processes *within* the organism, which are more or less vital for sustaining the integrity of the organism as a whole, keeping it operating and alive. These parts physiologically *interact* with other parts of the same organism or with ingested biotic and abiotic substances. Spatio-structurally, many of these parts are *fiat* entities, as for instance the human heart, which is delimited from its connecting veins and arteries by *fiat* boundaries. Physiologically, however, due to the functional role the heart has as a pumping organ maintaining the blood flow, it is a *bona fide* functional unit. The active center of an enzyme is another example of a *bona fide* functional unit of physiology that is spatio-structurally a *fiat* entity. Its spatio-structural demarcation involves *fiat* boundaries, which, however, rest upon *bona fide* landmarks (e.g. transmembrane domain, extracellular domain, transition between alpha helix and pleated sheet, coils, folds, indentations, grooves). Functional units of physiology exist independent of human mental or linguistic activities in the same way as functional units of locomotion.

3) **Physiological dispositions & functional units of ecology:** Organisms possess various parts that are bearers of the disposition to actively participate in causal processes involving *parts of other organisms* or material entities from their respective abiotic environment. In other words, every organism possesses structures with which it *interacts* with its biotic and abiotic *environment* and which are vital for sustaining the integrity of the organism as a whole. Because the respective functions of these structures involve a larger surrounding system, including the biotic and abiotic environment, they are ecologically relevant. Spatio-structurally, many of these interacting parts are *fiat* entities, as for instance the eye pits of gastropod species of *Patella*, or the pinhole eye of *Haliotis* species, the abalones, which represent specific concave regions of the epidermis, in which the epithelial cells are differentiated to photoreceptor cells but remain continuously connected to the rest of the epidermis. Although these eyes are spatio-structurally *fiat* entities, due to the functional role they take in for the organism as a whole in interacting with its environment, they are ecologically *bona fide* functional

units. Like the other types of functional units discussed above, functional units of ecology exist independent of human mental or linguistic activities.

2.3.2 Dispositions Involving Morphogenesis. Two different basic types of functional units that refer to dispositions that involve morphogenesis can be distinguished for biological material entities.

1) **Morphogenetic dispositions & functional units of development:** Biological objects originate, transform, mutate, merge, and differentiate. They are in a constant flux. The processes of genetically and environmentally induced changes of the spatio-structural composition and qualities of an organism are generally referred to as its development. All parts of an organism bear the disposition to develop. Each part has its own genetically determined developmental sequence of changes, called its morphogenesis. Often, the development of one part is causally dependent on the development of another part or both developments are controlled by the same genetic control mechanism. As a consequence, their development is causally linked and coordinated with one another—they cannot develop independently from each other. In other words, they form a functional unit of development.

For instance, the distinction of the three germ layers mesoderm, endoderm and ectoderm of undifferentiated tissue in early embryogenesis is the distinction of three functional units of development that can be characterized by the different morphogenetic dispositions they bear. Usually, the three germ layers cannot be differentiated on purely spatio-structural grounds. Despite their spatio-structural *flatness*, the layers are nevertheless *bona fide* functional units of development, because they exist independent of human mental or linguistic activities. Other examples are the 4 d cell in spiralian metazoans, which gives rise to the entire entomesoderm of these animals, the spermatogonia that gives rise to sperm cells, and the apical meristem in higher plants that gives rise to stem and leaves and their derivates.

2) **Morphogenetic dispositions & functional units of reproduction and propagation:** Organisms have the disposition to reproduce, either sexually or asexually. During the course of evolution, various structures (and behavioral strategies) have evolved that bear the function to facilitate the reproduction of the organism and the propagation of its offspring. These structures form functional units of reproduction and propagation. Typical examples for functional units of reproduction are sexual organs, as for instance flowers, and anatomical structures that attract the attention of potential partners for mating, as for instance the colorful plumage of males in some bird species, which are usually spatio-structurally demarcated from the rest of the organism by *fiat* boundaries. This holds for many functional units of propagation as well, as for instance fruits, the parachutes of *Taraxacum* species, and the hooks of the seed of *Arctium* species. Despite their spatio-structural *flatness*, these units nevertheless exist independent of human mental or linguistic activities and thus represent *bona fide* functional units of reproduction and propagation.

2.3.3 Structural Anatomy versus Functional Anatomy. The examples above demonstrate that entities can be demarcated not only exclusively on grounds of their intrinsic spatio-structural properties, but also on grounds of a combination of their spatio-structural properties and their dispositions. Whereas the former is important when delineating entities of *structural anatomy*, the latter is important for delineating entities of *functional anatomy* (cf. [31]). The boundaries of entities of structural anatomy are determined on spatio-structural grounds, the boundaries of entities of functional anatomy are determined in such a way that the entity delineated is a function bearer—a unit that, as a whole, bears the disposition to perform a certain function [31]. Entities of functional anatomy, however, often involve spatio-structurally *fiat* boundaries that are only *bona fide* from a functional point of view. Some functional units, as for instance metanephridial systems, even lack physical connectedness and are spatio-structurally discontinuous, because they are composed of a spatially separated group of material entities, i.e. the filtration site formed by podocytes at the blood vessels and the nephridial duct draining the coelomic space that store the filtrate. Instead of physical connectedness, they exhibit functional connectedness.

The identification of entities of functional anatomy often rests on the use of spatio-structural *bona fide* landmarks that are subparts of the entity, and the localization of the entity's boundary usually involves some degree of fuzziness. This fuzziness and indeterminacy, however, cannot be used as a categorial argument against their possible *bona fide* ontological nature, because the respective indeterminacy is in the same way granularity-dependent as the indeterminacy of spatio-structural *bona fide* boundaries at finer levels of granularity (see above).

From the fact that boundaries are not only granularity-dependent but also perspective-dependent, we conclude that for any biological organism one can always distinguish spatio-structural partitions from spatio-functional partitions. Moreover, as the examples from above indicate, several different spatio-functional partitions can be differentiated (e.g. locomotory, physiological, ecological, developmental, reproductive). What is seen as *fiat* and what as *bona fide* depends on the perspective: a given entity can be *fiat* from a spatio-structural point of view, but *bona fide* from a functional point of view. As a consequence, one has to distinguish a taxonomy of structural anatomy from a taxonomy of functional anatomy (see also [31]), or its more specific taxonomies of various types of functional anatomies.

Restricting the criteria for delimiting *fiat* from *bona fide* boundaries to intrinsic spatio-structural properties and ignoring all dispositional properties of an entity reflects the position of the proponents of *form* of the traditional opposition of *form* versus *function* (see the famous controversy between Etienne Geoffroy St. Hilaire and Georges Cuvier [32–34]). Hitherto, most ontology authors focused on form rather than function when discussing the ontological nature of boundaries and basic categories of material entity—they propagate, so to speak, Etienne Geoffroy St. Hilaire's position. We want to bring Georges Cuvier's position into the discussion.

2.4 Evidence for the Mind-Independence of some Non-Structural Boundaries: II. Bearers of Common Historical Traces *(Historical Relations—Past-Oriented 'Particular' Causality)*

2.4.1 Structural Integrity and Stability of an Entity over Time. Three different basic types of historical units can be distinguished that refer to a common causal history that is responsible for maintaining the structural integrity and stability of the respective biological material entities over time. (*Please note that we ignore all questions regarding temporal boundaries of the time of existence of material entities. Their discussion goes beyond the scope of this paper.*)

1) **Developmental relations & historical units of development:** Whereas a functional unit of development is a material entity that is delimited by bearing a specific developmental disposition, i.e. a promise to behave in a certain way in future developmental processes, a historical unit of development, in contrast, is a material entity that is delimited by the fact that all its parts *share the same developmental history*. Thus, the developmental processes have already occurred and certain particular structures shared the same developmental history. A historical unit of development is composed of those particular parts of a particular individual organism that have *proven in the past* that they have developed in concert, as a unit of development, thereby maintaining the spatio-structural integrity of the entity as a whole. In other words, the entity, with all its parts, is delimited on grounds of the fact that it *has acted as a functional unit of development in the past*. Its identity is thus based on a common morphogenetic history and not on morphogenetic dispositions.

One can, indeed, argue that the defining properties of a historical unit of development, i.e. their shared developmental origin, supervene on the morphogenetic dispositions and thus on the intrinsic functional properties of its corresponding functional unit of development, since the former is an *effect* caused by the latter. Epistemologically, however, when it comes to identifying morphogenetic dispositions and with them functional units of development, one must identify the respective historical units of development that serve as empirical evidence for the existence of morphogenetic dispositions in the first place. Anyhow, just like functional units of development, historical units of development exist independently of human mental or linguistic activities, although in many cases they possess spatio-structurally *fiat* boundaries.

2) **Heredity relations & historical units of heredity:** Whereas historical units of development relate to the integrity of particular parts of a particular organism during its individual development, population biologists also talk about other units that show integrity over time, thereby crossing the spatio-temporal boundaries of individual organisms (e.g. species as individuals, see [35]). Many bio-species or bio-populations, for instance, maintain a high degree of structural integrity or homogeneity of their member organisms over time. And so do genes or morphological traits that exhibit considerable stability within a population over a certain period of time. These structures are historical units of heredity that result from reproductive mechanisms that guarantee a high degree of similarity between the original morphological structure and its copies. And, again, also historical units of heredity exist independently of human mental or linguistic activities, although they in many cases possess spatio-structurally *fiat* boundaries.

3) **Heredity relations & historical units of evolution:** If we take a look at historical units of heredity at a coarser time-resolution, additional mechanisms must be in effect in order to still maintain structural integrity or stability. It requires stabilizing selection pressures for a given inheritable structure to maintain its spatio-structural integrity through an evolutionary period of time. The respective structures form historical units of evolution. For instance the "members" of a phylogenetic character state (i.e. a set of particular inheritable traits that are structurally identical throughout representatives of various species due to homology) are good examples for historical units of evolution (see also *modularity*

[36–39]). Historical units of evolution also exist independently of human mental or linguistic activities, although they possess in many cases spatio-structural *fiat* boundaries.

2.4.2 Lineages—Constituent Historical Relations of Entities distributed in Time and Space. Whereas the examples discussed above concern material entities that maintain their structural integrity and stability during some period of the lifespan of an organism or even across the spatio-temporal boundaries of single individuals, biological material entities can also historically relate to one another independently of any shared structural stability. When common historical origin is the only defining criterion, the respective sets of structures usually form spatio-temporally scattered groups of material entities whose delineation as a single material entity requires a principle of connectedness (i.e. principle of unity; see also [24]) other than the principles of physical connectedness, like electro-chemical bonds or physical junctions. The example of the functional connectedness of the scattered parts of a metanephridial system mentioned above already demonstrated that it is in principle possible to apply different perspectives for delineating scattered entities, i.e. groups of spatially separated material entities (for *groups* see [25]). Common historical origin may serve as another principle of connectedness that allows the delineation of groups of spatio-temporally scattered material entities. (*Again, please note that we ignore all questions regarding temporal boundaries.*)

1) **Relations of common developmental origin & developmental lineages:** When a cell divides into two daughter cells during a cell fission event, the two daughter cells share the same developmental origin. Although they may migrate to locations separated in space and may proliferate further into separate tissues (e.g. 4 d cell derivates), they nevertheless constitute, together with their parent cell from which they originated and which does not exist anymore, a developmental lineage. Developmental lineages, although spatio-structurally and spatio-temporally delimited by *fiat* boundaries, nevertheless exist independently of human mental or linguistic activities and are in this sense *bona fide* in nature.

2) **Relations of common kinship & genealogical lineages:** Whereas developmental lineages are restricted to the spatio-temporal boundaries of a single organism, the respective structures can be inherited to offspring and thereby constitute genealogical lineages that cross this boundary. In biology we are talking about gene lineages, lineages of morphological traits or even the kinship relations (i.e. parent-child relations) of families of individual organisms. They represent groups of entities that relate to one another across the spatio-temporal boundaries of a particular individual organism, based on common kinship. The phenomenon of infertile castes within some social insects, like for instance the worker caste of ants, demonstrates that these kinship relations can also have a biological impact (i.e. explaining the existence of infertile workers with otherwise zero fitness) and are thus not merely a construct of human mental or linguistic activities. Therefore, also genealogical lineages, although spatio-structurally and spatio-temporally delimited by *fiat* boundaries, nevertheless can be *bona fide* in nature.

3) **Relations of common ancestry/descent & evolutionary lineages:** If we take genealogical lineages to a coarser time scale and allow them to also cross the spatio-temporal boundaries of bio-populations and species, we are looking at evolutionary lineages. When we are talking about homologues, gene families, apomorphic characters and monophy-

letic groups of species, we are talking about evolutionary lineages. Evolutionary lineages are spatio-structurally and spatio-temporally delimited by *fiat* boundaries, but nevertheless exist independently of human mental or linguistic activities and are in this sense *bona fide* in nature.

2.4.3 Structural Anatomy versus Historical/Evolutionary Anatomy. The examples given above demonstrate that some material entities, be they scattered or not, can be demarcated on grounds of a combination of their historical relations and their intrinsic spatio-structural properties. Their delineation is relevant for *historical/evolutionary anatomy*. The boundaries of historical/ evolutionary entities of anatomy are determined in a way that the entity delineated is a whole, either spatio-temporally scattered or connected: every historical/evolutionary entity of anatomy is composed of parts which share the same historical/evolutionary origin, with no entity not belonging to it sharing the same historical origin.

Just like with entities of functional anatomy, historical/ evolutionary entities of anatomy often involve spatio-structurally *fiat* boundaries that are only *bona fide* from a historical/ evolutionary point of view. Their specification usually involves the use of spatio-structural *bona fide* landmarks of subparts of the entity, and their demarcation often involves some degree of fuzziness as well. While the determination of the boundaries of historical entities is often conducted by observation with the aid of specific instruments (e.g. video tracing, 4D microscopy), the actual determination of the boundary of evolutionary entities is usually the result of an extensive comparison of the spatio-structural properties of various structures and necessarily remains hypothetical.

Discussion

3.1 Boundaries Depend on Granular Perspective

Smith and Varzi [9,10] claimed that the distinction of *fiat* and *bona fide* is categorial, since it (i) exhaustively covers all possible cases of physical boundaries and (ii) unambiguously draws the line between mind-dependent and mind-independent reality. Moreover, the distinction is ontologically important, because it provides a clear criterion for distinguishing two foundational categories of material entity, i.e. *fiat* and *bona fide* entities, simply on the basis of the type of boundary that the entity possesses. With the examples given above we demonstrate that the distinction is not strictly categorial and that this prevailing view regarding the distinction between *fiat* and *bona fide* boundaries and entities is based on a rather static spatio-structural framework that completely ignores the dynamic nature of reality.

This is insofar unfortunate, as many scientific fields are not interested in purely spatio-structurally defined types of material entities. Instead, they focus on types of entities that *interact with* or *react to* other entities or specific basic conditions, or they are interested in entities that share a specific *history*. These types of entities cannot be characterized solely on spatio-structural grounds. Hence, their boundaries cannot be determined on purely spatio-structural grounds either. As the examples from above also demonstrate, some entities cannot be demarcated within a strictly spatio-structural framework. However, they may nevertheless be demarcated using other frameworks. Many of the entities delineated this way are *not* demarcated merely as a product of mental or linguistic activities, but as entities that reflect a mind-independent reality. Although their boundaries are often spatio-structurally indeterminate and *fiat*, these boundaries nevertheless delineate entities that exist independently of any human cognitive

act. Therefore, these entities are in the best sense *bona fide* entities. Their *fiatness* and indeterminacy thereby only concerns the spatio-structural aspects of their reality, but not their defining properties, be they functional or historical.

3.1.1 Granular Partitions and Basic Categories of Boundaries. So far, we have shown that the distinction between *fiat* and *bona fide* boundaries is not as straightforward as it is usually assumed. Instead, we are dealing with a variety of different foundational categories of boundaries that are independent of the distinction between *fiat* and *bona fide* itself: Before one can decide whether a given boundary is *fiat* or *bona fide*, one *first* has to specify from which perspective this distinction is being made. In other words, **the ontological status of a given boundary is perspective-dependent**.

In a series of papers, Smith and coauthors (e.g. [15,16,40,41] have introduced a formal theory of granular partitions and discussed its consequences and implications for various problems of reference and truth, including the abovementioned *Problem of the Many*. Smith and Brogaard argue that whenever we use an expression to refer to some real entity, this brings about "*a partition of reality into two domains: the foreground domain, within which the object of reference is located, and the background domain, which comprehends all entities left in the dark*" [16] (see also [15,41]). As a consequence, every partition has its granularity built in and is an artifact of perception, judgment or classification [41]. Judgments, at their turn, "*come along with partitions of reality of various sorts, whose type, granularity and scope depend upon the contexts in which our judgments are made*" [16].

Therefore, according to this theory of granular partitions, every partition is judgment-dependent and context-dependent (instead of the term 'context' we use the term 'perspective', which serves a similar function—we prefer 'perspective', because it refers to Keet's formal theory of granularity [42] and we already used it in [26,30]). A context, at its turn, is of a certain granularity, and it is understood to be "*a portion of reality associated with a given conversation or perceptual report and embracing also the beliefs and interests and background knowledge of the participants, their mental set, patterns of language use, ambient standards of precision, and so forth*" [16]. A context is thus "*a matter of what is paid attention to by participant speakers and hearers on given occasions*" [16]. Since granular partitions are context-dependent, a granular partition is "*a device for focusing upon what is salient and also for masking what is not salient*" [16]. As a consequence, a granular partition that puts a particular entity into its foreground will not necessarily recognize all the entity's parts and subparts: when a math teacher counts the number of students in her class, she will use a partition that does not recognize the students on the level of their cellular and subcellular composition. Accordingly, when she has to grade the students' performance in her class, she will use a partition that only judges their performance in math and not in English.

The theory of granular partitions can also manage vague boundaries: "*vagueness de dicto is captured at the partition level via multiple ways of projecting crisply*" [15] ('projection' here refers to the relation from judgment to reality, i.e. the relation from a partition cell to its corresponding portion of reality [40]). A granular partition is considered to be *crisp* if it projects onto reality in a single and unique manner, whereas it is considered to be *vague* if it involves a multitude of projections onto reality, thereby interpreting vagueness as a semantic property of names and predicates [15]. Thus, the theory of granular partitions introduced by Smith and coauthors provides an adequate formal framework for dealing with the problem of vague and indeterminate boundaries.

Moreover, in combination with Keet's formal theory of granularity [42], the theory of granular partitions also provides the formal framework for dealing with problems of the granularity-dependence and perspective-dependence of boundaries. Thereby one should note that every partition by itself is entirely fiat in nature [16]. However, as Smith and Brogaard argue, some partitions *"are coordinated with bona fide demarcations on the side of objects in reality and some of them merely with fiat demarcations which we ourselves have introduced into reality in our various dealings with nature"* [16]. Smith and Brogaard thus distinguish partitions that track *bona fide* boundaries in reality from partitions that track boundaries that only exist as a result of acts of human fiat. Unfortunately, Smith and coauthors do not provide any criteria for distinguishing *fiat* and *bona fide* boundaries—they merely discuss the possibilities of *fiat* and *bona fide* partitions.

The hitherto prevailing notion of boundaries restricted itself to the spatio-structural perspective that does not account for the role of time and thus processes in reality and how they affect the ontology of boundaries and the ontology of basic types of material entities. In other words, **the strictly spatio-structural perspective is ignorant of causality.** Above we have shown that there are other types of entities besides purely spatio-structurally defined ones, all of which are epistemologically relevant and empirically accessible. The main difference between them and spatio-structurally defined entities is that only spatio-structural entities are exclusively demarcated on grounds of their intrinsic structural properties, their qualities. In contrast, the other entities are defined either (i) in terms of dispositions and therefore involve types of processes that are *repeatable* in principle (i.e. based on a notion of *universal causality*), or (ii) they are defined in terms of historical relations and therefore involve a particular sequence of processes that *has taken place in the past* and the effects this sequence had on a collection of particular spatio-structural entities (i.e. based on a notion of *particular causality*). The resulting general distinction of foundational types of boundaries and of material entities thereby reflects the very basic distinction of scientific disciplines or aspects of empirical sciences, which can be classified into the following three basic categorial perspectives (i.e. context categories):

1. *Spatio-structural perspective*: the view on what *is given* now, at a particular point in time, i.e. what is *intrinsically inherent* in material entities; *descriptive* and inventory-oriented; restricted to passive observation; represents reality in a way that is analog to the painting of a still life (cf. Etienne Geoffroy St. Hilaire's structuralism position of priority of *form*).

2. *Predictive perspective*: the view on what *can happen* in the future, dealing with dispositional/functional aspects of reality and thus with *potentiality*; *predictive* and systems-oriented; involves experimentation and the active interference and manipulation of a human-independent reality by an investigator; represents reality as a dynamic system and describes material entities with a focus on their potential future interactions, i.e. models an entity's causal space of possible (inter-)action (cf. Georges Cuvier's functionalism position of priority of *function*).

3. *Retrodictive (diachronic) perspective*: the view on what *has happened* in the past, *retrodictive* and history-oriented; involves the observation and description of particular processes or the reconstruction of past processes by comparing present distribution patterns of spatio-structural properties and their bearers; represents reality as a dynamic system and describes material entities with a focus on their historical interactions and common origins (Karl Ernst von Baer's position of *embryology*; Charles Darwin's position of *evolution*).

Above, we have argued that one can meaningfully distinguish *fiat* from *bona fide* boundaries and *fiat* from *bona fide* entities within all three perspectives. In all three perspectives *fiatness* implies mind-dependent (i.e. purely epistemological/conceptual) delimitation, whereas *bona fideness* implies mind-independent (i.e. natural, ontological) delimitation. As a consequence, some entities are spatio-structurally demarcated by *fiat*, although functionally they are *bona fide*. One can argue that these spatio-structurally *fiat* boundaries are nevertheless mind-independent and, as a consequence, *bona fide*, but only from a non-spatio-structural point of view. Therefore, one must distinguish between mind-dependent and mind-independent spatio-structurally *fiat* boundaries.

Our analysis implies that if one does not restrict the distinction of *fiat* and *bona fide* boundaries to a purely spatio-structural framework, but, instead, focuses on the less restrictive and thus more general framework of mind-dependence and mind-independence, several different categories of *fiat* and *bona fide* boundaries must be distinguished. Our examples demonstrate that we should not talk about *fiat* and *bona fide* boundaries without specifying the perspective we use, because many biological objects can be partitioned into *bona fide* components in various different ways. Consequently, the distinction of *fiat* and *bona fide* material entities necessarily depends on the perspective as well (for a discussion of *granularity perspectives* cf. [42,30]).

Any perspective that itself is *mind-independent* provides a *principle of identity* (cf. [24]) that depends on the specific aspects of reality that are taken into account (i.e. intrinsic spatio-structural properties versus functional dispositions versus historical integrity or historically caused distribution patterns) and has its own two categories of boundaries, *fiat* and *bona fide* respectively. Therefore, one has to distinguish spatio-structural from functional and historical boundaries, and thus also spatio-structural from functional and historical types of material entities. This results in a taxonomy of spatio-structural entities alongside a taxonomy of functional entities and a taxonomy of historical entities. Moreover, in case one further differentiates these foundational perspectives into several more specific perspectives that are still ontologically independent from each other (i.e. perspectives must not supervene on each other), it results in even more taxonomies. This is the case for the different types of functional entities, i.e. locomotory, physiological, ecological, developmental, and reproductive entities, with each corresponding perspective resulting in its own specific taxonomy of material entities.

It should be noted that distinguishing types of boundaries in terms of *fiatness* versus *bona fideness* and at the same time in terms of different mind-independent perspectives results in the recognition of distinctions that are ontologically independent from each other and thus truly categorial. Muscles of multicellular animals, for instance, usually can be clearly delineated and distinguished from each other on grounds of their spatio-structurally *bona fide* boundaries. This is not necessarily the case when employing a locomotory perspective, since sometimes several muscles are innervated by the same nerve and thus cannot be moved independently from each other. As a consequence, at least from a locomotory perspective, not every individual muscle does constitute a *bona fide* functional unit of locomotion. Instead, individual muscles exist that are functionally *fiat* entities of locomotion. Thus, some spatio-functionally *fiat* entities are spatio-structurally *bona fide* objects, and vice versa, which demonstrates the ontological independence of the respective categories and their underlying perspectives.

3.2 Consequences for Ontology Design

We have shown that the distinction of *fiat* and *bona fide* boundaries and *fiat* and *bona fide* material entities is not only granularity-dependent, but also perspective-dependent. It is important to realize that the perspective-dependence does not imply a representational arbitrariness. The perspective-dependence of the distinction of *fiat* and *bona fide* boundaries and *fiat* and *bona fide* material entities does not merely result from human cognitive acts, but also depends on the entities themselves. As Smith and Brogaard point out: *"while partitions, and the cells by which they are constituted, are artefacts of our cognition, when once a given partition exists, it is, for each cell in the partition and for each object in reality, an objective matter whether or not that object is located in that cell"* [16]. There are *bona fide* partitions that correlate with reality. This is the case, because the foundation for the distinction of *fiat* and *bona fide* is the distinction of mind-dependent and mind-independent delimitations, independent of the perspective. Therefore, by differentiating several types of *fiat* and *bona fide* boundaries and allowing a material entity to be at the same time *fiat* and *bona fide*, we do not relativize the categorial distinction of *fiat* and *bona fide* in the sense that it would not represent a distinction that has a real correlate in nature. Instead, we rather point to the fact that the distinction of foundational types of boundaries has to be further differentiated, in order to accommodate additional perspectives into the ontological consideration besides the spatio-structural one.

What are the consequences for ontology design? Unfortunately, an already well known problem of ontology design becomes even worse: reality cannot be modeled within a single, universal and ontologically consistent taxonomy of top-level categories of material entity that can be easily organized within a universal single-inheritance tree (see also [26]). Considering ciliated and rhabdomeric light sensory cells as an example: should the category 'ciliated light sensory cell' be organized as a subcategory of 'ciliated cell' or as a subcategory of 'light sensory cell'? We are dealing with the lineup of intrinsic spatio-structural qualities on the one hand and functional dispositions on the other hand. Which one should we give preference in a universal single-inheritance taxonomy? Another example would be 'mesodermal muscle cell', which could be organized either as a subcategory of 'muscle cell' or as a subcategory of 'mesodermal cell'. Here, a structural-functional property (i.e. muscle) is lined up against a developmental property.

The situation for ontology design is somehow comparable to the problem of how to best organize and represent the system of chemical elements. With the periodic table, Dmitri Mendeleev came up with a good solution for the latter, in which he combined two very simple and ontologically independent but each other overlapping taxonomies within a single table. If we would only have to deal with 'ciliated light sensory cell', we could use the same approach in biology and combine a spatio-structural with a functional taxonomy, resulting in a hierarchical table of biological entities. However, as the second example, 'mesodermal muscle cell', and the other examples given in this paper indicate, in biology we are dealing with more than two categorial perspectives and thus more than two foundational taxonomies of different categories of material entity and, therefore, a tabular organization like the periodic table is not applicable.

The necessity to distinguish different taxonomies of material entity results from the perspective-dependence of boundaries and the existence of a multitude of ontologically independent perspectives. Which perspective is considered to be relevant thereby usually depends on the specific interests of the researchers and thus on a specific discipline or domain. Formal top-level ontologies must be compatible for all kinds of relevant scientific interests and thus must accommodate various perspectives. Therefore, and in order to guarantee the comparability and compatibility of different application and domain reference ontologies, formal top-level ontologies must provide a consistent

and uniform template of how these different taxonomies and their respective top-level categories must be organized.

The Basic Formal Ontology (BFO; http://www.ifomis.org/bfo; [11]) is such a scientific formal top-level ontology. In BFO, each category has exactly one single asserted parent class (except for the root category). This is the result of BFO having been developed according to the *single inheritance policy*, which requires all defined categories to be disjoint and exhaustive, i.e. categories must be mutually exclusive relative to a given level of granularity [43]. An important question that results from our findings is how to organize multiple, ontologically independent taxonomies within a single universal single-inheritance taxonomy?

Because every unit, independent of whether it is a developmental, ecological, evolutionary or physiological unit, necessarily possesses spatio-structural properties, the spatio-structural taxonomy can be used as backbone taxonomy: it is possible to classify the leaf categories of every non-spatio-structural taxonomy within the spatio-structural taxonomy according to their spatio-structural properties. As a consequence, leaf categories must be defined in reference to both spatio-structural properties *and* their specific defining dispositions or historical relations. Following this approach, the category 'ciliated light sensory cell' would be defined as:

A ciliated light sensory cell is a ciliated cell that is light-sensitive.

As such, 'ciliated light sensory cell' is a direct subcategory of the spatio-structural category 'ciliated cell'. However, it is at the same time also a direct subcategory of the physiological category 'light sensory cell'. Two questions immediately emerge: (i) where to place the category 'light sensory cell' within the spatio-structural taxonomy and (ii) how to specify the parent-child relation between 'light sensory cell' and 'ciliated light sensory cell' without violating the single inheritance principle?

The first question translates into the question of what spatio-structural properties do all instances of 'light sensory cell' have in common, which will specify the location of 'light sensory cell' within the spatio-structural taxonomy? This is not trivial, since new types of light sensory cells might be discovered in future, which would change the position of 'light sensory cell' within the spatio-structural taxonomy. We only know for certain that every light sensory cell necessarily is a cell and that it is necessarily light-sensitive. Thus, we could define 'light sensory cell' as:

A light sensory cell is a cell that is light-sensitive.

However, when comparing the direct functional child categories of 'light sensory cell' we might discover that we can provide a more specific spatio-structural definition. After all, light sensory cells must possess some light-sensitive proteins (i.e. opsin; e.g. [44]), which can help to identify a putative light sensory cell on purely spatio-structural grounds (e.g. by *in situ* hybridization of expressed genes; [45]). One could thus distinguish between a conservative part of the definition of 'light sensory cell' that is independent of its current composition of child categories (i.e. *a light sensory cell is a cell that is light-sensitive*) and a dynamic part that directly results from and thus depends on the comparison of spatio-structural features of all its child categories (i.e. *a light sensory cell is a cell that possesses some light-sensitive protein and that is light-sensitive*). Whenever a new child category is added, the dynamic part may change accordingly.

Regarding the second question of how to specify the parent-child relation between 'light sensory cell' and 'ciliated light sensory cell', we must introduce additional ontology relations for consistently organizing the various taxonomies and their interrelationships within a single universal single-inheritance taxonomy. For instance the transitive (i.e., for all A_i holds, if A_1 is-a A_2 and A_2 is-a A_3, then A_1 is-a A_3) 'is-a' relation for class-subclass relationships should be differentiated into for instance a 'is-structurally-a', 'is-physiologically-a', 'is-locomotory-a' and 'is-ecologically-a' relation, in order to differentiate the class-subclass relations of the different taxonomies and to allow to still organize them within a single-inheritance tree. In the same way one should distinguish different types of parthood relations as well (e.g. 'ecologically-fiat-part-of').

If we want to identify functional and historical *bona fide* entities that are spatio-structurally *fiat*, we also require properties such as 'ecological-unit' or 'developmental-unit' with a Boolean value space (i.e. yes/no). In case we want to distinguish spatio-structurally scattered units from units that are spatio-structurally connected (i.e. no part of the unit is spatially separated from the rest of the unit by a gap), a property such as 'ecological-unit' must be differentiated into the properties 'ecological-object-unit' for spatially connected units and 'ecological-group-unit' for spatially scattered units (for *groups* see [25]). Moreover, if we want to indicate that a spatio-structurally *bona fide* entity is a *fiat* unit, as it is for instance the case with some muscles that are not innovated individually but only as a muscle bundle, a property as for instance 'locomotory-fiat-unit' is required.

Except for lineages, all the different taxonomies discussed above could be organized within a single universal single-inheritance taxonomy for a scientific formal top-level ontology such as BFO. In case of lineages, however, we require ontology relations that indicate the respective historical relation between instances and spatio-structural categories of material entity, as for instance a symmetric (i.e., for all A_i holds, if A_1 is-homologues-with A_2, then A_2 is-homologues-with A_1) relation 'is-homologues-with'.

3.3 Conclusions

With various examples from biology we have demonstrated that the hitherto prevailingly used criteria for distinguishing *bona fide* from *fiat* physical boundaries are not unambiguously applicable. The ambiguity results from the combined use of two ontologically independent types of demarcation criteria: (i) intrinsic qualitative criteria (i.e. spatial/physical discontinuity, qualitative heterogeneity) and (ii) the criterion of mind-independence. Our examples demonstrate that in many cases physical boundaries are *bona fide* with respect to one type of criterion while *fiat* with respect to the other. Moreover, they demonstrate that the distinction of *fiat* and *bona fide* material entities is perspective-dependent. As a consequence, the distinction of *bona fide* and *fiat* boundaries itself is also perspective-dependent and thus cannot be categorial and absolute. Our examples also demonstrate that if the two types of demarcation criteria for boundaries both must be met, the distinction of *bona fide* and *fiat* boundaries is not exhaustive, because boundaries exist that only meet one type of criterion.

One possible solution would be dropping the criterion of mind-independence and confining the distinction of *bona fide* and *fiat* boundaries merely to spatio-structural criteria. This, however, would do no justice to the existence of the multitude of ontologically independent perspectives and the perspective-dependence of the distinction of *fiat* and *bona fide* material entities.

We therefore concluded that *fiat* and *bona fide* boundaries must be distinguished based on the criterion of mind-independence. The criterion of mind-independence, at its turn, seems to depend on the question whether the entity delimited by the boundary in question is a *bona fide* material entity, i.e. a spatio-structurally bound object or a functionally or historically bound causal unit. In

case it is, the respective boundary exists independent of any human cognitive acts and is *bona fide* in this sense. In other words, the ontological status of a boundary depends on the type of material entity it bounds and <u>not</u> vice versa. As a consequence, the distinction of *fiat* and *bona fide* material entities cannot rest on the type of boundary they possess, but must depend on criteria of causal unity—either spatio-structural or predictive or retrodictive (see the main perspectives discussed above).

If the defining criteria for distinguishing *bona fide* and *fiat* entities do not refer to boundary types, one can simplify the distinction of boundaries into those that bound *fiat* entities and those that bound *bona fide* entities, be they spatio-structural objects or functional or historical units. It seems as if the development of BFO 2.0 is currently taking a step towards this direction. In a draft that is currently (Aug. 2012) available from the BFO website (*Basic Formal Ontology 2.0 - Draft Specification and User's Guide*; http://ncorwiki.buffalo.edu/index.php/Basic_Formal_Ontology_2.0), objects are characterized as *natural units* that are *causally relatively isolated* entities, which means that they are structured through some *causal unity*, to which they are maximal (i.e. a maximal causally unified material entity). According to this draft, causal unity thereby includes unity through physical covering, internal physical forces or engineered assembly of components. According to the BFO 2.0 draft, however, this notion of objects still awaits a formal theory. It remains to be seen where the development of BFO 2.0 will finally lead to.

Regarding basic types of boundaries, the draft does not distinguish *bona fide* and *fiat* boundaries anymore, but treats all boundaries of material entities as *continuant fiat boundaries*. Although this notion follows, at least from an epistemic point of view, directly from the theory of granular partitions, which claims that every partition is *fiat* and thus every boundary is *fiat* too, it does not do justice to the underlying ontological nature of the entities bound: we can still distinguish between boundaries which bound entities that correlate with natural units and those which bound *fiat* entities—after all, the BFO 2.0 draft itself still distinguishes between object and *fiat* object part, and if objects are natural units than their boundaries must be natural as well. Therefore, if one wants to distinguish boundaries of different categories of material entities, one could distinguish for instance continuant *fiat* boundary$_{Object}$ from continuant *fiat* boundary$_{fiatObjectPart}$ and continuant *fiat* boundary$_{locomotoryUnit}$ or continuant *fiat* boundary$_{ecologicalObjectUnit}$.

We have shown that the problems resulting from the perspective-dependence of boundaries do not only affect very special and highly differentiated categories of material entity, which may be restricted to leaf classes of domain reference ontologies and thus to aspects of reality that are not relevant to formal top-level ontologies. Quite the contrary, they do apply already at the root of the top-level categories of 'material entity' and thus must be considered by any formal top-level ontology for the scientific domain, as for instance the BFO.

Unfortunately, the consequences make ontology design more complicated. However, they result from an attempt to meet the requirements of the perspectivalist position that a plurality of alternative perspectives on reality do exist that are ontologically equally legitimate. Apparently, reality is very complex and the problems that we face when organizing and categorizing reality seem to reflect the epistemic constraints and limitations that are inherent to our human cognitive devices when attempting to represent this reality within a consistent mental model. Obviously, we have fundamental problems to comprehend reality within a single universal perspective and therefore have to resort to the epistemic means of multiple perspectives.

Acknowledgments

We thank B. Smith for taking a look at an earlier draft of the MS early in 2012. We also thank two anonymous reviewers for giving valuable comments and suggestions. It goes without saying, however, that we are solely responsible for all the arguments and statements in this paper.

References

1. Rosse C, Kumar A, Mejino LV, Cook DL, Detwiler LT, et al. (2005) A Strategy for Improving and Integrating Biomedical Ontologies. In: AMIA 2005 Symposium Proceedings. pp. 639–643.
2. Brinkley JF, Suciu D, Detwiler LT, Gennari JH, Rosse C (2006) A framework for using reference ontologies as a foundation for the semantic web. In: AMIA 2006 Annual Symposium Proceedings. pp. 96–100.
3. Smith B, Kusnierczyk W, Schober D, Ceusters W (2006) Towards a Reference Terminology for Ontology Research and Development in the Biomedical Domain. In: Bodenreider O, editor. Proceedings of KR-MED 2006, Studies in Health Technology and Informatics, Vol. 124. IOS Press. pp. 57–66.
4. Smith B, Munn K, Papakin I (2004) Bodily systems and the spatial-functional structure of the human body. In: Pisanelli DM, editor. Medical Ontologies. Amsterdam: IOS Press. pp. 39–63.
5. Schulz S, Boeker M, Stenzhorn H, Niggemann J (2009) Granularity issues in the alignment of upper ontologies. Methods of Information in Medicine. 48(2): 184–189.
6. Smith B (1994) Fiat objects. In: Guarino N, Pribbenow S, Viue L, editors. Parts and Wholes: Conceptual Part-Whole Relations and Formal Mereology, 11th European Conference on Artificial Intelligence. Amsterdam: European Coordinating Committee for Artificial Intelligence. pp. 15–23.
7. Smith B (1995) On Drawing Lines on a Map. In: Frank AU, Kuhn W, Mark DM, editors. Spatial Information Theory: Proceedings in COSIT '95. Berlin/Heidelberg/Vienna/New York/London/Tokyo: Springer. pp. 475–484.
8. Smith B (2001) Fiat Objects. Topoi. 20(2): 131–148.
9. Smith B, Varzi AC (1997) Fiat and Bona Fide Boundaries: Towards an Ontology of Spatially Extended Objects. In: Spatial Information Theory: A Theoretical Basis for GIS. Berlin, Heidelberg: Springer. pp. 103–119.
10. Smith B, Varzi AC (2000) Fiat and Bona Fide Boundaries. Philosophy and Phenomenological Research. 60(2): 401–420.
11. Smith B, Kumar A, Bittner T (2005) Basic Formal Ontology for Bioinformatics. Journal of Information Systems. 1–16.
12. Schulz S, Johansson I (2007) Continua in Biological Systems. The Monist. 90(4): 23.
13. Brentano F (1988) Philosophical Investigations on Space, Time and the Continuum. London/Sydney: Croom Helm.
14. Varzi AC (2008) Boundary. The Stanford Encyclopedia of Philosophy (Fall 2008 Edition). Available: http://plato.stanford.edu/entries/boundary/. Accessed 2012 Oct 4.
15. Bittner T, Smith B (2001) Granular partitions and vagueness. In: Welty C, Smith B, editors. Formal Ontology in Information Systems. New York: ACM press. pp. 309–320.
16. Smith B, Brogaard B (2000) A unified theory of truth and reference. Logique et Analyse. 169–170: 49–93.
17. Smith B, Varzi AC (1997) The formal ontology of boundaries. The Electronic Journal of Analytic Philosophy. 5(5). Available: http://ejap.louisiana.edu/EJAP/1997.spring/smithvarzi976.html. Accessed 2012 Oct 4.
18. Tye M (1990) Vague Objects. Mind. 99: 535–557.
19. Cohn AG, Gotts NM (1996) A theory of spatial relations with indeterminate boundaries. In: Eschenbach C, Habel C, Smith B, editors. Topological Foundations of Cognitive Science. Hamburg: Graduiertenkolleg Kognitionswissenschaft. pp. 131–150.
20. Sorensen RA (1988) Blindspots. Oxford: Clarendon Press.
21. Williamson T (1994) Vagueness. London: Routledge.
22. Unger P (1980) The Problem of the Many. Midwest Studies in Philosophy. 5: 411–467.
23. Lewis D (1983) New Work for a Theory of Universals. Australasian Journal of Philosophy. 61: 343–377.
24. Smith B (1999) Agglomerations. In: Freska C, Mark DM, editors. Spatial Information Theory. Cognitive and Computational Foundations of Geographic Information Science, COSIT '99, LNCS 1661. Berlin, Heidelberg: Springer. pp. 267–282.
25. Vogt L, Grobe P, Quast B, Bartolomaeus T (2011) Top-Level Categories of Constitutively Organized Material Entities - Suggestions for a Formal Top-Level Ontology. PLoS ONE. 6(4): e18794.
26. Vogt L, Grobe P, Quast B, Bartolomaeus T (2012) Accommodating Ontologies to Biological Reality—Top-Level Categories of Cumulative-Constitutively Organized Material Entities. PLoS ONE. 7(1): e30004.
27. Vogt L, Bartolomaeus T, Giribet G (2010) The linguistic problem of morphology: structure versus homology and the standardization of morphological data. Cladistics. 26(3): 301–325.
28. Smith B, Mejino Jr JLV, Schulz S, Kumar A, Rosse C (2005) Anatomical information science. In: Cohn AC, Mark D, editors. Conference on Spatial Information Theory (COSIT 2005). Berlin: Springer. pp. 149–164.
29. Mejino JLV, Rosse C (2004) Symbolic modeling of structural relationships in the Foundational Model of Anatomy. In: Hahn U, Schulz S, Cornet R, editors. KR-MED Proceedings. Bethesda: AMIA. pp. 48–62.
30. Vogt L (2010) Spatio-structural granularity of biological material entities. BMC Bioinformatics. 11(289).
31. Johansson I, Smith B, Munn K, Tsikolia N, Elsner K, et al. (2005) Functional anatomy: A taxonomic proposal. Acta Biotheoretica. 53(3): 153–166.
32. Russell ES (1916) Form and Function: A Contribution to the History of Animal Morphology. London: John Murray.
33. Appel TA (1987) The Cuvier-Geoffrey Debate: French Biology in the Decades before Darwin. Monographs. Oxford: Oxford University Press.
34. Amundson RON (1998) Typology Reconsidered: Two Doctrines on the History of Evolutionary Biology. Biology and Philosophy. 13(2): 153–177.
35. Ghiselin MT (1974) A radical solution to the species problem. Systematic Zoology. 23: 536–544.
36. Wagner GP (1996) Homologues, natural kinds and the evolution of modularity. American Zoology. 36: 36–43.
37. Wagner GP, Altenberg L (1996) Complex Adaptations and the Evolution of Evolvability. Evolution. 50(3): 967–976.
38. Laubichler MD (2000) Homology in development and the development of the homology concept. American Zoologist. 40(4): 777–788.
39. Mitchell SD (2006) Modularity - More than a buzzword? Biological Theory. 1(1): 98–101.
40. Bittner T, Smith B (2003) A theory of granular partitions. In: Duckham M, Goodchild MF, Worboys, MF, editors. Foundations of geographic information science. London: Taylor & Francis Books. pp. 117–149.
41. Smith B (2004) Carving Up Reality. In: Gorman M, Sanford J, editors. Categories: Historical and Systematic Essays. Washington DC: Catholic University of America Press. pp. 225–237.
42. Keet CM (2008) A Formal Theory of Granularity: Toward enhancing biological and applied life sciences information system with granularity. PhD thesis, KRDB Dissertation Series DS-2008-01. Available: http://www.meteck.org/files/AFormalTheoryOfGranularity_CMK08.pdf. Accessed 2012 Oct 4.
43. Spear AD (2006) Ontology for the Twenty First Century: An Introduction with Recommendations. Science. 1–132.
44. Shichida Y, Matsuyama T (2009) Evolution of opsins and phototransduction. Philosophical Transactions of the Royal Society of London. Series B, Biological sciences. 364(1531): 2881–2895.
45. Arendt D, Hausen H, Purschke G (2009) The "division of labour" model of eye evolution. Philosophical transactions of the Royal Society of London. Series B, Biological sciences. 364(1531): 2809–2817.

Author Contributions

Conceived and designed the experiments: LV. Performed the experiments: LV. Analyzed the data: LV. Contributed reagents/materials/analysis tools: LV BQ PG TB. Wrote the paper: LV BQ PG TB.

8

A Spectrum of Pleiotropic Consequences in Development Due to Changes in a Regulatory Pathway

Ana E. Escalante[1,2,3]*, Sumiko Inouye[4], Michael Travisano[1,3]

1 Department of Ecology, Evolution and Behavior, University of Minnesota, St. Paul, Minnesota, United States of America, **2** Departamento de Ecología de la Biodiversidad, Instituto de Ecología, Universidad Nacional Autónoma de México, Mexico City, México, **3** Biotechnology Institute, University of Minnesota, St. Paul, Minnesota, United States of America, **4** Department of Biochemistry, Robert Wood Johnson Medical School, Piscataway, New Jersey, United States of America

Abstract

Regulatory evolution has frequently been proposed as the primary mechanism driving morphological evolution. This is because regulatory changes may be less likely to cause deleterious pleiotropic effects than changes in protein structure, and consequently have a higher likelihood to be beneficial. We examined the potential for mutations in *trans* acting regulatory elements to drive phenotypic change, and the predictability of such change. We approach these questions by the study of the phenotypic scope and size of controlled alteration in the developmental network of the bacterium *Myxococcus xanthus*. We perturbed the expression of a key regulatory gene (*fruA*) by constructing independent in-frame deletions of four *trans* acting regulatory loci that modify its expression. While mutants retained developmental capability, the deletions caused changes in the expression of *fruA* and a dramatic shortening of time required for completion of development. We found phenotypic changes in the majority of traits measured, indicating pleiotropic effects of changes in regulation. The magnitude of the change for different traits was variable but the extent of differences between the mutants and parental type were consistent with changes in *fruA* expression. We conclude that changes in the expression of essential regulatory regions of developmental networks may simultaneously lead to modest as well as dramatic morphological changes upon which selection may subsequently act.

Editor: Suzannah Rutherford, Fred Hutchinson Cancer Research Center, United States of America

Funding: Ana E. Escalante was supported by University of Minnesota and CONACyT postdoctoral Fellowship 126166, Sumiko Inouye by the Foundation of Medicine and Dentistry of New Jersey, and Michael Travisano by the US National Science Foundation (DEB-0918897, SES-0959134, MCB-1042335, DEB-1146463). The funders had no role in study design, data collection and analysis, decision to publish, or preparation of the manuscript.

Competing Interests: The authors have declared that no competing interests exist.

* E-mail: anaelena.escalante@gmail.com

Introduction

Heritable phenotypic change is a prerequisite for adaptive evolution. However, the process by which diversity in form originates, and the mechanisms that link phenotypic variation with genetic modification, are poorly understood. This is in part because the focus of traditional evolutionary theory has been on changes in gene frequencies and has not taken into account the complexity of biological systems that result in what we define as phenotypes (for a conceptual review see [1,2]). New approaches have emerged to fill this gap, combining knowledge from molecular biology and evolutionary theory in the search for mechanistic information about the origin and evolution of phenotypic traits [3–5]. In this respect, the study of the evolution of development has become paradigmatic [2], and progress is being made in understanding the evolution of phenotype by comparative genomic studies and experimental manipulation of biological laboratory models [5–7].

The foremost example of the integration of molecular biology and evolutionary theory in the study of phenotypic adaptation is the evolution of developmental networks [8]. A commonly stated hypothesis is that the evolution of *cis*-regulatory elements in developmental networks, is less likely to cause negative pleiotropic consequences in the phenotype than *trans*-acting regulatory factors [7]. *Cis*-regulatory elements are closely linked to the loci that they

affect, while *trans* regulators are either unlinked or distantly linked to the loci under their regulatory control. Because mutations in *cis*-regulatory elements impact closely linked loci, their effects are localized spatially or temporally, in contrast to mutations in *trans*-acting factors or structural genes that may affect global gene function [3]. The distinction among different mechanisms of regulatory control and structural genes has lead to the 'toolkit gene' concept [7], in which the localized expression of 'toolkit genes' can readily evolve via regulatory mutations. The prevalence of *cis*-acting regulatory elements within genomes provides a potential mechanism for decreased deleterious pleiotropy during evolution occurring via changes in development [7], which has been mainly evidenced by comparative genomic and expression studies [7,9]. Within this framework, it is expected that inframe deletions of *trans*-regulatory elements within a developmental network will have pleiotropic effects resulting in substantial phenotypic change. However, the importance of *trans*-regulation remains contentious as the potential for localized expression appears limited. In this paper, we explore the scale of changes in developmental time and place, heterochrony and heterotopy, arising via *trans*-regulation, and determine if such phenotypic changes could facilitate adaptive evolution.

Another hypothesis for the importance of developmental networks in evolution is the potential for large beneficial effect mutations. The observation of mutations having dramatic morphological consequences, such as losing the ability for development of complex structures [10] or due to changes in developmental timing (e.g., [11]) suggests that large beneficial effects are possible. The importance of large effect mutations has a long and contentious history in evolutionary biology [12–14]. Nevertheless, mechanisms promoting abrupt phenotypic evolution remain a topic of intense interest [15–17], in particular the structuring of developmental networks into modules. A module is a highly interconnected sub-developmental network that has relatively few connections with other modules. This hierarchical network structure potentially limits the pleiotropic consequences of mutations across an entire developmental network, facilitating large effects beneficial mutations and limiting their deleterious effects. However, the debate on the adaptive potential of alterations in development persists, in part, because it is difficult to integrate genomic and phenotypic information [18,19], and this is particularly true for complex traits. The lack of clarity on the mechanisms by which development affects evolutionary outcomes, undermine its utility in a reformulation of evolutionary theory [20].

We have initiated a research program to directly investigate developmental evolution using a model system that is both relatively simple to propagate and is genetically tractable, the free-living microbe *Myxococcus xanthus*. *M. xanthus* undergoes multicellular development as a social behavior via aggregation of vegetative cells and formation of fruiting bodies [21]. It is readily culturable in laboratory settings [22] and can be genetically manipulated with relative ease, allowing us to directly observe the developmental consequences of specific genetic changes [23]. Development in *M. xanthus* occurs when resources become scarce and individual cells migrate towards aggregation centers, gliding to form multicellular groups consisting of about 100,000 cells. These groups of cells (swarms) develop into fruiting bodies (FBs), each containing approximately 10,000 spores, after 24–72 hours via a series of temporally and spatially structured cellular activities [24,25]. Phenotypic variation resulting from variation in social traits (such as aggregation to form FBs), has been reported as prevailing in natural microbial populations [26–29] suggesting the value of such variation in competition and its potential for adaptation.

In this study we investigate the phenotypic consequences of regulatory changes in development, focusing on changes in developmental timing and outcome. What are the phenotypic consequences of mutations in *trans*-acting regulatory pathways? The pathway of signal dependence through *M. xanthus* development is an area of active research, but several key steps have been identified that involve both intra- and extracellular signaling [21,25]. One essential step is appropriate expression of the *fruA* gene, a key intracellular regulator of development that is required for cell aggregation and fruiting body maturation [30]. Regulation of *fruA* expression occurs via multiple signaling pathways that have been partially elucidated (Figure 1). We constructed in-frame deletions of four regulatory loci whose gene products have been tentatively identified as affecting *fruA* expression and fruiting body development [31]. Loci were chosen based on previous data indicating that their loss did not prevent development in *Myxococcus xanthus* [31–34], the importance of serine-threonine phosphate cascades in development [35], and their association with the *fruA* regulatory network. In-frame deletions were used as they have unambiguous effects on expression of a regulatory locus, and they have no or little effect on expression of upstream or downstream loci linked to the deleted gene. Although information exists on the pleiotropic effects of regulatory mutations in *M. xanthus* model [35], there have not been studies interpreting such observations

within an evolutionary framework. We designed assays to assess phenotypic change due to regulatory mutations relative to unmutated genotype and unexpected variance. This design provides a straightforward approach to compare traits and the scale of mutational effects. Thus, the present experimental and analytical approaches on the impact of regulatory mutations provides 1) direct information on their phenotypic effects that is otherwise difficult or impossible to achieve by other approaches such as comparative genomics and 2) evidence emphasizing the utility of a simple biological model in the study of phenotypic evolution.

The in-frame deletions caused dramatic shortening of time required for fruiting body development, consistent with the anticipated effects of loss of the four regulatory loci. We found phenotypic changes in the majority of traits measured, indicating pleiotropic effects of changes in regulation. The magnitude of the change for different traits was variable but the extent of phenotypic differences among the mutants and parental type were consistent with linear changes in *fruA* expression. These results show that multiple phenotypic changes in developmental traits can readily occur due to pleiotropy, via simple genetic changes affecting development in a predictable fashion [36].

Results

Changes in *fruA* Expression

Prior studies suggested that deletion of the *pktA2*, *pktC2*, *pktD1* and *pktD9* loci would alter *fruA* expression and thereby impact development and fruiting body formation [31–34]. We observed large changes in *fruA* expression during development of the mutant knockout strains. The observed differences in expression, between the knockouts and the parental strain, are consistent with the predicted regulatory structure of *fruA* (Figure 1). In particular, the mutants have an overall higher level of expression ($F_{1,10} = 9.8$, $p = 0.0107$, ANCOVA adj. $r^2 = 0.875$), as determined by a planned contrast of mutants versus parental strain. Temporal expression of *fruA* differs between the mutants and parental strain, as assessed by the interaction of genotypic state, mutant or parental, versus time ($F_{1,4} = 9.62$, $p = 0.0362$, ANCOVA adj. $r^2 = 0.915$). After 12 hours into development, the average expression of *fruA* for the mutants, is higher than the parental stain, and it is also maintained for longer time (Figure 2).

Phenotypic Traits

To determine the size and scope of phenotypic consequences of the developmental regulatory changes, we measured the rate of developmental progression, several fruiting body characteristics including size, variance in size, number, as well as spore number and viability. Exponential growth rate was our measure of the vegetative phenotype.

Parental and mutant strain development differs greatly (Table 1, Figures 3 and 4). Development, measured as progression rate into mature fruiting bodies, is accelerated in all mutants relative to the parental genotype DZF1. The largest amount of genetically determined phenotypic variation among the five genotypes is observed at 36 hours (analysis not shown), at which point all mutants have completed or nearly completed fruiting body development. Development is complete for all the mutants after 48 hours, while it is substantially slower for the parental genotype ($p = 2.2 \times 10^{-6}$) by an average of 19.2%. Even so, not all fruiting body size traits are affected by the in-frame regulatory deletions, and there were large differences in the phenotypic consequences depending upon the trait (Table 1, Figure 5). For example, fruiting body mean size is not significantly different between the mutants

A.

B.

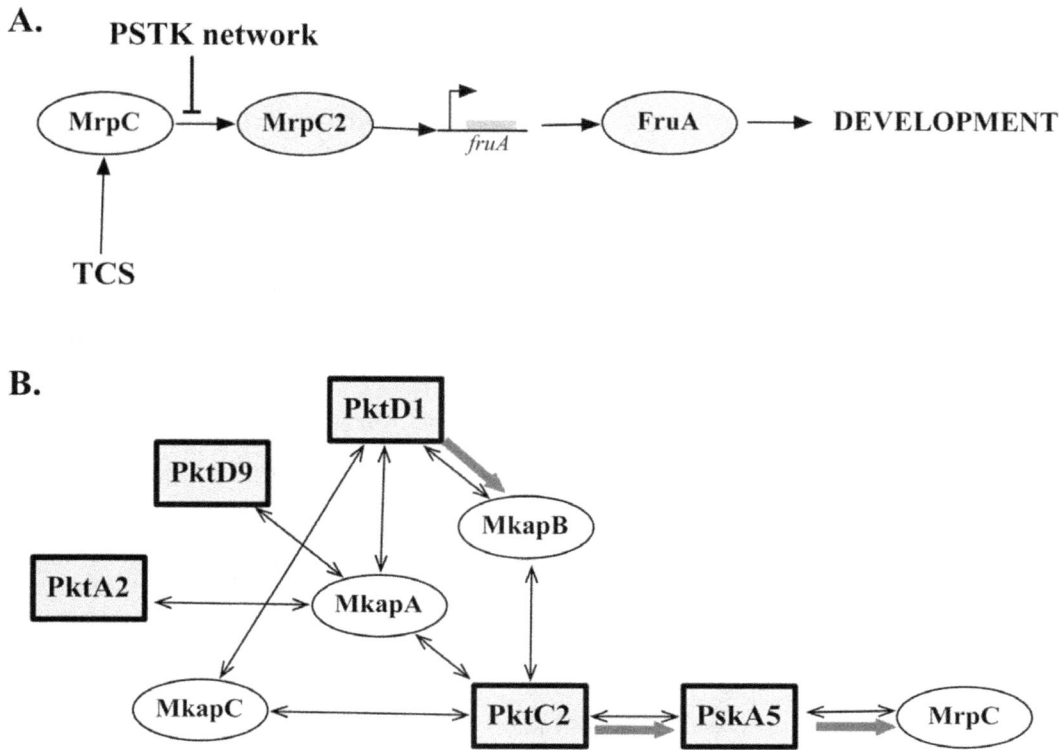

Figure 1. Regulatory network model of development in *Myxococcus xanthus*. A. Regulatory network model of *fruA* expression. Expression of *fruA* is a key step in the induction of developmental gene expression and the achievement of multicellular fruiting body formation and sporulation. During vegetative growth, *mrpC* is transcribed at low levels, using its own product (MrpC) as a transcription factor (positive feedback). When starvation signals trigger the expression of genes in a two component system (TCS) the newly synthesized MrpC is not phosphorylated, but is instead processed to MrpC2, which has a higher affinity for the *mrpC* and *fruA* promoter regions [34]. A PSTK network inhibits development by phosphorylating MrpC. B. The Protein Serine-Threonine Kinase (PSTK) network model. PSTK network is thought to consist of at least five kinases (squares) as well as three multikinase associated proteins (Mkaps). Double-headed arrows indicate interactions identified by yeast-two hybrid screens, while gray arrows are characterized phosphorylation pathways [35]. PskA5, a protein kinase activated by PktC2, phosphorylates MrpC, reducing its affinity for both *mrpC* and *fruA* promoter regions, and preventing untimely initiation of development. In this paper, PktA2, C2, D1 and D9 were deletion targets. The *pskA5* locus is closely linked to that of *mrpC/mrpC2* and therefore was not a candidate for in-frame deletion.

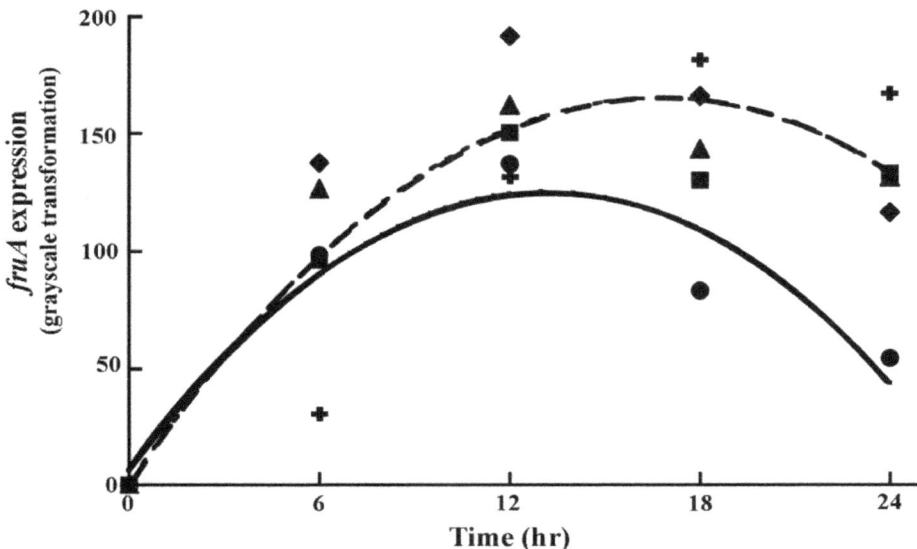

Figure 2. Expression of *fruA* over time for all strains. Code for strains are: DZF1 (circle), A2 (square), C2 (cross), D1 (triangle), and D9 (diamond). Solid and dashed lines correspond to the parental and knock-out mutant strains respectively.

and DZF1, but variance in size is far larger in the mutant strains compared with DZF1, which is notably homogeneous in the size of the mature fruiting bodies. Significant differences in the number and viability of resulting spores are observed between DZF1 and mutants, some mutants produce more spores than the parental strain and in all cases relative viability is diminished in the mutants. In this study we focus on the potential for phenotypic changes to occur, and spore production and viability are only considered as phenotypic traits that are assessed for change (rather than measures of fitness).

Far fewer differences among mutant strains were observed for phenotypic traits (Table 2). Statistically significant differences among mutants were observed for development ($F_{3,7} = 10.92$, $p = 0.005$), with mutant genotype D9 proceeding through development faster than A2. Mutants C2 nor D1 were indistinguishable from one another or A2 and D9. No other statistically significant differences were observed (analysis not shown).

Statistical Measurement of Scope and Size of Phenotypic Consequences

To simultaneously compare phenotypic effects across traits, we scaled each measure to their respective standard deviations [37]. This allowed us to determine the phenotypic variation that was attributable to differences between DZF1 and mutants and to gauge the size of the differences in the same units. This approach provides values that are in units of standard deviation (Table 3), so that comparable 95% confidence intervals can be generated (Figure 5). Confidence intervals excluding zero indicate statistically significant differences between the parental genotype (DZF1) and the mutants, not corrected for carrying out multiple simultaneous tests. Confidence intervals excluding 1 indicate the differences between DZF1 and mutants are greater than the non-genetic component of the phenotypic variation (Figure 5). The results show three things. First, the size of the change for different traits was variable and most traits measured were affected by the regulatory change in development. Second, 5 out of 6 traits have statistically significant phenotypic differences when comparing mutants with parental strain (DZF1). Finally, 3 out of 6 trait differences remain statistically significant even after carrying out sequential Bonferroni correction for multiple tests.

We evaluated the correlation of phenotypic changes and gene (*fruA*) expression changes (Figure 5, right hand column), noting a relationship between differences in phenotype between the mutant and parental strains with *fruA* expression. The differences among

trait responses is consistent with differences in *fruA* expression, as supported by a linear regression of SMD on the square root of the of the absolute correlation values for *fruA* expression and each phenotypic trait (slope = 2.26, $t_4 = 2.81$, $p = 0.048$, adj. $r^2 = 0.58$).

We also performed a principal components analysis of developmental traits, to assess the size and scope of statistically independent traits. While conclusions from the above analyses of individual traits are potentially limited, since the data for different traits were collected from same replicates and are therefore not independent, the structure of data collection allows for a simultaneous analysis of the developmental traits via a principal component analysis. Three components were statistically significant by a chi-square test ($p < 10^{-5}$, 10^{-4}, and 10^{-2}, respectively), accounting for total of 86.7% of the variation (40.0, 28.3, and 18.3, respectively). An ANOVA on the composite principal component trait values indicates that the genotypes are readily distinguished (Figure 6) for the first ($F_{4,11} = 6.78$, $p = 0.0053$) and second axes ($F_{4,11} = 4.40$, $p = 0.023$), but not the third ($F_{4,11} = 1.68$, $p = 0.224$). No statistically significant differences were detected among the mutants when considered alone, without the unaltered parental genotype. More importantly, the parental and mutant genotypes are statistically distinct, as determined by t-tests on the primary ($t_{11} = 5.195$, $p = 0.0003$) and secondary ($t_{11} = 3.295$, $p = 0.007$) axes. Their respective SMD are 1.92 and 1.23.

Growth Rate

A statistically significant difference between the parental genotype and the knockout mutants was observed ($F_{4,8} = 6.81$, $p = 0.011$), due to the decreased growth rate of one mutant (D1). No differences among genotypes for growth rate were observed when D1 was excluded from the analysis ($F_{3,6} = 1.9$, $p > 0.2$), and there was no evidence that the growth rate of parental genotype differed from the other three knockout mutants ($F_{1,8} = 0.81$, $p > 0.4$). The decreased growth rate of the D1 genotype indicates that there is 'crosstalk' by at least some regulatory elements to affect both development and vegetative growth.

Discussion

A major success of the Modern Synthesis was the abstraction of genetics. By focusing on the intersection of Mendelian genetics and Darwinian selection, a general evolutionary theory was developed. Nevertheless the limitations of the purely genetic

Table 1. Variation among genotypes for developmental traits.

Trait	Genotype[a]				Replicate[a]				Block[a]				Replicate x Block[b]				Error	
	MS	Df	F	p	MS	df	F	p	MS	df	F	p	MS	df	F	p	MS	df
Development[c]	0.0696	4	22.5	0.0004	0.0245	2	2.07	0.205	0.002	3	0.271	0.845	0.012	6	3.92	0.049	0.0031	7
FB CV	3184	4	3.61	0.318	2950	2	3.35	0.065	71.87	3	0.816	0.969	–	–	–	(0.99)	881	14
FB number[d]	0.064	4	3.57	0.033	0.026	2	1.43	0.271	0.027	3	1.51	0.255	–	–	–	(0.46)	0.18	14
FB size	20271	4	1.56	0.24	42523	2	3.28	0.068	17113	3	1.32	0.31	–	–	–	(0.15)	12961	14
Spore Count[d]	0.318	4	7.44	0.002	0.031	2	0.735	0.497	0.074	3	1.74	0.205	–	–	–	(0.22)	0.598	14
Spore Viability[d]	0.338	4	2.67	0.084	3.54	2	27.97	3×10^{-5}	0.139	3	1.10	0.388	–	–	–	(0.95)	0.126	12

[a]Random factor.
[b]Partial F-test values in parenthesis for inclusion of an interaction term in the analyses. In only one instance, for development, did a partial F-test indicate that including the interaction term statistically improved the analysis.
[c]Assessed at 36 hours.
[d]Analysis carried out on Log_{10} transformed data.

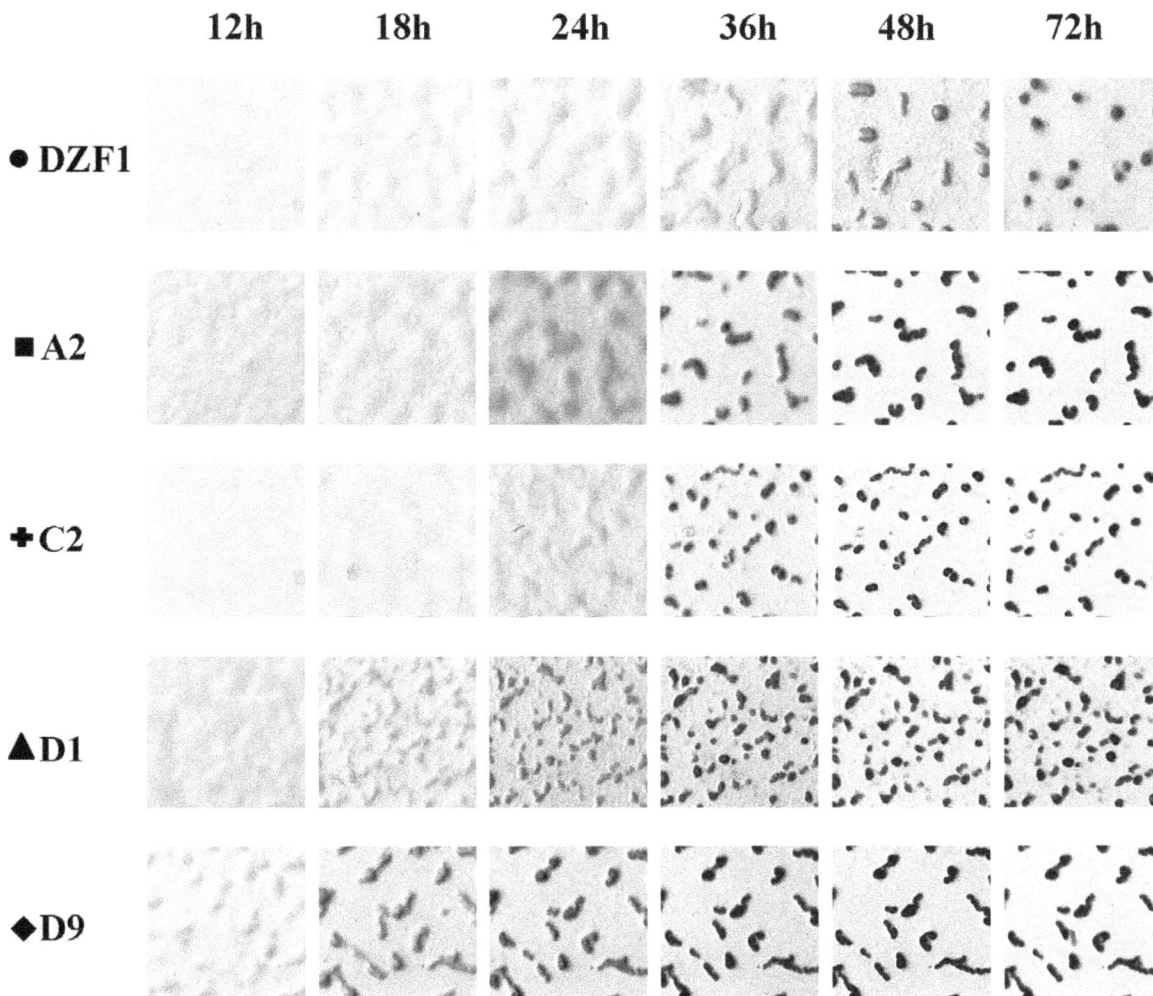

Figure 3. *Myxococcus xanthus* **parental strain and knock-out mutants fruiting body formation at 12, 18, 24, 36, 48 and 72 h development.** Micrographs were taken at 269.5 pixel/mm on TPM plates. Different mutations account for changes in developmental timing, which also has consequences in the final shape and distribution of fruiting bodies.

approaches have long been apparent [38,39], as they largely ignore the complexity of biological systems and fail to incorporate mechanistic details underlying phenotypic differences on evolu-

tion [40,41]. In the absence of mechanistic information, the evolutionary intricacies underlying complex phenotypes remain unclear.

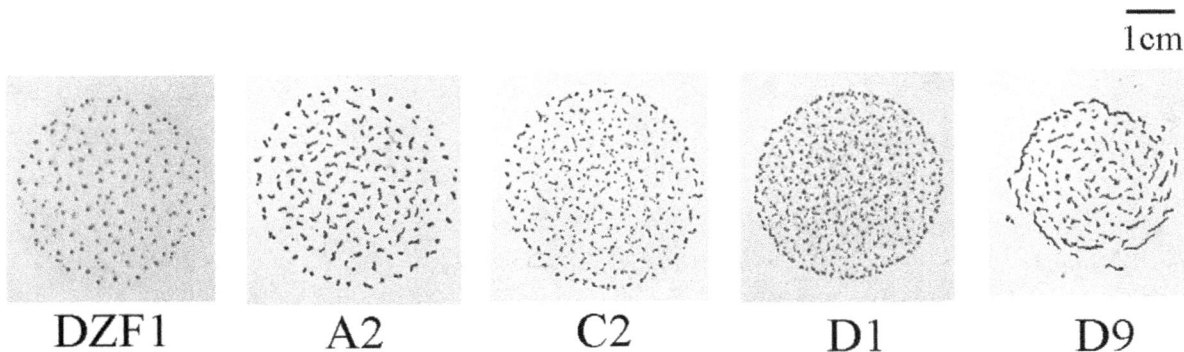

Figure 4. Phenotypic diversity after 72 h development of *Myxococcus xanthus* **parental strain and knock-out mutants.** Solid dark spots correspond to mature fruiting bodies containing myxospores after aggregation and differentiation of vegetative cells. Observed diversity results from knocking out genes associated with changes in developmental timing. Micrographs were taken at 269.5 pixel/mm on TPM plates.

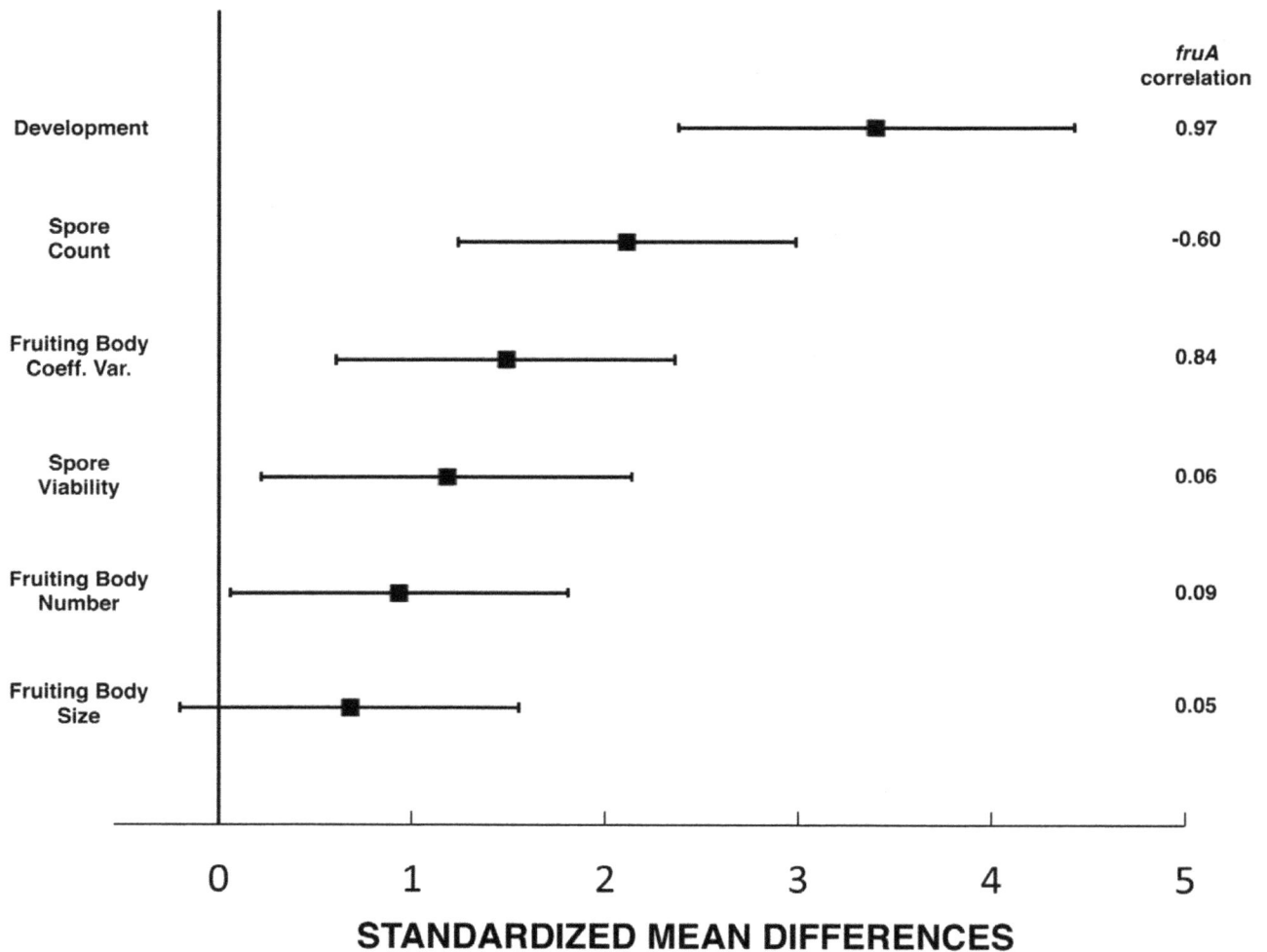

Figure 5. Standardized mean difference among mutants and parental strains. Error bars are 95% confidence intervals (CI) for the observed phenotypic variation. Statistical differences between parental strain and mutants are supported by exclusion of '0' within the confidence intervals. The right-hand column shows correlation coefficients for differences among the mutant strains with *fruA* expression.

This study is one step towards a functional synthesis [42] on the evolution of development. The approach taken combines the rigor of experimental molecular biology with the conceptual foundations provided by evolutionary biology (see also [43]). In this study, we investigated the potential for adaptation by developmental modification. We were interested in *trans*-regulators of development, as they seem to be critical for development, and there is relatively little quantitative information to determine if changes in

Table 2. Mutant trait means and 95% Confidence Intervals (CI).

Trait	Genotype			
	A2	C2	D1	D9
Development[a]	0.68±0.42[b]	0.82±0.1	0.74±0.3	0.92±0.22
FB CV	107.78±148.21	98.04±113.73	93.46±40.66	126.53±120.69
FB number[c]	2.36±0.40	2.49±0.23	2.57±0.43	2.37±0.54
FB Size	413.92±377.61	295.99±245.46	342.09±471.50	364.14±193.24
Spore Count[c]	6.22±0.30	6.02±0.36	5.94±0.23	6.09±0.51
Spore Viability[c]	5.08±0.90	4.77±1.36	4.18±1.03	4.77±2.17

[a]Assessed at 36 hours.
[b]95% Confidence intervals determined by a *t*-distribution with n −1 = 2 df.
[c]Analysis carried out on Log_{10} transformed data.

Table 3. Differences between the parental and mutant genotypes for developmental traits.

Trait	Mean Difference	t_s	df	p	SD_{pooled}	Standardized Mean Difference
Development	0.189	7.841	7	0.0001[a]	0.056	3.399
FB CV	44.01	3.632	14	0.0027[a]	29.68	1.483
FB number	0.124	2.281	14	0.039	0.134	0.931
FB size	77.15	1.66	14	0.119	113.8	0.678
Spore Count	0.436	5.17	14	0.0001[a]	0.207	2.109
Spore Viability	0.419	2.67	12	0.020	0.356	1.18

[a]Statistically significant after sequential Bonferroni correction for carrying out multiple simultaneous tests.

trans-regulation could give rise to the kind of phenotypic changes that would allow evolution to proceed. Our approach involved the investigation of scope and size of phenotypic consequences by alteration of a developmental network. This topic is relevant since it remains unclear how changes in developmental timing alter phenotype and adaptation as a consequence.

One model of developmental evolution states that morphological changes are more likely to occur through changes in the expression of "toolkit" loci via their promoter regions (*cis*-regulation). These "toolkit" loci encode functionally conserved proteins of mosaically pleiotropic influence within the vast regulatory network they control. Structural changes in them are presumed to be less tolerated because their large deleterious consequences in fitness [7]. Despite debate on the molecular nature of morphological change, it is clear that expression changes in these toolkit genes do occur and are associated with morphological modifications in animals. There is substantial evidence of their functional and sequence conservation across phylogenetic groups, like the Hox family of transcription factors. Nevertheless, the use of animal models in evolutionary development studies imposes inherent practical complications, as well as

potentially limited interpretation. Moreover, these studies have primarily focused on among species comparisons, unlike the within species differences examined in this study.

In the *Myxococcus* model, the importance of toolkit genes in the evolution of development is unclear, as many developmentally essential genes are not conserved across different Myxococcales species [44]. We perturbed the expression of an essential gene (*fruA*) by constructing independent in-frame deletions. The deleted loci were previously identified as associated with the *M. xanthus* developmental network, and were hypothesized to impede the onset of development by a phosphate cascade terminating in transcriptional regulation of *fruA* gene expression [31–34]. By generating precise in-frame deletions, we observed changes in developmental timing and altered *fruA* expression, verifying expectations. Moreover, the extent of change in other developmental traits was consistent with altered *fruA* expression, suggesting, that large changes in the timing of developmental networks can occur by proportional changes in the underlying mechanisms by which they occur. The observation that phenotypic variation is linearly associated with changes in gene expression of an essential gene shows the

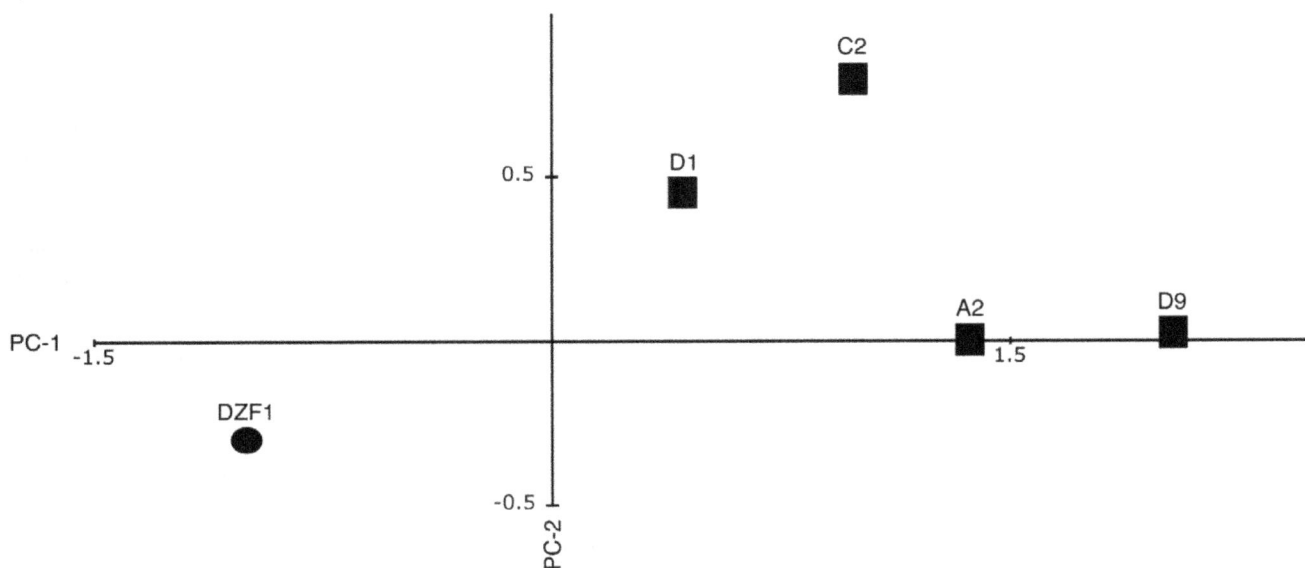

Figure 6. Principal Component Analysis of all strains. Traits included in the analysis are: $\log_{10}$ terminal fruiting body count, $\log_{10}$ terminal viability Count, $\log_{10}$ terminal spore count, 36 hour development, coefficient of variation for fruiting body size at 72 hours, and fruiting body size at 72 hours, PC-1 and PC-2 account for 40% and 28.3% of the phenotypic variance, respectively.

potential for diversification by simple mutations in *trans*-regulatory elements of a developmental network. Predictable, and potentially gradual, evolutionary change (see [36]) in developmental traits can proceed by changes in *trans* regulation without catastrophic consequences.

We devised planned comparisons between mutated and unmutated strains to specifically assess phenotypic change due to regulatory mutations, relative to the unexplained (error) variance in phenotype. This experimental design provides a straightforward approach to compare not only traits but to scale the effects of mutations.

Scope of Phenotypic Consequences

Statistically significant differences in five of six developmental traits, three after Bonferroni correction, were observed. Phenotypic consequences were evident across a range of traits, as seen by differences in spore count and fruiting body size variation within a developmental swarm. Moreover, substantial phenotypic effects were observed for the statistically independent first and second principal component axes, strongly supporting pleiotropic consequences. The observation of complex modes of developmental evolution is not restricted to our study, as discussed in a recent reevaluation of the Hox gene evolution [45]. A study by Liao and colleagues [5] suggests that genes associated with anatomical or morphological changes are in general more pleiotropic than genes involved with physiology. The results are part of a comparative genomics study predicting that morphological evolution more often should involve transcriptional regulation and gene expression changes. However, interpretation of these and results from other complex systems, such as human and mice, are necessarily difficult, in that assigning gene function and mode of activity is rarely unambiguous.

Size of Phenotypic Consequences

We measured the size of phenotypic changes by statistical analysis of morphological effects of mutations affecting developmental timing. We observed a gradient in the magnitude of consequences ranging from 0.68 to 3.4 standard deviations of within-genotype phenotypic variation. The large reduction in developmental timing illustrates the potential for dramatic alterations in developmental programs, without catastrophic consequences. The extent of phenotypic change in a trait was largely consistent with patterns of *fruA* gene expression. It is particularly interesting to note that one of the traits that differs between mutants and the parental strain is FB size variation (CV), and we hypothesize that this could be the result of heterotopy. Fruiting body formation in *Myxococcus* is affected by cell-cell communication and extracellular signals [46], that involve population density and nutrient availability [25]. Alteration of *fruA* gene expression changes the timing of development and thus may alter the consequences of extracellular signals. In our experimental system, heterochronic differences in expression of development-related genes across mutants lead to differences in fruiting body size. As stated by Carroll [7] "*changes in the spatial regulation of toolkit genes and the genes they regulate are associated with morphological divergence*". We hypothesize that heterochronic differences in *fruA* expression resulted in heterotopic differences in fruiting body size due to gradients in extracellular signals. These observations suggest that evolutionary change of phenotypic traits could be either gradual or discrete, depending on the traits, loci and mutations involved.

In addition to the planned comparison between mutants and unmutated parental genotype, variation among mutants may exist. For instance, mutant D9 appears to have faster development

(Figure 3), larger fruiting bodies (Figure 4), and the most variance in FB size (Figure 3). Despite these observations, statistical significance for trait differences among mutants was only observed for speed of development, between mutants D9 and A2. Developmental traits are highly environmental labile, and the primary focus of the study was contrasting the parental and mutant genotypes. Moving forward, it would be worthwhile to investigate the apparent differences among mutants that were not predicted, using our current results to frame expectations. These subsequent experiments would necessarily need to be carefully executed, and blocked accordingly, as our current three-fold measurements are insufficient to discriminate the mutant phenotypes with statistical rigor.

Genetic Basis for the Change in Form

A main goal in the study of morphology is to determine the underlying mechanistic bases for its evolution. Our observations suggest that there is a direct correlation of morphological changes of developmental phenotypes with changes in the expression of a key gene for development in *M. xanthus*. We also observe that there are differences in this correlation depending on the relative position of the gene affected in the network, as is the case for A1, D1, D9 compared to C2. These differences were not anticipated, despite differences in the hypothesized regulatory pathway (Figure 1, [35]). We anticipated similar roles of the four regulatory loci, with differences arising via magnitude of the effects. Future work disentangling the differences between these two groups will be helpful in further understanding the consequences of perturbing different parts of the regulatory network. Finally, we observed that the extent of phenotypic differences among the mutants and parental type are largely consistent with *linear* changes in *fruA* expression, indicating the major changes in the timing of developmental networks can occur by proportional changes in the underlying mechanism by which they occur. The finding of phenotypic variation correlated with changes in genetic expression, shows the potential for diversification by simple mutations in the regulatory network. Mutations altering social behavior and development potentially lead to evolutionary change in phenotype without catastrophic consequences. These observations may help future studies in looking for links between development evolution and natural patterns of phenotypic variation, such as those observed in populations of *M. xanthus* by Kraemer and colleagues [29].

Evolutionary Implications of Variation in Developmental Timing

Although most studies of developmental process in *M. xanthus* have been conducted with laboratory strains in controlled laboratory settings, there is documented evidence of the globally widespread presence of developmentally competent strains [47]. This widespread behavior is nonetheless rapidly lost if selection for social behavior is relaxed [22,48], indicating that social proficiency and development into FBs is highly beneficial in the wild. Moreover, work by Kraemer et al. [29], demonstrate natural variation in developmental timing of strains recovered from different sites, suggesting that variation in selective forces across different environments may contribute to the persistence of such variants. At this point is impossible to know if such forces act directly on developmental timing or in other pleiotropically linked trait(s) (such as social motility), but speculation on the selective advantage of developmental timing can be made. For example, if two different populations of strains are mixed, one being faster in developing than the other, it is possible to imagine that the faster developer might monopolize signaling molecules and exclude the slower developer from producing viable spores [29], this being a plausible mechanism for genetic differentiation and, potentially, speciation. If the trait under selection is, for example, social motility, which is linked with predatory efficiency and development, it is easy to imagine that slow-developing strains can have an advantage when resources are scarce, by using them for longer before going into development. This being said, the frequency with which *M. xanthus* goes into development in the wild is completely unknown, as is the contribution of such behavior to overall fitness or adaptation [49].

Conclusions

This study provides evidence that emphasize the utility of a simple microbial model for research on developmental evolution and the consequences of phenotypic diversity generation. This is of particular relevance because of the dependence on gene-phenotype mapping in the search for understanding the mechanisms underlying the origin and evolution of complex traits. The *Myxococcus xanthus* model for development used here provided detailed quantitative measurements of phenotypic consequences resulting from changes in the regulation of development. We observed that simple genetic perturbations of the signal cascade for development result in significant pleiotropic changes in phenotype that range in magnitude. Our results imply that changes in *trans* acting regulatory regions can potentially lead to predictable phenotypic evolution.

Materials and Methods

Strains and Mutant Construction

The strains used included the parental strain *Myxococcus xanthus* DZF1 [50] and 4 single PSTKs (Protein Serine/Threonine Kinases) in-frame deletion mutant strains. Mutants were constructed using the kanamycin resistant gene (kan) for positive screening and a galactokinase gene (galK) for negative screening. Briefly, two DNA fragments of approximately 600 bp in size were amplified by PCR using the genomic DNA as a template. Fragment 1 contained the 600-bp upstream region of the translation initiation codon with the first several amino acid codons and fragment 2 was the 600-bp downstream of the translation termination codon with several amino acid codons, also a unique six-base cutter restriction enzyme site was introduced upstream, and downstream the target sequence. This permitted construction of inframe-deletion mutants. Fragments 1 and 2 were cloned into pKO1kmr carrying the galK and kan genes. After the constructed plasmid was introduced into wild-type cells, the plasmid with the wild-type gene was eliminated by the addition of D-galactose in a medium [23]. The strains used were: Parental strain (DZF1), $\Delta pktA2$ (A2-1), $\Delta pktD1$ (D1-4), $\Delta pktC2$ (C2-2), and $\Delta pktD9$ (D9-2). Multiple vials for each isolate were frozen (20% glycerol) and stored at $-80°C$ until used. None of the four deleted loci *pktA2* (MXAN 1467), *pktD1* (MXAN 4017), *pktC2* (MXAN 1710), and *pktD9* (MXAN 6420) are adjacent to one another or to the *mrpC/mrpC2* (MXAN 5125) locus. The *pskA5* locus is nearby to the *mrpC/mrpC2* locus. MXAN number designations refer to the sequence annotations of the *M. xanthus* genome [50].

Microbiological Procedures

To revive strains from the frozen storage, stocks of each strain were thawed and 50 µl spotted into a CYE plate (1% Bacto Casitone, 10 mM Tris-HCl (pH 7.6), 0.5% yeast extract, 10 mM MOPS (pH 7.6) and 4 mM MgSO$_4$) [35]. Inoculated plates were incubated at 30°C for 3 days. After this, cells were picked with a loop and used for further experimentation.

All assays for vegetative phenotype were performed using CYE plates or broth and for developmental phenotype TPM plates or solution were always used: 10 mM Tris-HCl (pH 7.6), 1 mM K$_2$HPO$_4$, 8 mM MgSO$_4$ [43].

Quantification of Pleiotropic Effects

All the phenotypic measures for each strain were performed in triplicate and in blocks to give statistical support to the observed phenotypic measurements and to rigorously evaluate potential pleiotropic consequences in the resulting phenotypes.

Vegetative Phenotype

Growth rate. Growth on CYE broth is vegetative with no social predatory behavior or Fruiting Body (FB) development. For growth measurements, cultures were grown in 250-ml Erlenmeyer flasks with Klett tubes attached. All the inoculated flasks were incubated at 30°C with shaking (250 rpm) to keep the cultures well oxygenated. Growth was measured with a Klett-Summerson colorimeter [50], using amber filter (No. 66) with transmission 640 to 700 nm (Klett Mfg. Co., Inc).

Developmental Phenotype

We quantified development both as for timing and for final phenotypic results. For timing we analyzed darkness of FBs as a measure for cell aggregation and FB maturity [43], and *fruA* expression at different time points. We also assessed FB number, FB size, FB size variation, total spore counts and viable spore counts.

Gene expression assay. Protein samples for Western blot analysis were prepared from cells developing on TPM plates. 10 µl of the cell suspension prepared as described above were spotted at 64 spots per a square plate (8 cm × 8 cm). The developing cells were harvested from 2 plates at the indicated time points, suspended in ice cold 500 µl TM buffer (10 mM Tris-HCl: pH 7.6), 8 mM Mg$_2$SO$_4$), and precipitated. The precipitated cells were kept at $-80°C$ until used. The cells were solubilized in 100 µl sample loading buffer and heated for 5 min in boiling water with vigorous vortexing. Cell lysates were quantified using a Bradford assay (Bio-Rad Laboratories). Protein lysates (15 µg) were resolved by 12% SDS-PAGE and transferred to polyvinylidene difluoride (PVDF) membrane using semidry transfer apparatus (Bio-Rad Laboratories). Western blot analysis was performed using anti-FruA IgG, anti-CsgA (P17) IgG and anti-Tps polyclonal antibodies. Secondary goat anti-rabbit IgG-alkaline phosphatase (AP) conjugate (Bio-Rad Laboratories) was used according to the manufacture's protocol.

Development assay. Strains were propagated in CYE broth at 30°C and 250 rpm until an approximate optical density of 100 Klett units was reached (4×10^8 cells/ml). Cells were harvested at that moment by spinning them down using a microcentrifuge (6000 rpm × 10 min) and washing off remnant nutrients using TPM solution (10 mM Tris-HCl (pH 7.6), 1 mM K$_2$HPO$_4$, 8 mM MgSO$_4$) [51]. After washing, the cell pellet for each strain was resuspended in 1/10th of the original volume. To evaluate developmental behavior of each strain, 15 µl of the cell suspension was spotted onto TPM plates (1.5% agar). The plates were prepared 2 days in advance to avoid excess moisture and pre-warmed for 20 min at 30°C before each cell suspension is spotted. The spots were dried for 20 min and then incubated at 30°C for 4 days. While incubating the plates, FB formation (cell aggregation) was assayed by taking stereomicroscope photographs at different time points (0, 12, 18, 24, 36, 48, and 72 hours) using a Nikon SMZ1500 Zoom Stereo Microscope. All the images were saved in a digital format, processed, and analyzed using ImageJ software [52].

Developmental phenotype measurements

a. *fruA* expression. Digital images of the expression assay gels were processed to obtain quantitative measures. We used the Histogram Analysis tool implemented in Image J [52] and obtained the amount of 'black' in the image, as a direct correlate of the gene product (FruA). In this way, we were able to assess gene expression at different time points during development for the parental strain and mutants.

b. Fruiting body developmental timing. We used an increase in coefficient of variation between pixels as a measure of fruiting body (FB) maturation, and the time sequence of images was used to estimate the FB timing. As FBs develop, they darken and we measured the change in color over time for each mutant. Color change was measured using the Histogram Analysis function implemented in Image J [52] and this tool makes it possible to measure the black/white distribution of pixels in the image. The two extremes of the distribution are: 0 h time point when cells are first spotted onto the plate and 72 h time point when all genotypes have completed development. When the cells were first spotted they cannot be distinguished from the background and are homogenously distributed. At 72 hours, mature FBs contrast strongly with the background, and the variance in the color (black and while) across pixels will be >0. To standardize the variance values, each time point variance value was divided by the final 72 h value, so all samples had a variance of 1 at the time point 72 h. Finally this value was transformed into a coefficient of variance dividing it by the mean of the distribution. This was done for each time point to generate a FB developmental sequence. The coefficient of variation approach provides a method to assess developmental timing that does not depend upon determinations of either absolute or relative fruiting body color or size.

c. FB count and size. We processed the 72 h time point digital images by transforming them into a black/white binary image where each FB appeared as a black area in a white background. This transformation allows for estimation of final counts, sizes, and variation in size of FBs for each mutant. We used the coefficient of variation (CV) as a measure of heterogeneity in FB size in a developmental swarm.

d. Spore count and viability. We performed total spore counts and viable spore counts by flow cytometry and plating. Fruiting bodies were harvested, by taking a plug from the agar plates containing all the FBs that developed from a single inoculum of cells. The plug was forced into 13 mm diameter tube with a sterile wood applicator and washed with 2 ml of TPM solution by vortexing. The total volume of the wash was removed using a micropipette and then sonicated to disrupt the FBs and to obtain individual spores. The sonicated spores were then incubated at 60°C for 30 min in order to kill all the non-spore cells and possible contaminants that could remain in the solution.

Total spore counts were performed on 200 µl sample from each 2 ml spore solution via a flow cytometer (Benton Dickson FACS Calibur) using a 15 mW 488 nm argon laser. Since the spores are naturally refractile no staining was needed, and the counts were obtained for 15 s. Spore count ml^{-1} (C) of the original samples was determined by:

$$C = F/(t x r x D) \qquad (1)$$

Where F is the number of spores acquired (Forward Scattered count), t is the time in seconds of data acquisition, R is the flow rate in ml*s^{-1} of the cytometer and D is the dilution performed before running the sample.

Viable spore counts were made by mixing 1 ml of spore solution with 3 ml of soft CYE agar (40°C), vortexing the mix and pouring over CYE plates. The plates were incubated for 5 days at 30°C and the resulting colonies counted as viable spores.

Statistical analyses. We assessed *fruA* expression by a full factorial ANCOVA (genotype, time and time^2 as main effects), using *a priori* contrasts to compare the single knockout mutant strains versus their unmutated parental strain. Statistical significance for changes in timing of *fruA* expression was determined by a full factorial ANCOVA (mutant state, time and time^2) on the averages of the mutant strain values and those of the unmutated parental strain.

Statistical significance for developmental traits was assessed by ANOVA, with genotype, replicate and block as main effects. All were treated as random effects, as that is statistically conservative. A replicate *X* block interaction term was included in the ANOVA, when supported by partial F-tests for improved fitting [53]. An interaction term was not automatically included due to the large reduction in degrees of freedom associated with its inclusion.

Standard mean differences (SMD) were calculated from comparisons of parental genotype (DZF1) with the knockout mutants, as described below. For each trait, the difference between the parental genotype and the average of the knockouts was determined and a 95% confidence interval calculated based upon the standard error of the values (this analysis can be done as either a t-test or an ANOVA, since there is only one degree of freedom in the numerator). The SMD estimate was calculated by dividing the trait value by the standard deviation of the unexplained error, the square root of the Mean Square Error [37]. This provides values that are in units of standard deviation, so that the results can be compared across traits. The Confidence Intervals (CIs) were computed the same way (division by the standard deviation of the explained error). In other words, the SMD are in units of phenotypic standard deviation. Confidence intervals not overlapping with zero indicate statistical differences between the parental genotype (DZF1) and the knockouts. Confidence intervals not overlapping with 1 indicates that differences between DZF1 and mutants are greater than the unexplained phenotypic variation within genotypes (Figure 3).

A principal component analysis was conducted on correlations of the six developmental traits assessed for the five genotypes. A pre-planned contrast comparing the parental genotype and the mutants was computed for each value for each PCA trait (as determined by a chi-square test), using the same ANOVA structure as was performed for the individual traits. There were three main effects, genotype, replicate and block, with genotype as a fixed effect and replicate and block considered as random factors.

Acknowledgments

We are grateful to Alan C. Love and the Minnesota MicroPop Reading Group. During the course of final revision before submission, Prof. Sumiko Inouye, our good friend and coauthor died due to complications for a routine operation. Sumiko was intimately involved with the design and execution of the experiments and in writing the paper.

Author Contributions

Conceived and designed the experiments: AEE MT. Performed the experiments: AEE SI. Analyzed the data: AEE MT. Contributed reagents/materials/analysis tools: SI MT. Wrote the paper: AEE MT.

References

1. Pigliucci M (2009) An extended synthesis for evolutionary biology. Ann N Y Acad Sci 1168: 218–228.
2. Pigliucci M (2010) Genotype-phenotype mapping and the end of the "genes as blueprint" metaphor. Philos Trans R Soc, B 365: 557–566.
3. Carroll SB, Grenier JK, Weatherbee SD (2005) From DNA to Diversity: Molecular Genetics and the Evolution of Animal Design. Oxford: Blackwell Science. 258p.
4. Hoekstra HE (2006) Genetics, development and evolution of adaptive pigmentation in vertebrates. Heredity 97: 222–234.
5. Liao B-Y, Weng M-P, Zhang J (2010) Contrasting genetic paths to morphological and physiological evolution. Proc Natl Acad Sci USA 107: 7353–7358.
6. Jenner RA, Wills MA (2007) The choice of model organisms in evo-devo. Nat Rev Genet 8: 311–319.
7. Carroll SB (2008) Evo-devo and an expanding evolutionary synthesis: a genetic theory of morphological evolution. Cell 134: 25–36.
8. Rice SH (2008) Theoretical Approaches to the Evolution of Development and Genetic Architecture. Ann N Y Acad Sci 1133: 67–86.
9. Cañestro C, Yokoi H, Postlethwait JH (2007) Evolutionary developmental biology and genomics. Nat Rev Genet 8: 932–942.
10. Yu Y-TN, Yuan X, Velicer GJ (2010) Adaptive evolution of an sRNA that controls Myxococcus development. Science 328: 993.
11. Chuck G, Cigan AM, Saeteurn K, Hake S (2007) The heterochronic maize mutant Corngrass1 results from overexpression of a tandem microRNA. Nature Genetics 39: 544–549.
12. de Vries H (1906) Species and varieties, their origin by mutation. Chicago: The Open Court Publishing Co. 847p.
13. Goldschmidt R (1940) The material basis of evolution. New Haven: Yale University Press. 436p.
14. Barton NH, Charlesworth B (1984) Genetic Revolutions, Founder Effects, and Speciation. Annu Rev Ecol Syst 15: 133–164.
15. Rubinoff D, Le Roux JJ (2008) Evidence of repeated and independent saltational evolution in a peculiar genus of sphinx moths (Proserpinus: Sphingidae). PloS one 3: e4035.
16. Suzuki Y, Nijhout HF (2008) Genetic basis of adaptive evolution of a polyphenism by genetic accommodation. J Evol Biol 21: 57–66.
17. Minelli A, Chagas-Júnior A, Edgecombe GD (2009) Saltational evolution of trunk segment number in centipedes. Evol Dev 11: 318–322.
18. Lewontin RC (1974) The genetic basis of evolutionary change. New York: Columbia University Press. 346p.
19. Atallah J, Larsen E (2009) Genotype-phenotype mapping developmental biology confronts the toolkit paradox. Int Rev Cell Mol Biol 278 (119–148).
20. Hoekstra HE, Coyne JA (2007) The locus of evolution: evo devo and the genetics of adaptation. Evolution 61: 995–1016.
21. Shimkets LJ (1999) Intercellular signaling during fruiting-body development of Myxococcus xanthus. Annu Rev Microbiol 53: 525–549.
22. Velicer GJ, Kroos L, Lenski RE (1998) Loss of social behaviors by Myxococcus xanthus during evolution in an unstructured habitat. Proc Natl Acad Sci USA 95: 12376–12380.
23. Ueki T, Inouye S, Inouye M (1996) Positive-negative KG cassettes for construction of multi-gene deletions using a single drug marker. Gene 183: 153–157.
24. Wireman JW, Dworkin M (1977) Developmentally induced autolysis during fruiting body formation by Myxococcus xanthus. J Bacteriol 129: 798–802.
25. Zusman DR, Scott AE, Yang Z, Kirby JR (2007) Chemosensory pathways, motility and development in Myxococcus xanthus. Nat Rev Microbiol 5: 862–872.
26. Fortunato A, Strassman JE, Santorelli L, Wueller DC (2003) Co-ocurrence in nature of different clones in the social amoeba, Dyctiostelium discoideum. Mol Ecol 12: 1031–1038.
27. Davelos AL, Kinkel LL, Samac DA (2004) Spatial variation in frequency and intensity of antibiotic interactions hmong streptomycetes from prairie soil. Appl Environ Microb 70: 1051–1058.
28. Stefanic P, Mandic-Mulec I (2009) Social interactions and distribution of Bacillus subtilis phenotypes at microscale. J Bacteriol 191: 1756–1764.
29. Kraemer SA, Toups MA, Velicer GJ (2010) Natural variation in developmental life-history traits of the bacterium Myxococcus xanthus. FEMS Microbiol Ecol 73: 226–233.
30. Ueki T, Inouye S (2006) A novel regulation on developmental gene expression of fruiting body formation in Myxobacteria. Appl Microbiol Biotechnol 72: 21–29.
31. Nariya H, Inouye S (2006) A protein Ser/Thr kinase cascade negatively regulates the DNA-binding activity of MrpC, a smaller form of which may be necessary for the Myxococcus xanthus development. Mol Microbiol 60: 1205–1217.
32. Nariya H, Inouye S (2002) Activation of 6-phosphofructokinase via phosphorylation by Pkn4, a protein Ser/Thr kinase of Myxococcus xanthus. Mol Microbiol 46: 1353–1366.
33. Nariya H, Inouye S (2003) An effective sporulation of Myxococcus xanthus requires glycogen consumption via Pkn4-activated 6-phosphofructokinase. Molecular Microbiology 49: 517–528.
34. Nariya H, Inouye S (2005) Modulating factors for the Pkn4 kinase cascade in regulating 6-phosphofructokinase in Myxococcus xanthus. Mol Microbiol 56: 1314–1328.
35. Inouye S, Nariya H, Munoz-Dorado J (2008) Protein Ser/Thr kinases and phosphatases in Myxococcus xanthus. In: Whitworth D, editor. Myxobacteria: Multicellularity and Differentiation. Washington, DC: ASM Press. 191–210.
36. Stern DL, Orgogozo V (2008) The loci of evolution: how predictable is genetic evolution? Evolution 62: 2155–2177.
37. Whitlock MC, Schluter D (2009) The Analysis of Biological Data. Greenwood Village, Colorado: Roberts and Company. 700p.
38. Ohta T (1973) Slightly Deleterious Mutant Substitutions in Evolution. Nature 246: 96–98.
39. Dykhuizen DE, Dean AM (1990) Enzyme activity and fitness: Evolution in solution. Trends Ecol Evol 5: 257–262.
40. Travisano M, Lenski RE (1996) Long-term experimental evolution in Escherichia coli. IV. Targets of selection and the specificity of adaptation. Genetics 143: 15–26.
41. Müller GB (2007) Evo-devo: extending the evolutionary synthesis. Nat Rev Genet 8: 943–949.
42. Dean AM, Thornton JW (2007) Mechanistic approaches to the study of evolution: the functional synthesis. Nat Rev Genet 8: 675–688.
43. Queller DC, Ponte E, Bozzaro S, Strassmann JE (2003) Single-gene greenbeard effects in the social amoeba Dictyostelium discoideum. Science 299: 105–6.
44. Huntley S, Hamann N, Wegener-Feldbrügge S, Treuner-Lange A, Kube M, et al. (2011) Comparative genomic analysis of fruiting body formation in Myxococcales. Mol Biol Evol 28: 1083–1097.
45. Lemons D, McGinnis W (2006) Genomic evolution of Hox gene clusters. Science 313: 1918–1922.
46. Kroos L, Kaiser D (1987) Expression of many developmentally regulated genes in Myxococcus depends on a sequence of cell interactions. Genes Dev 1: 840–854.
47. Vos M, Velicer GJ (2008) Isolation by distance in the spore-forming soil bacterium Myxococcus xanthus. Curr Biol 18: 386–391.
48. Zhang YQ, Li YZ, Wang B, Wu ZH, Zhang GY, et al. (2005) Characteristics and living patterns of marine myxobacterial isolates. Appl Environ Microb 71: 3331–3336.
49. Velicer GJ, Hillesland KL (2008) Why cooperate? The ecology and evolution of myxobacteria. In: Whitworth D, editor. Myxobacteria: Multicellularity and Differentiation. Washington, DC: ASM Press. 17–40.
50. Morrison CE, Zusman DR (1979) Myxococcus xanthus mutants with temperature-sensitive, stage-specific defects: evidence for independent pathways in development. J Bacteriol 140: 1036–1042.
51. Higgs PI, Merlie JPJ (2008) Myxococcus xanthus: cultivation, motility and development. In: Whitworth D, editor. Myxobacteria: Multicellularity and Differentiation. Washington, DC: ASM Press. 465–478.
52. Abramoff MD, Magelhaes PJ, Ram SJ (2004) Image Processing with Image J. Biophotonics International 11: 36–42.
53. Zar JH (1999) Biostatistical Analysis. Upper Saddle River, New Jersey: Prentice Hall. 662p.

9

Testing Adaptive Hypotheses of Convergence with Functional Landscapes: A Case Study of Bone-Cracking Hypercarnivores

Zhijie Jack Tseng[1,2]*[¤]

1 Department of Biological Sciences, University of Southern California, Los Angeles, California, United States of America, 2 Department of Vertebrate Paleontology, Natural History Museum of Los Angeles County, Los Angeles, California, United States of America

Abstract

Morphological convergence is a well documented phenomenon in mammals, and adaptive explanations are commonly employed to infer similar functions for convergent characteristics. I present a study that adopts aspects of theoretical morphology and engineering optimization to test hypotheses about adaptive convergent evolution. Bone-cracking ecomorphologies in Carnivora were used as a case study. Previous research has shown that skull deepening and widening are major evolutionary patterns in convergent bone-cracking canids and hyaenids. A simple two-dimensional design space, with skull width-to-length and depth-to-length ratios as variables, was used to examine optimized shapes for two functional properties: mechanical advantage (MA) and strain energy (SE). Functionality of theoretical skull shapes was studied using finite element analysis (FEA) and visualized as functional landscapes. The distribution of actual skull shapes in the landscape showed a convergent trend of plesiomorphically low-MA and moderate-SE skulls evolving towards higher-MA and moderate-SE skulls; this is corroborated by FEA of 13 actual specimens. Nevertheless, regions exist in the landscape where high-MA and lower-SE shapes are not represented by existing species; their vacancy is observed even at higher taxonomic levels. Results highlight the interaction of biomechanical and non-biomechanical factors in constraining general skull dimensions to localized functional optima through evolution.

Editor: Andrew A. Farke, Raymond M. Alf Museum of Paleontology, United States of America

Funding: This research was supported by a National Science Foundation Graduate Research Fellowship and a Doctoral Dissertation Improvement Grant (DEB-0909807), American Society of Mammalogists Grant in Aid of Research, United States Fulbright Program, and a University of Southern California Zumberge Grant. The funders had no role in study design, data collection and analysis, decision to publish, or preparation of the manuscript.

Competing Interests: The author has declared that no competing interests exist.

* E-mail: jtseng@amnh.org

¤ Current address: Division of Paleontology, American Museum of Natural History, New York, New York, United States of America

Introduction

Convergent evolution is a prominent feature of mammalian evolution in the Cenozoic, so much so that many cases (e.g. convergently fossorial, arboreal, herbivorous, or carnivorous forms) have become textbook examples for the concept in evolutionary biology [1]. Morphological convergence is often interpreted as being adaptive for the very reason that they appeared in unrelated clades of species. This study addresses two questions about macroevolutionary morphological convergence in mammalian skull morphology: (1) Are morphologically convergent species actually convergent in functional capability? (2) If so, do those morphologies occupy local optimal peaks in a "functional" landscape? These questions are explored with theoretical morphology and finite element modeling in a case study of bone-cracking carnivorous mammals.

Adaptation, like convergent evolution, is a central concept in evolutionary biology. Studies of patterns and processes of adaptation on the macroevolutionary scale often rely on morphological characters, essentially those that are preserved in the fossil record. The concept of the fitness (or adaptive) landscape, as originally proposed to visualize possible evolutionary pathways of genetic interactions, has been adopted as a framework to examine

morphology in evolutionary and ecological contexts [2–5]. In a demonstration of the concept at its extremes, Kauffman [6] used simulations of hypothetical genetic interactions to create two fitness landscapes, one ("Fujiyama" landscape) with a single adaptive peak, and the other with a random distribution of equally adaptive peaks. Adaptive evolution is thought to proceed on intermediate landscapes between those extremes, with differentially elevated adaptive peaks, some of which act as "topological attractors" where examples of convergence can be sought [7,8].

In conventional morphometric studies, examples of convergent morphological evolution can be identified by macroevolutionary pathways that move toward each other in empirical morphospace, a morphospace built using existing, observed morphological diversity [8,9]. However, convergent morphologies can also evolve via parallel evolutionary pathways that do not exhibit obvious trends of such movement in empirical morphospace [10]. The complex craniodental system of vertebrates, particularly those of mammals with heterodont dentition, is subject to multiple functional demands not only of mastication and food acquisition, but also a range of sensory functions [11,12]. Understanding key evolutionary drivers of functional changes in such complex systems can be daunting, although there is some evidence of modularity to indicate that certain complex features evolved as integrated units [13,14]. To put the issue at hand as an analogy in engineering optimization theory, the number of possible designs of an engineered tool is proportional to the multiplicity of functions it is intended to serve; selective pressures on multi-tasking biological structures may similarly have resulted in equally fit morphologies on distinct (but comparable) adaptive peaks in a fitness landscape [15]. This phenomenon of "many-to-one" form-function relationship has been recognized as a major feature of adaptive evolution [16].

With the aid of computer-based simulation tools, questions that surround the functional aspect of morphological evolution can now be addressed with the creation of form-function landscapes based on hypothetical morphospace [8]. The bulk of previous work on theoretical morphospace has been done in studies of plants and invertebrate animals [8,17,18]. Complex mathematical models have been constructed to simulate growth patterns and possible (but sometimes non-existent) morphotypes in a variety of organismal groups. However, few studies have focused on constructing hypothetical morphospaces of vertebrates, particularly mammals (but see [19]). One factor in the paucity of such studies may lie in the large number of skeletal elements that exist in vertebrates, and the highly integrated functionality of many larger animals. Such emergent properties make parameterization of key morphological traits difficult. Nonetheless, the exploration of form and function using empirical morphospace and simulation of morphotypes that occur in different regions of such morphospace has already been proposed and explored in vertebrates [19,20]. A subset of functional simulations currently rely on finite element analysis (FEA), a technique which has gained wide use in the study of vertebrate biomechanics, particularly on the craniodental system [21,22]. However, FEA has mostly been applied to studies of existing or fossil morphology, and has not been used with emphasis on theoretical morphology. This study aimed to explore the union of functional simulations of craniodental function using FEA with the study of theoretical morphology using functional landscapes and hybrid morphospaces. A prominent example of convergent evolution in the Cenozoic record of mammals, that of bone-cracking hyaenids and borophagine canids, was used to demonstrate the utility of combining functional and theoretical approaches to study evolutionary (and potentially adaptive) changes in morphology. As this study attempts to demonstrate, the use of such a theoretical framework to test adaptive hypotheses regarding convergent morphologies, by comparing realized forms with a range of theoretically possible ones, provides new insights into the nature of constraint and adaptive function in the evolution of the carnivoran skull.

Finite element analysis

FEA was originally an engineering technique, used in the design process to conduct mechanical testing on simplified, discrete representations of real-world objects. The term was coined by Clough [23] for applications in the civil engineering field. In the past two decades, the application of FEA to studies of vertebrate functional morphology has seen a notable increase, particularly in the study of the craniodental system [22,24–26]. Application to mammalian craniodental biomechanics has been tested in a diverse range of research questions, from convergent evolution [27], ecological niche [28], conservation biology [29], bite force [30], to bone strain and model validation [31], among others.

The initial input to FEA is the morphology of interest, either derived from computer-generated models, photos of specimens, or more commonly, computed tomography (CT) images [21]. The representations of the morphology in question are modified and converted into element meshes, which are mathematical, geometric constructs of the original morphology. Material properties and boundary conditions are assigned to the mesh model with values derived from experiments, or in the case of extinct organisms, experimental values taken from closely related living taxa [21]. FE analysis software programs can then perform simulations of forces on the FE model, returning results in the form of stresses, strains, and bite force [32]. The process of improving models of actual species, usually by digitally repairing incomplete areas of the structure of interest, is amenable to manipulation and creation of non-existing, theoretical shapes that can then be tested in the same way as a model of an actual species.

Bone-cracking ecomorphology

Ecomorphologies are categories of ecological specialization, based on characteristic morphological features inferred to be associated with specific functional capabilities. The repetitive evolution of major ecomorphologies in carnivorous mammals is a key feature of this mammalian group throughout their Cenozoic evolution [33–36]. As in stereotypical cat-like and dog-like carnivorans, the hyena-like forms are hypercarnivores specialized in consumption of vertebrate flesh [34]. These hyena-like forms also have robust craniodental morphological features that are seen as adaptations for durophagy. Strong and bulbous cheek teeth, deep and often rounded foreheads, and large, rugose parietal areas for jaw muscle attachment are the main features of bone-cracking ecomorphologies [34,37]. These morphological features are associated with impressive bone-cracking capability in the extant spotted hyenas [38,39]. The generally large-bodied carnivorans that possess these morphological features have been identified in the fossil record in Hyaenidae [40], borophagine canids [41], and Percrocutidae [42–44].

Hyaenids and percrocutids are feliform carnivorans, with the majority of their evolutionary record in the Old World [40]. The earliest records of both groups are found in middle Miocene deposits of Eurasia; percrocutids did not survive the Miocene, whereas hyaenids are known today by four species, composing one of the smallest living carnivoran families [45]. Distinct morphological differences between percrocutids and hyaenids, which have been proposed to be sister groups, are established at their earliest occurrences [46]. The evolution of true hyaenids have been demonstrated to be quite gradual, with sequential appearance of six ecomorphological categories through their ~25 m.y. fossil record [40,47]. In contrast, the fragmentary fossil record of percrocutids is currently lacking a comprehensive phylogenetic framework. Nevertheless, it is clear that the most robust forms in either lineage, the hyaenines and the percrocutid *Dinocrocuta*, respectively, possessed full capability for bone-cracking comparable to, or exceeding, the modern spotted hyena (*Crocuta crocuta*) [43].

Canidae are North American natives, evolving into three subfamilies that represent some of the most common fossil carnivorans to be found in the Tertiary: Hesperocyoninae, Borophaginae, and Caninae [41,48,49]. All modern canids belong in Caninae, with no surviving species from the other two subfamilies [50]. Borophaginae contain the most hyena-like canids, some of which have long been considered ecological vicars of true Old World hyaenids [37,41,51]. Craniodental function in the most specialized borophagine canids has also been shown to resemble those of hyaenids in bone-cracking capability [52]. Furthermore, the bone-cracking ecomorphologies in the borophagine canids evolved derived craniodental morphology via parallel evolutionary pathways alongside the macroevlutionary patterns

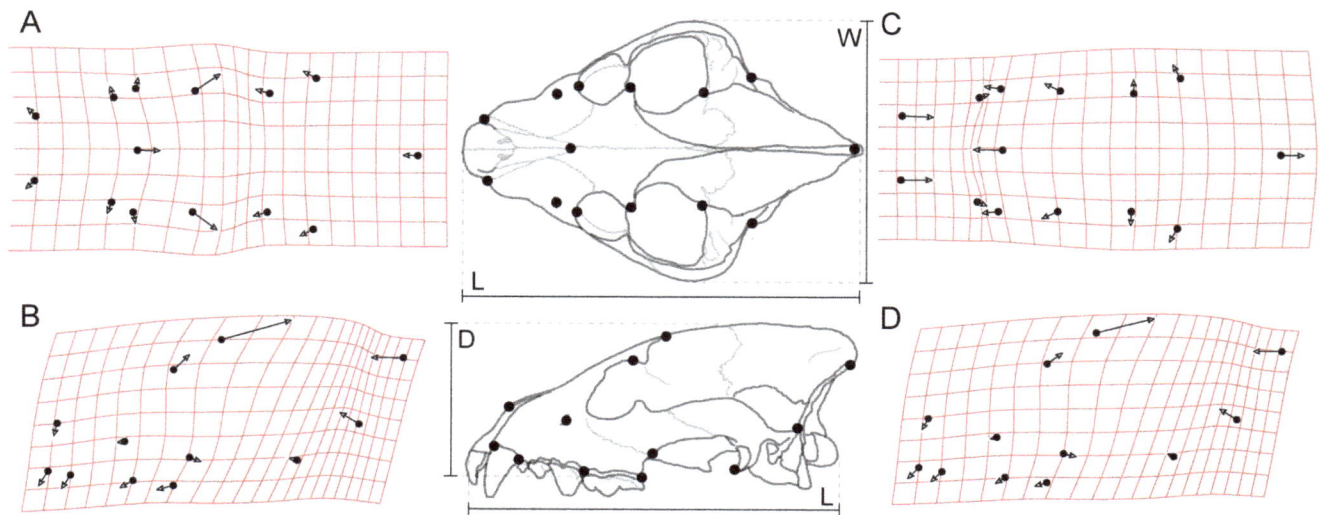

Figure 1. Convergent evolution of skull shapes in dogs and hyenas. Data for borophagine canids (A–B) and hyaenids (C–D) from two-dimensional geometric morphometric analyses in [10]. A, C, dorsal views; B, D, lateral views. Illustrations of skull show the measurements of width to length (W:L) and depth to length (D:L) taken from theoretical and actual skull shapes.

observed in hyaenids (Fig. 1, [10]). Such extensive convergence in craniodental morphology and inferred functional capability proceeded under a complex interplay of adaptation and constraint [10,37,53].

Taking the evolutionary patterns observed previously for borophagine canids, hyaenids, and percrocutids, I test the hypothesis that bone-cracking ecomorphologies were specialized forms that converged on identical or equivalent functional peaks on a simplified form-function landscape. Form and function are closely linked, so the functional pathways shared by bone-cracking ecomorphologies should reflect their parallel evolution in skull shape changes. Secondly, the convergently evolved specialist species in both lineages occupy optimal peaks in the theoretical morphospace containing a wide range of possible morphologies. A novel "functional" landscape built using principles of functional morphology and theoretical morphology is presented as a framework to test these hypotheses. The general utility of such approach is then demonstrated by tracking evolution of craniodental function, as inferred from FEA simulations, of actual fossil and extant species in the three carnivoran lineages discussed above.

Materials and Methods

No permits were required for the described study, which complied with all relevant regulations. All specimens, except for the skull of *Proteles cristata*, are in recognized museum collections listed in the Supplementary Information section. The dry skull of the extant aardwolf *Proteles cristata* was purchased from a natural history company (Necromance, 7220 Melrose Ave, Los Angeles, CA 90046) which sells specimens that are "legally obtained by-products and can be legally sold according to California state laws". The relevant California Penal Code 653o and 653p do not prohibit the import of *Proteles cristata*, which is listed by CITES Appendix III as a species of least concern in Botswana. In addition, the aardwolf is not listed under the federal Endangered Species Act foreign species list. The specimen, of unknown provenance other than "southern Africa", is being used as a destructive sample in a separate study, and was CT-scanned for

the current study prior to destruction. The raw CT dataset, which can be used to reconstruct the original morphology of the destructed specimen, is deposited online in Dryad (doi:10.5061/dryad.r2b1h). All theoretical and actual species models generated in this study are also available in Dryad at the above DOI address.

Hybrid morphospace

Strictly speaking, a theoretical morphospace, as defined by McGhee [8], is constructed without any morphometric input from actual specimens. The geometric shapes of organismal morphology are created using mathematical models, spanning a range that may encompass non-existent shapes [8]. In contrast, the hybrid morphospaces in this analysis were constructed with an actual ecomorphology: the jackal-like *Ictitherium* [40,47]. This fossil hyaenid provided a morphology that resembled less specialized forms of the convergent bone-cracking lineages, and therefore is a good starting point to examine how morphological evolution proceeded toward specialized forms. A two-dimensional morphospace was then used in conjunction with two functional properties (*sensu* [16]), described below, to create form-function landscapes. The morphological parameters were chosen to represent the main axes of evolutionary skull shape changes observed in both the Hyaenidae and the borophagine canids (Fig. 1), which exhibited parallel evolutionary pathways of change through time towards bone-cracking ecomorphologies [10]. These axes are relative skull width (width-to-length ratio, W:L) and relative skull depth (depth-to-length ratio, D:L). Even though the actual evolutionary patterns of skull shape change is complex, the morphospace constructed in this study used only simple overall cranial dimensions; more sophisticated methods of generating theoretical skull shapes are actively being developed to better characterize the observed variation [54]. Nevertheless, simple variation along the axes used generates a two-fold difference in the functional attributes tested.

During the evolution of the hyaenid and borophagine canid lineages, species evolved from relatively long-snouted, shallow- and narrow-skulled forms to short-snouted, deep- and wide-skulled robust forms [10,40,41]. These general skull shape changes are associated with the increased biomechanical capability of the larger and more robust species to consume hard foods

[43,52,55,56]. The exact causal links between incremental morphological changes and functional improvements are not known (and therefore the morphospaces created here are hypothetical in nature), but mechanical functions of specific craniodental features in bone-cracking ecomorphologies have been proposed [37,57,58]. Among these features are the development of a domed forehead and enlarged masticatory muscles, which are manifested in relatively deeper and wider skulls, respectively [59]. Accordingly, hybrid morphospaces were created to encompass and extend this range of observed evolutionary trends. Both morphological parameters were altered from the plesiomorphic state seen in the skull of *Ictitherium* by (1) increasing dorsoventral skull depth relative to skull length, and (2) increasing lateral skull width in increments of 25% up to 200% deviation from the *Ictitherium* specimen. To examine skull functionality that fall below this area, relative width and depth ratios of 75% from *Ictitherium* were also examined.

The two main axes of cranial shape change formulated were used to form a two-dimensional morphospace analogous to Raup's [17] classic "cube" of geometric parameters of shell coiling (Fig. 2). The cranial parameters used here, however, do not constitute theoretical morphospace in the strict sense; the direction of change along each morphological axis were chosen for analysis based on previous work using empirical morphospace [10]. This type of morphospace might also be referred to as a "combined" morphospace utilizing elements of both theoretical and empirical morphospaces [19].

Measures of function

Conventional adaptive landscapes rely conceptually on direct measures of survival and reproduction; such measures are dependent on environmental and ecological conditions at the specific temporal and spatial scale being examined [5,16]. The creation of a functional landscape, as defined here, aimed to measure more universal features of craniodental systems based on biomechanics. Bite force, regardless of the means for its estimation in living and extinct organisms, is one parameter that is crucial for vertebrates in both prey apprehension and mastication [60,61]. It is particularly important for bone-cracking ecomorphologies, as bite force is one direct determinant of the size of prey bone that can be consumed [39,62]. Thus, the bite force performance of fossil and living carnivorans is expected to be of major importance.

Similarly, it has been argued that skull strain energy, a measure of the work done in deformation of an object in FEA simulations, is a suitable measure of functional efficiency [32]. This argument is based on the logic that biological objects (e.g. skulls) with maximum stiffness for a given volume of material (i.e. low strain energy during deformation) should be favored by selective processes that maximize functionality [32]. Skull strain energy is used as a second axis of function in this study. The skulls of species in bone-cracking lineages are expected to be selected for increased stiffness per amount of skull bone, in order to perform the intensive bone-cracking behavior which places large amounts of stress and strain on the skull.

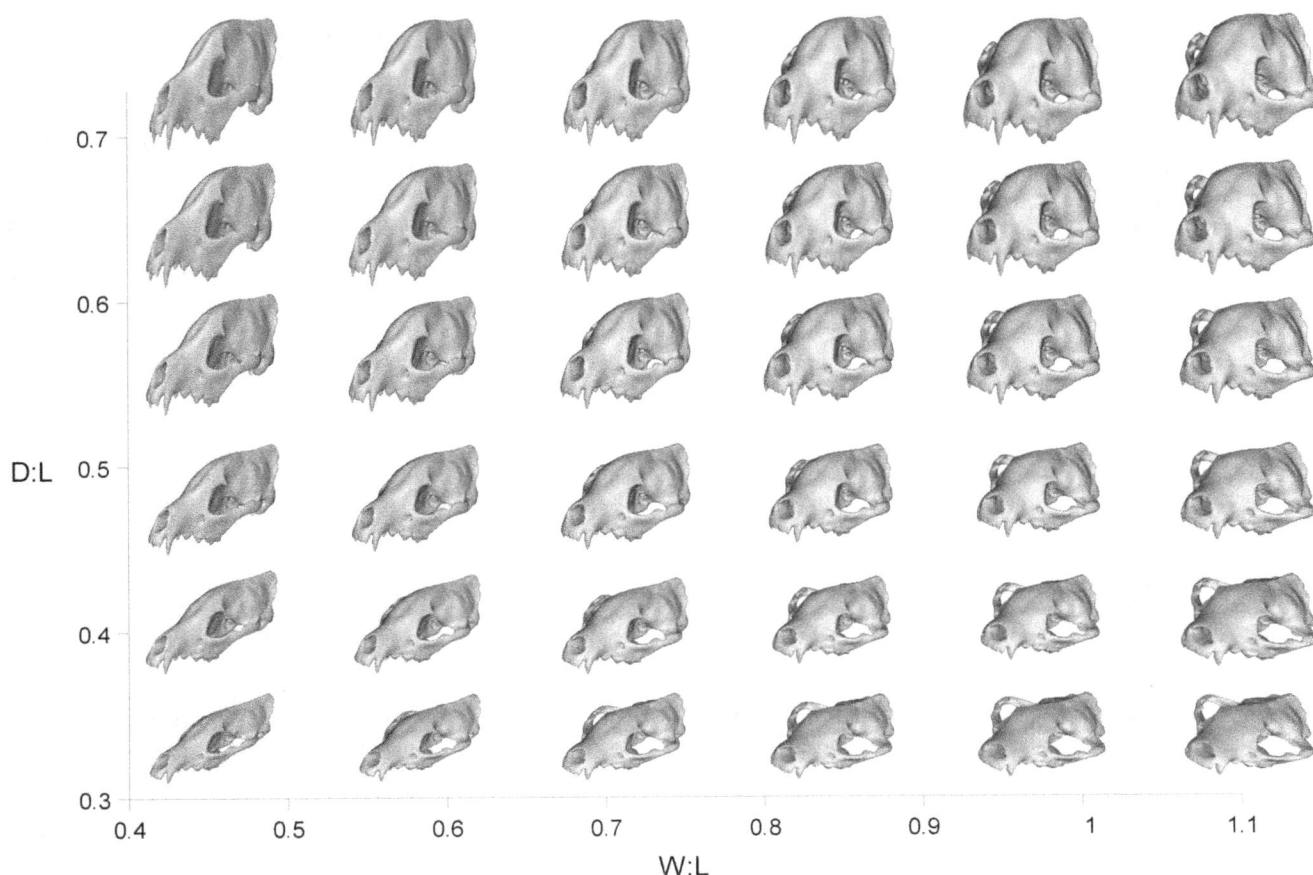

Figure 2. Theoretical models generated by geometric modification of an *Ictitherium* skull. The hybrid morphospace occupied by the 36 models spanned D:L ratios from 0.33 to 0.73 and W:L ratios from 0.42 to 1.11. Theoretical skull shapes are shown in rostral-lateral view.

Functional landscape

Analogous to an adaptive landscape, where the third-dimension is a fitness axis used to document adaptive peaks and valleys over a bivariate plot of morphological parameters [8], a functional landscape charts functional properties measured by a biomechanical axis over the bivariate plot of morphological parameters. The functional properties of bite force and skull strain energy are distinguished from measures of fitness because the former measure relatively narrow aspects of mastication, and not the overall organismal fitness or performance. Nevertheless, for lineages such as hyaenids and borophagine canids that experienced directional evolution towards bone-cracking ecomorphologies, the two functional properties analyzed were probably important for bone-cracking performance.

The incremental changes in morphological parameters create hypothetical morphotypes that show variation along the same directions as observed empirically in hyaenids and canids (Fig. 2). The association between parameters of skull shape and the ecological habits of extant carnivorans has been demonstrated in empirical morphospaces created by geometric morphometrics analyses [63,64]. Conceptually, the morphological parameters used can be supplemented with other functionally relevant parameters that are particular to the research question being addressed. Similarly, there may be other functional properties in addition to bite force and skull strain energy that are relevant to the specific type of functional morphology being examined. In its basic concept, the functional landscape is a functional manifestation of an adaptive landscape, its fitness axis (commonly the z-axis) having been modified to measure aspects of biomechanical function, which underlies organismal performance in relevant tasks.

Generation of theoretical models

Theoretical morphotypes representing incremental deviation of the two morphological parameters were generated by modification of an *Ictitherium* digital skull model. A complete and intact skull of the late Miocene hyaenid *Ictitherium* sp. (HMV 0163, Hezheng Paleozoology Museum, Gansu Province, China) was scanned using computed tomography (CT) at Lanzhou University Hospital No. 1 (Gansu Province, China) with a Siemens Somatom Sensation 64 scanner (120 KV, 304.00 mAs); images had a pixel size of 0.2578 mm, resolution of 512×512 pixels, and 0.36 mm interslice distance. Data were exported in the DICOM (Digital Imaging and Communications in Medicine) format. The cranium and mandible of the specimen were separated and digitized using the software program Mimics 13 (Materialise NV). Digital reconstructions, including internal morphology, were exported in the stereolithography format (*.stl). The files were then imported into Geomagic Studio 10 (Geomagic, Inc.) where generation of theoretical morphotypes took place.

Skull depth and width in theoretical morphotypes were changed by scaling the original digital model of *Ictitherium* in the respective axes by a set percentage (75%–200% of original). The axes of change were aligned so that depth increased along the line connecting the carnassial tooth and the top of the frontal dome (Figs. 1–2); width increased along the axis perpendicular to the long axis of the skull. The modified theoretical morphotypes were then exported into Strand7 2.3.7 finite element analysis software program (G+D Computer Pty Ltd), where finite element meshes were generated.

The finite element meshes representing different morphotypes were modeled with identical forces, material properties, and boundary conditions. Because the fourth premolar (carnassial tooth) represents a synapomorphy of Carnivora for shearing and masticating meat, all models simulated unilateral bites with the carnassial. Although specialized bone-cracking hyaenids and certain borophagine canids (e.g. *Aelurodon*) evolved robust P3 as the main bone-cracking tooth, other specialized canids and less specialized hyaenids do not equally emphasize the robustness of this tooth. Therefore, the carnassial tooth simulation provided a common point of comparison across convergent specialist and generalized species. Furthermore, previous findings indicate that both P3 and P4 exhibit higher mechanical advantage in the extant spotted hyena compared to gray wolf [43], thus adaptive signals in P3 bone crackers would be recorded in P4 simulations as well. Three jaw-closing muscle groups were modeled: temporalis, masseter, and pterygoid. The relative contributions of the muscle groups to total input force were set at 67% (Temporalis), 22% (Masseter), and 11% (Pterygoid); these values were based on wet weight of the relative muscles in modern *Crocuta crocuta* [65]. Proportions of 64%, 22%, and 11% have been reported for canids [28,66,67]; the small differences between hyaenids and canids were assumed to be negligible for the model results studied, and the construction of models from actual specimens (including canids) used the first set of percentages for consistency. Muscle activation on the balancing (non-biting) side cranium was adjusted to 60% of the total input force on the working (biting) side cranium; ratios across the muscle groups remained the same [68]. Force vectors within each muscle attachment area were divided evenly over the entire area, with adjustment for wrapping of musculature around the cranial muscle attachment sites using the Boneload program [69]. Muscle force vectors in the respective muscle groups were oriented toward centroids of each muscle group at the attachment sites on the corresponding dentaries. A gape of 30 degrees was simulated for all models, close to the optimal angle found in *Canis lupus dingo* [70]. A total of 39,820 N of input muscle force was simulated in all models, and the output bite force was calculated as mechanical advantage (output force/input force) with a maximum range of 0.0 (no output force) to 1.0 (output force = input force). All models were also adjusted to have identical total surface areas (1×10^6 mm^2), to allow comparison of performance variables among theoretical morphotypes as a function of shape changes, but not size [32]. This particular ratio of input force (39,820 N) to surface area ratio (1×10^6 mm^2) matched the ratio used by Tseng and Wang [52], which was derived from the force-surface area ratio in their *Canis lupus* model that was validated by maximal measured bite force in *Canis familiaris* [71].

Three nodal constraints were placed on the cranium models: the left and right temporomandibular joints (TMJ), and the unilateral bite point. The bite point was modeled as a nodal constraint fixed from all translational and rotational movements. The TMJ was modeled as a single nodal constraint in the middle of each glenoid fossa, fixed from all but rotational movement in the sagittal plane. All models were given a single set of material properties, representing typical values for mammalian cortical bone. All analyses were linear and static, therefore only two material parameters were required: Young's (Elastic) modulus = 20 GPa, and Poisson's ratio = 0.3. Heterogeneous models that contain multiple material properties have been shown to have higher stresses and bite forces compared to identical models made with a single set of material properties; such differences in results are acknowledged, but they are assumed to have no great effect on the comparative context being pursued in this study [25,52,55].

Bite force output is graphed as mechanical advantage (MA) and skull strain energy (SE) values recorded in Joules. Both were plotted against bivariate plots of the two morphological parameters (D:L and W:L ratios) as wireframe plots, upon which

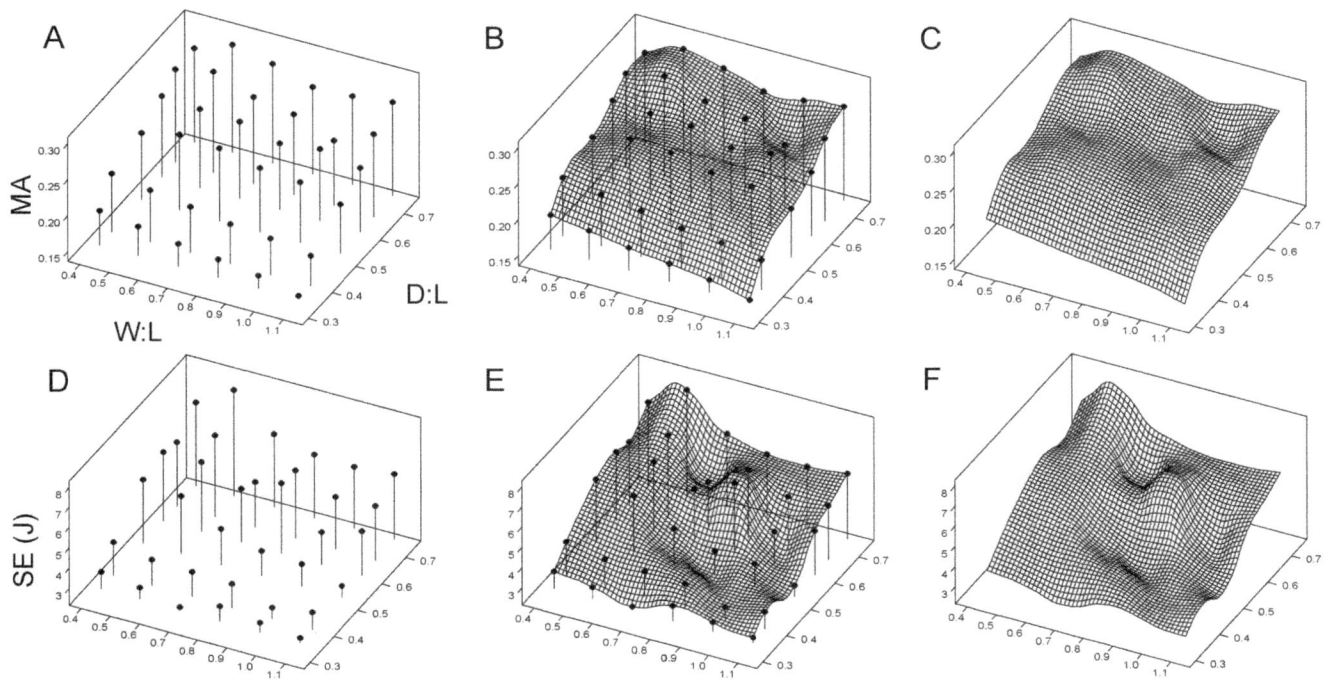

Figure 3. Construction of the functional landscape from theoretical morphologies. W:L and D:L are plotted on the x- and y-axes, respectively. The functional properties mechanical advantage (MA) and skull strain energy (SE, in joules) are plotted on the z-axis. A, D, three-dimensional plots of the data points from analysis of theoretical models; B, E, the wire frame mesh overlaid and interpolated using the theoretical models; C, F, the theoretical models removed, leaving the mesh representing the functional landscapes for MA and SE, respectively. For data values see Table S5.

simulation results from models of actual hyaenid and borophagine species were plotted (Fig. 3).

Skull dimensions of actual species

To use the functional landscape in predictions of functional evolution in actual lineages, the W:L and D:L ratios of actual hyaenids and canid species were measured from specimen photos. The dataset of fossil and extant hyaenids and canids from Tseng and Wang [10] was used. Nine hyaenid species and 15 canid species were complete enough to be measured. Means were used where sample size >1, and specimens that were used in FEA (see below) were plotted individually (Table 1, S1, S2).

Models of actual species

FE models of actual fossil and living species of Hyaenidae and borophagine canids were constructed as described above for the theoretical models. Most of the models were existing ones taken from previous studies [43,52,55,65]. Bite force and skull strain energy values were obtained from analyses after all models were standardized so that the ratios of total muscle input force to total model surface area were kept constant across all models [32]. Such standardization allowed the absolute size of models to be removed, and comparisons of skull shape and function measured. This type of comparisons are desired in this case because the functional landscape is constructed from morphological parameters that approximate evolutionary shape changes, most of which are not allometric in hyaenids and canids [10]. Also, body size increased dramatically over the course of evolution in the two carnivoran groups examined, so that bite force would show increases even in absence of biomechanical adaptations. Therefore, comparisons solely based on skull shape appeared to be the most appropriate.

To examine evolutionary trends predicted by the functional landscape, a series of FE models that represent different degrees of specialization for bone-cracking in the hyaenid and borophagine lineages, respectively, were used. The hyaenids *Proteles cristata* (J050607T02, ZJT comparative collection, prepared dry skull), *Ictitherium* sp. (HMV 0163), *Chasmaporthetes lunensis* [55,72], *Ikelohyaena abronia* [65], *Parahyaena brunnea* (MVZ 117842, Museum of Vertebrate Zoology, University of California, Berkeley), and *Crocuta crocuta* [55] were analyzed. The fossil and modern canids analyzed included *Mesocyon coryphaeus*, *Microtomarctus conferta*, *Epicyon haydeni*, *Borophagus secundus*, and *Canis lupus* from Tseng and Wang [52], and *Lycaon pictus* from Tseng and Stynder [65]. In addition, the percrocutid *Dinocrocuta gigantea*, a feliform carnivoran that convergently evolved bone-cracking morphology independent of hyaenids or canids, was included in the analysis using the model from Tseng [43]. All specimens, except for *P. cristata*, are deposited in the museum collections listed above and described in the relevant publications cited. The raw CT data for *P. cristata* are archived online in Dryad (doi:10.5061/dryad.r2b1h). A total of 13 models of actual fossil and extant species were used.

In addition to MA and SE values, the stress distributions on the skulls of actual species were also visualized. Values of von Mises stress, which approximate materials that fail under a ductile mode of fracture, were used [26,73]. As FEA conducted on models of actual species were scaled in a similar manner to the theoretical models, the distribution of von Mises stress on the skull represents relative levels of stress that can be directly compared across species. High levels of stress under such comparisons can therefore be interpreted as likely areas of material failure.

Table 1. List of actual models and species measurements of hyaenids and canids used in the study.

Actual models:	
Hyaenidae	**Canidae**
Crocuta crocuta	Lycaon pictus
Parahyaena brunnea	Canis lupus
Ikelohyaena abronia[†]	Borophagus secundus[†]
Chasmaporthetes lunensis[†]	Epicyon haydeni[†]
Ictitherium sp.[†]	Microtomarctus conferta[†]
Proteles cristata	Mesocyon coryphaeus[†]
Additional measurements:	
Crocuta crocuta (n = 45)	Canis dirus (n = 1)[†]
Adcrocuta eximia (n = 3)[†]	Borophagus secundus (n = 1)[†]
Hyaena hyaena (n = 1)	Epicyon haydeni (n = 1)[†]
Hyaenictitherium wongi (n = 1)[†]	Epicyon saevus (n = 1)[†]
Proteles cristata (n = 3)	Aelurodon ferox (n = 3)[†]
	Aelurodon mcgrewi (n = 1)[†]
Percrocutidae	Aelurodon taxoides (n = 1)[†]
Dinocrocuta gigantea (n = 3)[†]	Protomarctus optatus (n = 1)[†]
	Phlaocyon leucosteus (n = 1)[†]
	Desmocyon matthewi (n = 1)[†]
	Paraenhydrocyon josephi (n = 2)[†]
	Hesperocyon gregarius (n = 1)[†]

For list of specimen numbers see Tables S1, S2.
[†]extinct taxon.

Results

The hybrid morphospace comprised 36 theoretical models, onto which a wire mesh was interpolated to create the functional landscapes (Figs. 2–3). Two separate landscapes were created, one for mechanical advantage (MA; Fig. 3A–C) and the other for skull strain energy (SE; Fig. 3D–F). The landscapes showed predictable trends of variation. Increasing skull depth, regardless of the starting skull width, translated into higher MA and higher SE (Fig. 3). Increasing skull width generated progressively lower MA and SE at shallower skull depths, but the patterns became more complex at higher skull depths (Fig. 3). Peaks in MA are found at skull depth-to-length (D:L) ratio of >0.7 and width-to-length (W:L) ratios of 0.4–0.7 (Fig. 3A–C). Lowest MA values are found at D:L<0.4 and W:L>0.7 (Fig. 3A–C).

D:L and W:L ratios of actual hyaenid and borophagine canid species overlapped extensively in their distribution on the functional landscape (Fig. 4). Furthermore, the species followed an evolutionary pathway from D:L 0.3–0.4 and W:L 0.5–0.6 to D:L ~0.5 and W:L ~0.7 (Fig. 4A, D). This pathway showed a continuous climb towards higher elevation on the MA landscape (Fig. 4B), and a path into an adaptive valley on the SE landscape (Figs. 4E). The pathways occupied by actual hyaenids and canids traversed a region of increasing MA and moderately low SE, which is bordered at the bottom right with a large region of low MA and low SE theoretical shapes (Fig. 5). In the upper regions are high MA and high SE shapes; both of these regions represent suboptimal areas (Fig. 5).

The regions of the MA vs. SE plot occupied by actual species represent a pathway from low MA (~0.18) towards higher MA

(~0.25) at or below the fitted curve (SE = 174.74*MA^2 −47.04*MA+5.8609, r^2 = 0.8363) for the theoretical models (Fig. 6). The MA values of models of actual species covered a slightly larger range than predicted by theoretical models, from ~0.16 to ~0.27 (Fig. 6A). The exception is the myrmecophagous hyaenid *Proteles cristata*, which has an MA of ~0.12, lower than all actual species and theoretical models. SE values of actual species followed the overall trend predicted by the theoretical models, but do not follow the theoretical pathways exactly. To test for potential differences created by scaling factor, models of actual species were re-analyzed after scaling by total volume, total muscle attachment surface area, or total skull length (condylobasal length). Volume- and muscle-scaled models returned essentially identical results as the total surface area method (Fig. 6A). Scaling by skull length returned similar results, except that MA values for the derived hyaenids *Crocuta crocuta* and *Parahyaena brunnea* were lower, and SE values for *Ictitherium* and *Chasmaporthetes lunensis* were also lower (Fig. 6B). Such differences did not change the overall trends, however.

Von Mises stress distributions on the actual models showed a general trend of increasingly stressed fronto-parietal regions in hyaenids (Fig. 7A–G). The canid models showed no such trend, and in general had moderate levels of von Mises stress spread over the dorsal cranium, except for elevated stress levels in *Canis lupus* and decreased levels in *Epicyon haydeni* (Fig. 7H–M). The fronto-parietal region in *Ikelohyaena abronia* and *Canis lupus* showed the highest peak stress, and in all models the temporomandibular joints tend to have elevated stress levels (Fig. 7).

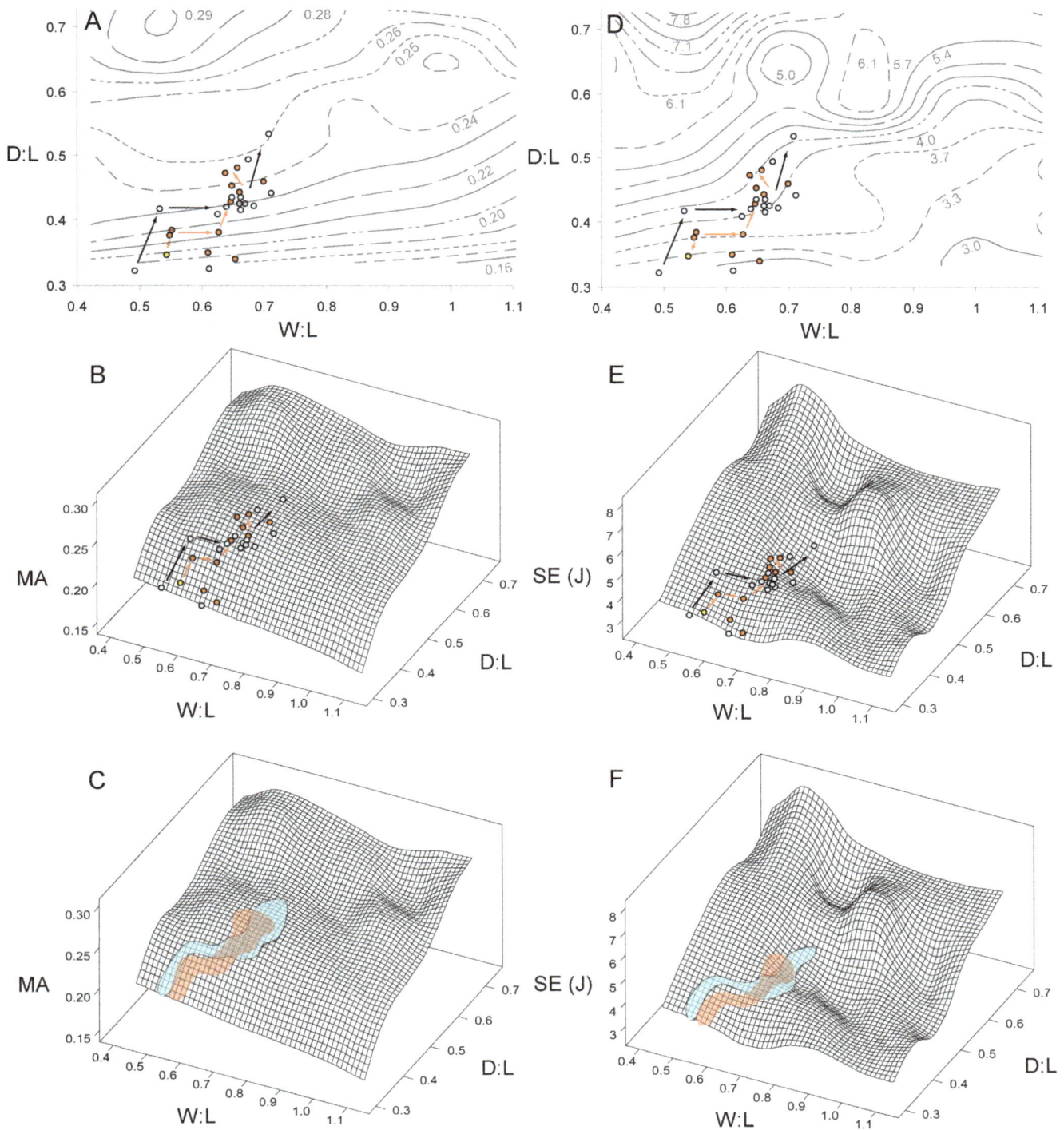

Figure 4. Distribution of actual species on the functional landscapes. A, D, distribution of hyaenids (dark circles) and fossil canids (light circles) on two-dimensional contour plots of MA and SE, respectively. Lines are isoclines. B, E, distribution of hyaenid and canid species on the three-dimensional functional landscapes for MA and SE, respectively. C, F, the pathways occupied by the hyaenid (shaded) and canid (outlined) lineages on the MA and SE landscapes, respectively. Sequential arrows indicate directions of change from less derived, earlier species to more derived, younger species. Note continuous climb on the MA landscape and shifting towards shallower slopes on the SE landscape. For data values of species models see Table S6.

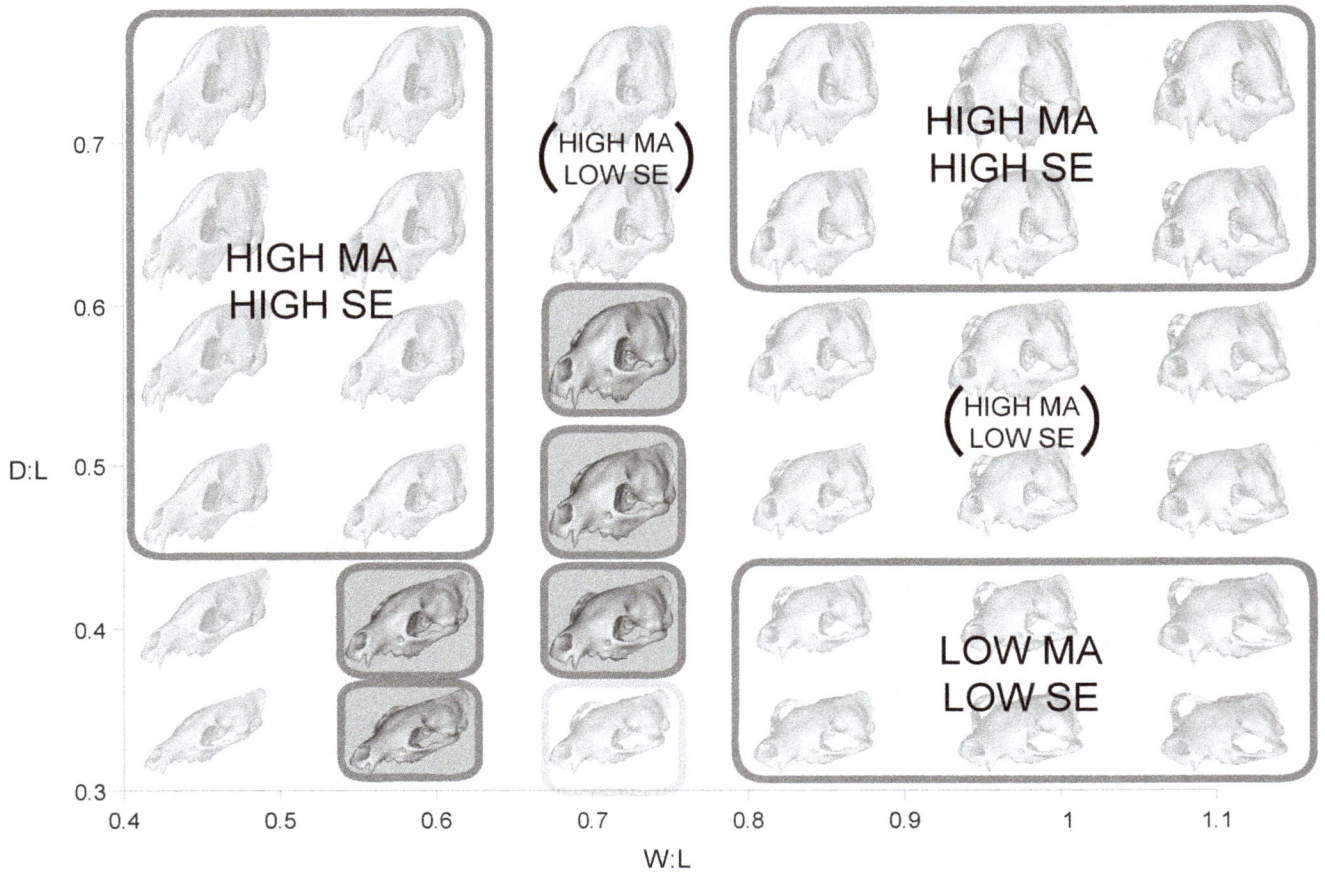

Figure 5. Locations of optimal functional capability in the hybrid morphospace. Small shaded squares represent theoretical models matched by existing hyaenid and canid species (small unshaded square shows position of insectivorous *Proteles cristata*). Suboptimal regions are shown in larger squares. Regions marked by parenthesized labels represent optimal areas not occupied by actual species.

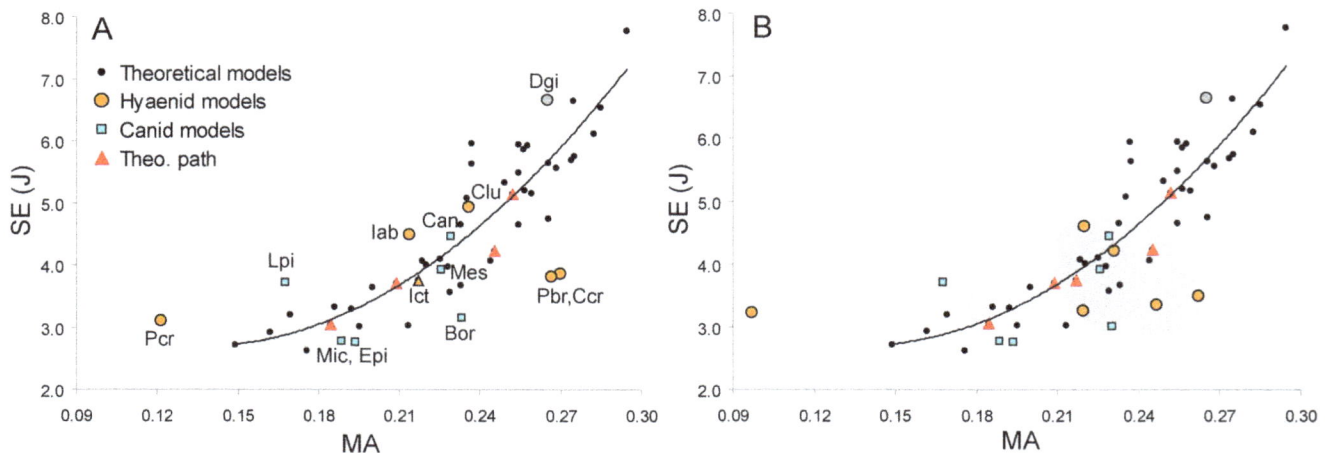

Figure 6. Theoretical and actual MA and SE values. A. distribution of theoretical models overlaid with values from FE models of actual species, all scaled by total surface area. B, distribution of theoretical and actual models, the latter scaled by condylobasal length of the skull. Red triangles indicate the theoretical pathway traveled by actual species on the functional landscape. The positions of hyaenid (darker shade) and canid (lighter shade) groupings are shown as ovals in (B). Species abbreviations (hyaenids): Ccr, *Crocuta crocuta*; Hlu, *Chasmaporthetes lunensis*; Iab, *Ikelohyaena abronia*; Ict, *Ictitherium sp.*; Pbr, *Parahyaena brunnea*; Pcr, *Proteles cristata*. Canids: Bor, *Borophagus secundus*; Can, *Canis lupus*; Epi, *Epicyon haydeni*; Lpi, *Lycaon pictus*; Mes, *Mesocyon coryphaeus*; Mic, *Microtomarctus conferta*. Percrocutid: Dgi, *Dinocrocuta gigantea*.

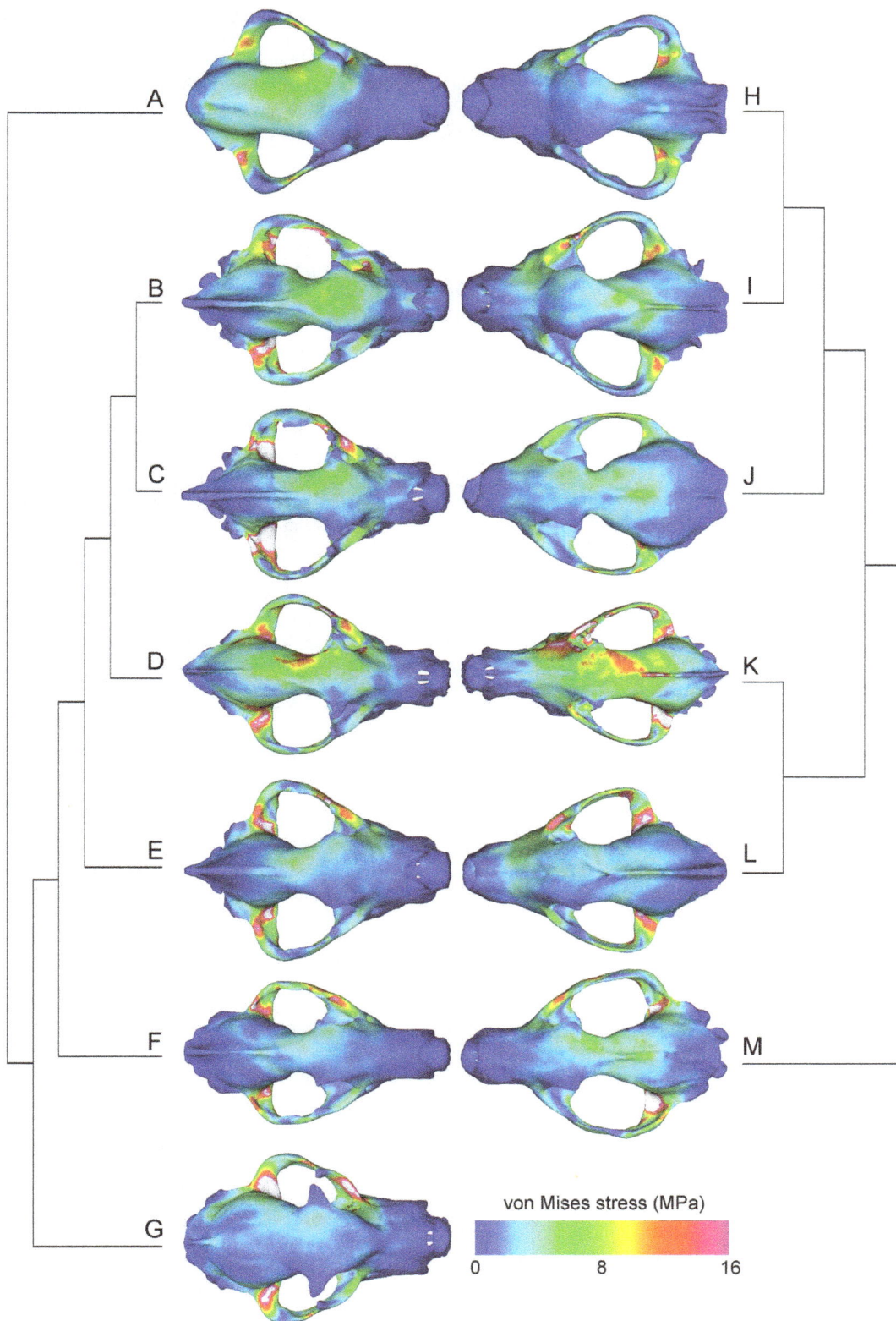

Figure 7. Stress distributions on the FE skull models of actual fossil and extant species. A, *Dinocrocuta gigantea*; B, *Crocuta crocuta*; C, *Parahyaena brunnea*; D, *Ikelohyaena abronia*; E, *Chasmaporthetes lunensis*; F, *Ictitherium sp.*; G, *Proteles cristata*; H, *Epicyon haydeni*; I, *Borophagus secundus*; J, *Microtomarctus conferta*; K, *Canis lupus*; L, *Lycaon pictus*; M, *Mesocyon coryphaeus*. Phylogenetic relationships for hyaenids (A–G) based on Werdelin and Solounias (1991), and for canids (H–M) based on Wang (1994), Wang et al. (1999), and Tedford et al. (2009).

Discussion

The generation of simplified functional landscapes was used to test the hypothesis that convergent morphological evolution in canids and hyaenids can be explained in terms of functional evolution towards more optimal bone-cracking capability. Theoretical skull shapes showed a general increase in mechanical advantage (MA) with higher D:L ratios, although the narrower (lower W:L) skulls had largest levels of strain energy (Fig. 3). Actual hyaenid and canid species showed a steady climb up the MA landscape, at the same time moving along topological isoclines in SE. Such a pattern of evolution is consistent with optimization theory; in this case two functions of the skull, maximizing MA and minimizing SE, are optimized by traveling upslope on the MA landscape and moving along topological isoclines on the SE landscape (Fig. 4). Therefore, the hypothesis that convergent morphologies shared convergent functional capability is supported.

Hyaenid and canid models are adjacent to each other on the MA vs. SE plots, with more derived species having higher MA (Fig. 6). As predicted by the landscape model, the path from less derived species to more specialized species tend to occur downward (i.e. smaller SE for a given MA) or rightward (i.e. larger MA for a given SE) on the plot (Fig. 6). Such distribution is expected under an optimization model. The functional correlation of the MA and SE distributions is further supported by the fact that *Proteles cristata*, a specialized insectivorous hyaenid, does not crack bones, and accordingly has very low MA towards the bottom left corner of the MA vs. SE plot (Fig. 6). This exception also supports the interpretation that there is an adaptive coupling of functional attributes and ecology in this sample of carnivorans.

Overall, among the theoretical models used to construct the functional landscape, only a small number overlapped with actual morphologies (Fig. 5). In the upper left and upper right regions high MA is coupled with high SE, making those morphologies suboptimal; those areas are accordingly not occupied by actual species (Fig. 5). The bottom right corner is marked by both low MA and low SE, and is similarly not optimal. The actual path taken by canids and hyaenids constitutes a route of increasing MA at relatively small cost in SE increase (Fig. 5). Therefore, the functional landscape distributions of actual species predict skull MA to be maximized relative to increase in skull SE through evolution. The distribution of MA versus SE values for the 13 actual skull models show an intermediate position within the range of theoretical morphologies, overlapping the regions predicted by the functional landscape, therefore supporting theoretical predictions (Fig. 6).

The remaining unoccupied regions in the functional landscape, however, indicate that the hypothesis predicting the derived morphotypes occupying functional optima on the landscape was not supported. Contrary to expectation, the most optimized theoretical shapes in the functional landscape are not occupied by actual species (Fig. 5). Skull shapes with D:L = 0.7, W:L = 0.7 and D:L = 0.5, W:L = 0.8–1.1 tend to have high MA and relatively low SE, making them more suitable for generating large bite forces than shapes toward the central and bottom left regions of the landscape, where canids and hyaenids are located (Fig. 5). There appears to be no visible barriers or functional valleys on the MA landscape, or prohibitively high SE peaks on the SE landscape, to explain the lack of actual species in those regions (Fig. 4). To check whether this bias in distribution is a function of similarly restricted skull shape changes specific to hyaenids and borophagine canids, the modern carnivoran dataset composed of 37 North American and East African carnivoran species from Tseng and Wang [10]

was plotted onto the functional landscapes (Fig. 8, Table 2, S3, S4). The pathways taken by canid and hyaenids species overlapped with the skull shapes observed among the major representatives of modern carnivoran families (Fig. 8A–C). Only the cheetah, *Acinonyx jubatus*, was distinct from all other carnivorans by much higher W:L ratios that placed the species on an SE peak (Fig. 8).

The extensive overlap of modern carnivorans with the evolutionary sequence of canids and hyaenids indicates constraint in skull shape disparity within the hybrid morphospace of theoretical possible shapes (Fig. 2). Presence of such a higher-level constraint created a limitation on the evolution of functionally more optimized skull shapes in bone-cracking carnivorans (Fig. 8). Of course, the complex suite of functions that the mammalian skull plays in mastication, food acquisition, and sensory reception meant that constraints on the realized skull shapes are more complex than just biomechanical ones. In the absence of additional non-biomechanical constraints, one optimal path to maximize MA and minimize SE would be to travel along topological lines at D:L between 0.3 and 0.4 towards higher W:L ratios (Fig. 8C–D). At higher W:L ratios, SE increases more slowly and therefore those shapes have relatively higher MA. In reality, there appears to be a constraint in increasing W:L ratio across the modern carnivorans analyzed, and consequently skull shape evolved towards higher D:L with an upper limit of W:L ~0.7 to instead increase MA on local (but not global) optima (Fig. 8).

The high-MA high-SE skull of *Dinocrocuta gigantea* also demonstrates the presence of others factors in determining performance in addition to the two functional properties examined. The largest bone-cracking carnivorans examined in this study, *Epicyon haydeni* and *Dinocrocuta gigantea*, share similarities in skull shape but not in biomechanical attributes (Fig. 7). *Epicyon* has a low-MA and low-SE skull, in contrast to the high-MA high-SE skull of *Dinocrocuta*. This seemingly contradictory result might be explained by the one-to-one form to function property of mechanical advantage [16]. Mechanical advantage by itself is a scale-free measure of force generation, but in fact a system with high MA and low absolute muscle force can generate the same resulting bite force as a system with low MA and large absolute bite force. Therefore, the disparate distributions of *Dinocrocuta* and *Epicyon* can in fact represent similar performing morphologies that converge along another axis of evolutionary change, namely body size. A large body size would allow *Epicyon* to generate bite forces required to crack bones to a comparable degree as smaller, more shape-adapted skulls of *Crocuta* and *Borophagus*. On the other hand, the large body size of *Dinocrocuta* would allow a smaller muscle input to generate sufficient bone-cracking bite forces, therefore not producing the high-SE predicted at its maximum capability (Fig. 6). Such alternatives to evolutionary changes in skull shape can be further coupled with behavior, in which bones of smaller prey are cracked and consumed, and bones of larger prey intentionally left alone. With this interpretation, body size increase in bone-cracking carnivorans as a masticatory adaptation would be analogous to larger body size in ungulates as a defense mechanism, in that both increases in body size alone constitutes an adaptation. Whether "body-size" specialists should constitute a distinct sub-category of bone-cracking ecomorphology is a fascinating issue that remains to be explored. Archaic mammals such as creodonts and condylarths, for example, evolved dental morphology and body size approaching the larger carnivoran bone-crackers, even though the skulls of most creodonts do not share the suite of morphological features seen in carnivorans [35].

A concept intimately associated with adaptive landscapes is the macroevolutionary ratchet, which has been studied in carnivorans

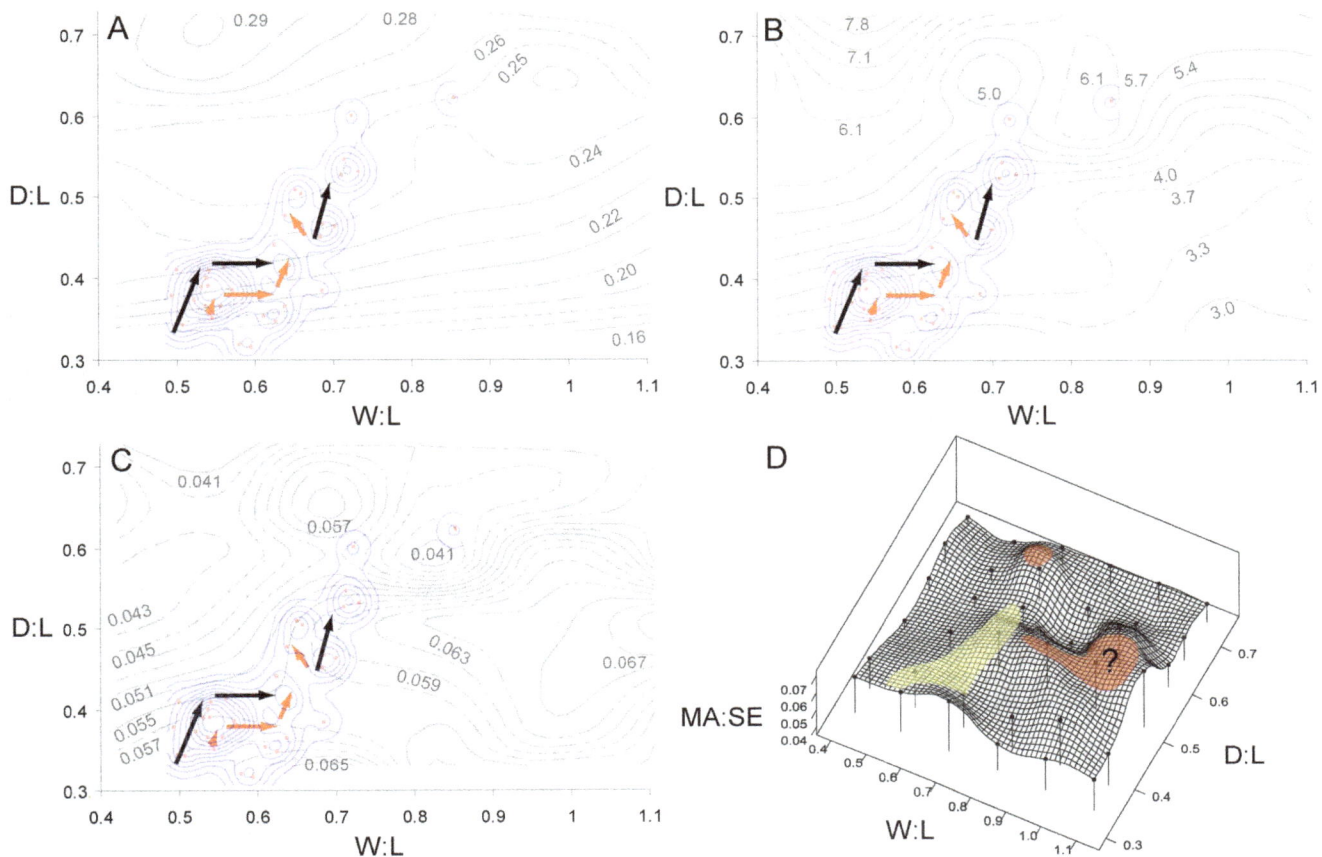

Figure 8. Distributions of modern North American and East African carnivoran species on the functional landscapes. Distributions are plotted on the MA (A), SE (B), and MA:SE (C–D) landscapes. Arrows indicate pathways of evolution for hyaenids (light arrows) and borophagine canids (dark arrows). Species distributions of modern carnivoran species are plotted as solid contours. Peaks on the MA:SE landscape (D) represent optimized theoretical skull shapes that are either realized (light shade) or unoccupied (dark shade, with question mark). (D) corresponds with Figure 5.

[53,74]. The limited number of alternative means of morphological specialization is associated with decrease in morphological disparity in repeatedly specialized lineages, which affected the long-term fitness of those lineages [34,53]. In this context, generalist species are located at lower elevations of the adaptive landscape, and specialists are higher up adaptive peaks; the macroevolutionary ratchet can be visualized as the evolutionary process of moving up in elevation on the landscape [75]. Catastrophic, sometimes even localized, events may shift the position of those adaptive peaks, causing the demise of specialists by their very inability to move or survive in other regions of the fitness landscape [75]. Others argue for the mobility and dynamic nature of adaptive peaks through time, which may imply a different mode of adaptation and specialization of organisms that involves more evolutionary "adjustment" to current peaks [8]. The fact that convergent canids and hyaenids evolved via pathways within the overall distribution of modern carnivorans indicates that a general constraint on skull depth and width ratios is present, perhaps as a more general phenomenon than caused by specific factors in a macroevolutionary ratchet model for bone-cracking specialists (Fig. 8). Nevertheless, it would be interesting to further explore whether the pathways on the functional landscape are "one-way streets", and if the distance already traveled by a particular lineage may indeed represent the macroevolutionary ratchet in action.

The proxy for functionality used in this study, namely measures of bite force and skull strain energy, are biomechanical function indicators, arguably not a very complete measure of fitness (using a definition of the organismal ability to both survive and reproduce). However, the fact that terminal members of the lineages studied represent the best examples of *Crocuta*-equivalent bone-cracking ecomorphologies in the Cenozoic, and that their evolutionary processes show overwhelming trend towards robust craniodental features, suggest that in this case the functional properties likely would have been quite important in their evolution. Furthermore, carnassial mechanical advantage is a common selective parameter for all carnivorans, and measures of its biomechanical function are directly linked to mastication and food intake. One can also argue that plotting phylogenetic trends onto the static functional landscape is not greatly affected by the possibility of shifting adaptive peaks in other types of landscapes which are contingent upon environmental variations [8]; biomechanical function underlies the capability of different species to utilize harder food, which existed in the form of prey skeletal remains regardless of their taxonomic identity or the surrounding environment. In other words, the same selective pressures for masticatory capability would exist independently of environmental changes, as long as larger vertebrate prey are present. Thus, performance measures based on physical principles such as mechanical advantage are suitable rulers to test specific form-function hypotheses in ecomorphological contexts.

Table 2. List of modern North American and East Africa carnivoran species used to construct contour of carnivoran distribution.

North America	n	East Africa	n
Alopex lagopus	10	Acinonyx jubatus	2
Canis latrans	10	Atilax paludinosus	5
Canis lupus	11	Bdeogale crassicauda	5
Gulo gulo	5	Canis aureus	8
Lynx canadensis	5	Caracal caracal	1
Lynx rufus	4	Civettictis civetta	4
Martes pennanti	9	Crocuta crocuta	45
Mephitis mephitis	4	Felis sylvestris	7
Mustela frenata	7	Genetta rubiginosa	13
Neovison vison	2	Herpestes sanguineus	8
Procyon lotor	3	Hyaena hyaena	1
Puma concolor	9	Ichneumia albicauda	2
Taxidea taxus	6	Ictonyx striatus	3
Urocyon cinereoargenteus	1	Lycaon pictus	8
Ursus americanus	10	Mellivora capensis	1
Ursus arctos	9	Nandinia binotata	5
Vulpes vulpes	11	Otocyon megalotis	7
		Panthera leo	24
		Panthera pardus	7
		Proteles cristata	3

For specimen numbers see Tables S3, S4 and Tseng and Wang [10].

Regardless of the simplicity of a two-dimensional framework, the resulting distribution of actual species on the functional landscape shows a remarkable consistency of maintaining MA:SE ratios throughout the region occupied by bone-cracking canids, hyaenids, and the corresponding modern faunas (Fig. 8). Such a pattern indicates an overarching selection for the maintenance of strong skulls and efficient bites across Carnivora, attributes which are principal in both active hunting and passive scavenging behaviors. Despite the outstanding morphological features of the skull and teeth in specialized bone-cracking ecomorphologies, the functional properties of those derived ecomorphs still operated within the bounds of the carnivoran distribution. Again, the notable exception in the modern east African fauna is the cheetah, *Acinonyx jubatus*. Skull shape in the cheetah has fallen off the tall ridges on the functional landscape, and is located in a valley with low MA. The strict requirements for speed may have overridden the base functional demands of mastication, demonstrating that such deviation from the major trend is nevertheless feasible (Fig. 8).

Among metazoan animals, redundancy in body segments has been proposed to enhance evolutionary potential for differentiation in functions [16]. An analogous explanation can be applied to the plesiomorphically homodont dentition of vertebrates, which evolved into highly heterodont dentition in mammals. Carnivores exhibit fine examples of diversified function of heterodont teeth [76]. The shallow, slashing bites of pursuit predators are made using the anterior incisors and canines, and the crushing bites of omnivores are made with the posterior bunodont molars [77]. Such differentiation in dental function is shared by all carnivorans on a more general level, indicating the presence of multiple axes of functional properties for different tooth positions alone. Carnassial

function, the focus of the current study, for example, should be supplemented with study of functional properties in other teeth in order to more fully characterize the potential selective forces that shape craniodental morphology. Such integration requires more complex mathematical formulations of a multi-dimensional problem, of which the current study represents a two-dimensional first step that is easily visualized.

The fact that the functional landscapes predicted movement of canid and hyaenid species through a more or less isoclinal ridge in the D:L and W:L morphospace of MA:SE ratios suggests there are other important factors besides general skull dimensions in the functional evolution of bone-cracking ecomorphologies. Movement on the landscape towards deeper and wider skulls also allows more masticatory musculature to be present in the parietal region, which was not adjusted in the theoretical shapes analyzed here. In addition, the relative proportions of the rostrum and the braincase, and also the positions of the dentition relative to the masticatory muscles both affect mechanical advantage. Such changes require more sophisticated theoretical models, and more fine-tuned variations in FE skull models, which might be generated using algorithms derived from geometric morphometrics analysis [78]. Nevertheless, the usage of simple morphological parameters to create theoretical skull shapes was shown to be informative in discovering potential biomechanical and non-biomechanical constraints on overall skull shape in convergent evolution of adaptive morphologies in carnivorous mammals. Many more studies are needed to explore the begging questions and to improve the completeness of such theoretical frameworks.

In sum, a functional landscape framework constructed from theoretical morphologies showed the presence of functional peaks that are not attained by actual species. The pathways that actual species traversed, however, were nevertheless local optima of relatively high mechanical advantage and moderate skull strain energy. Predictions from the functional landscape are supported by results obtained using models of actual species, showing a clear link between form and function in the evolution of bone-cracking ecomorphologies. The restricted region occupied by a wider sampling of modern carnivorans on the functional landscape indicates higher-level phylogenetic constraint as an explanation for the unoccupied optimal peaks. The combination of theoretical morphology and functional modeling with FEA has been shown to be an informative approach to test adaptive hypotheses regarding morphological convergence, and has implications for applications in broader taxonomic contexts.

Conclusions

An analytical framework combining biomechanical analysis of three-dimensional theoretical morphologies and functional landscapes to evaluate the evolutionary trends in actual lineages represents a novel approach to the study of convergent evolution. Modeling approaches such as finite element analysis not only permits the incorporation of fossil species into biomechanical simulations, but also provides comparative data that inform the robustness of previously hypothesized form-function relationships. Given an asymmetrical understanding of morphological disparity relative to its functional significance, many more such studies are needed, especially for extinct lineages. The case study of convergent, bone-cracking hypercarnivores showed that both biomechanical and broader-scale factors act to shape the observed morphologies of hyaenids and dogs, and that the existing morphological disparity in those two lineages likely represent only local optima in functional morphology. More sophisticated theoretical and functional frameworks will continue to shed light on mechanisms that underlie such prominent examples of evolutionary convergence.

Supporting Information

Table S1 Cranium ratio measurements of fossil canids. Institutional abbreviations: AMNH, American Museum of Natural History, New York; F:AM, Frick Collection, American Museum of Natural History, New York; HMV, Hezheng Paleozoology Museum, Gansu, China; IVPP, Institute of Vertebrate Paleontology and Paleoanthropology, Beijing, China; LACM, Natural History Museum of Los Angeles County, California; MCZ, Museum of Comparative Zoology, Harvard University, Massachusetts; MVZ, Museum of Vertebrate Zoology, University of California, California; PPHM, Plains-Panhandle Museum, Texas; UAMZ, University of Alberta Museum of Zoology, Alberta, Canada; UCMP; University of California Museum of Paleontology, Berkeley, California. Other Abbreviations: D:L, skull depth to length ratio; W:L, skull width to length ratio.

Table S2 Cranium ratio measurements of fossil hyaenids and percrocutids. For abbreviations see Table S1 legend.

Table S3 Cranium ratio measurements of extant North American carnivorans. For abbreviations see Table S1 legend.

Table S4 Cranium ratio measurements of extant east African carnivorans. For abbreviations see Table S1 legend.

Table S5 Theoretical models and their parameters. D:L, skull depth to length ratio; W:L, skull width to length ratio; elements: number of four-noded tetrahedral finite elements in model; SE, skull strain energy (in Joules); adjSE, strain energy adjusted by model volume; Fout, output bite force (in Newtons); MA, mechanical advantage; S.T., solution time required for FEA

(in minutes). Model files are deposited in Dryad (doi:10.5061/dryad.r2b1h).

Table S6 Finite element model parameters of actual fossil and extant carnivorans analyzed in the study. The number of elements were kept as close to 1,000,000 elements as possible for both actual and theoretical models. Three models had lower counts of elements (*P. brunnea*, *P. cristata*, *C. lupus*), but no correlation of MA or SE values to lower element counts was detected. Abbreviations as in Table S5. Model files are deposited in Dryad (doi:10.5061/dryad.r2b1h).

Acknowledgments

I thank my Ph.D. advisor X. Wang and my committee members for their input on previous versions of this paper. J. Liu provided much emotional and intellectual support without which this study could not have been completed. G. Xie helped with CT scanning of *Ictitherium*. B. Van Valkenburgh and the Digimorph project (UT Austin) provided CT images of *Parahyaena* and *Crocuta*. M. Antón, W. Binder, M. Salesa, and D. Stynder contributed collaborative efforts on previously published computer models used in this paper. The editor A. Farke, reviewers, and G. Marroig provided constructive comments on the paper that improved its content and breadth. The curators and staff of the collections from which the materials used in the paper originated: S. Chen, W. He (Hezheng Paleozoology Museum); Z. Qiu (Institute of Vertebrate Paleontology and Paleoanthropology, Chinese Academy of Sciences); J. Indeck (Plains-Panhandle Museum); J. Meng, J. Galkin (American Museum of Natural History). A large portion of this study was completed while the author was a visitor at the University of Alberta; M. Wilson and A. Murray provided access to collections under their care, research space, and also contributed their time in discussion.

Author Contributions

Conceived and designed the experiments: ZJT. Performed the experiments: ZJT. Analyzed the data: ZJT. Contributed reagents/materials/analysis tools: ZJT. Wrote the paper: ZJT.

References

1. Futuyma DJ (1997) Evolutionary Biology. Sunderland: Sinauer Associates, Inc. 763 p.
2. McGhee GR, Jr. (1980) Shell form in the biconvex articulate Brachiopoda: a geometric analysis. Paleobiology 6: 57–76.
3. Wright S (1932) The roles of mutation, inbreeding, crossbreeding and selection in evolution. Proceedings of the XI International Congress of Genetics 1: 356–366.
4. Simpson GG (1944) Tempo and mode in evolution. New York: Columbia University Press. 237 p.
5. Arnold SJ (2003) Performance surfaces and adaptive landscapes. Integrative and Comparative Biology 43: 367–375.
6. Kauffman SA (1995) At Home in the Universe. Oxford: Oxford University Press. 336 p.
7. Thomas RDK, Reif W-E (1993) The skeleton space: a finite set of organic designs. Evolution 47: 341–360.
8. McGhee GR, Jr. (1999) Theoretical Morphology: The concept and its applications. New York: Columbia University Press. 316 p.
9. Stayton CT (2006) Testing hypotheses of convergence with multivariate data: morphological and functional convergence among herbivorous lizards. Evolution 60: 824–841.
10. Tseng ZJ, Wang X (2011) Do convergent ecomorphs evolve through convergent morphological pathways? Cranial shape evolution in fossil hyaenids and borophagine canids (Carnivora, Mammalia). Paleobiology 37: 470–489.
11. Savage RJ (1977) Evolution in carnivorous mammals. Palaeontology 20: 237–271.
12. Greaves WS (1985) The generalized carnivore jaw. Zoological Journal of the Linnean Society 85: 267–274.
13. Goswami A (2006) Cranial modularity shifts during mammalian evolution. The American Naturalist 168: 270–280.
14. Goswami A (2006) Morphological integration in the carnivoran skull. Evolution 60: 169–183.
15. Niklas KJ (1997) The Evolutionary Biology of Plants. Chicago: University of Chicago Press. 470 p.
16. Wainwright PC (2007) Functional versus morphological diversity in macroevolution. Annual Review of Ecology, Evolution, and Systematics 38: 381–401.
17. Raup DM (1966) Geometric analysis of shell coiling: general problems. Journal of Palaeontology 40: 1178–1190.
18. Raup DM (1967) Geometric analysis of shell coiling: coiling in ammonoids. Journal of Palaeontology 41: 43–65.
19. Figueirido B, MacLeod N, Krieger J, Di Renzi M, Perez-Claros JA, et al. (2011) Constraint and adaptation in the evolution of carnivoran skull shape. Paleobiology 37: 490–518.
20. O'Higgins P, Cobb SN, Fitton LC, Groning F, Phillips R, et al. (2011) Combining geometric morphometrics and functional simulation: an emerging toolkit for virtual functional analyses. Journal of Anatomy 218: 3–15.
21. Rayfield EJ (2007) Finite element analysis and understanding the biomechanics and evolution of living and fossil organisms. Annual Review of Earth and Planetary Science 35: 541–576.
22. Ross CF (2005) Finite element analysis in vertebrate biomechanics. The Anatomical Record Part A 283A: 253–258.
23. Clough RW (1960) The Finite Element Method in Plane Stress Analysis. Proceedings of 2nd ASCE Conference on Electronic Computation. Pittsburgh, PA.
24. Rayfield EJ, Norman DB, Jorner CC, Horner JR, Smith PM, et al. (2001) Cranial design and function in a large theropod dinosaur. Nature 409: 1033–1037.
25. McHenry C, Wroe S, Clausen PD, Moreno K, Cunningham E (2007) Supermodeled sabercat, predatory behavior in *Smilodon fatalis* revealed by high-resolution 3D computer simulation. Proceedings of the National Academy of Sciences 104: 16010–16015.

26. Dumont ER, Piccirillo J, Grosse IR (2005) Finite-element analysis of biting behavior and bone stress in the facial skeletons of bats. The Anatomical Record Part A 283A: 319–330.

27. Wroe S, Clausen PD, McHenry C, Moreno K, Cunningham E (2007) Computer simulation of feeding behaviour in the thylacine and dingo as a novel test for convergence and niche overlap. Proceedings of the Royal Society B: Biological Sciences 274: 2819–2828.

28. Slater GJ, Dumont ER, Van Valkenburgh B (2009) Implications of predatory specialization for cranial form and function in canids. Journal of Zoology 278: 181–188.

29. Slater GJ, Figueirido B, Louis L, Yang P, Van Valkenburgh B (2010) Biomechanical consequences of rapid evolution in the polar bear lineage. PLoS ONE 5: e13870. doi:13810.11371/journal.pone.0013870.

30. Davis JL, Santana SE, Dumont ER, Grosse I (2010) Predicting bite force in mammals: two-dimensional versus three-dimensional lever models. Journal of Experimental Biology 213: 1844–1851.

31. Ross CF, Berthaume MA, Dechow PC, Iriarte-Diaz J, Porro LB, et al. (2010) In vivo bone strain and finite-element modeling of the craniofacial haft in catarrhine primates. Journal of Anatomy 218: 112–141.

32. Dumont ER, Grosse I, Slater GJ (2009) Requirements for comparing the performance of finite element models of biological structures. Journal of Theoretical Biology 256: 96–103.

33. Van Valkenburgh B (1999) Major patterns in the history of carnivorous mammals. Annual Review of Earth and Planetary Science 27: 463–493.

34. Van Valkenburgh B (2007) Déjà vu: the evolution of feeding morphologies in the Carnivora. Integrative and Comparative Biology 47: 147–163.

35. Werdelin L (1996) Chapter 17. Carnivoran ecomorphology: a phylogenetic perspective. In: Gittleman JL, editor. Carnivore behavior, ecology, and evolution. New York: Cornell University Press. pp. 582–624.

36. Van Valkenburgh B (1988) Trophic diversity in past and present guilds of large predatory mammals. Paleobiology 14: 155–173.

37. Werdelin L (1989) Constraint and adaptation in the bone-cracking canid Osteoborus (Mammalia: Canidae). Paleobiology 15: 387–401.

38. Kruuk H (1972) The spotted hyena: a study of predation and social behavior; Shaller GB, editor. Chicago: The University of Chicago Press. 335 p.

39. Binder WJ, Van Valkenburgh B (2000) Development of bite strength and feeding behaviour in juvenile spotted hyenas (Crocuta crocuta). Journal of the Zoological Society of London 252: 273–283.

40. Werdelin L, Solounias N (1991) The Hyaenidae: taxonomy, systematics and evolution. Fossils and Strata 30: 1–104.

41. Wang X, Tedford RH, Taylor BE (1999) Phylogenetic systematics of the Borophaginae (Carnivora: Canidae). Bulletin of the American Museum of Natural History 243: 1–391.

42. Qiu Z, Xie J, Yan D (1988) Discovery of the skull of Dinocrocuta gigantea. Vertebrata PalAsiatica 26: 128–138.

43. Tseng ZJ (2009) Cranial function in a late Miocene Dinocrocuta gigantea (Mammalia: Carnivora) revealed by comparative finite element analysis. Biological Journal of the Linnean Society 96: 51–67.

44. Tseng ZJ, Binder WJ (2010) Mandibular biomechanics of Crocuta crocuta, Canis lupus, and the late Miocene Dinocrocuta gigantea (Carnivora, Mammalia). Zoological Journal of the Linnean Society 158: 683–696.

45. Nowak RM (1999) Walker's carnivores of the world. Baltimore: The Johns Hopkins University Press. 313 p.

46. Chen GF, Schmidt-Kittler N (1983) The deciduous dentition of Percrocuta Kretzoi and the diphyletic origin of the hyaenas (Carnivora, Mammalia). Palaontologische Zeitschrift 57: 159–169.

47. Turner A, Antón M, Werdelin L (2008) Taxonomy and evolutionary patterns in the fossil Hyaenidae of Europe. Geobios 41: 677–687.

48. Wang X (1994) Phylogenetic systematics of the Hesperocyoninae (Carnivora: Canidae). Bulletin of the American Museum of Natural History 221: 1–207.

49. Tedford RH, Wang X, Taylor BE (2009) Phylogenetic systematics of the North American fossil Caninae (Carnivora: Canidae). Bulletin of the American Museum of Natural History 325: 1–218.

50. Wang X, Tedford RH, Antón M (2008) The Dog Family, Canidae, and Their Evolutionary History. New York: Columbia University Press. 219 p.

51. Van Valkenburgh B, Sacco T, Wang X, editors (2003) Pack hunting in Miocene borophagine dogs: evidence from craniodental morphology and body size. New York: American Museum of Natural History. 147–162 p.

52. Tseng ZJ, Wang X (2010) Cranial functional morphology of fossil dogs and adaptation for durophagy in Borophagus and Epicyon (Carnivora, Mammalia). Journal of Morphology 271: 1386–1398.

53. Holliday JA, Steppan SJ (2004) Evolution of hypercarnivory: the effect of specialization on morphological and taxonomic diversity. Paleobiology 30: 108–128.

54. Parr WCH, Wroe S, Chamoli U, Richards HS, McCurry MR, et al. (2012) Toward integration of geometric morphometrics and computational biomechanics: New methods for 3D virtual reconstruction and quantitative analysis of Finite Element Models. Journal of Theoretical Biology 301: 1–14.

55. Tseng ZJ, Antón M, Salesa MJ (2011) The evolution of the bone-cracking model in carnivorans: Cranial functional morphology of the Plio-Pleistocene cursorial hyaenid Chasmaporthetes lunensis (Mammalia: Carnivora). Paleobiology 37: 140–156.

56. Figueirido B, Tseng ZJ, Martín-Serra A (2013) Skull shape evolution in durophagous carnivorans. Evolution doi: 10.1111/evo.12059.

57. Rensberger JM, Stefen C (2006) Functional differentiations of the microstructure in the upper carnassial enamel of the spotted hyena. Palaeontographica Abt A 278: 149–162.

58. Joeckel RM (1998) Unique frontal sinuses in fossil and living Hyaenidae (Mammalian, Carnivora): description and interpretation. Journal of Vertebrate Paleontology 18: 627–639.

59. Tanner JB, Zelditch M, Lundrigan BL, Holekamp KE (2010) Ontogenetic change in skull morphology and mechanical advantage in the spotted hyena (Crocuta crocuta). Journal of Morphology 271: 353–365.

60. Wroe S, McHenry C, Thomason JJ (2005) Bite club: comparative bite force in big biting mammals and the prediction of predatory behaviour in fossil taxa. Proceedings of the Royal Society of London, Series B 272: 619–625.

61. Meers MB (2002) Maximum bite force and prey size of Tyrannosaurus rex and their relationships to the inference of feeding behavior. Historical Biology 16: 1–12.

62. Binder WJ, Thompson EN, Van Valkenburgh B (2002) Temporal variation in tooth fracture among Rancho La Brea dire wolves. Journal of Vertebrate Paleontology 22: 423–428.

63. Wroe S, Milne N (2007) Convergence and remarkably consistent constraint in the evolution of carnivore skull shape. Evolution 61: 1251–1260.

64. Meloro C, Raia P, Piras P, Barbera C, O'Higgins P (2008) The shape of the mandibular corpus in large fissiped carnivores: allometry, function and phylogeny. Zoological Journal of Linnean Society 154: 832–845.

65. Tseng ZJ, Stynder D (2011) Mosaic functionality in a transitional ecomorphology: skull biomechanics in stem Hyaeninae compared to modern South African carnivorans. Biological Journal of the Linnean Society 102: 540–559.

66. Turnbull WD (1970) Mammalian masticatory apparatus. Fieldiana: Geology 18: 149–356.

67. Davis D (1955) Masticatory apparatus in the spectacled bear Tremarctos ornatus. Fieldiana: Zoology 37: 25–46.

68. Dessem D (1989) Interactions between jaw-muscle recruitment and jaw-joint forces in Canis familiaris. Journal of Anatomy 164: 101–121.

69. Grosse I, Dumont ER, Coletta C, Tolleson A (2007) Techniques for modeling muscle-induced forces in finite element models of skeletal structures. The Anatomical Record 290: 1069–1088.

70. Bourke J, Wroe S, Moreno K, McHenry C, Clausen PD (2008) Effects of gape and tooth position on bite force and skull stress in the dingo (Canis lupus dingo) using a 3-dimensional finite element approach. PLoS ONE 3: e2200.

71. Ellis JL, Thomason JJ, Kebreab E, France J (2008) Calibration of estimated biting forces in domestic canids: comparison of post-mortem and in vivo measurements. Journal of Anatomy 212: 769–780.

72. Antón M, Turner A, Salesa MJ, Morales J (2006) A complete skull of Chasmaporthetes lunensis (Carnivora, Hyaenidae) from the Spanish Pliocene site of La Puebla de Valverde (Teruel). Estudios Geológicos 62: 375–388.

73. Nalla RK, Kinney JH, Ritchie RO (2003) Mechanistic failure criteria for the failure of human cortical bone. Nature Materials 2: 164–168.

74. Van Valkenburgh B, Wang X, Damuth J (2004) Cope's Rule, hypercarnivory, and extinction in North American canids. Science 306: 101–104.

75. Strathman RR (1978) Progressive vacating of adaptive types during the Phanerozoic. Evolution 32: 907–914.

76. Van Valkenburgh B (1989) Carnivore dental adaptations and diet: a study of trophic diversity within guilds In: Gittleman JL, editor. Carnivore behavior, ecology, and evolution. New York: Cornell University Press. pp. 410–436.

77. Van Valkenburgh B (1996) Feeding behavior in free-ranging, large African carnivores. Journal of Mammalogy 77: 240–254.

78. Stayton CT (2009) Application of thin-plate spline transformations to finite element models, or, how to turn a bog turtle into a spotted turtle to analyze both. Evolution 63: 1348–1355.

10

On the Relationship between the Macroevolutionary Trajectories of Morphological Integration and Morphological Disparity

Sylvain Gerber*

Department of Biology and Biochemistry, University of Bath, Bath, England

Abstract

How does the organization of phenotypes relate to their propensity to vary? How do evolutionary changes in this organization affect large-scale phenotypic evolution? Over the last decade, studies of morphological integration and modularity have renewed our understanding of the organizational and variational properties of complex phenotypes. Much effort has been made to unravel the connections among the genetic, developmental, and functional contexts leading to differential integration among morphological traits and individuation of variational modules. Yet, their macroevolutionary consequences on the dynamics of morphological disparity–the large-scale variety of organismal designs–are still largely unknown. Here, I investigate the relationship between morphological integration and morphological disparity throughout the entire evolutionary history of crinoids (echinoderms). Quantitative analyses of interspecific patterns of variation and covariation among characters describing the stem, cup, arm, and tegmen of the crinoid body do not show any significant concordance between the temporal trajectories of disparity and overall integration. Nevertheless, the results reveal marked differences in the patterns of integration for Palaeozoic and post-Palaeozoic crinoids. Post-Palaeozoic crinoids have a higher degree of integration and occupy a different region of the space of integration patterns, corresponding to more heterogeneously structured matrices of correlation among traits. Particularly, increased covariation is observed between subsets of characters from the dorsal cup and from the arms. These analyses show that morphological disparity is not dependent on the overall degree of evolutionary integration but rather on the way integration is distributed among traits. Hence, temporal changes in disparity dynamics are likely constrained by reorganizations of the modularity of the crinoid morphology and not by changes in the variability of individual traits. The differences in integration patterns explain the more stereotyped morphologies of post-Palaeozoic crinoids and, from a broader macroevolutionary perspective, call for a greater attention to the distributional heterogeneities of constraints in morphospace.

Editor: Richard J. Butler, Ludwig-Maximilians-Universität München, Germany

Funding: This research was supported by a Leverhulme Trust research grant (Grant F/00 351/Z). The funders had no role in study design, data collection and analysis, decision to publish, or preparation of the manuscript.

Competing Interests: The author has declared that no competing interests exist.

* E-mail: s.gerber@bath.ac.uk

Introduction

Heterogeneous patterning of morphospaces (quantitative state space representations of taxa relative to an underlying set of possibilities for morphological variation) has been frequently documented in clade-wide temporal studies, and is now widely acknowledged as a prominent feature of phenotypic macroevolution [1–7]. These heterogeneities are expressed in the spread and spacing of taxa in morphospace, as revealed by statistical measures of morphological disparity [8,9]. Morphospace and disparity patterns may variously be the expression of functional factors, developmental constraints, historical contingency and/or stochasticity influencing the waxing and waning of taxa over the evolutionary dynamics of clades [10].

Although morphological disparity analyses have been undertaken primarily as a means to globally characterize patterns of stability and change of realized morphospace during the long-term history of clades (the magnitude of disparity), disparity arguably also has an underlying, non-trivial structure. This structure potentially reflects aspects of the hierarchical organization of

phenotypes into quasi-independent units of evolutionary transformation, i.e., evolutionary modules [11–14]. This near-decomposability of morphological phenotypes, as can be observed or inferred when quantifying morphological changes within evolving lineages, underlines patterns of differential integration within and among suites of phenotypic traits influenced by pleiotropic effects, developmental pathways and functional factors [15,16].

In a macroevolutionary context, how phenotypic integration and modularity may actually be related to morphological disparity is an important but still largely unexplored question [17,18]. For instance, might changes in morphological disparity characteristically result from the interplay of parcellation and integration of phenotypic organization (decrease or loss of correlation within primarily integrated set of traits leading to increased modularity and vice-versa)? Or might they result instead from intrinsic changes in the variational potential of a relatively constant number of modules? When are integration and disparity likely to correlate? Analogously, might disparity be operationally used as a meaningful proxy for modularity?

Here, I address some of these questions by quantifying the temporal trajectories of clade-wide measures of morphological integration in the Class Crinoidea (Echinodermata) over the Phanerozoic. The evolutionary history of crinoids is marked by two distinct radiations, occurring firstly in the early Palaeozoic (mainly Ordovician, ~500–435 Myr ago) and secondly in the Triassic-Early Jurassic, as part of the recovery from the end-Permian mass extinction (~251 Myr ago). Both radiations are characterized by rapid morphological diversifications at relatively low taxonomic diversity. Nevertheless, Foote [19] showed that post-Palaeozoic crinoids were morphologically less disparate than their Palaeozoic counterparts and also occupied a different, non-overlapping region of the morphospace. This distinct and more limited array of morphological designs perhaps suggests a different set of ecological opportunities [20] or internal constraints on the evolvability of crinoids. This case study, spanning more than 400 millions years of morphological evolution, enables one to portray macroevolutionary patterns of morphological integration and to contrast them with disparity profiles.

Given the temporal scale, the taxonomic level and the degree of morphological resolution, the temporal changes in the overall degree of integration do not focus on patterns comparable to those that are generally described at low taxonomic levels and concerned with small-scale aspects of organismal organization and variation. Rather, the evolutionary dynamics of integration quantified here is more closely related to the dimensionality of the crinoid morphospace *itself*, reflecting the highest levels of the hierarchical embedding of evolutionary modules within the crinoid body plan.

Materials and Methods

The morphological dataset used in the present study has been compiled and regularly augmented by Foote [19,21–25]. Its quality and adequacy for documenting evolutionary patterns in crinoids have been evaluated through numerous sensitivity analyses testing for potential biases induced by character selection and weighting, missing data, taxon sampling protocols, unequal time interval duration, morphospace dimensionality and disparity measures [19,21]. The use of a different character-coding scheme applied to early Palaeozoic crinoids has also been tested and it provided results consistent with previous accounts [26]. The dataset includes 1032 species representing one species per genus per time interval. Each species is described by 90 discrete morphological characters offering a comprehensive coverage of the stem, cup, arm, and tegmen parts of the crinoid body (14, 40, 28, and 8 characters, respectively). See Foote [19] for further details on character definition and coding (data available in Appendix S1).

I followed two complementary approaches in order to allow the use of different measures of morphological disparity and integration. The first approach treats discrete characters directly and is hereafter referred to as the discrete character space approach; the second approach consists in extracting a dissimilarity matrix from the discrete character space using the mean character difference as the measure of morphological dissimilarity between two species [27] and then carrying out a principal coordinate analysis (PCoA) of this dissimilarity matrix. The first ten principal coordinates provide a fair representation of among-species dissimilarities and define the principal coordinate space explored in subsequent analyses.

With these two approaches, morphological disparity is measured as the mean pairwise dissimilarity and as the sum of univariate variances respectively, which are both standard indices

of disparity relatively insensitive to sample size [28]. For discrete characters, I measured integration as the relative mean mutual compatibility. Two characters are said to be compatible if their state combinations do not necessarily imply homoplasy (e.g., for binary characters, not all four possible character state combinations 00, 01, 10 and 11 are found, so they can be mapped onto a tree without requiring convergence or reversal) [29]. In phylogenetics, compatibility analysis can be used to avoid overweighted correlated suites when selecting characters. For each time interval, I constructed a matrix of mutual compatibility, where the mutual compatibility of two characters i and j is defined as the total number of characters compatible with both i and j [30]. I then calculated the mean mutual compatibility and divided it by the maximum possible number of mutual compatibilities (i.e., total number of characters minus two). Hence, this measure of integration ranges from zero to one, respectively corresponding to low and high levels of correlation among characters.

For the continuous variables obtained via PCoA, I used the relative standard deviation of the eigenvalues of the correlation matrix proposed by Pavlicev et al [31] as a measure of morphological integration. This index also ranges from zero to one. If morphological integration is important, only a few dimensions are necessary to summarize most of the observed variation and the standard deviation of eigenvalues will be high because of the marked differences among them. Conversely, if morphological traits are weakly integrated, the standard deviation will be low because all eigenvalues will be roughly similar. These integration indices are therefore unrelated to the magnitude of disparity but instead describe its structure, that is, the dimensionality of the distribution of taxa in the morphospace.

The temporal partitioning of the morphospace into successive time intervals often leads to the extraction of matrices with more variables than individuals from the total morphological dataset. This "small n, large p" problem makes the sample correlation matrix an unreliable estimator of the population correlation matrix. Indeed, when the number of individuals becomes too small compared to the number of variables, the sample correlation matrix loses its full-rank and positive definiteness, thereby biasing the distribution of its eigenvalues and, in the present context, the measures of morphological integration. In addition, it has been shown analytically that the lower bound of the range of the standard deviation of eigenvalues for finite sample correlation matrices varies as $(1/n)^{1/2}$ [32]. If not accounted for, this sample size effect can thus mislead the interpretation of temporal changes in integration, because the range of the index will vary as a function of taxonomic diversity. In order to circumvent these problems, I derived estimates of correlation matrices from a shrinkage procedure using the R package *corpcor* [33]. This approach allows one to obtain accurate, well-conditioned, and positive definite estimates of correlation matrices even for small sample sizes [34]. Based on simulations of random matrices of uncorrelated variables, I found that it also maintains the lower bound of Pavlicev et al.'s index close to zero down to sample sizes of about 15. Therefore, I chose to discard six time intervals with sample sizes lower than 15 to avoid any spurious estimates of integration: Early Ordovician, Late Permian, Triassic (two intervals) and Cenozoic (two intervals). The remaining time intervals have an average p/n ratio of 2.73 with a maximum of 6 (second time interval of the Cretaceous).

Comparison between morphological disparity and integration cannot be made directly because of the potential effect of trends and serial correlation inherent to most time series. To circumvent these effects, I used the generalized differencing approach [35], which consists of first detrending the time series by regressing their

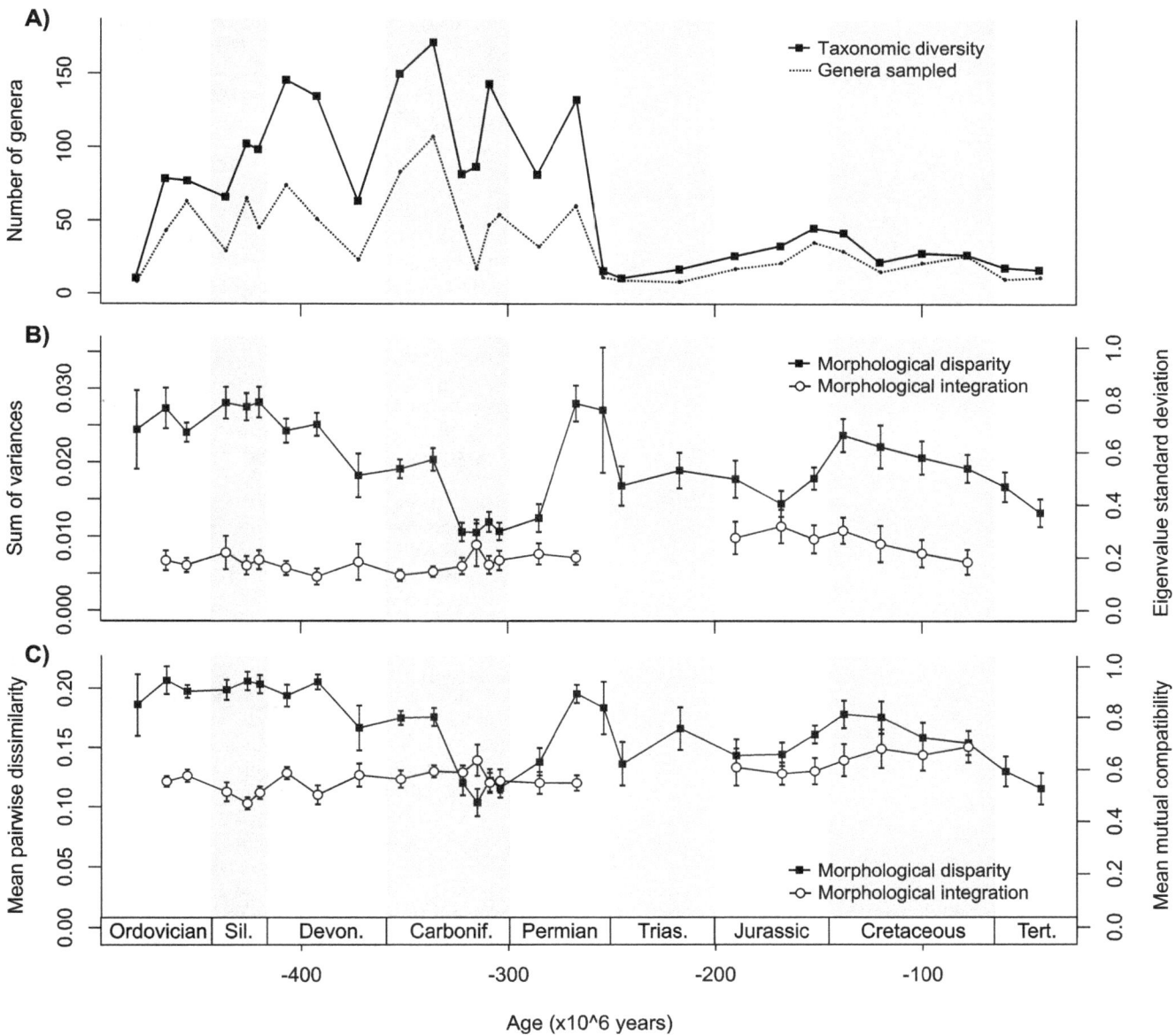

Figure 1. The temporal trajectories of taxonomic diversity, morphological disparity and morphological integration of Phanerozoic crinoids. (A) Number of genera known and number of species sampled per genus per stratigraphic interval. (B) Disparity measured as the sum of univariate variances; integration measured as the relative standard deviation of the eigenvalues of the correlation matrix. (C) Disparity measured as the mean character dissimilarity; integration measured as mean mutual compatibility. Error bars are bootstrapped standard errors. Because low sample sizes prevent from deriving reliable estimates of correlation matrices (see text), integration values are not presented for Early Ordovician, Late Permian, Triassic, and Cenozoic data. Whether based on the analysis of discrete or continuous variables, variations in the overall degree of integration do not appear to be associated with concomitant changes in disparity.

values against numerical time and then correcting for serial correlation by taking first differences (differences between adjacent values) modulated by the serial correlation coefficient (lag-1 coefficient). Correlation analyses between integration and disparity are performed on their generalized differences.

Finally, to trace the evolution of patterns of integration in greater details, I also used the metric recently proposed by Mitteroecker and Bookstein [36], the square root of the summed squared log relative eigenvalues, which provides a measure of distance between two covariance (or correlation) matrices. I computed all pairwise distances between the shrinkage estimates of correlation matrices corresponding to each time interval and then

performed a principal coordinate analysis of the distance matrix obtained. This method enables to visualize the temporal trajectory of patterns of integration in the space of correlation matrices.

Assessing patterns of correlation and compatibility among characters can be hindered by the fact that species are not independent entities but parts of a hierarchically structured phylogeny resulting from branching evolution [37]. Unfortunately, in the absence of detailed phylogenetic hypothesis, it is not possible to correct for the non-independence of species by applying phylogenetic comparative methods. Nevertheless, in order to evaluate the potential effect of phylogenetic autocorrelation on estimates of integration, I applied the permutation-compatibility

A) All characters

B) Taxonomically non-significant characters

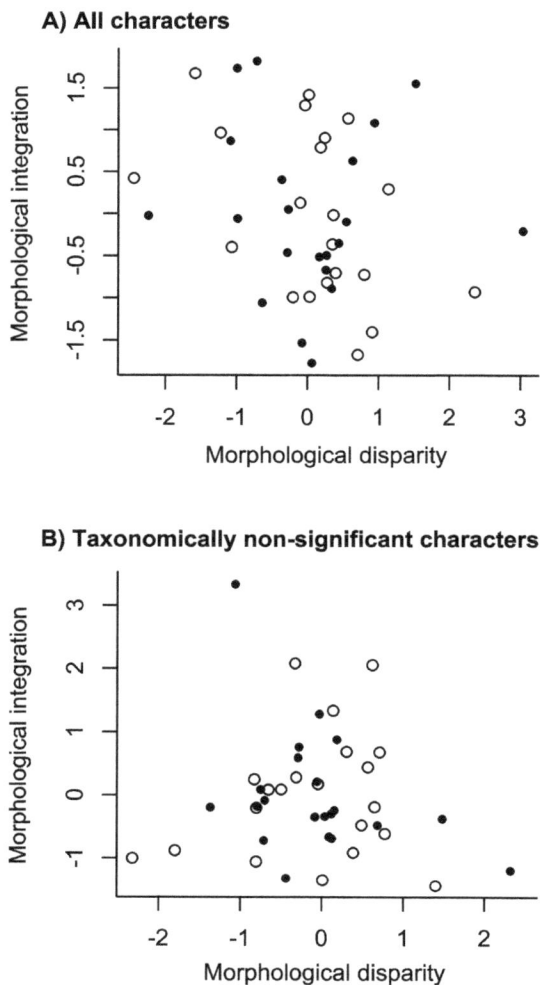

Figure 2. Correlation between temporal changes in disparity and integration. Spearmann's rank correlation between level of morphological disparity and degree of morphological integration. (A) Generalized differences of morphological integration versus disparity for the PCoA-based approach (black circles; $r = -0.118$, $P = 0.609$) and the discrete character approach (open circles; $r = -0.449$, $P = 0.042$) when all characters are considered. (B) Generalized differences of morphological integration versus disparity for the PCoA-based approach (black circles; $r = -0.340$, $P = 0.131$) and the discrete character approach (open circles; $r = 0.118$, $P = 0.609$) when only taxonomically non-significant characters are considered (see text). In general, the amount of morphological disparity displayed by crinoids is not significantly correlated with the overall degree of integration among morphological traits.

test for hierarchic structure in discrete character matrix [38]. This test compares the observed number of compatible character pairs with the null distribution obtained by permuting the original character matrix. If the observed compatibility is within the range of permuted matrices, then there is no (or little) phylogenetic signal in the data, and the observed patterns of integration are more likely to reflect secondary signals of correlated character changes.

Even though characters provide ecological and functional information about crinoid morphology, some also enable taxonomic distinctions within and among higher taxa [21]. To further ensure the robustness of the conclusions, all the above analyses have been run for the total morphological dataset, but also on a subset of 27 characters that are not taxonomically relevant (i.e.,

not used for diagnosing subclasses and orders; characters 1–2, 4–15, 21, 30, 47–48, 55, 57, 60, 62, 64, 69–71, 77), and should therefore not bear a strong phylogenetic signal. All statistical analyses were programmed and carried out in R (functions available in Appendix S2).

Results

Figure 1 provides the curves of taxonomic diversity, morphological disparity and morphological integration for crinoids over the Phanerozoic, so as to examine the relative behaviours of these metrics, each emphasizing different aspects of biodiversity dynamics. Two complementary approaches are used in order to draw estimates of disparity and integration from both continuous and discrete character variables (Fig. 1A and 1B). As reported previously [19], morphological disparity shows marked variations over the period studied, most of them being decoupled from the rises and drops in taxonomic diversity. Contrastingly, indices of morphological integration measured as the relative standard deviation of eigenvalues and as the relative mean mutual compatibility appear to be fairly stable. Most increases and decreases in the overall degree of integration are not significant and do not appear associated with similar changes in the level of disparity. However, post-Palaeozoic crinoids on average display a higher degree of morphological integration than Palaeozoic crinoids for both measures of integration ($P < 0.01$ in both cases with a Mann-Whitney U test).

I further investigate the relationships between degree of correlation among traits and level of morphological variety by calculating the correlation between the generalized differences of integration and disparity estimates (Fig. 2). Only the correlation between the mean mutual compatibility and the mean pairwise dissimilarity is significant (Spearmann's $r = -0.449$, $P = 0.042$; Fig. 2A). Nevertheless, a permutation-compatibility test [38] detects a significant hierarchic structure in the dataset (as a whole and within individual time intervals), suggesting a phylogenetic signal potentially biasing estimates of integration (phylogenetic autocorrelation). I reran the same analysis on a subset of 27 characters of putatively low phylogenetic significance ([21] and see methods) and for which the permutation-compatibility test does not reveal significant underlying phylogenetic signal (Fig. 2B). Whether based on continuous or discrete character approaches, no significant correlation is observed between changes in disparity and integration (time series available in Appendix S3).

Finally, I computed the pairwise distances among the trait correlation matrices associated with each geologic time interval so as to ordinate and visualize patterns of integration within the space of correlation matrices (Fig. 3). The temporal trajectory of correlation matrices follows a non-random pathway in this space, reflecting progressive but non-regular changes in patterns of integration across the Phanerozoic. The most striking feature of the distribution of these integration patterns is the clear separation of Palaeozoic and post-Palaeozoic patterns along the first principal coordinate of the space. The location of most Palaeozoic patterns in the vicinity of the identity matrix is indicative of homogeneously structured correlation matrices (i.e., all off-diagonal elements are of comparable magnitude), whereas post-Palaeozoic matrices tend to be more heterogeneously structured (unequal values of off-diagonal elements delineating blocks of variables; Fig. 4). The average distance among post-Palaeozoic patterns is significantly greater than that of Palaeozoic patterns ($P < 0.001$; Mann-Whitney U test) despite the roughly equivalent duration separating successive intervals. This suggests greater magnitudes of transition between successive patterns of integration in post-Palaeozoic

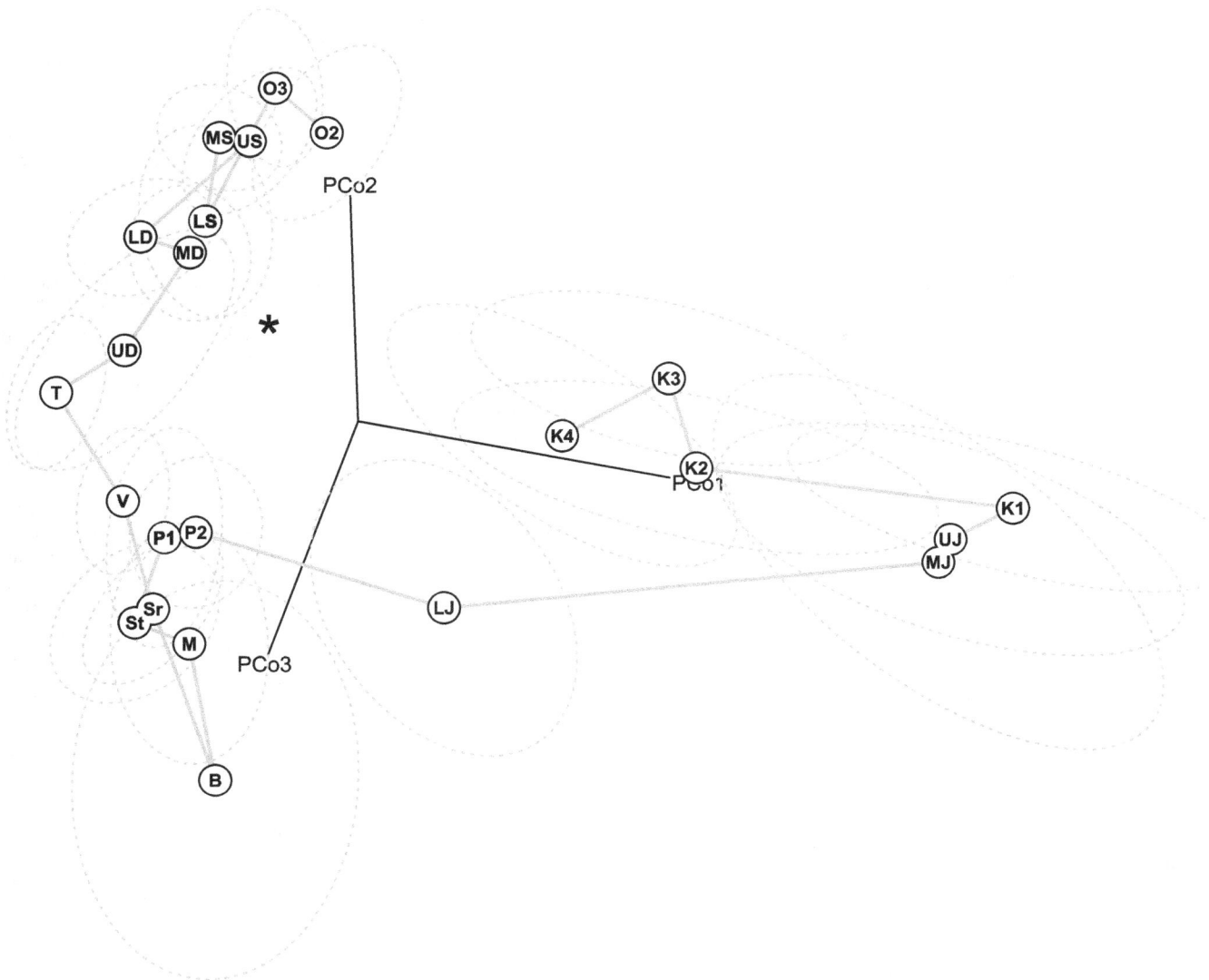

Figure 3. The temporal trajectory of integration patterns of Phanerozoic crinoids. The plot shows the first three principal coordinates of the space of correlation matrices. Each point corresponds to the correlation matrix of crinoids within a given geologic time interval (The correlation between pairwise Euclidean distances in the space of the first three principal coordinates and the actual distances between correlation matrices is 0.85). The grey line represents the temporal trajectory of correlation matrices from the Ordovician (O2) to the end of the Cretaceous (K4), and the asterisk gives the location of the identity matrix (i.e., a matrix with no integration among traits). Dotted lines are 68% confidence ellipses based on bootstrap resampling. Labels: O2 = Llanvirnian to lower Caradocian, O3 = remainder of Ordovician, LS = Lower Silurian, MS = Middle Silurian, US = Upper Silurian, LD = Lower Devonian, MD = Middle Devonian, UD = Upper Devonian, T = Tournaisian (Carboniferous, Mississippian), Sr = Serpukhovian (Carboniferous, Mississippian), B = Bashkirian (Carboniferous, Pennsylvanian), M = Moscovian (Carboniferous, Pennsylvanian), St = Stephanian (Carboniferous, Pennsylvanian), P1 = Asselian-Sakmarian (Permian), P2 = Artinskian-Kungurian (Permian), LJ = Lower Jurassic, MJ = Middle Jurassic, UJ = Upper Jurassic, K1 = Neocomian (Cretaceous), K2 = Barremian-Aptian (Cretaceous), K3 = Albian-Turonian (Cretaceous), K4 = Senonian (Cretaceous). The first principal coordinate separates Palaeozoic from post-Palaeozoic forms. The distribution of most Palaeozoic correlation matrices near the identity matrix emphasizes their homogeneous structure (roughly similar pairwise correlation among traits), whereas post-Palaeozoic correlation matrices display individuated blocks of correlated traits.

crinoids. Similar results are obtained when the space of correlation matrices is built from the set of taxonomically non-significant characters or from a drastic culling of data preserving only characters with less than five percent of missing data.

Discussion

The present work examined large-scale patterns of evolutionary integration among morphological traits in crinoids and tested if and how changes in these patterns were associated with

concomitant changes in the level of morphological disparity expressed by the clade. The analyses reveal relatively stable measures of the overall degree of integration despite marked temporal variations in taxonomic diversity and morphological disparity. Correlation analyses accounting for and limiting the effect of phylogenetic autocorrelation did not detect a significant one-to-one relationship between integration and disparity. Nevertheless, significant differences in the degree and pattern of integration are observed between Palaeozoic and post-Palaeozoic crinoids. Post-Palaeozoic crinoids have a higher overall degree of

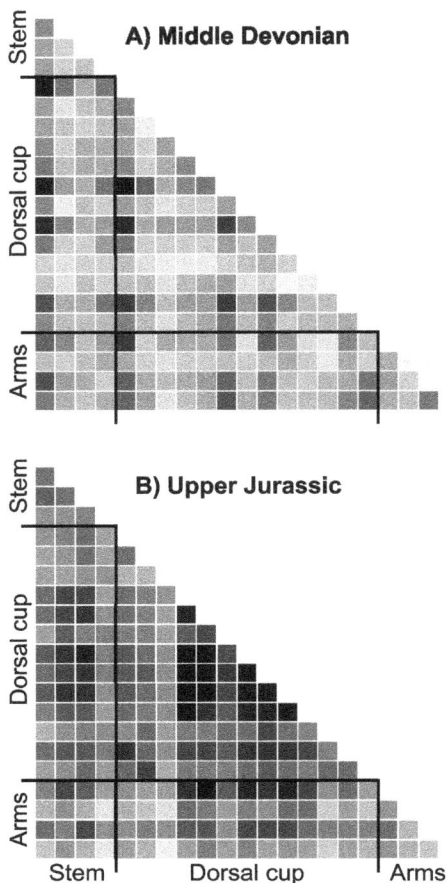

Figure 4. Matrices of mutual compatibility for Palaeozoic and post-Palaeozoic crinoids. These two matrices exemplify the differences in patterns of integration between (A) Palaeozoic (Middle Devonian, MD) and (B) post-Palaeozoic crinoids (Upper-Jurassic, UJ). The choice for these two time intervals has been driven by their location in the space of correlation matrix (separation along PCo1; see Figure 3) and the comparability of their sets of applicable characters (number and distribution over the whole character matrix). The gray-scale correlates with the strength of mutual compatibility (~correlation): the darker the gray, the higher the compatibility. The comparison of these two matrices shows the overall stronger integration among characters within the post-Palaeozoic matrix and its heterogeneous structure with larger blocks of compatible characters (e.g., stem and dorsal cup characters, arm and dorsal cup characters).

integration and occupy a different region of the space of correlation matrices. Their location indicates heterogeneously structured correlation matrices with larger blocks of correlated traits, which could explain the less disparate and more stereotyped post-Palaeozoic morphologies reported previously [19].

Hence, if the amount of morphological disparity does not appear to be conditional upon any given degree of overall integration, the results suggest that disparity is related to the modular nature of the correlation matrix, that is, to its pattern of organization into evolutionarily quasi-independent blocks of integrated traits. With regards to the two competing hypotheses presented in the introduction, the temporal trajectory of morphological disparity in crinoids would then be tied to changes in the pattern of correlation among traits rather than to changes in their individual variability.

In a study comparing the disparity levels of ecological and non-ecological (developmental) characters before and after mass extinctions, Ciampaglio [20] concluded that crinoid disparity patterns were mainly driven by the increasing structuring of ecological guilds rather than by developmental constraints. Nevertheless, his model of developmental constraints was focusing on upper limits for the level of disparity and not on biases in the spatial deployment of taxa in morphospace. Yet, developmental integration of traits and their dedication to specific functions generate evolutionary patterns of association and covariation among them, which shape the distribution of taxa in morphospace and the potential for evolutionary change [39]. The propensity of modular phenotypes to vary depends on the match between their developmental and functional modularity (i.e., the alignment of the genotype-phenotype map with the phenotype-fitness map; [40]). Specifically, if the pattern of developmental integration among traits coincides with their association to perform adaptive functions, evolvability is enhanced. Post-Palaeozoic crinoids derived from one family of Palaeozoic cladids [41] and their evolution has been characterized by an increased frequency of traits required for passive and active motility [42]. This has been interpreted as a response to increased interactions with benthic predators such as cidaroid sea-urchins [42,43]. It is possible that these changes in predatory pressures may be responsible for the redeployment of traits into novel or modified functional complexes (increased aggregation of traits here). Then, the differences in evolutionary modularity documented between Palaeozoic and post-Palaeozoic crinoids potentially indicate a modification of the match between developmental and functional integration and the restricted range for trait covariation could explain the lower propensity to vary of post-Palaeozoic crinoids. Nevertheless, it is important to stress that current statistical indices of disparity are measures of observed macroevolutionary variation and therefore do not necessarily reflect the full potential to vary.

In summary, morphological disparity should be seen as more than a mere summary statistic of the amount of morphospace occupied. On the one hand, disparity reflects the building-up of the genealogical hierarchy over long timescales, with for instance the changing taxonomic composition of clades and the signature of mass and background extinctions. On the other hand, the behaviour of morphological disparity in face of these macroevolutionary phenomena is tied to the dimensionality of phenotypic variation constrained by the apportionment of variability among units of evolutionary transformation. Hence, the distribution and dynamics of taxa in morphospace should provide insights into the architecture of phenotypes and the constraints on their evolvability. This challenges the frequent conceptualization of morphospace as a homogeneous state-space. Such an interpretation of morphospace is unlikely to hold at the level of macroevolutionary phenotypic variation where development imposes a strong structure on the evolutionary accessibility of phenotypes (e.g. [44–46]). To different locations in morphospace are attached different sets of constraints and opportunities for phenotypic change in terms of probability, magnitude and directionality of evolutionary transitions. The morphospace is said to be structured [47], that is, patterns of phenotypic change are constrained by the location in the morphospace. This can be critical when comparing and interpreting the evolutionary dynamics of lineages originating in different regions of the morphospace. It does not mean that natural selection does not play any role at this scale, but rather that selection plays with a non-randomly distributed set of developmentally possible options in the vicinity of the evolving lineage [1,48]. Further work is required to assess the relative role of selective pressures and developmental constraints in shaping patterns of diversification in the crinoid morphospace, for instance by conducting similar analyses at different temporal scales and taxonomic levels, in combination with an improved knowledge of crinoid development (e.g., [49,50]). Even if developmental data are not directly obtainable for some groups and might imply hypotheses from comparisons with extant relatives, a greater attention to the organizational and variational properties of morphological phenotypes is necessary when constructing, exploring, and interpreting morphospaces. This is an important step to refine our understanding of the evolutionary history of higher taxa and of the processes driving macroevolutionary change.

Supporting Information

Appendix S1   Crinoid data (Foote 1999). The file includes details of the stratigraphic intervals used, a description of the 90 discrete morphological characters, and the coding of these characters for the crinoid species retained in the analyses. See M. Foote, 1999. Paleobiology Memoir 1:1–115 (supplement to Paleobiology vol. 25, number 2) for additional details (doi: 10.1666/0094-8373(1999)25[1:MDITER]2.0.CO;2.).

Appendix S2   *R* functions. The file includes the *R* functions for running disparity and integration analyses as described in the main text. Crinoid data are available in Appendix S1, *R* can be

downloaded from http://www.r-project.org/. The shrinkage estimators of correlation matrices were obtained using the *R* package *corpcor* (http://strimmerlab.org/software/corpcor/). For additional details or questions: s.gerber@bath.ac.uk.

Appendix S3   Time series for morphological disparity and integration. The file provides the numerical values of the temporal trajectories of disparity and integration throughout the Phanerozoic as displayed in Figure 1. It also includes the results when only the subset of taxonomically non-significant characters is used (see main text).

Acknowledgments

I am especially grateful to M. J. Foote for kindly letting me use his crinoid data. For discussions and comments, I thank G. J. Eble, M. J. Foote, M. A. Wills, Y. Savriama, A. Haber, J. Rudkin, K. Davis, M. Hughes, A. O'Connor, R. C. P. Mounce, V. Hinman, and P. Oliveri. Two anonymous reviewers provided helpful reviews of the manuscript.

Author Contributions

Conceived and designed the experiments: SG. Performed the experiments: SG. Analyzed the data: SG. Contributed reagents/materials/analysis tools: SG. Wrote the paper: SG.

References

1. Alberch P (1980) Ontogenesis and morphological diversification. Am Zool 20: 653–667. doi: 10.1093/icb/20.4.653.
2. Maynard-Smith J, Burian R, Kauffman S, Alberch P, Campbell J, et al. (1985) Developmental constraints and evolution. Q Rev Biol 60: 265–287. doi:10.1086/414425.
3. Raup DM (1987) Neutral models in paleobiology. In: Nitecki MH, Hoffman A, editors. Neutral Models in Biology. Oxford, UK: Oxford University Press. pp. 121–132.
4. Arthur W (1997) The Origin of Animal Body Plans. Cambridge, UK: Cambridge University Press. 357 p.
5. Foote MJ (1997) The evolutionary history of morphological diversity. Annu Rev Ecol Syst. 28: 129–152. doi: 10.1146/annurev.ecolsys.28.1.129.
6. Gould SJ (2002) The Structure of Evolutionary Theory. Cambridge, MA: Harvard University Press. 1464 p.
7. Erwin DH (2007) Disparity: Morphological pattern and developmental context. Palaeontology 50: 57–73. doi: 10.1111/j.1475–4983.2006.00614.x.
8. Foote MJ (1993) Discordance and concordance between morphological and taxonomic diversity. Paleobiology 19: 185–204.
9. Wills MA, Briggs DEG, Fortey RA (1994) Disparity as an evolutionary index: A comparison of Cambrian and recent arthropods. Paleobiology 20: 93–130.
10. Jablonski D (2000) Micro- and macroevolution: scale and hierarchy in evolutionary biology and paleobiology. Paleobiology 26 (Suppl. to No.4): 15–52. doi: 10.1666/0094–8373(2000)26[15:MAMSAH]2.0.CO;2.
11. Olson EC, Miller RL (1958) Morphological Integration. Chicago, IL: University of Chicago Press (reprinted 1999). 376 p.
12. Lewontin RC (1978) Adaptation. Sci Am 239: 156–169.
13. Wagner GP (1996) Homologues, natural kinds and the evolution of modularity. Am Zool 36: 36–43. doi: 10.1093/icb/36.1.36.
14. Raff RA (1996) The Shape of Life. Chicago, IL: University of Chicago Press. 544 p.
15. Schlosser G, Wagner GP (2004) Modularity in Development and Evolution. Chicago, IL: University of Chicago Press. 600 p.
16. Klingenberg CP (2008) Morphological integration and developmental modularity. Annu Rev Ecol Evol S. 39: 115–132. doi: 10.1146/annurev.ecolsys.37.091305.110054.
17. Chernoff B, Magwene PM (1999) Morphological Integration: Forty Years Later. In: Olson EC, Miller RL, editors. Morphological Integration. Chicago, IL: University of Chicago Press. pp. 319–353.
18. Eble GJ (2004) The macroevolution of phenotypic integration. In: Pigliucci M, Preston K, editors. Phenotypic Integration: Studying the Ecology and Evolution of Complex Phenotypes. Oxford, UK: Oxford University Press. pp. 253–273.
19. Foote MJ (1999) Morphological diversity in the evolutionary radiation of Paleozoic and post-Paleozoic crinoids. Paleobiology 25 (sp1): 1–116. doi: 10.1666/0094–8373(1999)25[1:MDITER]2.0.CO;2.
20. Ciampaglio CN (2002) Determining the role that ecological and developmental constraints play in controlling disparity: examples from the crinoid and blastozoan fossil record. Evol Dev 4: 170–188. 10.1046/j.1525-142X.2002.02001.x.
21. Foote MJ (1994a) Morphological disparity in Ordovician-Devonian crinoids and the early saturation of morphological space. Paleobiology 20: 320–344.
22. Foote MJ (1994b) Morphological disparity in Ordovician-Devonian crinoids. Contrib Mus Pal Univ of Michigan 29: 1–39.
23. Foote MJ (1995a) Morphological diversification of Paleozoic crinoids. Paleobiology 21: 273–299. doi: 10.2307/2401167.
24. Foote MJ (1995b) Morphology of Carboniferous and Permian crinoids. Contrib Mus Pal Univ of Michigan 29: 135–184.
25. Foote MJ (1996) Ecological controls on the evolutionary recovery of post-Paleozoic crinoids. Science 274: 1492–1495. doi: 10.1126/science.274.5292.1492.
26. Deline B, Ausich WI (2011) Testing the plateau: a reexamination of disparity and morphologic constraints in early Paleozoic crinoids. Paleobiology 37: 214–236. doi: 10.1666/09063.1.
27. Sneath PHA, Sokal RR (1973) Numerical taxonomy. The principles and practice of numerical classification. San Francisco, CA: H. Freeman and Co. 588 p.
28. Foote MJ (1992) Rarefaction analysis of morphological and taxonomic diversity. Paleobiology 18: 1–16. doi: 10.2307/2400977.
29. Camin JH, Sokal RR (1965) A method for deducing branching sequences in phylogeny. Evolution 19: 311–326. doi: 10.2307/2406441.
30. O'Keefe FR, Wagner PJ (2001) Inferring and testing hypotheses of cladistic character dependence by using character compatibility. Syst Biol 50: 657–675. doi: 10.1080/106351501753328794.
31. Pavlicev M, Cheverud JM, Wagner GP (2009) Measuring morphological integration using eigenvalue variance. Evol Biol 36: 157–170. doi: 10.1007/s11692–008–9042–7.
32. Wagner GP (1984) On the eigenvalue distribution of genetic and phenotypic dispersion matrices: Evidence for a nonrandom organization of quantitative character variation. J Math Biology 21: 77–95.
33. Schäfer J, Strimmer K (2005) A shrinkage approach to large-scale covariance matrix estimation and implications for functional genomics. Stat Appl Genet Mol Biol 4: 32. doi:10.2202/1544–6115.1175. Available: http://strimmerlab.org/software/corpcor/. Accessed 2013 Apr 25.
34. Ledoit O, Wolf M (2004) A well-conditioned estimator for large-dimensional covariance matrices. J Mult Anal 88: 365–411. doi: 10.1016/S0047-259X(03)00096–4.
35. McKinney ML, Owen CW (1989) Causation and nonrandomness in biological and geological time series: Temperature as a proximal control of extinction and diversity. Palaios 4: 3–15.
36. Mitteroecker P, Bookstein F (2009) The ontogenetic trajectory of the phenotypic covariance matrix, with examples from craniofacial shape in rats and humans. Evolution 63: 727–737. doi: 10.1111/j.1558–5646.2008.00587.x.
37. Felsenstein J (1985) Phylogenies and the comparative method. Am Nat 125: 1–15. doi: 10.1086/284325.
38. Alroy J (1994) Four permutation tests for the presence of phylogenetic structure. Syst Biol 43: 430–437. doi:10.1093/sysbio/43.3.430.

39. Gerber S, Hopkins MJ (2011) Mosaic heterochrony and evolutionary modularity: The trilobite genus *Zacanthopsis* as a case study. Evolution 65: 3241–3252. doi: 10.1111/j.1558-5646.2011.01363.x.

40. Altenberg L (2005) Modularity in evolution: some low-level questions. In: Callebaut W, Rasskin-Gutman D, editors. Modularity: Understanding the Development and Evolution of Natural Complex System. Cambridge, MA: The MIT Press. pp. 100–128.

41. Simms MJ, Sevastopulo GD (1993) The origin of articulate crinoids. Palaeontology 36: 91–109.

42. Baumiller TK (2008) Crinoid ecological morphology. Annu Rev Earth Pl Sc. 36: 221–249. doi: 10.1146/annurev.earth.36.031207.124116.

43. Baumiller TK, Salamon MA, Gorzelak P, Mooi R, Messing CG, et al. (2010) Post-Paleozoic crinoid radiation in response to benthic predation preceded the Mesozoic marine revolution. Proc Natl Acad Sci USA 107: 5893–5896. doi: 10.1073/pnas.0914199107.

44. Kaufman SA (1983) Developmental constraints: internal factors in evolution. In: Goodwin BC, Wylie CC, editors. Development and Evolution. Cambridge, UK: Cambridge University Press. pp. 195–225.

45. Stadler BMR, Stadler PF, Wagner GP, Fontana W (2001) The topology of the possible: formal spaces underlying patterns of evolutionary change. J Theor Biol 213: 241–274. doi: 10.1006/jtbi.2001.2423.

46. Rasskin-Gutman D (2005) Modularity: jumping forms within morphospace. In: Callebaut W, Rasskin-Gutman D, editors. Modularity: Understanding the Development and Evolution of Natural Complex System. Cambridge, MA: The MIT Press. pp. 207–219.

47. McShea DW (1998) Dynamics of diversification in state space. In: McKinney ML, Drake JA, editors. Biodiversity Dynamics. New York: Columbia University Press. pp. 91–108.

48. Arthur W (2001) Developmental drive: an important determinant of the direction of phenotypic evolution. Evol Dev 3: 271–278.

49. Hara Y, Yamaguchi M, Akasaka K, Nakano H, Nonaka M, et al. (2006) Expression patterns of Hox genes in larvae of the sea lily *Metacrinus rotundus*. Dev Genes Evol. 216: 797–809.

50. Shibata TF, Sato A, Oji T, Akasaka K (2008) Development and growth of the feather star *Oxycomanthus japonicus* to sexual maturity. Zoolog Sci. 25: 1075–1083. doi: 10.2108/zsj.25.1075.

11

Repeated Origin and Loss of Adhesive Toepads in Geckos

Tony Gamble[1,2], Eli Greenbaum[3¤], Todd R. Jackman[3], Anthony P. Russell[4], Aaron M. Bauer[3*]

1 Department of Genetics, Cell Biology and Development, University of Minnesota, Minneapolis, Minnesota, United States of America, **2** Bell Museum of Natural History, University of Minnesota, St. Paul, Minnesota, United States of America, **3** Department of Biology, Villanova University, Villanova, Pennsylvania, United States of America, **4** Department of Biological Sciences, University Department of Calgary, Calgary, Canada

Abstract

Geckos are well known for their extraordinary clinging abilities and many species easily scale vertical or even inverted surfaces. This ability is enabled by a complex digital adhesive mechanism (adhesive toepads) that employs van der Waals based adhesion, augmented by frictional forces. Numerous morphological traits and behaviors have evolved to facilitate deployment of the adhesive mechanism, maximize adhesive force and enable release from the substrate. The complex digital morphologies that result allow geckos to interact with their environment in a novel fashion quite differently from most other lizards. Details of toepad morphology suggest multiple gains and losses of the adhesive mechanism, but lack of a comprehensive phylogeny has hindered efforts to determine how frequently adhesive toepads have been gained and lost. Here we present a multigene phylogeny of geckos, including 107 of 118 recognized genera, and determine that adhesive toepads have been gained and lost multiple times, and remarkably, with approximately equal frequency. The most likely hypothesis suggests that adhesive toepads evolved 11 times and were lost nine times. The overall external morphology of the toepad is strikingly similar in many lineages in which it is independently derived, but lineage-specific differences are evident, particularly regarding internal anatomy, with unique morphological patterns defining each independent derivation.

Editor: Jose Castresana, Institute of Evolutionary Biology (CSIC-UPF), Spain

Funding: This research was supported by grants DEB 0515909 and DEB 0844523 from the National science Foundation (www.nsf.gov) and by Discovery Grant 9745-2008 from the Natural Sciences and Engineering Research Council of Canada (www.nserc-crsng.gc.ca). The funders had no role in study design, data collection and analysis, decision to publish, or preparation of the manuscript.

Competing Interests: The authors have declared that no competing interests exist.

* E-mail: aaron.bauer@villanova.edu

¤ Current address: Department of Biological Sciences, University of Texas at El Paso, El Paso, Texas, United States of America

Introduction

Repeated evolution, also called convergent or parallel evolution, is the independent emergence of similar traits in separate evolutionary lineages and is typically seen as evidence of adaptation through natural selection or of developmental constraints that limit or bias morphological evolution [1,2,3,4,5]. Examining instances of repeated evolution serves as an important means of studying evolutionary processes and is analogous to studying multiple experimental replicates [6]. Indeed, each case of convergent or parallel evolution reveals the degree of common response to some fundamental biological challenge. As a result, extensive effort has been devoted to identifying instances of repeated evolution. To do this effectively, an accurate phylogeny is required for the "mapping" of traits and to permit examination of whether similarity is the result of shared ancestry or represents true independent derivation [3]. Many aspects of vertebrate body form related to locomotion have evolved repeatedly, being both gained and lost many times over. This includes functionally significant traits such as wings as aerodynamic devices, and limb reduction or elimination associated with burrowing [7,8,9]. Likewise, adhesive toepads employed in climbing have evolved several times in vertebrates, including multiple lineages of treefrogs, *Anolis* lizards, *Prasinohaema* skinks and, perhaps most notably in geckos [10,11].

The key component of the adhesive apparatus in lizards is the presence of setae, microscopic hair-like outgrowths of the superficial layer of the subdigital epidermis (the Oberhäutchen),

which promote adhesion via van der Waals forces and complex frictional interactions [12,13,14,15]. Setae evolved from the microscopic spinules that are typical of the outer epidermis of all limbed gekkotans and some other squamates [15,16,17,18], and are hypothesized to aid in skin shedding [16,19]. A hierarchy of anatomical specializations have evolved to govern the adhesive properties of the setae, and dynamic interactions with the substrate depend on numerous morphological adaptations and behaviors that facilitate control of the adhesive mechanism during locomotion [13,20,21,22,23]. Collectively, these specializations permit effective and rapid application and removal of the setae with reference to the substrate and constitute a functionally integrated complex [13,24].

Geckos are among the most species-rich and geographically widespread of terrestrial vertebrate lineages, with ~1450 described species in 118 genera, and comprise 25% of all described lizard species [25]. They are the likely sister group of all other lizards and snakes, excluding the limbless dibamids, having diverged from other squamates 225–180 MY ago [26,27]. The gekkotan adhesive system has been present since at least the mid-Cretaceous, as revealed by scansorial pads preserved in amber-embedded gecko fossils [28,29]. Approximately 60% of gecko species possess adhesive toepads, whereas the remainder lack functional adhesive toepads (or lack limbs altogether, in the case of the Australian pygopodid geckos) [7]. Geckos with adhesive toepads can easily scale vertical or even inverted surfaces, and these extraordinary clinging abilities have long attracted scientific attention [16,30,31].

Recently, interest has focused on mimicking the gecko adhesive mechanism to develop bio-inspired technologies [32,33,34]. Biomimetic studies have concentrated largely on adhesion at the molecular level, but functional control of adhesive toepads requires integration across a hierarchy of systems operating at different scales. These complex interactions – from molecular bonds to the locomotor control of the entire organism – are incorporated across seven orders of magnitude of size in geckos [13].

The form and structure of adhesive toepads in geckos have been used historically for taxonomic purposes, chiefly for assigning species to genera [35,36,37]. Traditional views of gecko evolution presupposed a single [38], or at most two [22], origins of the adhesive apparatus. These views were inferred from phylogenetic hypotheses that used few characters and sparse taxon sampling, and that placed the padless eublepharid geckos as sister to all remaining geckos, a position refuted by recent molecular phylogenies [26,27,39,40]. Reconstructing the evolution of gekkotan adhesive toepads, therefore, requires a comprehensive phylogeny derived from an independent data source, i.e., molecular genetic data. Here we estimate the phylogenetic relationships among nearly all recognized gecko genera using a multilocus dataset. We optimize the evolution of adhesive toepads on this phylogeny and reveal extensive homoplasy both in toepad morphology and in patterns of toepad loss. Our approach provides an appropriate framework for investigating broader functional and ecological questions that are associated with the origin, diversification and secondary loss of adhesive toepads. Being able to focus upon evolutionary events in different parts of the gekkotan phylogeny will permit more specific questions to be explored. In this contribution we provide exemplars of such phenomena, and consider the environmental circumstances that may have triggered particular transitions. Further explorations of similar transitions in other parts of the phylogeny will ultimately lead to potential generalizations about the form, function and adaptive significance of adhesive pad configuration in its various guises.

Methods

Phylogenetic Analyses

We estimated phylogenies using approximately 4,100 aligned bases of nucleotide data, from 244 gekkotan taxa and 14 outgroups (Table S1). The dataset was mostly complete, with only about 3% missing data. This included exemplars from 107 of 118 recognized gekkotan genera. Several recently described or elevated genera [41,42] were not sampled, but these new taxa are invariant in digital morphology in comparison to related taxa that are represented in our phylogenetic analyses. DNA sequence data consisted of fragments of five nuclear protein-coding genes: *RAG1*, *RAG2*, *C–MOS*, *ACM4*, and *PDC*; and one mitochondrial gene: *ND2* and associated tRNAs. Primers, PCR conditions, and sequencing conditions are detailed elsewhere [43,44]. Sequence data have been deposited in GenBank (Table S1). We aligned sequences using T-Coffee [45] with default parameters and fine-tuned alignments by hand to ensure insertions and deletions did not disrupt the translation of DNA sequence into amino acids. Protein-coding sequences were translated into amino acids using MacClade 4.08 [46] to confirm alignment and gap placement. Alignment gaps were treated as missing data and nuclear gene sequences were unphased. We estimated phylogenetic relationships among taxa using Maximum Likelihood (ML) in RAxML 7.2.6 [47] and Bayesian analysis in MrBayes 3.1.2 [48]. Data in both analyses were divided into seven partitions; first by genome (nDNA and mtDNA) and then by codon, with a separate partition

for tRNAs. This partitioning scheme contains fewer parameters than the preferred partitioning strategy used in previous phylogenetic analyses of the same nuclear loci (partitioning by both gene and codon), but with far fewer taxa [43,49]. The more parameter-rich strategy resulted in convergence problems in the Bayesian analysis of this taxon-rich dataset, likely due to low phylogenetic signal in the smaller partitions; these problems were resolved by reducing the number of partitions. Model selection was based on AIC scores using the software jModeltest [50], which recovered either the GTR + I + G or the GTR + G models for each partition (Table S2). The GTR + G model was used for all partitions in the ML analysis, which is the only model implemented in RAxML due to problematic interactions between the I and G parameters [51,52]. Bayesian analyses were run with multiple MCMC chains for 40 million generations, sampling every 1000th generation. Post burn-in convergence was checked by visual inspection of likelihood values by generation using Tracer 1.5 [53] and comparing split frequencies between runs using AWTY [54].

Comparative Analyses

We categorized digital morphologies in all sampled taxa as a binary character, coding species lacking a functional digital adhesive mechanism as 0 and species with a functional digital adhesive mechanism as 1 (Table S1). Morphological data were gathered from the literature as well as our personal examination of museum specimens representing 95% of described gecko species. Methods summarizing the collection of paraphalangeal data have been detailed elsewhere [21].

We estimated the number of independent gains and losses of the gekkotan digital adhesive mechanism using ancestral state reconstruction under parsimony and Maximum Likelihood in Mesquite [55], and Bayesian reconstruction in Bayestraits [56]. We incorporated phylogenetic uncertainty into our ancestral state reconstructions by summarizing ancestral states over a random subsample of 5,000 post burn-in trees from the Bayesian phylogenetic analyses onto the ML tree [57]. To investigate whether gains and losses of a functional digital adhesive mechanism occurred at the same rate in geckos, we compared the 1–rate MK1 model [58] to the asymmetric 2–rate model [59,60] with the likelihood ratio test in both the ML and Bayesian reconstructions.

Ancestral state reconstruction methods can be positively misleading if the trait in question influences diversification rates [61,62]. To correct for this artefact we used the binary-state speciation and extinction (BiSSE) model [61] to simultaneously estimate transition rates between binary characters (q01 and q10) and state-specific extinction (mu0 and mu1) and speciation rates (lambda0 and lambda1). We accounted for the incomplete species sampling of our phylogeny (~10% of described gekkotan species) by converting our ML phylogeny into a terminally-unresolved generic-level tree that could accommodate all unsampled taxa [63]. We pruned our phylogeny to 107 terminal taxa, roughly equivalent to genera, to which we could unambiguously assign all 1,452 described gecko species. There were several instances where multiple genera were grouped together for convenience, as well as several instances where genera were split into multiple groups due to the revelation of generic paraphyly (see results). In all cases, there were no changes in the presence or absence of adhesive toepads among impacted clades, so any influence of this taxonomic assignment on our results should be negligible. The ML phylogeny was made ultrametric using penalized likelihood in APE 2.7 [64,65] with the root arbitrarily scaled to 100. We calculated BiSSE model parameters from the ultrametric ML tree using maximum likelihood in the software Diversitree [63]. We also tested several hypotheses regarding the evolution of the digital adhesive mechanism using a range of constrained BiSSE models. We calculated parameters for the unconstrained, six-parameter model and then sequentially constrained each of the model parameters, alone and in combination, to yield a single rate for each parameter (e.g., mu0 = mu1, lambda0 = lambda1, q01 = q10) to determine if constrained models provided a better fit to the data than did the unconstrained model. We also explored whether models that restricted transitions between character states provided a realistic evaluation of our data. We did this by constraining q01 = 0, where a functional digital adhesive mechanism evolved just once; and q10 = 0, where once gained, a functional digital adhesive mechanism is never lost. We used AIC scores to determine which model provided the best fit to our data. Bayesian posterior distributions of BiSSE model parameters were also estimated using Markov Chain Monte Carlo analyses with the terminally unresolved generic-level ML tree in Diversitree [63]. Priors for each parameter used an exponential distribution, and estimated ML model parameters were used as a starting point. We combined results from two separate MCMC chains run for 10,000 generations each, with the first 10% of each run discarded as burn-in.

Results

Molecular phylogenies recover patterns of interfamilial relationships consistent with previous molecular studies (Fig. 1, Figs. S1-S2) [26,39,40,43]. This includes well-supported monophyly of all seven gekkotan families (Table 1, Figs. S1-S2), with both Bayesian and maximum-likelihood trees concordant at well-supported nodes. Portions of the phylogeny with short internal branches are generally poorly supported, making it difficult to resolve phylogenetic relationships among many genera. This is the case at the base of Gekkonidae, Phyllodactylidae and Sphaerodactylidae. Several recognized genera are recovered with strong support as either para- or polyphyletic: *Afrogecko*, *Cnemaspis*, *Cyrtodactylus*, *Gekko*, *Rhacodactylus* and *Saurodactylus*.

Comparative analyses using multiple methodologies reveal repeated gains and losses of adhesive toepads (Fig. 1, Figs. S3, S4, S5 and S6). Phylogenetic uncertainty, due to short internodes, makes unambiguous ancestral state reconstructions difficult in some parts of the tree, particularly within the Gekkonidae (Fig. S4). Even so, well-resolved, strongly supported nodes across the phylogeny provide clear evidence of independent gains and losses. Reconstructing ancestral character states with parsimony (Fig. S5) across a selection of trees from the Bayesian phylogenetic analysis results in 20 transitions, with an average of 11 gains (min = 3, max = 17) and 9 losses (min = 3, max = 18). Indeed, gains and losses occur at about the same rate in all of our analyses (Fig. S6). A 1–rate transition model yields results that are not significantly different from an asymmetric 2–rate model for both maximum likelihood reconstructions (likelihood ratio test; $P = 0.4394$) and Bayesian reconstructions (Fig. S3). Similarly, the distribution of character transition rates shows considerable overlap in credibility intervals using a Bayesian implementation of the BiSSE model (Fig. 2) [61,63]. This extends to overlapping diversification rates (calculated as trait-specific speciation - extinction) among padded and padless lineages (Fig. 2). Comparing the full and constrained maximum likelihood BiSSE models (Table 2) reveals that constraints five and six best fit the data, although AIC differences among most models are small. Constraints five and six both have equal transition rates (q01 = q10) and constrain either equal speciation rates (constraint 5, lambda0 = lambda1) or equal extinction rates (constraint 6, mu0 = mu1). Models that restrict

transitions between character states (i.e., constraints eight and nine where q01 = 0, q10 = 0), provide a significantly worse fit to the data than the unconstrained and remaining constrained models. All of the comparative analyses indicate that the most recent common ancestor of all geckos lacked adhesive pads. Many padless lineages retain this ancestral state (e.g., Carphodactylidae and Eublepharidae), but in many others this condition is secondarily derived (e.g., *Homonota*, *Garthia* and *Gymnodactylus*).

Discussion

Phylogenetic comparative analyses recover multiple gains and losses of adhesive toepads in geckos. This contrasts with previous hypotheses that suggest one, or at most two origins of toepads in geckos [22,38]. This rampant convergence and parallelism in digital design helps explain the generally poor performance of superficial digital characters for systematic purposes, particularly at higher levels of inclusiveness [22,43,66,67]. Morphological evidence for gekkotan relationships exists, but a high noise-to-signal ratio among the relatively few morphological characters that have been exploited in gecko systematics to date has hampered both phylogenetic reconstruction and the study of character evolution. Recent work using molecular systematic approaches reveals that many gecko genera, originally defined by toepad morphology, are polyphyletic [39,68,69,70]. Here we identify three more polyphyletic genera: *Afrogecko*, *Cnemaspis* and *Rhacodactylus*. The genera *Gekko* and *Cyrtodactylus* are rendered paraphyletic by *Ptychozoon* and *Geckoella*, respectively. These results indicate that additional work at the generic level is necessary to ensure that gecko taxonomy is isomorphic with phylogeny.

The BiSSE model co-estimates character transition rates and trait-specific speciation and extinction rates, which allows for the estimation of diversification rates (speciation - extinction, Fig. 2) for lineages with and without adhesive toepads. Whereas diversification rates in gecko lineages with toepads are higher than in lineages lacking toepads, these differences are small, and there is overlap in the Bayesian posterior distributions of BiSSE diversification parameters. Therefore, the presence of adhesive toepads, on its own does not appear to have directly influenced the number of species in different gecko lineages. The lack of a direct relationship between adhesive toepads and diversification rates in geckos highlights the complicated relationship between the evolution of complex traits, speciation and extinction. The success of geckos has been linked to possessing many derived traits including nocturnality, visual and olfactory prey discrimination, and shifts in diet, as well as adhesive toepads [71,72,73]. That adhesive toepads do not, on their own, explain gecko diversification rates should therefore come as no surprise. Uncovering the patterns and processes that explain the great diversity of geckos overall, as well as the disparities in species richness among gekkotan clades, is a rich source for further research that will be greatly facilitated by the comprehensive phylogeny presented here.

An unambiguous gain of adhesive toepads from a padless ancestor is exemplified by the globally distributed genus *Hemidactylus*. The modular construction of the adhesive mechanism is evident when detailed digital morphology is compared to that of related padless genera, and when comparing the elaboration of specialized components from unspecialized precursors (Fig. 3). The likely key initial modification of the digit in *Hemidactylus*, indeed the minimum requirement necessary to possess a functional adhesive mechanism, involves the elaboration of the subdigital spinules into setae with multi-spatulate tips. Because a spinulate epidermis seems ubiquitous among limbed geckos [15,16,17,18], a setal precursor does not need to evolve *de novo* each time the

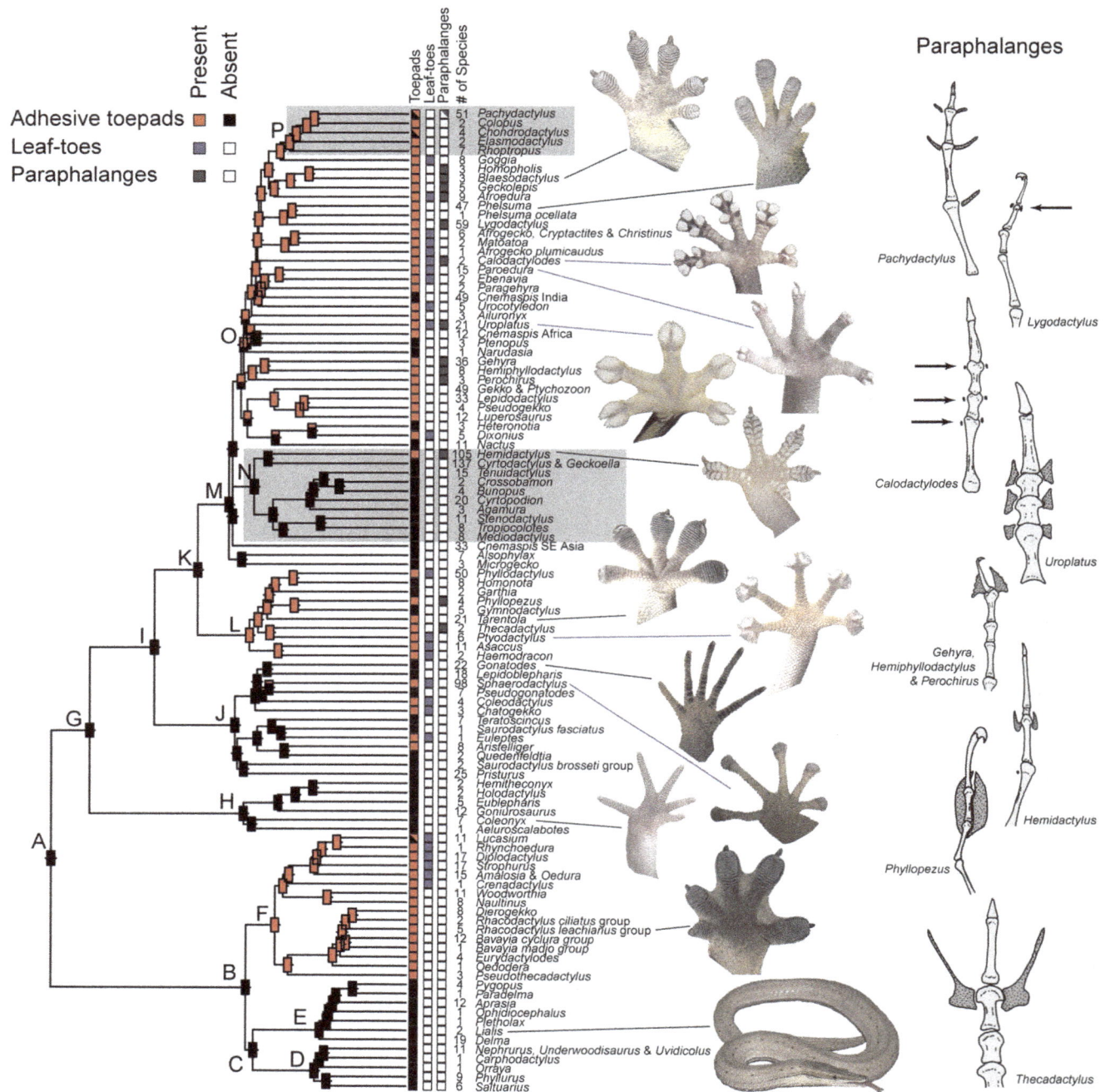

Figure 1. Gecko phylogeny and the evolution of adhesive toepads. Maximum likelihood tree showing phylogenetic relationships among gecko genera. Toepad traits, including the presence of adhesive toepads, toepad shape and the presence of paraphalanges, are indicated by colored squares on the tips of the branches (squares with two colors indicate polymorphism within the clade). Rectangles at internal nodes represent ancestral presence or absence probabilities of adhesive toepads inferred using the 6-parameter binary-state speciation and extinction (BiSSE) model. Details for lettered clades are presented in Table 1. Representative images illustrate a variety of gecko toepad morphologies. Single digits from representative gecko species illustrating the morphological diversity of paraphalangeal elements (in gray with stippling) are shown on the right. Clades enclosed in gray boxes are shown in greater detail in Figures 3 and 4.

adhesive mechanism evolves. Elongation of the epidermal spinules, initially likely involved in the enhancement of traction [74], influenced the ability of the integumentary outgrowths to interact with the substrate *via* van der Waals forces, promoting further setal elaboration and the subsequent integration of associated morphological traits that control the elaborated setae as a directional

adhesive complex [22]. These associated morphological traits in *Hemidactylus*, and indeed all padded gecko lineages, include a broadened subdigital surface (scansors), and modified tendons and muscles to control these scansors. Other modifications specific to *Hemidactylus*, and a few other padded lineages, include a raised penultimate phalanx resulting in a claw that is free of the

Table 1. Nodal support and ancestral states for key nodes of the gecko phylogeny.

Node	Clade Name	P(toepads)	ML bootstrap	Bayesian PP	Age (mya)
A	Gekkota	0.014 (0.000–0.035)	100	1.00	118–167
B	Pygopodoidea	0.233 (0.063–0.386)	100	1.00	66–102
C	unnamed	0.034 (0.000–0.136)	52	0.71	59–95
D	Carphodactylidae	0.000 (0.000–0.002)	100	1.00	20–46
E	Pygopodidae	0.000 (0.000–0.000)	100	1.00	28–44
F	Diplodactylidae	0.999 (0.999–1.00)	100	1.00	47–78
G	Gekkomorpha	0.020 (0.001–0.005)	92	1.00	113–157
H	Eublepharidae	0.001 (0.000–0.002)	100	1.00	60–98
I	Gekkonoidea	0.194 (0.031–0.386)	100	1.00	96–132
J	Sphaerodactylidae	0.008 (0.001–0.017)	100	1.00	85–117
K	unnamed	0.908 (0.775–0.997)	100	1.00	82–114
L	Phyllodactylidae	0.999 (0.998–1.00)	100	1.00	63–93
M	Gekkonidae	0.205 (0.008–0.523)	100	1.00	73–101
N	unnamed	0.020 (0.008–0.034)	100	1.00	60–87
O	Afro-Malagasy Clade	0.994 (0.973–1.00)	22	0.99	73–100
P	*Pachydactylus* Clade	0.998 (0.995–0.999)	100	1.00	41–69

Node labels refer to Figure 1. Posterior probabilities of the presence of toepads, P(toepads), calculated from the Bayesian comparative analysis. Nodal support values include maximum likelihood bootstrap values and Bayesian posterior probabilities. Node ages are from [39].

expanded pad, and neomorphic skeletal structures, the para-phalanges, which aid in the support of the scansors.

Adhesive toepads were lost nearly as many times as they originated, and a padless morphology is secondarily derived in many lineages. Unequivocal losses occurred in several lineages of Phyllodactylidae, within the diplodactylid genus *Lucasium* and within the gekkonid genera *Pachydactylus* and *Chondrodactylus* (Fig. 4). The latter three losses are associated with habitat shifts away from a rupicolous lifestyle to burrowing in loose sand [75], and highlight the adaptive significance of toepad morphology. The padless *Chondrodactylus angulifer*, for example, still retains skeletal, muscular and tendinous structures in the digits similar to those of related species that possess a functional adhesive mechanism [76]. The secondary loss of adhesive toepads results in a more highly derived morphology and, consistent with Dollo's law [77], does not simply reverse to the ostensibly primitive state. This pattern of reduction demonstrates that the adhesive system, once fully assembled, becomes reduced as a functionally integrated structural module [78,79] that remains fully intact but diminished in size, rather than displaying disassembly and dissolution. This pattern can be seen in six additional species in the genera *Rhoptropus* and *Pachydactylus* that have independently transitioned to terrestriality and show reductions (but not complete loss as seen in *C. angulifer* and *P. rangei*) in the number of scansors and in setal length [75,80].

Geckos show many lineage-specific morphological traits associated with the repeated gains and losses of adhesive toepads. These traits (which include modifications of the integument, digital skeleton, paraphalanges, musculo-tendinous system, and the vascular sinus network.), when re-examined in light of the hypothesis presented here, allow us to distinguish among most gecko lineages with independently derived adhesive systems as well as identify primitively padless lineages [13,21,22,76]. Two morphological traits associated with the digital adhesive mechanism show multiple independent origins and highlight lineage-specific differences among geckos with adhesive toepads. The first trait is toepad form. Toepads have traditionally been classified either as "leaf-toed," having divided, expanded scansors at the distal end of the digit, or "basal," having scansors distributed either proximally or along the entire length of the digit [22]. The leaf-toed morphology evolved in parallel 13–15 times and occurs in all of the major pad-bearing lineages (Fig. 1). Some leaf-toed lineages are independent derivations from a padless ancestor (e.g., *Euleptes*), whereas others are derived from a pad-bearing ancestor with close relatives having basal pads, implying that transitions between pad types are possible (e.g., *Goggia*; the Australian diplodactylids – *Crenadactylus*, *Oedura*, *Strophurus*, *Rhynchoedura*, *Diplodactylus* and *Lucasium*). Thus, the leaf-toed morphology has originated more often than adhesive pads as a whole, indicating the prevalence of transitions between pad types. The second trait is paraphalanges, cartilaginous or bony neomorphic structures associated with interphalangeal joints and thought to aid in support of the digital scansors or interdigital webbing [21,75]. Paraphalanges evolved nine times independently in geckos (Fig. 1). In almost every case their morphology is unique and easily distinguishable from those derived in other lineages. Parapha-langes exemplify complex characters that, when interpreted in a morphologically naïve context (e.g., a single binary character), may be seen as highly homoplastic, but if considered in light of specific structure and function (Fig. 1), reveal that each instance is unique.

The repeated gains and losses of the digital adhesive mechanism illustrate the importance of digital morphology in substrate interactions. Adhesive toepads enable animals that posses them to exploit vertically structured habitats, thereby allowing enhanced partitioning of the spatial niche [71,72]. The ability to adapt to specific substrates, for both digits with and without adhesive toepads, is also an important characteristic of geckos, and regions typified by geologic and topographic heterogeneity have been linked to increased diversity of gecko species [81]. Further research into the gekkotan adhesive mechanism should provide extensive material conducive to the study of the evolution of adaptive, complex phenotypes and partitioning of the spatial niche. Results

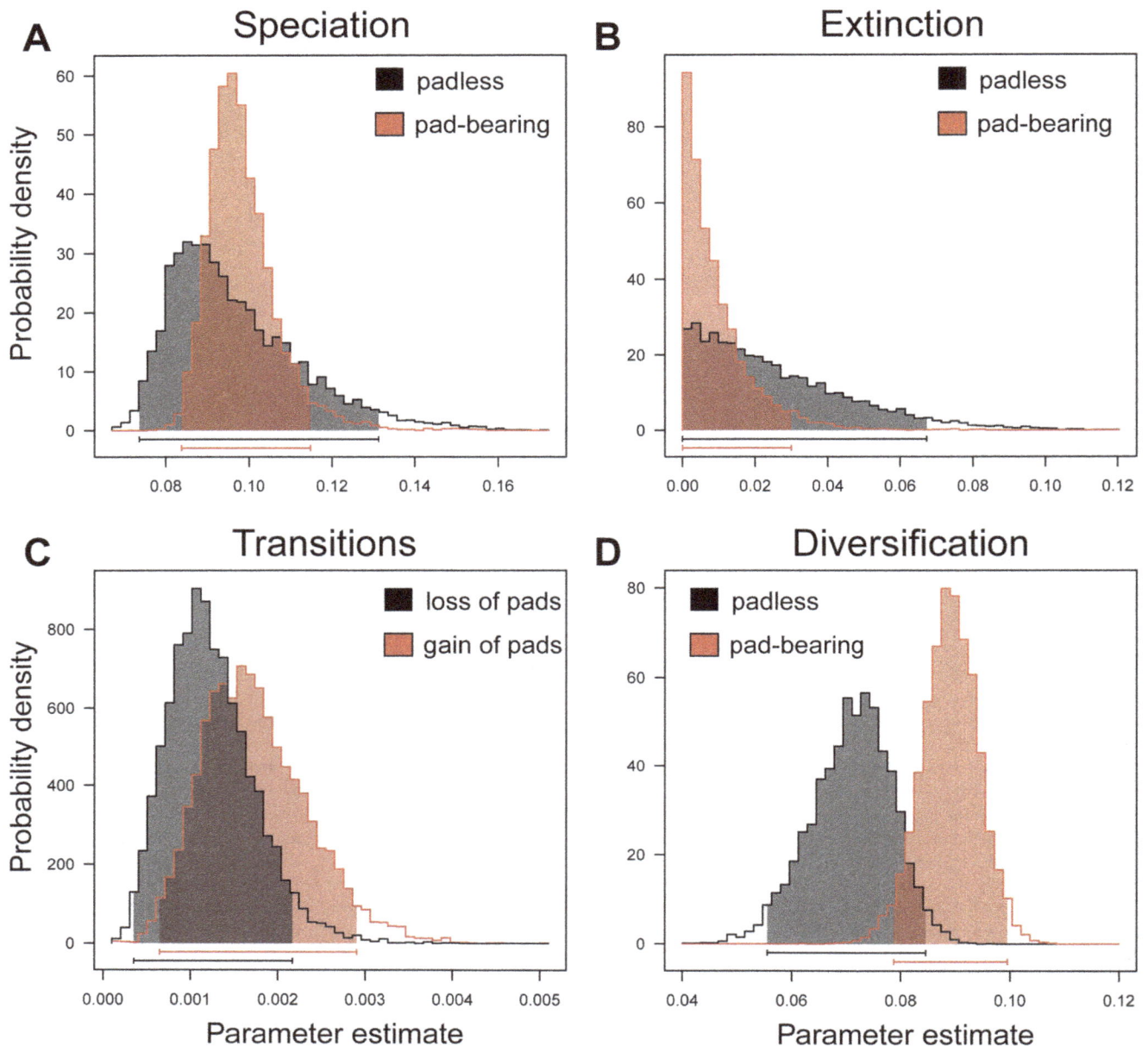

Figure 2. Bayesian parameter estimates inferred using the 6-parameter binary-state speciation and extinction (BiSSE) model. Estimates of: A. trait-specific speciation rates (lambda); B. trait-specific extinction rates (mu); C. transition rate parameters (q01 = gain of adhesive toepads, q10 = loss of adhesive toepads); D. net diversification rates calculated as the difference between speciation (lambda) and extinction (mu) rates for genera with and without adhesive toepads. The 95% credibility intervals for each parameter are shaded and indicated by bars along the x-axis.

presented here will prove useful in fostering additional research by identifying lineages with uniquely derived adhesive toepad morphology, and in differentiating between ancestrally padless lineages and species that are secondarily padless. The repeated evolution of adhesive toepads in the diverse and ancient geckos therefore, like the well-studied Caribbean *Anolis* ecomorphs [82], provides an outstanding resource for the understanding of mechanisms that drive phenotypic evolution, the balance between predictable evolutionary outcomes and historical contingency, and the relative influence of adaptation and developmental constraint on convergent and parallel evolution [2,3,5,83]. The sorts of questions that might arise from these considerations relate to

particular regions of the phylogeny, rather than to the synthetic bigger picture. Our broad-scale approach characterized adhesive toepads as essentially being present or absent. It does not explore, except for the exemplar taxa chosen, any of the variations in expression of the anatomical components [13,76] of the adhesive system. Aspects such as the significance of adhesive pad size [84] within and between gekkotan lineages, the manifestation of particular morphological patterns [22,76] or the environmental circumstances associated with the reduction or loss of the adhesive system [80] necessitate a finer scale of focus. For example, the relative size and configuration of adhesive toepads within lineages requires detailed examination at the species level in association

Table 2. Comparison of full and constrained maximum likelihood binary-state speciation and extinction (BiSSE) models.

Model	constraints	lambda0	lambda1	mu0	mu1	q01	q10	parameters	lnLik	AIC
full	None	0.0919287	0.0916504	0.0196976	0.0000042	0.0015639	0.0011354	6	−775.51	1563.0
constraint 1	lambda0 = lambda1	0.0917114	0.0917114	0.0195089	0.0000005	0.0015538	0.0011399	5	−775.51	1561.0
constraint 2	mu0 = mu1	0.0782689	0.0915912	0.0000147	0.0000147	0.0015438	0.0011227	5	−775.59	1561.2
constraint 3	q01 = q10	0.0917787	0.0917094	0.0197393	0.0000002	0.0013342	0.0013342	5	−775.61	1561.2
constraint 4	lambda0 = lambda1, mu0 = mu1	0.0858881	0.0858881	0.0000149	0.0000149	0.0015973	0.0010989	4	−777.70	1563.4
constraint 5	lambda0 = lambda1, q01 = q10	0.0917360	0.0917360	0.0197087	0.0000034	0.0013375	0.0013375	4	−775.61	**1559.2**
constraint 6	mu0 = mu1, q01 = q10	0.0781298	0.0916917	0.0000117	0.0000117	0.0013121	0.0013121	4	−775.75	**1559.5**
constraint 7	lambda0 = lambda1, mu0 = mu1, q01 = q10	0.0857970	0.0857970	0.0000003	0.0000003	0.0013039	0.0013039	3	−777.92	1561.8
constraint 8	q01 = 0	0.0815919	0.1616154	0.0000076	0.0960942	0.0000000	0.0019102	5	−778.31	1566.6
constraint 9	q10 = 0	0.1898625	0.0943962	0.1408593	0.0000000	0.0036345	0.0000000	5	−820.50	1651.0

Trait 0 lacks adhesive toepads; trait 1 possesses adhesive toepads. Lambda = trait specific speciation rates; mu = trait specific extinction rates; q = transition rate parameters. Constrained models are compared using the Akaike Information Criterion (AIC). The models with the lowest AIC scores are in bold.

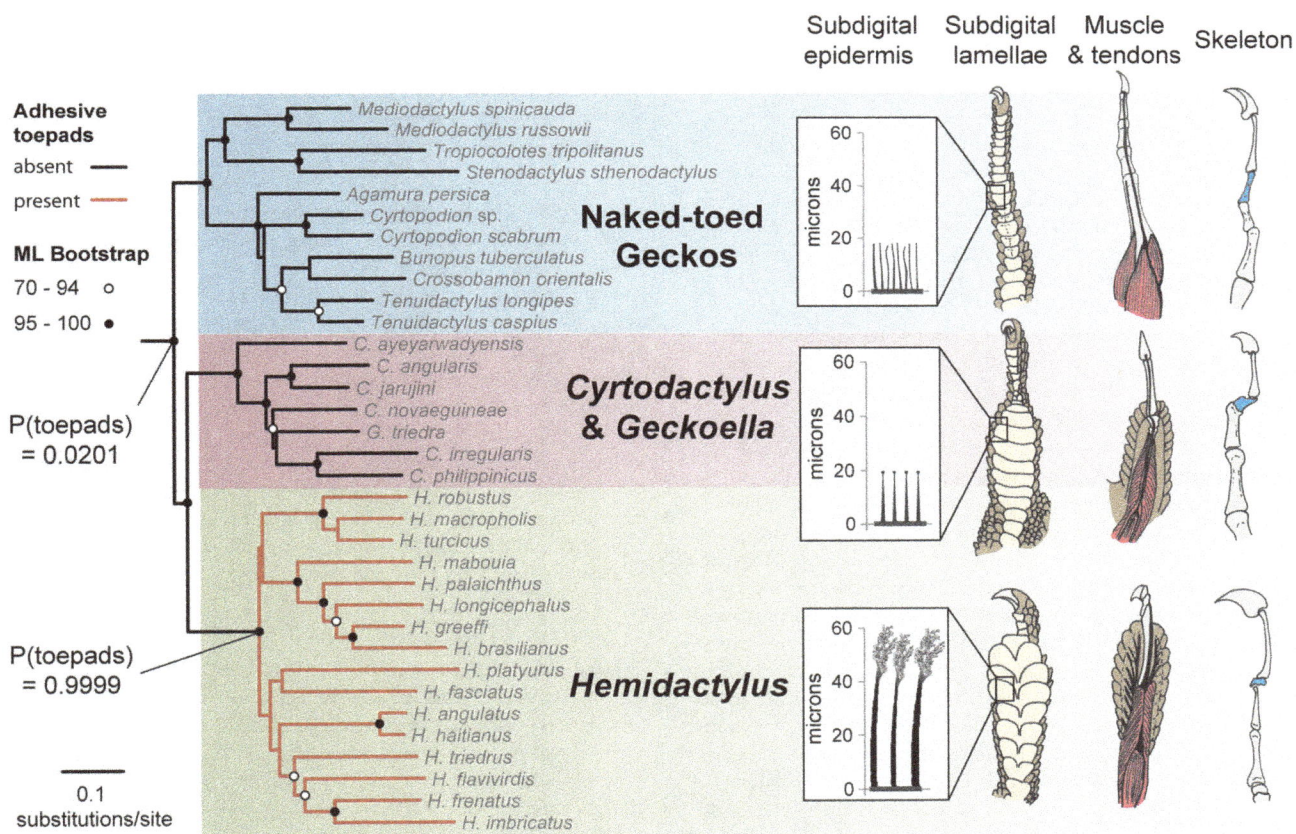

Figure 3. An unambiguous gain of adhesive toepads in house geckos (*Hemidactylus*). Maximum likelihood tree of included *Hemidactylus* species and their close relatives, the padless "naked-toed" geckos and the *Cyrtodactylus* + *Geckoella* clade. Circles at nodes indicate bootstrap support. Bayesian posterior probabilities of the presence of toepads are shown for two key nodes. Selected morphological components that comprise the digital adhesive mechanism are illustrated for each major clade. All three clades share spinules on the subdigital epidermis although only in *Hemidactylus* are they fully elaborated as setae. In the *Cyrtodactylus* + *Hemidactylus* clade: the subdigital lamellae are broadened; the antepenultimate phalanx of the digit (in blue) is reduced and, together with the penultimate phalanx and the claw, forms a raised arc; and the dorsal (extensor) musculature is expanded distally along the digit. The transition to fully functional toepads occurs in *Hemidactylus*, which incorporate the tendinous system that controls individual scansors, and possesses epdidermal spinules that are of increased length and that are multi-spatulate, enhancing functional adhesive surface area. These are recognizable as setae.

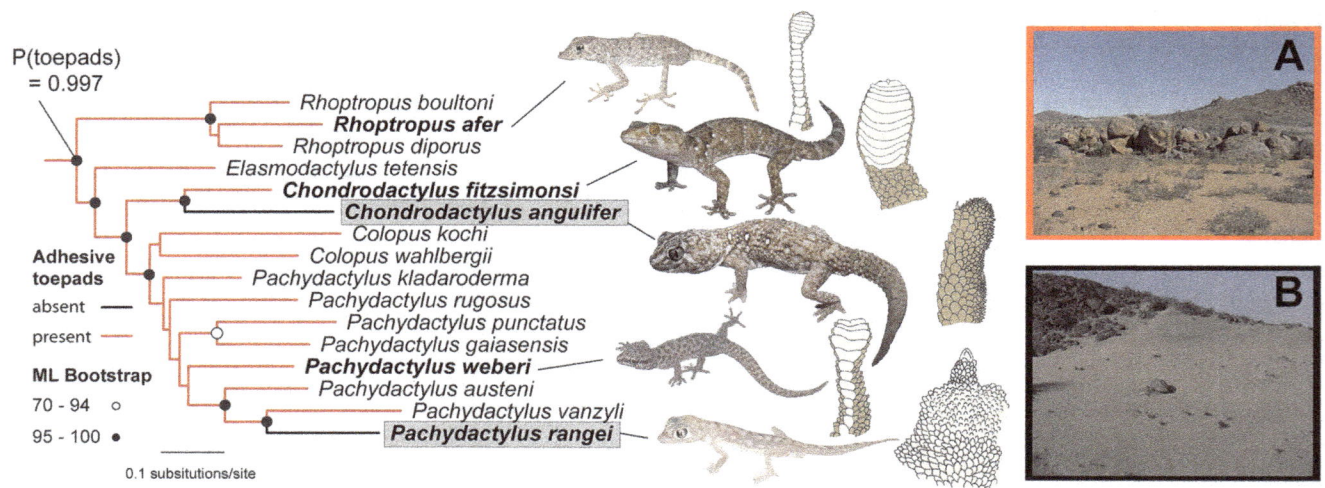

Figure 4. Two unambiguous losses of adhesive toepads in south African geckos. Maximum likelihood tree illustrating two independent losses of the digital adhesive mechanism in the southern African geckos *Chondrodactylus angulifer* and *Pachydactylus rangei* (in shaded boxes). Circles at nodes indicate bootstrap support. Bayesian posterior probabilities of the presence of toepads are shown for the most recent common ancestor of the included lineages, clearly indicating that the ancestor of this group possessed toepads. Representative species and their associated digital morphologies are illustrated. (**A**) Rupicolous habitat where padded members of this clade typically occur. (**B**) Sand dune habitat where the padless *Chondrodactylus angulifer* and the web-footed *Pachydactylus rangei* typically occur.

with study, at the microscopic scale, of the locomotor surfaces that they exploit. Such approaches have been conducted for a limited number of taxa [85,86], and can now be expanded to other parts of the phylogeny to test for congruence in observed patterns. Likewise, localized radiations within the phylogeny can be explored for circumstances related to adhesive pad reduction and loss. Increasing aridity and the exploitation of terrestrial habitats have been associated with such trends in southern Africa and the interior of Australia [75,76,87,88]. Additionally, the evolution of adhesive pad form (leaf-toed versus basal toepad patterns) can now be investigated in detail by pinpointing instances in the phylogeny in which each pattern has arisen independently, and in which transitions from leaf-toed to basal toepad expression have occurred [76], enabling questions about functional and mechanical effectiveness to be investigated.

The diversity of adhesive toepads in geckos holds enormous potential for biomimicry research, not only at the molecular level but also across the entire range of size scales at which geckos operate [12]. Repeated evolution of adhesive toepads can provide the foundation for understanding what is necessary and sufficient to make the "natural" adhesive system operable and functional. That foundation will allow the phylogenetic variation to be stripped away so that basic assembly rules can be understood, which will make formulation of biomimetic approaches more logical. Rather than selecting one exemplar gecko to copy, identifying distinct morphological modules from an array of separate evolutionary origins will permit a simpler and more directed approach to understanding how this functionally integrated complex operates.

Supporting Information

Figure S1 Phylogenetic relationships among sampled gecko species estimated using partitioned maximum likelihood. Bootstrap values from 100 rapid bootstrap replicates are shown at nodes.

Figure S2 Phylogenetic relationships among sampled gecko species estimated using partitioned Bayesian analysis. Bayesian posterior probabilities are shown at nodes.

Figure S3 Gecko phylogeny and the evolution of adhesive toepads estimated using Bayesian methods. A. Bayesian posterior distributions of the presence of toepads for key nodes across the gecko phylogeny estimated using Bayestraits over 5,000 trees from the Bayesian phylogenetic analysis. Numbers refer to node labels in panel B. B. Maximum likelihood tree showing phylogenetic relationships among gecko genera. The presence (red) or absence (black) of adhesive toepads is illustrated by colored squares on the tips of the branches (squares with two colors indicate polymorphism within the clade). Numbered nodes refer to Bayesian posterior distributions in panel A. C. Transition rate parameters from the Bayestraits analyses for the one rate model (in blue) and the two rate model where q01 = gain of adhesive toepads (in red) and q10 = loss of adhesive toepads (in black).

Figure S4 Phylogenetic relationships among sampled gecko species and the evolution of adhesive toepads estimated using maximum likelihood. Maximum likelihood tree showing phylogenetic relationships among sampled gecko species. Node color indicates ancestral states reconstructed using the mk1 model, summarized across a sample of 5,000 trees from the Bayesian phylogenetic analysis.

Figure S5 Phylogenetic relationships among sampled gecko species and the evolution of adhesive toepads estimated using parsimony. Maximum likelihood tree showing phylogenetic relationships among sampled gecko species. Node color indicates ancestral states reconstructed using parsimony (one of 114 equally parsimonious reconstructions).

Figure S6 The number of transitions between the gain and loss of adhesive toepads in geckos. Number of toepad gains (0 ->1) and losses (1 ->0) calculated using parsimony for 5,000 trees sampled from the Bayesian posterior distribution. Treescore = 20.

Table S1 Details of material examined.

Table S2 Summary of DNA sequence partitions.

References

1. Gompel N, Prud'homme B (2009) The causes of repeated genetic evolution. Developmental Biology 332: 36–47.
2. Losos JB, Jackman TR, Larson A, De Queiroz K, Rodriguez-Schettino L (1998) Contingency and determinism in replicated adaptive radiations of island lizards. Science 279: 2115–2118.
3. Wake DB, Wake MH, Specht CD (2011) Homoplasy: From detecting pattern to determining process and mechanism of evolution. Science 331: 1032–1035.
4. Haldane JBS (1932) The Causes of Evolution. London: Longmans, Green and Co.
5. Losos JB (2011) Convergence, adaptation, and constraint. Evolution 65: 1827–1840.
6. Kopp A (2009) Metamodels and phylogenetic replication: A systematic approach to the evolution of developmental pathways. Evolution 63: 2771–2789.
7. Pianka ER, Vitt LJ (2003) Lizards: Windows to the Evolution of Diversity. Berkeley, CA: University of California Press.
8. Vermeij GJ (2006) Historical contingency and the purported uniqueness of evolutionary innovations. Proceedings of the National Academy of Sciences of the United States of America 103: 1804–1809.
9. Wiens JJ, Brandley MC, Reeder TW (2006) Why does a trait evolve multiple times within a clade? Repeated evolution of snakelike body form in squamate reptiles. Evolution 60: 123–141.
10. Irschick DJ, Austin CC, Petren K, Fisher RN, Losos JB, et al. (1996) A comparative analysis of clinging ability among pad-bearing lizards. Biological Journal of the Linnean Society 59: 21–35.
11. Green DM (1981) Adhesion and the toe-pads of treefrogs. Copeia 1981: 790–796.
12. Autumn K, Sitti M, Liang YCA, Peattie AM, Hansen WR, et al. (2002) Evidence for van der Waals adhesion in gecko setae. Proceedings of the National Academy of Sciences of the United States of America 99: 12252–12256.
13. Russell AP (2002) Integrative functional morphology of the gekkotan adhesive system (Reptilia: Gekkota). Integrative and Comparative Biology 42: 1154–1163.
14. Autumn K, Dittmore A, Santos D, Spenko M, Cutkosky M (2006) Frictional adhesion: A new angle on gecko attachment. Journal of Experimental Biology 209: 3569–3579.
15. Maderson PFA (1964) Keratinized epidermal derivitives as an aid to climbing in gekkonid lizards. Nature 203: 780–781.
16. Maderson PFA (1970) Lizard glands and lizard hands: Models for evolutionary study. Forma et Functio 3: 179–204.
17. Peattie AM (2008) Subdigital setae of narrow-toed geckos, including a eublepharid (*Aeluroscalabotes felinus*). The Anatomical Record 291: 869–875.
18. Bauer AM, Russell AP (1987) Morphology of gekkonid cutaneous sensilla, with comments on function and phylogeny in the Carphodactylini (Reptilia: Gekkonidae). Canadian Journal of Zoology 66: 1583–1588.
19. Alibardi L, Maderson PFA (2003) Observations on the histochemistry and ultrastructure of regenerating caudal epidermis of the tuatara *Sphenodon punctatus* (Sphenodontida, Lepidosauria, Reptilia): A contribution to an understanding of the lepidosaurian epidermal generation and the evolutionary origin of the squamate shedding complex. Journal of Morphology 256: 134–145.
20. Russell AP (1975) A contribution to the functional analysis of the foot of the Tokay, *Gekko gecko* (Reptilia: Gekkonidae). Journal of Zoology 176: 437–476.
21. Russell AP, Bauer AM (1988) Paraphalangeal elements of gekkonid lizards - a comparative survey. Journal of Morphology 197: 221–240.
22. Russell AP (1979) Parallelism and integrated design in the foot structure of gekkonine and diplodactyline geckos. Copeia 1979: 1–21.
23. Pianka ER, Sweet SS (2005) Integrative biology of sticky feet in geckos. Bioessays 27: 647–652.
24. Russell AP, Higham TE (2009) A new angle on clinging in geckos: Incline, not substrate, triggers the deployment of the adhesive system. Proceedings of the Royal Society B–Biological Sciences 276: 3705–3709.
25. Uetz P (2010) The original descriptions of reptiles. Zootaxa 2334: 59–68.
26. Vidal N, Hedges SB (2009) The molecular evolutionary tree of lizards, snakes, and amphisbaenians. Comptes Rendus Biologies 332: 129–139.
27. Townsend TM, Larson A, Louis E, Macey JR (2004) Molecular phylogenetics of Squamata: The position of snakes, amphisbaenians, and dibamids, and the root of the squamate tree. Systematic Biology 53: 735–757.
28. Arnold EN, Poinar G (2008) A 100 million year old gecko with sophisticated adhesive toe pads, preserved in amber from Myanmar. Zootaxa 1847: 62–68.
29. Bauer AM, Böhme W, Weitschat W (2005) An early Eocene gecko from Baltic amber and its implications for the evolution of gecko adhesion. Journal of Zoology 265: 327–332.
30. Aristotle (1910) The History of Animals. Thompson DW, translator. Oxford: Clarendon Press.
31. Mahendra BC (1941) Contributions to the bionomics, anatomy, reproduction and development of the Indian house gecko, *Hemidactylus flaviviridis* Rüppel. Part II. The problem of locomotion. Proceedings of the Indian Academy of Sciences 13: 288–306.
32. Mahdavi A, Ferreira L, Sundback C, Nichol JW, Chan EP, et al. (2008) A biodegradable and biocompatible gecko-inspired tissue adhesive. Proceedings of the National Academy of Sciences of the United States of America 105: 2307–2312.
33. Kim S, Spenko M, Trujillo S, Heyneman B, Santos D, et al. (2008) Smooth vertical surface climbing with directional adhesion. IEEE Transactions on Robotics 24: 65–74.
34. Geim AK, Grigorieva SVDIV, Novoselov KS, Zhukov AA, Shapoval SY (2003) Microfabricated adhesive mimicking gecko foot-hair. Nature Materials 2: 461–463.
35. Fitzinger L (1843) Systema Reptilium (Amblyglossae). Vindobonae (Vienna): Braumüller et Seidel Bibliopolas. 106 p.
36. Vanzolini PE (1968) Geography of the South American Gekkonidae (Sauria). Arquivos de Zoologia (São Paulo) 17: 85–112.
37. Loveridge A (1947) Revision of the African lizards of the family Gekkonidae. Bulletin of the Museum of Comparative Zoology 98: 1–469.
38. Underwood G (1954) On the classification and evolution of geckos. Proceedings of the Zoological Society of London 124: 469–492.
39. Gamble T, Bauer AM, Colli GR, Greenbaum E, Jackman TR, et al. (2011) Coming to America: Multiple origins of New World geckos. Journal of Evolutionary Biology 24: 231–244.
40. Han D, Zhou K, Bauer AM (2004) Phylogenetic relationships among gekkotan lizards inferred from *C-mos* nuclear DNA sequences and a new classification of the Gekkota. Biological Journal of the Linnean Society 83: 353–368.
41. Fujita MK, Papenfuss TJ (2011) Molecular systematics of *Stenodactylus* (Gekkonidae), an Afro-Arabian gecko species complex. Molecular Phylogenetics and Evolution 58: 71–75.
42. Nielsen SV, Bauer AM, Jackman TR, Hitchmough RA, Daugherty CH (2011) New Zealand geckos (Diplodactylidae): Cryptic diversity in a post-Gondwanan lineage with trans-Tasman affinities. Molecular Phylogenetics and Evolution 59: 1–22.
43. Gamble T, Bauer AM, Greenbaum E, Jackman TR (2008) Out of the blue: A novel, trans-Atlantic clade of geckos (Gekkota, Squamata). Zoologica Scripta 37: 355–366.
44. Jackman TR, Bauer AM, Greenbaum E, Glaw F, Vences M (2008) Molecular phylogenetic relationships among species of the Malagasy-Comoran gecko genus *Paroedura* (Squamata : Gekkonidae). Molecular Phylogenetics and Evolution 46: 74–81.
45. Notredame C, Higgins DG, Heringa J (2000) T-Coffee: A novel method for fast and accurate multiple sequence alignment. Journal of Molecular Biology 302: 205–217.
46. Maddison WP, Maddison DR (1992) MacClade, analysis of phylogeny and character evolution. 3.0 ed. Sunderland, MA: Sinauer.
47. Stamatakis A (2006) RAxML-VI-HPC: Maximum likelihood-based phylogenetic analyses with thousands of taxa and mixed models. Bioinformatics 22: 2688–2690.
48. Huelsenbeck JP, Ronquist F (2001) MrBayes: Bayesian inference of phylogenetic trees. Bioinformatics 17: 754–755.

Acknowledgments

We thank D. J. Irschick and an anonymous reviewer for valuable comments that greatly improved the quality of the manuscript; R. FitzJohn, K. Kozak, A. M. Simons and D. B. Wake for comments on an earlier draft of the manuscript and help with analyses; the museums and colleagues that generously provided tissues and access to specimens; J. Marais and A. Captain for photos.

Author Contributions

Conceived and designed the experiments: TG AMB TRJ. Performed the experiments: TG EG. Analyzed the data: TG EG. Contributed reagents/materials/analysis tools: AMB APR TRJ. Wrote the paper: TG APR AMB.

49. Gamble T, Bauer AM, Greenbaum E, Jackman TR (2008) Evidence for Gondwanan vicariance in an ancient clade of gecko lizards. Journal of Biogeography 35: 88–104.

50. Posada D (2008) jModelTest: Phylogenetic model averaging. Molecular Biology and Evolution 25: 1253–1256.

51. Sullivan J, Swofford DL, Naylor GJP (1999) The effect of taxon sampling on estimating rate heterogeneity parameters of maximum-likelihood models. Molecular Biology and Evolution 16: 1347–1356.

52. Yang Z (2006) Computational Molecular Evolution. Oxford, England: Oxford University Press.

53. Rambaut A, Drummond AJ (2007) Tracer. 1.5 ed: Distributed by authors.

54. Nylander JAA, Wilgenbusch JC, Warren DL, Swofford DL (2008) AWTY (are we there yet?): A system for graphical exploration of MCMC convergence in Bayesian phylogenetics. Bioinformatics 24: 581–583.

55. Maddison WP, Maddison DR (2008) Mesquite: A modular system for evolutionary analysis. 2.5 ed.

56. Pagel M, Meade A, Barker D (2004) Bayesian estimation of ancestral character states on phylogenies. Systematic Biology 53: 673–684.

57. Lutzoni F, Pagel M, Reeb V (2001) Major fungal lineages are derived from lichen symbiotic ancestors. Nature 411: 937–940.

58. Lewis PO (2001) A likelihood approach to estimating phylogeny from discrete morphological character data. Systematic Biology 50: 913–925.

59. Pagel M (1999) Inferring the historical patterns of biological evolution. Nature 401: 877–884.

60. Schluter D, Price T, Mooers AØ, Ludwig D (1997) Likelihood of ancestor states in adaptive radiation. Evolution 51: 1699–1711.

61. Maddison WP, Midford PE, Otto SP (2007) Estimating a binary character's effect on speciation and extinction. Systematic Biology 56: 701–710.

62. Goldberg EE, Igic B (2008) On phylogenetic tests of irreversible evolution. Evolution 62: 2727–2741.

63. FitzJohn RG, Maddison WP, Otto SP (2009) Estimating trait-dependent speciation and extinction rates from incompletely resolved phylogenies. Systematic Biology 58: 595–611.

64. Sanderson MJ (2002) Estimating absolute rates of molecular evolution and divergence times: A penalized likelihood approach. Molecular Biology and Evolution 19: 101–109.

65. Paradis E, Claude J, Strimmer K (2004) APE: Analyses of Phylogenetics and Evolution in R language. Bioinformatics 20: 289–290.

66. Kluge AG (1983) Cladistic relationships among gekkonid lizards. Copeia 1983: 465–475.

67. Russell AP, Bauer AM (2002) Underwood's classification of the geckos: a 21st century appreciation. Bulletin of The Natural History Museum (Zoology) 68: 113–121.

68. Bauer AM, Good DA, Branch WR (1997) The taxonomy of the southern African leaf–toed geckos (Squamata: Gekkonidae), with a review of Old World "*Phyllodactylus*" and the description of five new genera. Proceedings of the California Academy of Sciences 49: 447–497.

69. Gamble T, Daza JD, Colli GR, Vitt LJ, Bauer AM (2011) A new genus of miniaturized and pug-nosed gecko from South America (Sphaerodactylidae: Gekkota). Zoological Journal of the Linnean Society 163: 1244–1266.

70. Oliver PM, Bauer AM, Greenbaum E, Jackman T, Hobbie T (2012) Molecular phylogenetics of the arboreal Australian gecko genus *Oedura* Gray 1842 (Gekkota: Diplodactylidae): Another plesiomorphic grade? Molecular Phylogenetics and Evolution 63: 255–264.

71. Vitt LJ, Pianka ER, Cooper WE, Schwenk K (2003) History and the global ecology of squamate reptiles. American Naturalist 162: 44–60.

72. Vitt LJ, Pianka ER (2005) Deep history impacts present-day ecology and biodiversity. Proceedings of the National Academy of Sciences of the United States of America 102: 7877–7881.

73. Losos JB (2010) Adaptive radiation, ecological opportunity, and evolutionary determinism. American Naturalist 175: 623–639.

74. Russell AP, Johnson MK, Delannoy SM (2007) Insights from studies of gecko-inspired adhesion and their impact on our understanding of the evolution of the gekkotan adhesive system. Journal of Adhesion Science and Technology 21: 1119–1143.

75. Lamb T, Bauer AM (2006) Footprints in the sand: Independent reduction of subdigital lamellae in the Namib-Kalahari burrowing geckos. Proceedings of the Royal Society B–Biological Sciences 273: 855–864.

76. Russell AP (1976) Some comments concerning interrelationships amongst gekkonine geckos. In: Bellairs AD, Cox CB, editors. Morphology and Biology of Reptiles. London: Academic Press. pp 217–244.

77. Gould SJ (1970) Dollo on Dollo's law: Irreversibility and the status of evolutionary laws. Journal of the History of Biology 3: 189–212.

78. Von Dassow G, Munro E (1999) Modularity in animal development and evolution: elements of a conceptual framework for EvoDevo. Journal of Experimental Zoology 285: 307–325.

79. Wagner GP (1996) Homologues, natural kinds and the evolution of modularity. American Zoologist 36: 36–43.

80. Johnson MK, Russell AP (2009) Configuration of the setal fields of *Rhoptropus* (Gekkota:Gekkonidae): Functional, evolutionary, ecological and phylogenetic implications of observed pattern. Journal of Anatomy 214: 937–955.

81. Bauer AM (1999) Evolutionary scenarios in the *Pachydactylus* group geckos of southern Africa: New hypotheses. African Journal of Herpetology 48: 53–62.

82. Losos JB (2009) Lizards in an Evolutionary Tree: Ecology and Adaptive Radiation of Anoles. Berkeley, CA: University of California Press.

83. Gould SJ (2002) The Structure of Evolutionary Theory. Cambridge, MA, USA: Belknap Press.

84. Webster NB, Johnson MK, Russell AP (2009) Ontogenetic scaling of scansorial surface area and setal dimensions of *Chondrodactylus bibronii* (Gekkota: Gekkonidae): Testing predictions derived from cross-species comparisons of gekkotans. Acta Zoologica 90: 18–29.

85. Russell AP, Johnson MK (2009) The gecko effect: Design principles of the gekkotan adhesive system across scales of organization. In: Favret EA, Fuentes NO, editors. Functional Properties of Bio-inspired Surfaces: Characterization and Technological Applications. Singapore: World Scientific Publishing Co. pp 103–132.

86. Russell AP, Johnson MK (2007) Real-world challenges to, and capabilities of, the gekkotan adhesive system: Contrasting the rough and the smooth. Canadian Journal of Zoology 85: 1228–1238.

87. Higham TE, Russell AP (2010) Divergence in locomotor performance, ecology, and morphology between two sympatric sister species of desert-dwelling gecko. Biological Journal of the Linnean Society 101: 860–869.

88. Johnson MK, Russell AP, Bauer AM (2005) Locomotor morphometry of the *Pachydactylus* radiation of lizards (Gekkota : Gekkonidae): A phylogenetically and ecologically informed analysis. Canadian Journal of Zoology 83: 1511–1524.

12

Landmarking the Brain for Geometric Morphometric Analysis

Madeleine B. Chollet[1]*, Kristina Aldridge[2], Nicole Pangborn[1], Seth M. Weinberg[3], Valerie B. DeLeon[1]

1 Center for Functional Anatomy and Evolution, Johns Hopkins University School of Medicine, Baltimore, Maryland, United States of America, **2** Department of Pathology and Anatomical Sciences, University of Missouri School of Medicine, Columbia, Missouri, United States of America, **3** Center for Craniofacial and Dental Genetics, University of Pittsburgh School of Dental Medicine, Pittsburgh, Pennsylvania, United States of America

Abstract

Neuroanatomic phenotypes are often assessed using volumetric analysis. Although powerful and versatile, this approach is limited in that it is unable to quantify changes in shape, to describe how regions are interrelated, or to determine whether changes in size are global or local. Statistical shape analysis using coordinate data from biologically relevant landmarks is the preferred method for testing these aspects of phenotype. To date, approximately fifty landmarks have been used to study brain shape. Of the studies that have used landmark-based statistical shape analysis of the brain, most have not published protocols for landmark identification or the results of reliability studies on these landmarks. The primary aims of this study were two-fold: (1) to collaboratively develop detailed data collection protocols for a set of brain landmarks, and (2) to complete an intra- and inter-observer validation study of the set of landmarks. Detailed protocols were developed for 29 cortical and subcortical landmarks using a sample of 10 boys aged 12 years old. Average intra-observer error for the final set of landmarks was 1.9 mm with a range of 0.72 mm–5.6 mm. Average inter-observer error was 1.1 mm with a range of 0.40 mm–3.4 mm. This study successfully establishes landmark protocols with a minimal level of error that can be used by other researchers in the assessment of neuroanatomic phenotypes.

Editor: Jean-Claude Baron, INSERM, France

Funding: This study was supported by grants F31 DE021302-01 and 5 R01 DE014399-05 from the National Institutes of Dental and Craniofacial Research. Publication of this article was funded in part by the Open Access Promotion Fund of the Johns Hopkins University Libraries. The funders had no role in study design, data collection and analysis, decision to publish, or preparation of the manuscript.

Competing Interests: The authors have declared that no competing interests exist.

* E-mail: mchollet@wustl.edu

Introduction

An examination of brain morphology, as defined by the size and shape of the brain as a whole and of individual structures within the brain, is one of the cornerstones of neuropsychiatric research and diagnosis. For example, age-related changes in brain morphology have been used to provide insight into the processes that underlie cognitive development throughout childhood and adolescence [1–2] and into those that underlie cognitive decline in senescence [3]. It has also been shown that brain morphology is altered in a variety of diseases (*e.g.*, type 2 diabetes [4]; major depressive disorder [5]; schizophrenia [6]; autism [7]) and thus has a practical role in clinical care and therapeutic research. Similarly, differences in brain morphology across species have been used to provide clues into human evolutionary history [8].

Traditionally, measures of size have been used to evaluate brain morphology. The use of these measures is based upon the idea that the size, and in particular the volume, of any given structure within the brain is determined by the functional requirements of that structure. Thus, a structure will be larger if it requires greater processing capacity and will be smaller if it requires less, such that form follows function [9–13]. However, there are notable exceptions to this tenet, including changes in size associated with certain pathological conditions. In these cases, a structure may undergo pathological enlargement rather than reduction in order to overcome functional deficits or as a result of the underlying

disease process. For example, larger brain volume in children with autism has been attributed to alterations in the biochemistry governing apoptosis and synaptic density, abnormally enlarged neurons, and reduced synaptic density [14]. Measures of size have other limitations as well. As a univariate measure, volume does not provide the information necessary to determine whether an effect is global or local, whether and if so how regions within the brain are interrelated, or to quantify changes in shape that may be distinct even when size is not [15–18]. It is in these areas that an analysis of shape, rather than size, provides more extensive and appropriate information for study and comparison.

'Shape' is defined as the set of geometric properties of an object that are independent of position, size, and orientation [15]. A variety of techniques exist to characterize the shape of the brain (*e.g.*, outline analysis [19–20], deformation-based morphometry [21–23], surface-based morphometry for cortical folding patterns [24–25]), many of which are still evolving. The focus of this study is landmark-based statistical shape analysis. Landmark-based statistical shape analysis is a technique that has been used widely in the fields of anthropology, genetics, and evolutionary biology and has more recently emerged as a tool to assess brain shape. With regard to the brain, the strengths of landmark-based shape analysis are two-fold: (1) landmarks can be placed throughout the brain, creating a three-dimensional spatial map consisting of both cortical and subcortical structures; and (2) a variety of independent methods have been developed to analyze landmark data, permitting one to visualize and interpret data in a variety of ways [26–28]. The first point is critical to the utility of statistical shape analysis as a methodological technique in the evaluation of the brain, since the brain is believed to consist of a collection of networks running in series and parallel to achieve specific functions. Changes in the spatial arrangement of the components of these networks, and thus the shape of the brain either as a whole or regionally, likely reflect changes in functional capacity and execution [13,17,18,29–32].

However, there are also limitations of landmark-based shape analysis. Landmarks can only be placed at locations that can be identified reliably on every individual under study. In some forms (or regions within a form), landmarks do not exist because there are not distinguishing features that reliably identify a particular point; thus, this technique may leave some regions underrepresented with potential overrepresentation of other regions. In addition, methods based on inferential statistics require that the number of variables does not exceed the degrees of freedom. Therefore, the number of landmarks that can be analyzed under these methods is limited by the sample size.

A more practical limitation of landmark-based statistical shape analysis of the brain is that, to date, none of the studies that have used this technique have published protocols for landmark identification and most have not published intra- or inter-observer error studies. Error studies were only available in the literature for three studies [33–35].

Of note, a variety of automated brain registration and cortical mapping programs have been validated that employ computer algorithms to delineate landmarks rather than rely upon manual landmark placement [36–42]. Automated methods for landmark placement have the benefit of removing inter-rater error and are especially useful for large sample sizes in which manual placement of landmarks would be grossly time consuming; however, as Pantazis et al. [43] has shown, automated methods may be less accurate in aligning occipital and frontal regions of the brain and these methods can be more susceptible to error when anatomical variation is high. Many automated methods are restricted to cortical landmarks, which limits their utility in the assessment of gross brain shape. Also, knowledge of different computational platforms is necessary to execute automated protocols and can be an impediment to novice researchers.

The purpose of the current study was to create a set of validated landmark protocols for the assessment of brain shape via landmark-based statistical shape analysis. Specifically, the primary aims were (1) to precisely and clearly define a set of landmarks that provide a biologically meaningful representation of brain shape, and (2) to evaluate the accuracy and repeatability of these landmarks, and by proxy, to optimize the landmark protocols in a multi-institutional intra- and inter-observer error study.

Materials and Methods

This study was approved by the University of Iowa institutional review board and the Johns Hopkins institutional review board. Written informed consent was obtained from a parent or guardian for all children enrolled in the study. All participants signed an informed consent approved by the University of Iowa review board and were compensated for their participation in the study.

This study was completed using magnetic resonance images (MRIs) from ten healthy, right-handed, white males, age 12. Participants were originally recruited from the community by the University of Iowa. Exclusion criteria included braces and diagnosis of a major medical, neurologic, or psychiatric illness.

This tightly constrained sample was chosen to limit variation among individuals.

Images were obtained using a 1.5-T Signa magnetic resonance scanner (General Electric, Milwaukee, Wisconsin) using a T1-weighted sequencing protocol. Voxel size was 1 mm × 1 mm × 1 mm. Post-acquisition processing was completed by technicians at the University of Iowa using the software BRAINS (Brain Research: Analysis of Images, Networks, and Systems) [44–47]. Post-acquisition processing consisted of brain extraction, in which the neural tissue was extracted from the surrounding skull and soft tissues of the face and scalp using an automated edge detection algorithm, followed by AC-PC alignment. The BRAINS software uses automated detection of the AC centroid, PC centroid, four ventricles, and mid-sagittal plane as defined by the interhemispheric fissure for spatial alignment. Details of linear alignment and post-processing are described elsewhere [44–48]. The use of the same verified automated processing technique across individuals minimized any innate error in the alignment process and thus would not be expected to have a significant impact on the reliability of manual landmark placement as was measured in this study.

As a baseline, landmarks were chosen according to (1) the frequency of their use in the literature and (2) the distribution of the landmark set with the goal of describing gross brain shape. In terms of frequency, landmarks were tabulated from the literature [17,18,33,34,49–54] (**Table 1**). "Common landmarks" were defined as those landmarks that have been used by at least three separate research teams in publication. "Uncommon landmarks" from the literature or novel landmarks were also included when the contributors agreed they were vital to the description of gross brain shape or of particular interest to the contributors. The final set of twenty-nine [29] landmarks (**Tables 2–3**) was determined by consensus among the contributors. Landmark data were collected for the left side of the brain only in order to limit the number of landmarks that needed to be collected, while still maintaining the diversity of landmarks being tested. The detailed landmark protocols established in this study are available in **Figure S1**.

Three of the contributors participated in data collection for the error study. Familiarity with neuroanatomy and brain landmarks varied among the raters from a complete novice brain landmarker (Rater 3) to an advanced brain landmarker (Rater 1). This spectrum of expertise was intentional, because it captured the range of likely users of the landmark definitions and of the protocols under study.

Landmark coordinate data were collected from three-dimensional reconstructions of brain imaging data in eTDIPS (http://www.cc.nih.gov/cip/software/etdips/) [55–56]. Raters initially conducted 3 trials per individual with each trial separated by at least 24 hours (10 individuals × 3 raters × 3 trials × 29 landmarks). A rater was not allowed to return to any trial once it had been completed. Landmark precision was calculated using the following formula:

$$\bar{P}_L = \sqrt{\frac{\left\{\sum_{k=10} \sum_{i=3} [d(L_{ki}, \bar{L}_{kt})]^2\right\}}{(ki-1)}}$$

where $(\bar{P}_L)$ = estimate of placement error for any given landmark, $d(L_{ki}, \bar{L}_{kt})$ = distance from replicate landmark location to mean landmark location, k = individual, i = trial. Landmark precision for a given rater was thus calculated as the square root of the squared distance of a landmark trial to the mean position of that landmark for an individual subject, summed across individuals and divided

Table 1. Brain landmarks tabulated from the literature.

Landmark	Aldridge	DeQuardo	Gharaibeh	Maudgil	Weinberg
amygdala	X				
anterior cingulate s/superior rostral s				X	
anterior commissure	X				
calcarine s./parieto-occipital s.				X	
caudate nucleus	X				
central s./lateral s.	X			X	X
central s.				X	
cerebellum - lateral pole					X
cerebellum - midsagittal inferior		X			X
cerebellum - midsagittal posterior					X
cerebellum - midsagittal superior		X	X		X
cerebral aqueduct/4th ventricle	X				
cingulate s.				X	
cingulate s./superior rostral s.				X	
corpus callosum - genu, anterior	X	X	X		X
corpus callosum - genu, posterior		X			
corpus callosum - midbody, inferior		X			
corpus callosum - midbody, sup		X	X		X
corpus callosum - splenium, ant		X	X		
corpus callosum - splenium, inf		X			
corpus callosum - splenium, post		X	X		X
fourth ventricle	X	X	X		
frontal pole	X		X		
inferior colliculus	X				
inferior frontal s./precentral s.	X			X	
lateral s./precentral s.				X	
lateral s./postcentral s.				X	
lateral s. – posterior termination	X				
lateral ventricle - anterior horn	X				
lateral ventricle - inferior horn	X				
lateral ventricle - posterior horn	X				
mammillary body	X				X
occipital pole	X				X
optic chiasm		X	X		
orbito-triangular s.					
parietooccipital s.				X	
pons - inferior	X	X	X		X
pons - superior	X	X	X		X
posterior commissure	X				
precentral s.				X	X
precentral s./superior frontal s.	X			X	
preoccipital notch				X	
superior colliculus	X	X	X		
thalamus	X				
temporal pole					X

Landmarks were tabulated from published studies where the primary methodology was landmark-based shape analysis of the brain in order to determine each landmark's frequency of use. Column headings indicate the source of the landmarks: (1) Aldridge [17,49,50]. (2) DeQuardo [51,52]. (3) Gharaibeh [53]. (4) Maudgil [33,54]. (5) Weinberg [34].

Table 2. Average intra-observer error and inter-observer error measured for landmarks.

#	LM	Intra-observer Error			Inter-observer Error		
		P1	P2	ΔP	P1	P2	ΔP
1	Frontal pole	1.5	-	-	0.88	-	-
2	Occipital pole	0.75	-	-	0.42	-	-
3	Temporal pole	1.2	-	-	0.71	-	-
4	Central s./Lateral s.	4.9	3.1	−1.8	3.1	1.9	−1.2
5	Central s. – superior point	4.0	3.2	−0.8	2.6	2.2	−0.4
6	Pre-central s./Superior frontal s.	4.6	4.9	+0.3	2.6	3.1	+0.5
7	Pre-central s./Inferior frontal s.	5.8	5.6	−0.2	3.3	3.4	+0.1
8	Ascending ramus lateral s.	3.3	2.7	−0.6	1.8	1.5	−0.3
9	Horizontal ramus lateral s.	2.9	2.8	−0.1	1.7	1.6	−0.1
10	Superior temporal s.	7.3	4.1	−3.2	4.1	2.4	−1.7
11	Parieto-occipital s.	2.9	2.8	−0.1	2.0	1.9	−0.1
12	Cerebellum – lateral pole	1.1	-	-	0.72	-	-
13	Cerebellum – inferior pole	1.9	1.0	−0.9	1.1	0.55	−0.55
14	Cerebellum – posterior pole	2.6	1.5	−1.1	1.5	0.84	−0.66
15	Cerebellum – superior pole	1.1	-	-	0.60	-	-
16	Fourth ventricle	0.95	-	-	0.54	-	-
17	Amygdala	1.9	1.7	−0.2	1.1	0.96	−0.14
18	Caudate nucleus	1.3	-	-	0.74	-	-
19	Thalamus	1.5	-	-	0.85	-	-
20	Corpus callosum – genu	0.89	-	-	0.50	-	-
21	Corpus callosum – midbody	1.5	-	-	0.94	-	-
22	Corpus callosum – splenium	0.78	-	-	0.44	-	-
23	Anterior commissure	0.76	-	-	0.44	-	-
24	Pons - inferior	0.93	-	-	0.53	-	-
25	Pons - superior	0.87	-	-	0.49	-	-
26	Superior colliculus	1.1	-	-	0.60	-	-
27	Left ventricle - anterior horn	0.72	-	-	0.40	-	-
28	Left ventricle - inferior horn	1.9	2.0	+0.1	1.2	1.2	0
29	Left ventricle posterior horn	4.1	1.7	−2.4	2.4	0.97	−1.4

In the landmark name, the backslash (/) indicates that the landmark is located at the intersection of the two sulci and s. is an abbreviation for sulcus. P1 is the imprecision (mm) for each landmark that was assessed in the first round of analysis. P2 is the imprecision (mm) for each landmark that was assessed in the second round of analysis using the modified protocols. The hyphen (-) indicates that the landmark was not reassessed in the second round of analysis because the error was less than 1.5 mm. ΔP is the difference between P2 and P1.

by the total number of measurements minus one. Similarly, intra-rater error was calculated as the deviation of each landmark trial from the average landmark position *for that rater*, summed across trials, across raters, and across individuals, and divided by the total number of measurements minus one. Inter-rater error was calculated as the deviation of the average landmark position for each rater from the average position *across raters*, summed across raters and across individuals, and divided by the total number of measurements minus one.

Any landmark with average intra-observer error greater than 1.5 mm (when rounded to two significant digits) was reassessed in a second round of error analysis. For the second round, the protocols of these "problem landmarks" were modified to provide additional clarity. The same three raters completed three additional trials per individual using the new (modified) protocols with each trial separated by at least 24 hours (15 landmarks×10 individuals×3 trials×3 raters). Intra-observer and inter-observer

errors were calculated as described above. Raw coordinate data for all trials are available in **Table S1**.

Results

Twenty-nine cortical and subcortical brain landmarks (**Table 1; Figure 1**) were assessed in this study (see Materials and Methods for selection criteria and definition of protocols). Using the initial set of protocols devised by the study team, average intra-observer error ranged from 0.72 mm (left ventricle anterior horn) – 7.3 mm (superior temporal sulcus) (**Table 2; Figures 1–2**). Sixteen of the 29 landmarks (55%) had an average intra-observer error of less than or equal to 1.5 mm, six (21%) had an average intra-observer error of 1.6–3.0 mm, and seven (24%) had an average intra-observer error greater than 3 mm. Inter-observer error ranged from 0.40 mm (left ventricle anterior horn) – 4.1 mm (superior temporal sulcus) (**Table 2; Figures 1, 3**). Twenty of the 29 landmarks (69%) had an average inter-observer error of less than

Table 3. Average intra-observer error by rater.

#	LM	Rater 1		Rater 2		Rater 3	
		P1	P2	P1	P2	P1	P2
1	Frontal pole	1.2	-	1.3	-	2.1	-
2	Occipital pole	0.93	-	0.61	-	0.72	-
3	Temporal pole	1.8	-	0.48	-	1.3	-
4	Central s./Lateral s.	6.6	4.8	0.74	0.99	7.4	3.6
5	Central s. – superior point	7.3	6.8	2.7	1.1	2.1	1.8
6	Pre-central s./Superior frontal s.	6.3	9.0	4.6	3.5	3.1	2.3
7	Pre-central s./Inferior frontal s.	5.2	9.2	4.0	3.6	8.1	3.9
8	Ascending ramus lateral s.	3.3	3.6	2.4	2.4	4.2	2.0
9	Horizontal ramus lateral s.	4.3	4.0	1.5	1.8	2.8	2.5
10	Superior temporal s.	8.3	5.3	5.1	2.4	8.3	4.7
11	Parieto-occipital s.	5.9	5.8	1.3	0.87	1.4	1.6
12	Cerebellum – lateral pole	2.1	-	0.48	-	0.74	-
13	Cerebellum – inferior pole	2.6	1.1	1.3	0.98	1.9	0.96
14	Cerebellum – posterior pole	2.2	1.3	1.7	1.6	3.8	1.7
15	Cerebellum – superior pole	0.96	-	0.76	-	1.4	-
16	Fourth ventricle	0.92	-	0.64	-	1.3	-
17	Amygdala	2.3	2.1	1.5	1.4	1.9	1.8
18	Caudate nucleus	1.5	-	0.92	-	1.5	-
19	Thalamus	1.8	-	1.1	-	1.6	-
20	Corpus callosum – genu	1.1	-	0.59	-	0.96	-
21	Corpus callosum – midbody	2.7	-	0.73	-	0.99	-
22	Corpus callosum – splenium	0.95	-	0.55	-	0.85	-
23	Anterior commissure	1.1	-	0.45	-	0.74	-
24	Pons - inferior	1.2	-	0.57	-	0.98	-
25	Pons - superior	1.2	-	0.63	-	0.81	-
26	Superior colliculus	1.4	-	0.68	-	1.1	-
27	Left ventricle - anterior horn	0.92	-	0.69	-	0.54	-
28	Left ventricle - inferior horn	3.0	2.6	0.73	1.1	2.0	2.3
29	Left ventricle posterior horn	5.8	1.2	4.0	1.5	2.5	2.4

In the landmark name, the backslash (/) indicates that the landmark is located at the intersection of the two sulci. P1 is the imprecision (mm) for each landmark in the first round of analysis. P2 is the imprecision (mm) for each landmark in the second round of analysis using the modified protocols. The hyphen (-) indicates that the landmark was not reassessed in the second round of analysis.

or equal to 1.5 mm, six (21%) had an average inter-observer error of 1.6–3.0 mm, and three (10%) had an average inter-observer error greater than 3 mm. The range of error and the landmarks associated with the greatest error were consistent among raters (**Table 3**). The range of error for each rater for the initial set of protocols was: Rater 1 = 0.92–8.30 mm, Rater 2 = 0.45–5.1 mm, and Rater 3 = 0.54–8.3 mm.

Midsagittal landmarks tended to have the least amount of error, while cortical and ventricular landmarks had the greatest error. Visual inspection of landmark placement revealed that most of the error for cortical landmarks was due to occasional misidentification of the central and pre-central sulci and the opercular and triangular sulci. Error in the inferior and posterior horns of the ventricles seemed to reflect the amount of CSF in the ventricles. Larger ventricles, filled with less radio-opaque CSF, tended to be associated with less landmark error, because the boundaries were more clearly defined. The terminations of narrower, more tapered ventricles were much less distinct. Moderate error in cerebellar

landmarks resulted from confusion about whether landmarks should be placed on the vermis or lobar tissue. These issues were addressed by clarifying protocols and by providing additional information on how to identify cortical sulci.

The thirteen (13) landmarks that had an average intra-observer error of greater than 1.5 mm were re-collected using modified protocols with greater specificity in landmark definition. Average intra-observer error of the new data ranged from 1.0 mm (cerebellum – inferior pole) – 5.6 mm (pre-central s./inferior frontal s. intersection) (**Table 2**). Two (15%) of the 13 landmarks had an average intra-observer error of less than or equal to 1.5 mm, six (46%) had an average intra-observer error of 1.6–3.0 mm, and five (38%) had an average intra-observer error greater than 3 mm. Inter-observer error ranged from 0.55 mm (cerebellum – inferior pole) – 3.4 mm (pre-central s./inferior frontal s. intersection). Six of the 13 landmarks (46%) had an average inter-observer error of less than or equal to 1.5 mm, five (38%) had an average inter-observer error of 1.6–3.0 mm, and two

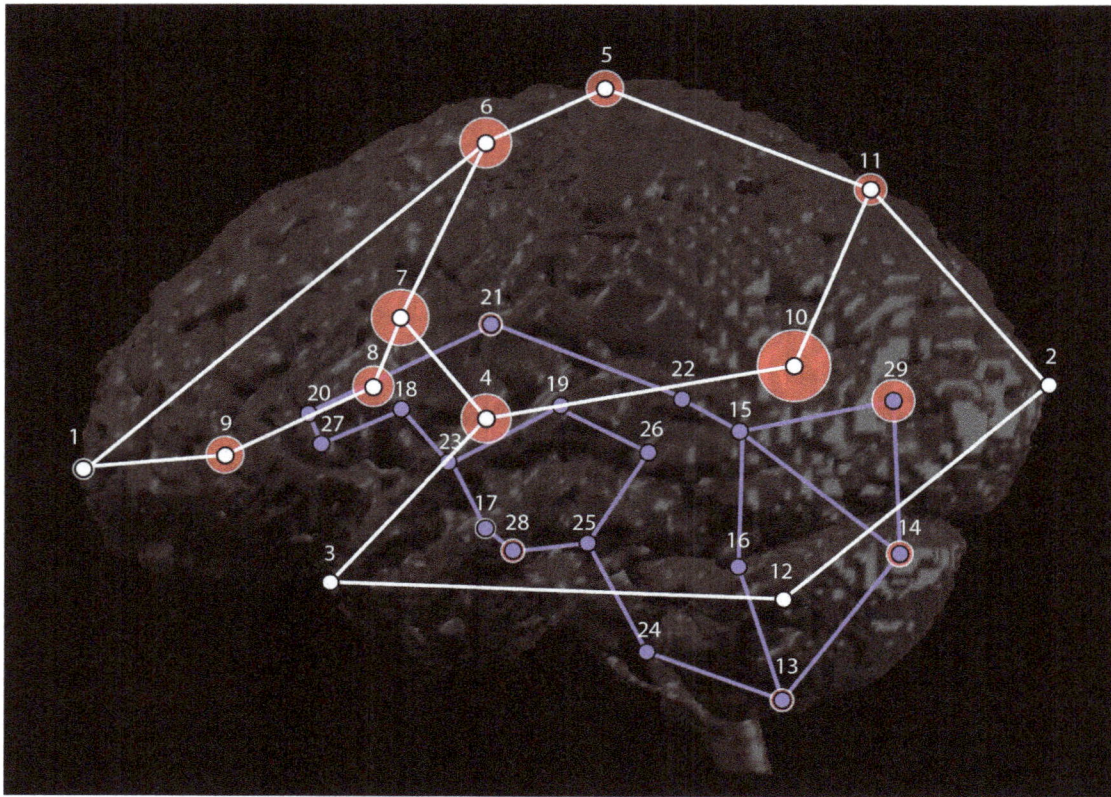

Figure 1. Landmarks and the associated error analyzed in this study. Left lateral view of a 3D reconstruction of the brain (anterior is to the left). Projected positions of landmarks are shown with numbers corresponding to Table 2. Cortical surface landmarks are white with white wireframe; subcortical landmarks are purple with purple wireframe. The size of the pink ellipses around each landmark indicate the magnitude of average precision (error) at anatomic scale. Landmarks for which no ellipse is visible had average error less than the 1.5 mm radius of the landmark marker. Note that the greatest magnitudes of error were associated with cortical surface landmarks.

(15%) had an average inter-observer error greater than 3 mm. Average intra-observer error decreased for eleven out of the thirteen landmarks using the modified protocols. Intra-observer error increased by 0.3 mm for the intersection of the precentral sulcus with the superior frontal sulcus and increased by 0.1 mm for the inferior horn of the lateral ventricle. Inter-observer error decreased for every landmark but two – pre-central s./inferior

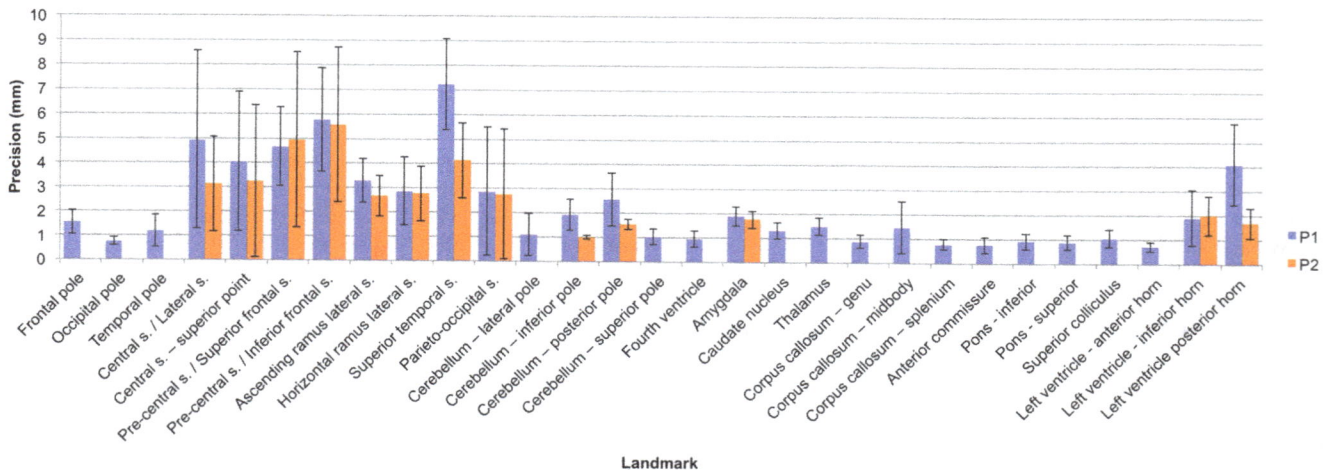

Figure 2. Histogram of the intra-observer precision of each landmark. This histogram indicates the level of intra-observer precision associated with each landmark using the original (P1) and modified (P2) protocols. The error bar is equal to one standard deviation above and below the mean. Landmark numbers correspond with the landmark numbers in Table 2.

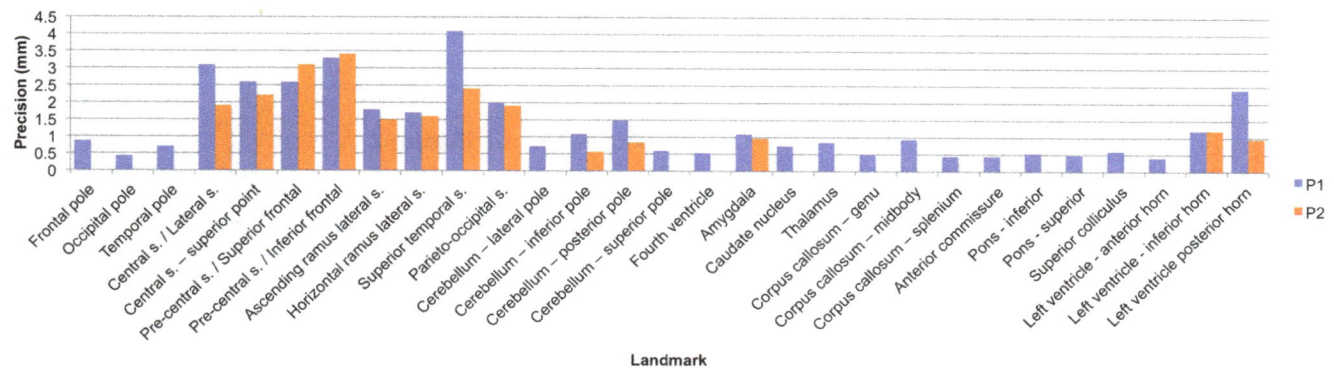

Figure 3. Histogram of the inter-observer precision of each landmark. This histogram indicates the level of inter-observer precision associated with each landmark using the original (P1) and modified (P2) protocols. Landmark numbers correspond with the landmark numbers in Table 2.

frontal s. intersection (+0.1 mm) and pre-central s./superior frontal s. intersection (+0.5 mm) – using the modified protocols.

The original landmark protocols, rather than the modified protocols, for the precentral s./superior frontal s. and the lateral ventricle-inferior horn were thus included in the final landmark protocol set. All other modified protocols were included in the final landmark set. Overall, the landmarks with the greatest intra- and inter-observer error measures were located at the superior temporal sulcus and at the intersection of the pre-central sulcus with the superior frontal sulcus and the inferior frontal sulcus. The range of error and the landmarks associated with the greatest error were consistent between Raters 2 and 3, but Rater 1, who was also the most experienced rater, had consistently greater error for cortical landmarks (**Table 3**). The range of error for the final set of landmarks for each rater was: Rater 1 = 0.92–9.2 mm, Rater 2 = 0.45–4.6 mm, and Rater 3 = 0.54–4.7 mm.

Discussion

This study established protocols for collecting three-dimensional coordinate data for 29 cortical and subcortical brain landmarks. The average intra-observer error was 1.9 mm with a range of 0.72 mm–5.6 mm, and the average inter-observer error was 1.1 mm with a range of 0.4 mm–3.4 mm. Error was particularly high for landmarks located at the intersection of the pre-central sulcus and inferior frontal sulcus, the intersection of the pre-central sulcus and inferior frontal sulcus, and the superior temporal sulcus. The increased level of error for these landmarks should be taken into account both when deciding which landmarks to collect for a given study and, if these landmarks are employed, during data analysis. Notably, error can be further minimized by completing multiple landmarking trials, calculating the average landmark coordinates, and then conducting statistical analyses using these averaged data [57].

Intra- and inter-observer error was consistent with, and often better than, previous studies using landmark-based statistical shape analysis of the brain. Maudgil et al. [33] reported a mean intra-rater precision of 3.7 mm and inter-rater precision of 6.0 mm for 12 cortical landmarks. Aldridge [35] reported a mean intra-observer precision of 2.17 mm, with error ranging from 0.61–7.51 mm. Weinberg et al. [34] reported intra-class correlation coefficients of 0.86–1.0, which are not directly comparable to results in this study. Although the error was consistent with what has been reported in the literature, it remains unclear what the functional implications of this level of error are. The functional boundaries of cortical regions are diffuse, and even well circumscribed anatomical structures within the brain such as the thalamus or caudate nucleus are not limited to a single function. It is thus possible that an error of even 1.5 mm could impact a study's results.

One of the overarching goals of this study was to create a resource that could be used by both students and researchers who are new to the field of brain landmarking and advanced landmarkers as a reference source. In alignment with this goal, the aim was to create protocols that could be collected using existing morphometric software packages and not rely upon cortical parcellation or functional brain mapping before landmark placement. This meant creating a set of protocols that went beyond a single line definition of the landmark and included step-by-step specifications with associated images. When assessing the validity of these protocols, Rater 3 was chosen because she only had cursory knowledge of brain structure and had never before landmarked the brain. The only guidance she was given before execution of the project was a single 2-hour review session on the location of cortical and subcortical structures. As the ranges of intra-rater error indicate (Rater 1 = 1.07–9.19 mm, Rater 2 = 0.45–5.11 mm, and Rater 3 = 0.54–8.34 mm), researchers who have limited knowledge of brain landmarks can successfully follow these protocols.

The primary limitation of this study is that the sample size was limited to ten subjects. A sample size of ten was chosen to provide a sufficient level of evaluation while also making it possible to complete each landmarking trial in a single sitting. In addition, it is notable that this study was completed on a set of 12-year-old Caucasian males. It is possible that the protocols established in this study are not as accurate for individuals of a different age due to subtle changes in brain morphology with aging, but unlikely considering that most landmarks were defined by stable boundaries such as the intersection of two sulci or the centroid of a subcortical structure.

In summary, this study established detailed protocols with a minimal level of error for a set of twenty-nine subcortical and cortical landmarks. Future work includes the definition of additional landmarks relevant to hypotheses about brain shape, establishment and testing of protocols for these landmarks, and continued refinement of existing protocols in response to documented anatomical variation at landmark sites.

Supporting Information

Figure S1 Landmark guide. The landmark guide includes detailed step-by-step directions for the location of the 29 landmarks assessed in this study.

Table S1 Raw landmark coordinate data. This table includes the raw coordinate data for all of the landmark trials for all three raters.

References

1. Casey BJ, Tottenham N, Liston C, Durston S (2005) Imaging the developing brain: what have we learned about cognitive development? Trends Cogn Sci 9: 104–110.
2. Giedd JN, Rapoport JL (2010) Structural MRI of pediatric brain development: what have we learned and where are we going? Neuron 67: 728–734.
3. Raz N, Rodrigue KM (2006) Differential aging of the brain: patterns of cognitive correlates and modifiers. Neurosci Biobehav Rev 30: 730–748.
4. Gold SM, Dziobek I, Sweat V, Tirsi A, Rogers K, et al. (2007) Hippocampal damage and memory impairments as possible early brain complications of type 2 diabetes. Diabetologia 50:711–719.
5. Lorenzetti V, Allen NB, Fornito A, Yucel M (2009) Structural brain abnormalities in major depressive disorder: a selective review of recent MRI studies. J Affect Disord 117: 1–17.
6. Olabi B, Ellison-Wright I, McIntosh AM, Wood SJ, Bullmore E, et al. (2011) Are there progressive brain changes in schizophrenia? A meta-analysis of structural magnetic resonance imaging studies. Biol Psychiatry 70:88–96.
7. Courchesne E, Campbell K, Solso S (2011) Brain growth across the life span in autism: age-specific changes in anatomical pathology. Brain Res 1380: 138–145.
8. Sherwood CC, Rilling JK, Holloway RL, Hof PR (2009) Evolution of the brain in humans: specializations in a comparative perspective. In: Binder MD, Hirokawa N, Windhorst U, Hirsch MC, editors. Springer-Verlag: Encyclopedia of Neuroscience. pp 1334–1338.
9. Jerison HJ (1973) Evolution of the brain and intelligence. New York: Academic Press.
10. Ringo JL (1991) Neuronal interconnection as a function of brain size. Brain Behav Evol 38: 1–6.
11. Purves D, While L, Zheng D, Andrews T, Riddle D (1996) Brain size, behavior, and the allocation of neural space. In: Magnussen D, editor. Individual development over the lifespan: biological and psychological perspectives. Cambridge: Cambridge University Press. pp 162–178.
12. Gerhart J, Kirschner M (1997) Cell, embryos, and evolution. London: Blackwell Science.
13. Caviness VS Jr, Lange NT, Makris N, Herbert MR, Kennedy DN (1999) MRI-based brain volumetrics: emergence of a developmental brain science. Brain Dev 21: 289–295.
14. Bauman ML, Kemper TL (2005) Neuroanatomic observations of the brain in autism: a review and future directions. Int J Dev Neurosci 23: 183–187.
15. Bookstein F (1991) Morphometric tools for landmark data: geometry and biology. Cambridge: Cambridge University Press.
16. Lele S, Richtsmeier J (2001) An invariant approach to the statistical analysis of shapes. Boca Raton: Chapman & Hall/CRC.
17. Aldridge K, Marsh JL, Govier D, Richtsmeier JT (2002) Central nervous system phenotypes in craniosynostosis. J Anat 201: 31–39.
18. Aldridge K (2011) Patterns of differences in brain morphology in humans as compared to extant apes. J Hum Evol 60: 94–105.
19. Bookstein FL (1997) Landmark methods for forms without landmarks: morphometrics of group differences in outline shape. Med Image Anal 1: 225–243.
20. Joshi SH, Narr KL, Philips OR, Nuechterlein KH, Asarnow RF, et al. (2013) Statistical shape analysis of the corpus callosum in Schizophrenia. Neuroimage 64: 547–559.
21. Ashburner J, Hutton C, Frackowiak R, Johnsrude I, Price C, et al. (1998) Identifying global anatomical differences: deformation-based morphometry. Hum Brain Mapp6: 348–357.
22. Joseph J, Warton C, Jacobson SW, Jacobson JL, Molteno CD, et al. (2012) Three-dimensional surface deformation-based shape analysis of hippocampus and caudate nucleus in children with fetal alcohol spectrum disorders. Hum Brain Mapp doi: 10.1002/hbm.22209.
23. Ceyhan E, Beg MF, Certiglu C, Wang L, Morris JC, et al. (2012) Metric distances between hippocampal shapes indicate different rates of change over time in nondemented and demented subjects. Curr Alzheimer Res 9:972–981.
24. Mangin JF, Riviere D, Cachia A, Duchesnay E, Cointepas Y, et al. (2004) A framework to study the cortical folding patterns. Neuroimage 23: S129–S138.
25. Nordahl CW, Dierker D, Mostafavi I, Schumann CM, Rivera SM, et al. (2007) Cortical folding abnormalities in autism revealed by surface-based morphometry. J Neurosci 27: 11725–11735.
26. Dryden IL, Mardia KV (1998) Statistical shape analysis. Chichester: Wiley.
27. Adams DC, Rohlf FJ, Slice DE (2004) Geometric morphometrics: Ten years of progress following the 'revolution'. Ital J Zool 71: 5–16.
28. Mitteroecker P, Gunz P (2009) Advances in geometric morphometrics. Evol Biol 36: 235–247.
29. Deacon T (1990) Rethinking mammalian brain evolution. Am Zool 30: 629–705.
30. Harvey PH, Krebs JR (1990) Comparing brains. Science 249: 140–146.
31. Aboitiz F (1996) Does bigger mean better? Evolutionary determinants of brain size and structure. Brain Behav Evol 47: 225–245.
32. Keverne EB, Fundele R, Narasimha M, Barton SC, Surani MA (1996) Genomic imprinting and the differential roles of parental genomes in brain development. Brain Res Dev Brain Res 92: 91–100.
33. Maudgil DD, Free SL, Sisodiya SM, Lemieux L, Woermann FG, et al. (1998) Identifying homologous anatomical landmarks on reconstructed magnetic resonance images of the human cerebral cortical surface. J Anat 193: 559–571.
34. Weinberg SM, Andreasen NC, Nopoulos P (2009) Three-dimensional morphometric analysis of brain shape in nonsyndromic orofacial clefting. J Anat 214: 926–936.
35. Aldridge K (2004) Organization of the human brain: development, variability, and evolution. Johns Hopkins University: Dissertation.
36. Lohmann G, von Cramon Y (2000) Automatic labeling of the human cortical surface using sulcal basins. Med Image Anal 4:179–188.
37. Cachia A, Mangin JF, Riviere D, Kherif F, Boddaert N, et al. (2003) A primal sketch of the cortex mean curvature: a morphogenesis based approach to study the variability of the folding patterns. IEEE Trans Med Imaging 22:754–765.
38. Thompson PM, Hayashi KM, Sowell ER, Gogtay N, Giedd JN, et al. (2004) Mapping cortical change in Alzheimer's disease, brain development, and schizophrenia. NeuroImage 23:S2–S18.
39. Van Essen D (2005) A population-average, landmark- and surface-based (pals) atlas of human cerebral cortex. NeuroImage 28:635–662.
40. Shattuck DW, Joshi AA, Pantazis D, Kan E, Dutton RA, et al. (2009) Semi-automated technique for delineation of landmarks on models of the cerebral cortex. J Neurosci Methods 178:385–392.
41. Im K, Jo HJ, Mangin J, Evans AC, Kim SL, et al. (2010) Spatial distribution of deep sulcal landmarks and hemispherical asymmetry on the cortical surface. Cereb Cortex 20:602–611.
42. Zhong J, Phua DY, Qiu A (2010) Quantitative evaluation of LDDMM, FreeSurfer, and CARET for cortical surface mapping. NeuroImage 52:131–141.
43. Pantazis D, Joshi A, Jiang J, Shattuck DW, Bernstein LE, et al. (2010) Comparison of landmark-based and automatic methods for cortical surface registration. NeuroImage 49:2479–2493.
44. Andreasen NC, Cohen G, Harris G, Cizadlo T, Parkkinen J, et al. (1992) Image processing for the study of brain structure and function: problems and programs. J Neuropsych Clin Neurosci 4: 125–133.
45. Andreasen NC, Cizadlo T, Harris G, Swayze V, O'Leary DS, et al. (1993) Voxel processing techniques for the antemortem study of neuroanatomy and neuropathology using magnetic resonance imaging. J Neuropsych Clin Neurosci 5: 121–130.
46. Andreasen NC, Harris G, Cizadlo T, Arndt S, O'Leary DS, et al. (1994) Techniques for measuring sulcal/gyral patterns in the brain as visualized through magnetic resonance scanning: BRAINPLOT and BRAINMAP. Proc Natl Acad Sci 91: 93–97.
47. Magnotta V, Harris G, Andreasen NC, O'Leary DS, Yuh WT, et al. (2002) Structural MR image processing using the BRAINS2 toolbox. Comput Med Imaging Graph 26: 251–264.
48. Pierson R, Johnson H, Harris G, Keefe H, Paulsen JS, et al. (2011) Fully automated analysis using BRAINS: AutoWorkup. NeuroImage 54:328–336.
49. Aldridge K, Kane AA, Marsh JL, Panchal J, Boyadjiev SA, et al. (2005a) Brain morphology in nonsyndromic unicoronal craniosynostosis. Anat Rec A Discov Mol Cell Evol Biol 285: 690–698.
50. Aldridge K, Kane AA, Marsh JL, Yan P, Govier D, et al. (2005b) Relationship of brain and skull in pre- and postoperative sagittal synostosis. J Anat 206: 373–385.
51. DeQuardo JR, Bookstein FL, Green WDK, Brunberg JA, Tandon R (1996) Spatial relationships of neuroanatomic landmarks in schizophrenia. Psychiatry Res 67: 81–95.

Acknowledgments

Special thanks to Dr. Peg Nopoulos at the University of Iowa for providing the magnetic resonance images used in this project.

Author Contributions

Conceived and designed the experiments: MBC VBD SW KA NP. Performed the experiments: MBC KA NP. Analyzed the data: MBC VBD. Contributed reagents/materials/analysis tools: MBC VBD KA. Wrote the paper: MBC KA NP SW VBD.

52. DeQuardo JR, Keshavan MS, Bookstein FL, Bagwell WW, Green WDK, et al. (1999) Landmark-based morphometric analysis of first-episode schizophrenia. Biol Psychiatry 45: 1321–1328.

53. Gharaibeh WS, Rohlf FJ, Slice DE, DeLisi LE (2000) A geometric morphometric assessment of change in midline brain structural shape following a first episode of schizophrenia. Biol Psychiatry 48: 398–405.

54. Free SL, O'Higgins P, Maudgil DD, Dryden IL, Lemieux L, et al. (2001) Landmark-based morphometrics of the normal adult brain using MRI. Neuroimage 13: 801–813.

55. Mullick R, Venkataraman S, Warusavithana S, Nguyen HT, Raghavan R (1998) eTDIPS: 2D/3D image processing system for volume rendering and telemedicine. Annual Meeting of the Society for Computer Applications in Radiology

56. Mullick R, Warusavithana SV, Shalini V, Pang P (1999) Plug-ins: a software model for biomedical imaging and visualization research. Biomedical Imaging Symposium: Visualizing the Future of Biology and Medicine, National Institutes of Health (NIH).

57. Valeri CJ, Cole TM, Lele S, Richtsmeier JT (1998) Capturing data from three-dimensional surfaces using fuzzy landmarks. Am J Phys Anthropol 107:113–124.

13

A Reporter Assay in Lamprey Embryos Reveals Both Functional Conservation and Elaboration of Vertebrate Enhancers

Hugo J. Parker[1¤a], **Tatjana Sauka-Spengler**[2¤b], **Marianne Bronner**[2], **Greg Elgar**[1]*

1 Division of Systems Biology, Medical Research Council National Institute for Medical Research, London, United Kingdom, **2** Division of Biology, California Institute of Technology, Pasadena, California, United States of America

Abstract

The sea lamprey is an important model organism for investigating the evolutionary origins of vertebrates. As more vertebrate genome sequences are obtained, evolutionary developmental biologists are becoming increasingly able to identify putative gene regulatory elements across the breadth of the vertebrate taxa. The identification of these regions makes it possible to address how changes at the genomic level have led to changes in developmental gene regulatory networks and ultimately to the evolution of morphological diversity. Comparative genomics approaches using sea lamprey have already predicted a number of such regulatory elements in the lamprey genome. Functional characterisation of these sequences and other similar elements requires efficient reporter assays in lamprey. In this report, we describe the development of a transient transgenesis method for lamprey embryos. Focusing on conserved non-coding elements (CNEs), we use this method to investigate their functional conservation across the vertebrate subphylum. We find instances of both functional conservation and lineage-specific functional evolution of CNEs across vertebrates, emphasising the utility of functionally testing homologous CNEs in their host species.

Editor: Sylvie Rétaux, CNRS, France

Funding: This work was supported by MRC core funding (U117597141) to GE and grants GM090049 and DE017911 to MEB. The funders had no role in study design, data collection and analysis, decision to publish, or preparation of the manuscript.

Competing Interests: The authors have declared that no competing interests exist.

* E-mail: gelgar@nimr.mrc.ac.uk

¤a Current address: Stowers Institute for Medical Research, Kansas City, Missouri, United States of America
¤b Current address: Weatherall Institute of Molecular Medicine, University of Oxford, John Radcliffe Hospital, Oxford, United Kingdom

Introduction

The sea lamprey, *Petromyzon marinus*, is a member of the jawless fish lineage (agnathans), the only extant vertebrate sister group to the jawed vertebrates (gnathostomes) [1]. As such, it provides a unique window into early vertebrate history, enabling inference of the ancestral states and evolutionary origins of vertebrate characters. For example, investigations into lamprey genetics and embryogenesis have shed light on the evolution of the jaw [2], [3], paired fins [4], neural crest [5], [6], pharynx [7], immune system [8], sympathetic nervous system [9], forebrain [10], [11], and hindbrain [12]. Whilst the restricted summer breeding season presents some practical difficulties for studying lamprey embryology, the availability of copious embryos during this period has enabled the establishment of a suite of developmental biology techniques for lamprey, including *in-situ* hybridisation, morpholino-mediated gene knockdown and cell labelling [5], [7], [13] [14]. Furthermore, the recent genome assembly has significantly enhanced its utility as a model organism [15].

The emergence and elaboration of *cis*-regulatory elements in early vertebrates are likely to have contributed significantly to the evolution of the vertebrate body plan. Whilst embryonic gene expression patterns can be informative as to the developmental changes underlying morphological evolution, the changes at the *cis*-regulatory level that are responsible for divergent gene expression have received less attention. There are a handful of studies where early vertebrate *cis*-regulatory evolution has been addressed through testing amphioxus or lamprey genomic sequences for enhancer activity in gnathostomes [16–18]. These elements provide evidence for ancient conserved mechanisms of gene regulation across chordates. However, in order to gain a more complete picture of the contribution of changes in both *cis*- and *trans*- regulatory state during vertebrate evolution, it is important also to test such elements in their host species [19]. As increasing numbers of lamprey *cis*-regulatory elements are uncovered [20], [21], an efficient reporter assay methodology for functionally testing them in lamprey is sorely needed.

Shared conserved non-coding elements (CNEs), have been identified through comparisons between lamprey and gnathostome genomic sequences, representing a collection of putative *cis*-regulatory elements that were present in the last common ancestor of all vertebrates [20], [21], [22]. As the vast majority of these elements are not identifiable in invertebrate genomes, they are a vertebrate-specific character [23]. Many CNEs have been shown to function as developmental enhancers by reporter assay within their host species (e.g. [23], [24]), yet the degree to which the regulatory functions of CNEs are conserved between species has been investigated less frequently (e.g. [25]). Recent findings from zebrafish and mouse assays suggest that the regulatory functions of orthologous CNEs in their respective species are frequently well conserved, but on occasion can differ considerably [19], [26], which echoes findings from testing paralogous pairs of CNEs in a zebrafish assay [27], [28]. For the instances where homologous CNEs show functional divergence, these elements could provide a route through which divergence in gene expression patterns can be directly linked to specific sequence changes within *cis*-regulatory elements (e.g. [29]). Thus, lamprey-gnathostome CNEs could be informative as to the functional elaboration of homologous *cis*-regulatory elements across the vertebrate sub-phylum.

Kusukabi *et al.* [30] conducted the first application of transgenesis in an agnathan, using a close relative of the sea lamprey – the Japanese lamprey, *Lampetra japonica*. In their study, circular plasmid constructs containing the GFP coding sequence downstream of either a viral promoter or upstream regulatory regions of medaka actin genes were injected into fertilised eggs before the first cleavage, resulting in mosaic GFP expression. Interestingly, muscle-specific actin promoters from medaka were able to drive GFP expression in the developing muscle of lamprey embryos. Whilst the long lamprey life-cycle impinges upon the likelihood of generating transgenic lines, this study illustrated the feasibility of assaying CNEs for enhancer activity in lamprey.

In a study focusing on putative Hox-regulated enhancers conserved between jawed vertebrates and lamprey, we presented reporter expression patterns driven by two lamprey elements in lamprey embryos, verifying that ancient CNEs have broad functional conservation across vertebrates [31]. In this report, we expand upon that by describing the development of the I-SceI meganuclease-mediated transient transgenesis approach for lamprey embryos. We add versatility to the assay by identifying multiple minimal promoters that are functional in lamprey. As an example of its application, we perform reciprocal cross-species comparisons of CNE enhancer activity, finding evidence for both functional conservation and divergence across vertebrate taxa.

Materials and Methods

Zebrafish transgenesis

CNEs were amplified from genomic DNA by PCR (primer sequences are given in Figures S1 and S2) and transferred into the pGW_cfosEGFP vector by Gateway recombination via the pCR8/GW/TOPO TA entry vector (Invitrogen). Transgenesis was carried out according to the protocol of Fisher *et al.* [32] using the wild-type QMWT zebrafish strain.

Lamprey transgenesis

For linearised plasmid injection, the pm3285_cfos_EGFP plasmid, consisting of the lamprey homolog of CNE 3285 cloned into the pGW_cfosEGFP vector, was linearised with KpnI (NEB), purified with a Qiagen PCR purification kit and eluted in distilled water. Lamprey embryos were obtained as described previously [6] and injected with approximately 2–3 nl of linearised plasmid at

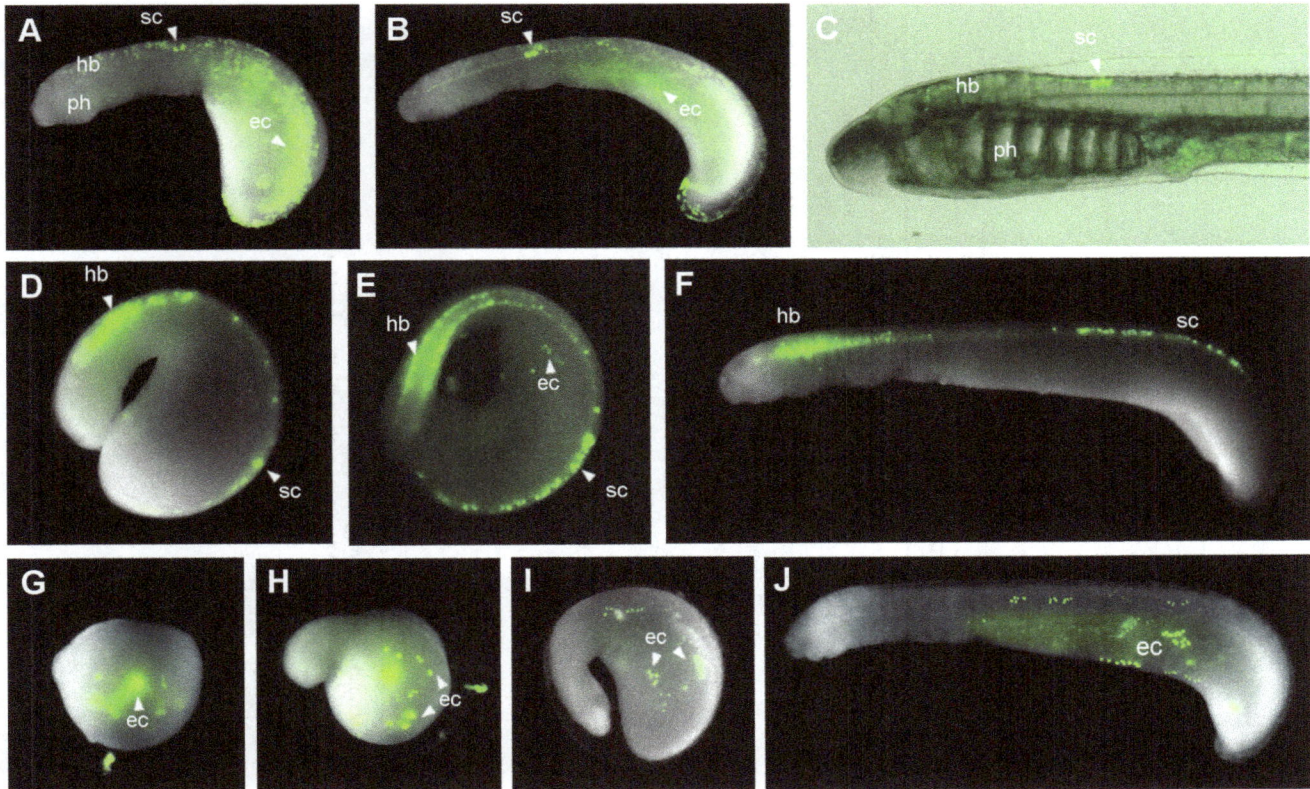

Figure 1. I-scel meganuclease-mediated transgenesis in lamprey embryos. (A–C) Transient transgenic lamprey generated through injection of the linearised pm3285_cfos_EGFP plasmid showing mosaic GFP expression in the ectoderm and in neurons of the spinal cord at stage 23 (**A**), stage 25 (**B**) and at the ammocoete larval stage (**C**). **(D–F)** Lamprey embryo with GFP expression in the hindbrain and spinal cord at stage 23 (**D**), stage 24 (**E**) and stage 25 (**F**), obtained using the meganuclease assay with the pm3285 enhancer. **(G–J)** 'Background' GFP expression in the ectoderm of lamprey embryos at stage 19 (**G**), 21 (**H**), 23 (**I**) and 25 (**J**), driven by the cfos_Iscel_EFGP plasmid with no enhancer using the meganuclease method (embryos shown are different individuals). Abbreviations: ec, ectoderm; hb, hindbrain; ph, pharynx; sc, spinal cord.

Table 1. A comparison of lamprey transgenesis methods.

Method	Element	Promoter	Plasmid conc. /ngul^{-1}	Embryos injected	Survivors (% of injected)	Ectodermal expression (% of survivors)	Neuronal expression (% of survivors)
Linearised plasmid	pm3285	cfos	100	500	18 (4%)	10 (56%)	5 (28%)
Linearised plasmid	pm3285 (repeat)	cfos	100	500	17 (3%)	11 (65%)	2 (12%)
Meganuclease	no enhancer	cfos	20	350	232 (66%)	203 (88%)	0
Meganuclease	pm3285	cfos	20	600	220 (37%)	not counted	35 (16%)
Meganuclease	pm3285 (repeat)	cfos	20	550	422 (77%)	166 (39%)	83 (20%)
Circular plasmid	pm3285	cfos	50	550	120 (22%)	0	0
Circular plasmid	pm3285 (repeat)	cfos	50	500	72 (14%)	0	0
Meganuclease (same embryo batch as for circular vector above)	pm3285	cfos	20	500	139 (28%)	not counted	22(16%)
Meganuclease	pm3299	cfos	20	700	302 (43%)	not counted	56 (19%)
Meganuclease	dr3285	cfos	20	600	217 (36%)	not counted	29 (13%)
Meganuclease	dr3299	cfos	20	650	467 (72%)	not counted	49 (10%)
Meganuclease	pm3299	hsp70	20	600	124 (21%)	not counted	44 (35%)

Results are shown for linearised plasmid injection and meganuclease-mediated transgenesis. Ectodermal background expression was not counted for the injections of pm3285 and pm3299 with the meganuclease method, but the proportion of embryos with this background expression was in keeping with that found for the cfos_I-Scel_EGFP construct.

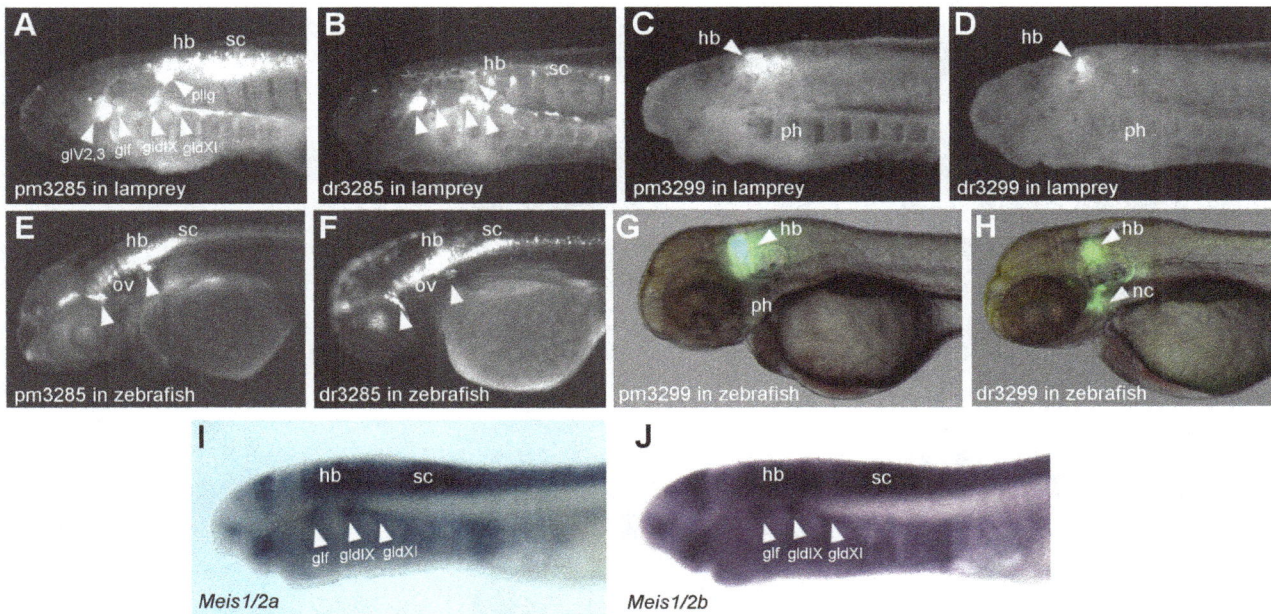

Figure 2. Reporter expression driven by CNEs 3285 and 3299 in lamprey and zebrafish embryos. (A–B) GFP fluorescence in stage 26 transient transgenic lamprey embryos, generated by meganuclease-mediated transgenesis with the pm3285 (A) and dr3285 enhancers (B). GFP expression is seen in the cranial ganglia (arrowheads), hindbrain and spinal cord. (C–D) GFP fluorescence in the rostral hindbrain in stage 26 transient transgenic lamprey embryos generated by meganuclease-mediated transgenesis with the pm3299 (C) and dr3299 (D) enhancers. GFP expression driven by pm3285 (E) and dr3285 (F) in 54hpf F1 transgenic zebrafish embryos, created by Tol2 transgenesis. Expression in the cranial ganglia is seen for the elements from both zebrafish and lamprey (arrowheads). Expression is also seen in primary neurons of the hindbrain and spinal cord. (G–H) 54hpf F1 transgenic zebrafish embryos showing GFP expression driven by the pm3299 (G) and dr3299 enhancers (H). GFP expression in the hindbrain is driven by both elements, whilst only the zebrafish element up-regulates GFP in the neural crest. (I–J) Expression of *Meis1/2a* (I) and *b* (J) genes in stage 25 lamprey embryos, revealed by *in-situ* hybridisation. Arrowheads highlight expression in cranial ganglia. Abbreviations: glV2,3, trigeminal ganglion; gldIX, epibranchial ganglion of the glossopharyngeal nerve; gldX1, epibranchial ganglion of the vagus nerve; glf, facial ganglion; hb, hindbrain; nc, neural crest; ov, otic vesicle; ph, pharynx; pllg, posterior lateral line ganglion; sc, spinal cord.

a concentration of 100 ngμl^{-1} during the first cell division. Circular plasmid injection was performed at a concentration of 50 ngμl^{-1}. I-SceI meganuclease-mediated transgenesis was based on the protocol of Ogino *et al.* [33]. Putative enhancers were cloned into the cfos_I-SceI_EGFP lamprey reporter plasmid, engineered for this study, upstream of the mouse *c-Fos* promoter. 20 μl restriction digests containing 15 units I-SceI enzyme (NEB), 2 μl 10X I-SceI buffer, 0.2 μl 100X BSA and 400 ng of reporter plasmid (final concentration 20 ngμl^{-1}), were incubated at 37°C for 40 minutes prior to being immediately injected into lamprey embryos during their first cell division (drop volume approximately 2–3 nl).

Lamprey in-situ hybridisation

performed according to published protocols [14] using the *meis1/2 a* and *b* probes [6].

Results and Discussion

Identification of a minimal promoter, c-Fos, that functions in lamprey

We sought to develop a functional assay to test the ability of lamprey homologs of two CNEs, pm3285 and pm3299, associated with the *meis2* locus, to drive reporter expression in lamprey embryos. These CNEs are highly conserved between jawed vertebrates and lamprey [20] (Multiple sequence alignments are provided in Figures S1 and S2) and drive specific patterns of GFP expression in a zebrafish reporter assay. pm3285 drives expression

in the cranial ganglia, hindbrain and spinal cord, and pm3299 in the anterior hindbrain [31]. These patterns are consistent with the endogenous expression of *meis2* in zebrafish, which is broadly expressed in the central nervous system, particularly in the hindbrain [34], [35].

In order to test these elements for enhancer activity in lamprey embryos, we sought a minimal promoter that can function in lamprey. We tested the suitability of the mouse *c-Fos* minimal promoter, which we used in the zebrafish assay, by injecting the pm3285_cfos_EGFP plasmid into lamprey embryos. The plasmid was linearised to increase the chance of genomic integration. Injection of this construct during the first cleavage resulted in a high death rate immediately post-injection, as well as during gastrulation, such that the frequency of injected embryos surviving through gastrulation was very low (Table 1) and they were often deformed. Injecting the plasmid at lower concentration (50 ngμl^{-1}) produced surviving embryos but none were expressing GFP. In a proportion of the survivors from the 100 ngμl^{-1} injections, GFP was expressed in a mosaic manner in the cranial ganglia and neurons of the spinal cord (Table 1), in agreement with the reporter expression driven by the same construct using the Tol2 assay in zebrafish. One of these embryos is shown in Figure 1 A–C, exhibiting GFP expression in neurons of the spinal cord, as well as broad reporter expression in the ectoderm from an early stage. Interestingly, the neuronal expression persisted up to the ammocoete stage in this embryo (Figure 1C). The successful generation of embryos exhibiting GFP expression mediated by this

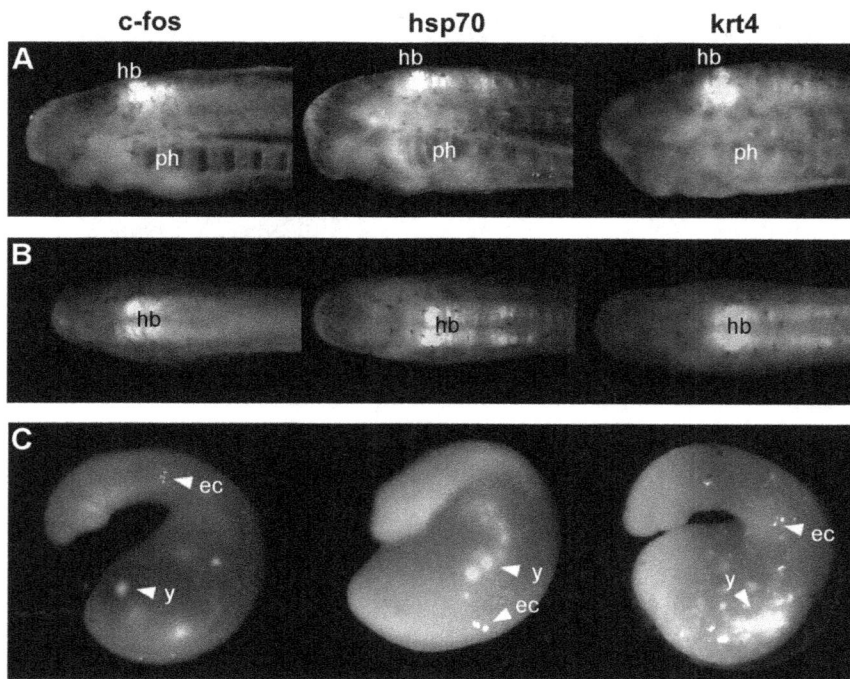

Figure 3. Identifying minimal promoters that are functional in lamprey. (**A–B**) Lateral (**A**) and dorsal (**B**) views of stage 26 transient transgenic lamprey embryos, showing reporter expression in the anterior hindbrain driven by different minimal promoters in conjunction with the pm3299 enhancer using the I-SceI meganuclease approach. (**C**) Lateral views of stage 23 lamprey embryos injected with the same constructs as those in **A** and **B**, showing early expression in ectoderm and yolk, similar to the background expression driven by the control cfos_I-SceI_EGFP construct (see Figure 1I). The identity of the minimal promoter used in each case is indicated above the top panel. Abbreviations: ec, ectoderm; hb, hindbrain; ph, pharynx; y, yolk.

construct suggested that the mouse *c-Fos* minimal promoter is functional in lamprey embryos.

The effect of I-SceI meganuclease-mediated transgenesis on mosaicism

In an effort to reduce mosaicism and increase embryo survival rate, we applied the I-SceI meganuclease-mediated transgenesis method to lamprey embryos. We reasoned that by increasing the probability of early genomic integration of the injected construct, the amount of DNA injected could also be lowered, lessening the toxic effect of exogenous DNA whilst decreasing mosaicism. This method utilises the rare-cutting I-SceI meganuclease and requires a construct in which the DNA fragment to be integrated is flanked by I-SceI recognition sites. The construct is digested with the meganuclease enzyme *in-vitro* and the reaction mix is then injected immediately into fertilised eggs. Whilst the integration mechanism is unclear, it is unlikely that the enzyme cuts genomic DNA, as its 18 bp recognition site is predicted to occur once in every 7×10^{10} bp of genomic sequence [36] (compared to the lamprey genome size of 2.3×10^9 bp). Rather, the continued association of the enzyme with the digested construct may prevent its degradation or concatamerisation, thus increasing the probability of genomic integration [36].

When pm3285 was tested using the meganuclease assay, the embryo survival was considerably higher than with linearised plasmid (37% survival post-gastrulation compared to 3.6%) and 35 embryos showed expression in the hindbrain or spinal cord, with 10 of these 35 also expressing GFP in the cranial ganglia (Table 1). Repeat injections confirmed that the increased embryo survival achieved through the meganuclease approach is reproducible

(Table 1). The pattern of GFP expression obtained using the meganuclease method is in the same domains as that obtained by the injection of the pm3285_cfos linearised plasmid (Figure 1A–C compared to D–F). The earliest neural expression is seen at stage 21 in a low number of cells, becoming more expansive at later stages, whilst most embryos also display mosaic ectodermal expression.

The transient transgenic embryos obtained through the meganuclease approach display a range of mosaicism with respect to GFP expression in neurons. To compare mosaicism between the two transgenesis approaches, we categorised transient transgenics into two groups depending on whether they had more than or less than 50 GFP-positive neurons at stage 24. For the linearised plasmid injection, 0/5 embryos with neuronal GFP expression exhibited this expression in more than 50 neurons. In contrast, for the meganuclease approach, 9/35 embryos with neuronal expression had more than 50 neurons expressing GFP (examples of transient transgenic embryos obtained from each method are shown in Figure S3). This suggests that the meganuclease approach yields transient transgenic embryos with decreased mosaicism compared to linearised plasmid injection. However, due to the low number of survivors obtained through linearised plasmid injection, it is unclear whether its apparently high mosaicism is a general trend. Nevertheless, in this instance, the meganuclease approach was beneficial in generating an appreciable number of GFP-expressing embryos that showed lower mosaicism than those obtained through linearised plasmid injection. We were unable to generate any GFP-expressing lamprey embryos through injecting the pm3285_cfos_I-SceI_EGFP construct as a circular plasmid without the I-sceI meganuclease enzyme, an approach that had been used by Kusukabi *et al.* (2003) [30] with a different construct (Table 1). Thus, for our construct, the meganuclease transgenesis approach represents an improvement over circular plasmid injection in terms of the frequency of obtaining transient transgenic embryos, and an improvement over linearised plasmid injection in terms of the balance of reporter expression and embryo survival. Importantly, the promoter control – the cfos_I-SceI_EGFP construct with no enhancer – when injected using the meganuclease method, resulted in a large proportion (87.5%) of the survivors showing mosaic GFP expression in the ectoderm and yolk, but not in any other domains (Table 1, Figure 1G–I). As this expression is driven by the construct in the absence of an enhancer, we consider it to be 'background' expression.

Effects of CNE sequence divergence on transgene expression

To address whether CNE enhancer function has diverged between vertebrate lineages, we have focused upon the reporter expression patterns driven by the zebrafish (dr) and lamprey (pm) CNEs when tested in zebrafish and lamprey embryos. As the same CNE sequences were used in each assay, the expression patterns driven by them are directly comparable [28]. For each CNE, at least two independent transgenic lines were generated and their expression domains were compared. We observed GFP expression from pm3299 in 3 lines, dr3299 in 3 lines, pm3285 in 2 lines and dr3285 in 2 lines. The tissue specific domains of reporter expression that we highlight were conserved between lines for each enhancer. Expression in the cranial ganglia driven by pm3285 and dr3285 (Figure 2A,B), is in agreement with the reporter expression seen in transgenic zebrafish lines generated with these elements, in which GFP is seen in clusters of cranial ganglia both anterior and posterior to the otic vesicle (Figure 2E,F). Two lamprey genes showing homology to jawed vertebrate *meis* genes have been identified and named *pmMeis1/2a* and *b* [6]. The expression patterns of these two genes at this developmental stage are very similar to each other, with both showing clear cranial ganglia expression (Figure 2I,J). However, the expression of these genes does not entirely overlap with the GFP expression driven by pm3285, which extends to more anterior cranial ganglia. Both pm3285 and dr3285 also drive expression in neurons of the hindbrain and spinal cord in lamprey embryos (Figure 2A,B), expression domains that are also seen for these elements in zebrafish (Figure 2E,F). The reporter expression patterns driven by these elements when tested in zebrafish and lamprey embryos suggests that, with respect to these elements, *cis*- and *trans*-regulatory state are broadly conserved between lamprey and gnathostomes.

The hindbrain expression driven by pm3299 in lamprey has an anterior limit consistent with the expression pattern of the two lamprey *meis* genes (Figure 2C,I,J). In transgenic zebrafish lines, the expression driven by dr3299 differs from that of its lamprey homolog in two regards: firstly, dr3299 drives expression in the neural crest cells settling in the hyoid arch, whilst pm3299 does not (Figure 2G,H); secondly it is restricted to a smaller domain of the hindbrain, whilst pm3299 drives broader hindbrain expression (Figure 2G,H). These patterns led us to speculate that this enhancer may have been elaborated in gnathostomes relative to lamprey, such that it gained a new expression domain in the neural crest and its expression in the hindbrain became more restricted. No observable neural crest expression is driven by pm3299 in lamprey embryos (Figure 2C), and the broad pattern of hindbrain expression (Figure 2C) is in line with the reporter expression driven by pm3299 in zebrafish (Figure 2G). The zebrafish element, dr3299, when tested in lamprey, drives reporter expression in the hindbrain but no reporter expression is seen in the pharyngeal neural crest (Figure 2D). This suggests that the neural crest expression that is driven by the zebrafish enhancer when tested in zebrafish is a consequence of differences in both *cis* and *trans* between zebrafish and lamprey.

Effects of different promoters on transgene expression

To add further versatility to the assay, we have tested a selection of promoter elements with the I-SceI meganuclease method, using lamprey CNE 3299 as the enhancer. We selected the mouse *β-globin* minimal promoter and three zebrafish minimal promoters from the genes *hsp70*, *klf4* and *krt4* (see Table S1 for sequences), which have previously been shown to display low background activity and high interactivity with a variety of enhancers in zebrafish [37]. No GFP-expressing lamprey embryos were obtained using the *β-globin* or *klf4* promoters with CNE 3299. When tested in conjunction with lamprey CNE 3299, the zebrafish *hsp70*, *krt4* and mouse *c-Fos* promoters all up-regulate GFP expression in a consistent manner in the hindbrain and spinal cord (Figure 3). The consistent GFP expression domains driven by lamprey CNE 3299 with three different minimal promoters confirm the validity of this expression pattern by eliminating the possibility of promoter bias. Each of these three promoters also drives mosaic expression in the ectoderm. The *hsp70* and *krt4* promoters were not tested in the absence of an enhancer, so we cannot definitively pronounce this to be background expression for these two promoters. However, we consider it likely to be background expression as it is in the same domain as that driven by the *c-Fos* promoter control and as described previously for other constructs in lamprey [30].

Conclusion

We have identified an improved method for lamprey transgenesis. The increased numbers of transient transgenic embryos obtained through the I-SceI meganuclease assay in lamprey makes it an improvement over simple linearised plasmid injection. Given the utility of lamprey as a model organism for investigating early vertebrate evolution and the completion of the lamprey genome assembly, this is both important and timely. This assay will make it possible to probe the lamprey gene regulatory architecture by characterising lamprey enhancers and testing jawed vertebrate regulatory elements for activity in lamprey embryos. Lamprey transgenesis also offers scope for experimental attempts to re-create evolutionary transitions, such as the acquisition of the jaw or paired limbs. In these approaches, jawed vertebrate enhancers could be used to layer novel gene expression domains upon the putatively ancestral embryonic plan of the sea lamprey. The effects of inducing novel genetic cascades in these territories could provide insight into the regulatory changes underlying the evolution of jawed vertebrate characters.

Focusing upon ancient vertebrate CNEs, we have used our lamprey reporter assay to demonstrate conservation and elaboration of CNE function between two distantly related extant vertebrates. Our data from zebrafish and lamprey support the notion that, to a certain degree, ancient conserved enhancer sequences are indicative of core developmental programs that are common to all vertebrates. Nevertheless, these elements also appear to have been susceptible to lineage-specific evolutionary tinkering, with changes in *cis* and *trans* contributing to modification of their regulatory output.

Supporting Information

Figure S1 Multiple sequence alignment of CNE 3285 from vertebrate genomes. Primers used for amplification are highlighted in red.

Figure S2 Multiple sequence alignment of CNE 3299 from vertebrate genomes. Primers used for amplification are highlighted in red.

Figure S3 Comparison of mosaicism from linearised plasmid and I-SceI meganuclease transgenisis approaches. Examples of stage 24–25 transient transgenic embryos obtained through each approach using the pm3285 enhancer are shown.

Table S1 Minimal promoters tested in the lamprey reporter assay.

Acknowledgments

We thank Natalya Nikitina, Benjamin Uy, Melinda Modrell and Marcos Simoes-Costa for advice and assistance on lamprey husbandry. We thank Heather Callaway for zebrafish maintenance.

Author Contributions

Conceived and designed the experiments: HP GE TSS MB. Performed the experiments: HP TSS. Analyzed the data: HP. Contributed reagents/materials/analysis tools: GE MB. Wrote the paper: HP GE.

References

1. Blair JE, Hedges SB (2005) Molecular phylogeny and divergence times of deuterostome animals. Mol Biol Evol 22: 2275–2284.
2. Shigetani Y, Sugahara F, Kawakami Y, Murakami Y, Hirano S, et al. (2002) Heterotopic shift of epithelial-mesenchymal interactions in vertebrate jaw evolution. Science. 296: 1316–1319.
3. Cerny R, Cattell M, Sauka-Spengler T, Bronner-Fraser M, Yu F, et al. (2010) Evidence for the prepattern/cooption model of vertebrate jaw evolution. Proc Natl Acad Sci U S A. 107: 17262–17267.
4. Freitas R, Zhang G, Cohn MJ (2006) Evidence that mechanisms of fin development evolved in the midline of early vertebrates. Nature. 442: 1033–1037.
5. McCauley DW, Bronner-Fraser M (2003) Neural crest contributions to the lamprey head. Development. 130: 2317–2327.
6. Sauka-Spengler T, Meulemans D, Jones M, Bronner-Fraser M (2007) Ancient evolutionary origin of the neural crest gene regulatory network. Dev Cell. 13: 405–420.
7. McCauley DW, Bronner-Fraser M (2006) Importance of SoxE in neural crest development and the evolution of the pharynx. Nature. 441: 750–752.
8. Pancer Z, Amemiya CT, Ehrhardt GR, Ceitlin J, Gartland GL, et al. (2004) Somatic diversification of variable lymphocyte receptors in the agnathan sea lamprey. Nature. 430: 174–180.
9. Häming D, Simoes-Costa M, Uy B, Valencia J, Sauka-Spengler T, et al. (2011). Expression of sympathetic nervous system genes in Lamprey suggests their recruitment for specification of a new vertebrate feature. PloS one, 6(10), e26543.
10. Guérin A, d'Aubenton-Carafa Y, Marrakchi E, Da Silva C, Wincker P, et al. (2009) Neurodevelopment genes in lampreys reveal trends for forebrain evolution in craniates. PLoS One. 4: e5374.
11. Sugahara F, Aota S, Kuraku S, Murakami Y, Takio-Ogawa Y, et al. (2011) Involvement of Hedgehog and FGF signalling in the lamprey telencephalon: evolution of regionalization and dorsoventral patterning of the vertebrate forebrain. Development. 138: 1217–1226.
12. Murakami Y, Pasqualetti M, Takio Y, Hirano S, Rijli FM, et al. (2004) Segmental development of reticulospinal and branchiomotor neurons in lamprey: insights into the evolution of the vertebrate hindbrain. Development. 131: 983–995.
13. Murakami Y, Ogasawara M, Sugahara F, Hirano S, Satoh N, et al. (2001). Identification and expression of the lamprey Pax6 gene: evolutionary origin of the segmented brain of vertebrates. Development. 128: 3521–3531.
14. Nikitina N, Bronner-Fraser M, Sauka-Spengler T (2009) The sea lamprey Petromyzon marinus: a model for evolutionary and developmental biology. Cold Spring Harb 15. Protoc. 1: pdb-mo113.
15. Smith JJ, Kuraku S, Holt C, Sauka-Spengler T, Jiang N, et al. (2013). Sequencing of the sea lamprey (Petromyzon marinus) genome provides insights into vertebrate evolution. Nature genetics, 45(4), 415–421.
16. Carr JL, Shashikant CS, Bailey WJ and Ruddle FH (1998) Molecular evolution of Hox gene regulation: cloning and transgenic analysis of the lamprey HoxQ8 gene. J Exp Zool. 280: 73–85.
17. Manzanares M, Wada H, Itasaki N, Trainor PA, Krumlauf R, et al. (2000) Conservation and elaboration of Hox gene regulation during evolution of the vertebrate head. Nature. 408: 854–857.
18. Yu JK, Meulemans D, McKeown SJ, Bronner-Fraser M (2008) Insights from the amphioxus genome on the origin of vertebrate neural crest. Genome Res. 18: 1127–1132.
19. Ritter DI, Li Q, Kostka D, Pollard KS, Guo S, et al. (2010) The importance of being cis: evolution of orthologous fish and mammalian enhancer activity. Mol Biol Evol. 27: 2322–2332.
20. McEwen GK, Goode DK, Parker HJ, Woolfe A, Callaway H, et al. (2009) Early evolution of conserved regulatory sequences associated with development in vertebrates. PLoS Genet. 5: e1000762.
21. Kano S, Xiao JH, Osório J, Ekker M, Hadzhiev Y, et al. (2010) Two lamprey Hedgehog genes share non-coding regulatory sequences and expression patterns with gnathostome Hedgehogs. PLoS One. 5: e13332.
22. Irvine SQ, Carr JL, Bailey WJ, Kawasaki K, Shimizu N, et al. (2002) Genomic analysis of Hox clusters in the sea lamprey Petromyzon marinus. J Exp Zool. 294: 47–62.
23. Woolfe A, Goodson M, Goode DK, Snell P, McEwen GK, et al. (2005) Highly conserved non-coding sequences are associated with vertebrate development. PLoS Biol. 3: e7.
24. Pennacchio LA, Ahituv N, Moses AM, Prabhakar S, Nobrega MA, et al (2006) In vivo enhancer analysis of human conserved non-coding sequences. Nature. 444: 499–502.
25. de la Calle-Mustienes E, Feijóo CG, Manzanares M, Tena JJ, Rodríguez-Seguel E, et al. (2005) A functional survey of the enhancer activity of conserved non-coding sequences from vertebrate Iroquois cluster gene deserts. Genome Res. 15: 1061–1072.
26. Navratilova P, Fredman D, Hawkins TA, Turner K, Lenhard B, et al. (2009) Systematic human/zebrafish comparative identification of cis-regulatory activity around vertebrate developmental transcription factor genes. Dev Biol. 327: 526–540.
27. McEwen GK, Woolfe A, Goode D, Vavouri T, Callaway H, et al. (2006) Ancient duplicated conserved noncoding elements in vertebrates: a genomic and functional analysis. Genome Res. 16: 451–465.
28. Goode DK, Callaway HA, Cerda GA, Lewis KE, Elgar G (2011) Minor change, major difference: divergent functions of highly conserved cis-regulatory elements subsequent to whole genome duplication events. Development. 138: 879–884.
29. Prabhakar S, Visel A, Akiyama JA, Shoukry M, Lewis KD, et al. (2008). Human-specific gain of function in a developmental enhancer. Science. 321: 1346–1350.
30. Kusakabe R, Tochinai S, Kuratani S (2003) Expression of foreign genes in lamprey embryos: an approach to study evolutionary changes in gene regulation. J Exp Zool B Mol Dev Evol. 296: 87–97.
31. Parker HJ, Piccinelli P, Sauka-Spengler T, Bronner M, Elgar G (2011) Ancient Pbx-Hox signatures define hundreds of vertebrate developmental enhancers. BMC Genomics, 12: 637.
32. Fisher S, Grice EA, Vinton RM, Bessling SL, Urasaki A, et al. (2006) Evaluating the biological relevance of putative enhancers using Tol2 transposon-mediated transgenesis in zebrafish. Nat Protoc. 1: 1297–1305.
33. Ogino H, McConnell WB, Grainger RM (2006) High-throughput transgenesis in Xenopus using I-SceI meganuclease. Nat Protoc. 1: 1703–1710.
34. Biemar F, Devos N, Martial JA, Driever W, Peers B (2001) Cloning and expression of the TALE superclass homeobox Meis2 gene during zebrafish embryonic development. Mech Dev 109: 427–431.
35. Zerucha T, Prince VE (2001) Cloning and developmental expression of a zebrafish meis2 homeobox gene. Mech Dev. 102: 247–250.
36. Thermes V, Grabher C, Ristoratore F, Bourrat F, Choulika A, et al. (2002) I-SceI meganuclease mediates highly efficient transgenesis in fish. Mech Dev. 118: 91–98.
37. Gehrig J, Reischl M, Kalmár E, Ferg M, Hadzhiev Y, et al. (2009) Automated high-throughput mapping of promoter-enhancer interactions in zebrafish embryos. Nat Methods. 6: 911–916.

14

Bigger Helpers in the Ant *Cataglyphis bombycina*: Increased Worker Polymorphism or Novel Soldier Caste?

Mathieu Molet[1,2]*, **Vincent Maicher**[1,2], **Christian Peeters**[1,2]

1 Laboratoire Ecologie & Evolution – Unité Mixte de Recherche 7625, Université Pierre et Marie Curie, Paris, France, **2** Laboratoire Ecologie & Evolution – Unité Mixte de Recherche 7625, Centre National de la Recherche Scientifique, Paris, France

Abstract

Introduction: The mechanisms by which development favors or constrains the evolution of new phenotypes are incompletely understood. Polyphenic species may benefit from developmental plasticity not only regarding ecological advantages, but also potential for evolutionary diversification. For instance, the repeated evolution of novel castes in ants may have been facilitated by the existence of alternative queen and worker castes and their respective developmental programs.

Material and Methods: *Cataglyphis bombycina* is exceptional in its genus because winged queens and size-polymorphic workers occur together with bigger individuals having saber-shaped mandibles. We measured seven body parts in more than 150 individuals to perform a morphometric analysis and assess the developmental origin of this novel phenotype.

Results: Adults with saber-shaped mandibles differ from both workers and queens regarding the size of most body parts. Their relative growth rates are identical to workers for some pairs of body parts, and identical to queens for other pairs of body parts; critical sizes differ in all cases.

Conclusions: Big individuals are a third caste, i.e. soldiers, not major workers. Novel traits such as elongated mandibles are combined with a mix of queen and worker growth rates. We also reveal the existence of a dimorphism in the queen caste (microgynes and macrogynes). We discuss how novel phenotypes can evolve more readily in the context of an existing polyphenism. Both morphological traits and growth rules from existing queen and worker castes can be recombined, hence mosaic phenotypes are more likely to be viable. In *C. bombycina*, such a mosaic phenotype appears to function both for defense (saber-shaped mandibles) and fat storage (big abdomen). Recycling of developmental programs may have contributed to the morphological diversity and ecological success of ants.

Editor: Jesus E. Maldonado, Smithsonian Conservation Biology Institute, United States of America

Funding: This work was funded by Agence Nationale de la Recherche (project ANTEVO ANR-12-JSV7-0003-01) and by Laboratoire Ecologie & Evolution UMR7625 (Université Pierre et Marie Curie – Centre National de la Recherche Scientifique). The funders had no role in study design, data collection and analysis, decision to publish, or preparation of the manuscript.

Competing Interests: The authors have declared that no competing interests exist.

* E-mail: mathieu.molet@upmc.fr

Introduction

During animal ontogeny, organs follow distinct growth rules that produce an integrated phenotype. Two parameters are crucial: growth rate defines the speed at which each organ grows, and critical size specifies the overall size of the individual at which growth stops and development is complete [1]. By modifying growth rules (growth rate and critical size), new phenotypes can be produced. Phenotypic plasticity relies on such modifications to generate variation. Changing critical size without altering growth rates results in new phenotypes that are in the continuity of existing ones. Potential changes are thus limited. In contrast, modifying growth rates can lead to dramatically different phenotypes. Here we explore how growth rules are modified to generate new phenotypes by using one of the most polyphenic animal taxon as a model: ants.

In ants, phenotypic plasticity generates two morphologically distinct female castes: workers and queens. The growth rules of these adult phenotypes differ in both growth rates and critical size (when a larva reaches critical size, pupation is triggered) [1]. Queens are generally large, with a complex articulated winged thorax, and a specialized reproductive apparatus, whereas workers are smaller, with a simplified wingless thorax, and a reduced reproductive apparatus ([2]; Fig. 1). These two castes are adapted for distinct functions, namely independent colony foundation and reproduction for queens, and foraging, brood care, nest building and nest defense for workers.

This elementary dichotomy constitutes the groundplan for all ants, but it does not reflect the diversity of female phenotypes across the 13.000 extant species. The diversification of lifestyles has selected for changes in colony life history, phenotypes of colony members and their underlying growth rules. Indeed, the degree of dimorphism between queen and worker castes has increased considerably in many taxa, and this has created an empty morphospace associated with potential new functions. For instance, size polymorphic workers can be more efficient in

Figure 1. *Cataglyphis bombycina* **ants are highly polyphenic.** Colonies contain one dealate queen (left), numerous workers (middle and bottom), and fewer large individuals with saber-shaped mandibles 'ISM' (soldiers, right). Photo © P. Landmann.

resource acquisition or defense, ultimately improving colony growth [3]. The production of small queens -microgynes- can be more economical for dependent colony foundation, i.e. when young queens do not fly away but disperse on foot with nestmate workers [4]. Such female phenotypes result from changes in critical size relative to the ancestral phenotypes, but not in growth rates. Indeed, major workers are distributed along the growth curve of workers, with some traits becoming enlarged relative to others due to allometry only ([1]; Fig. 2). Similarly, microgynes in some species are 'isometric reductions' of macrogynes [5]. In contrast, other phenotypes result from changes in both critical size and growth rates, such as soldiers (e.g. [6–12]; Fig. 2) and ergatoid -permanently wingless- queens (reviewed in [13]). Their production also enhances colony success, respectively for survival/growth [7] and reproduction [14]. They can be seen as functional equivalents of major workers and microgynes respectively, although the mechanisms that produce them differ. Importantly the phenotypic outcomes of such new growth rules are potentially much more diverse than those based on similar growth rates (Fig. 2).

Cataglyphis ants are characteristic insects of arid regions distributed around the Mediterranean Basin and reaching into Central Asia [15]. Out of 90 species, most have continuously polymorphic workers. In contrast, *C. bombycina* (Roger) has colonies with two types of non-reproductive females: workers with variable sizes (reaction norm, i.e. continuous variation) and invariably big individuals with saber-shaped mandibles (discrete size; Fig. 1). The latter are morphologically striking and occur only in this species and *C. kurdistanicus* [16]. Their function in colonies is poorly understood. In *C. bombycina*, Délye [17] reported that they are rarely seen outside, but when the nest is disturbed they run around with open mandibles and bite.

We aimed to test whether individuals with saber-shaped mandibles ('ISM') are the product of allometric growth along the worker developmental curve (same growth rates but distinct critical sizes) or whether they are the outcome of completely different growth rules (new growth rates and new critical size). The first case would correspond to a major worker subcaste, whereas the second case would be a novel soldier caste. Accordingly, we compared the size, morphology and growth rules of 'ISM' with workers and queens. We also assessed behavior and physiology.

We discuss why the production of new phenotypes using existing growth rules is relatively risk-free in terms of viability and functionality but limits the number of possible outcomes. In contrast, the production of novel phenotypes using new growth rules can generate high diversity at the price of higher failure rate. We argue that ants can escape this cost by recombining worker and queen growth rules to produce novel phenotypes [18].

Materials and Methods

Colony Collection

Cataglyphis bombycina is a dominant species in the sand dunes of north Africa; nests are huge and deep, with multiple entrances distant by several meters. One colony (#1) was partly excavated in November 2010 by C. Peeters in Remlia, SE of Merzouga, Morocco (30.71°, −4.40°); only surface chambers (<1 m deep) could be sampled due to collapse of the rapidly drying sand. This yielded about 1000 adult ants including workers and individuals with saber-shaped mandibles, but no gynes (virgin winged queens). Four colonies (#2–5) were sampled by Serge Aron in Amerzgane (31.05°, −7.21°) [19] during nuptial flights in April 2011, and this yielded workers and gynes. No specific permissions were required to collect ants as these sand dunes are public land. Our field study did not involve endangered or protected species. Voucher specimens have been deposited in the California Academy of Sciences (see www.antweb.org for images, specimen IDs CASENT0906666, CASENT0906667, CASENT0906668).

Data Collection

We used 48 workers (40 random, 4 of the smallest and 4 of the largest) and 40 'ISM' (individuals with saber-shaped mandibles)

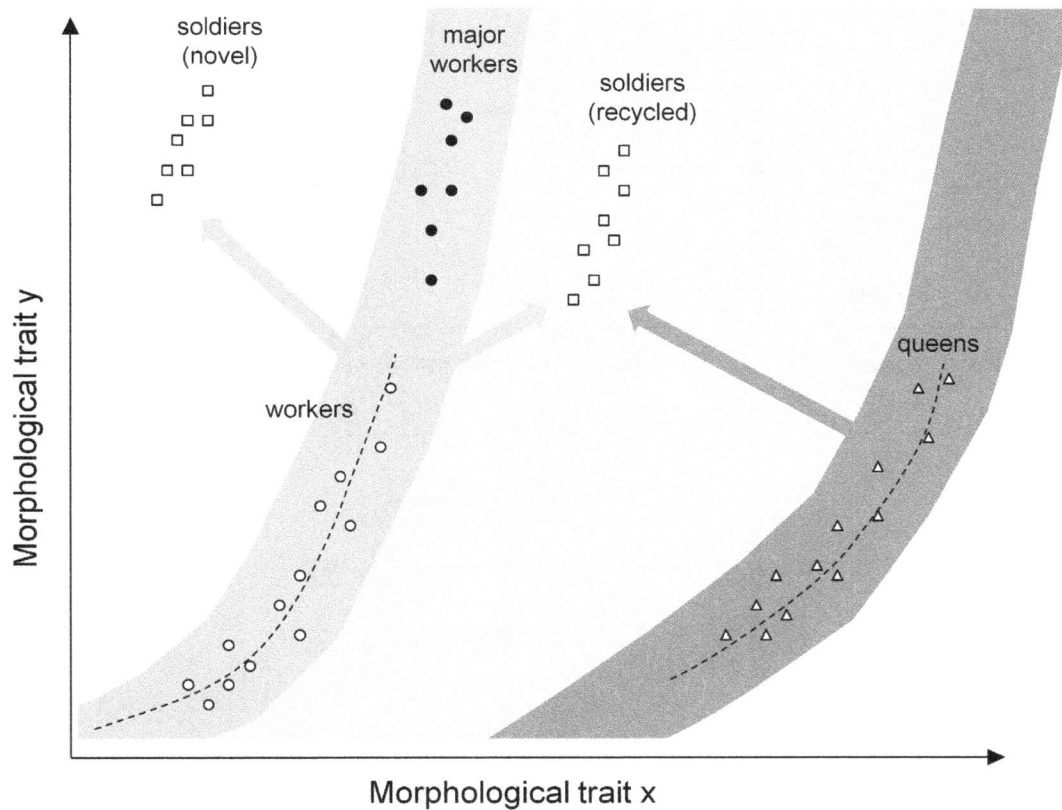

Figure 2. Distinct adult castes exhibit distinct growth rules. Growth rules of the worker (empty circles) and queen (empty triangles) castes for a theoretical pair of morphological traits 'x' and 'y' are illustrated by dashed lines. Increasing critical size of the worker growth rule leads to the production of adults that are larger and have a different shape due to allometry, i.e. major workers (black circles). However the range of possible phenotypes that can be produced is limited (grey area surrounding the workers' curve). Alternatively, modifying both critical size and growth rate leads to the production of novel adult phenotypes that are outside of this range and accordingly do not belong to the worker caste, i.e. soldiers (empty squares). We suggest that this can be done either by combining parameters of the growth rules of existing worker and queen castes ('recycled' soldiers) or by evolving brand new growth rules ('novel' soldiers).

from colony #1, and 12 random workers and 56 gynes from colonies #2–5. We measured seven body parts: head width, palp length, mandible length, area of the interior side of the mandible, tibia length, thorax volume and cross-sectional area of the first gaster segment (abdominal segment III) using ImageJ software (http://rsb.info.nih.gov/ij) following Molet et al. [20]. Since workers and gynes originated from different colonies, we first checked that the size of the various body parts did not differ between colonies. Kruskal-Wallis rank tests revealed no colony effect on either worker or queen body parts (P-values respectively 0.74 and 0.06 for head width, 0.19 and 0.72 for palp length, 0.60 and 0.06 for mandible length, 0.53 and 0.06 for area of the interior side of the mandible, 0.50 and 0.88 for tibia length, 0.51 and 0.06 for thorax volume and 0.21 and 0.69 for cross-sectional area of the first gaster segment. We also took scanning electron microscope photographs of all female types.

Statistical Analyses

In order to contrast growth rules among female types, we compared the sizes of body parts and the growth rates between body parts. Sizes were compared using Kruskal-Wallis rank tests followed by pairwise comparisons using Wilcoxon rank tests with Bonferroni correction. Growth rates, also known as allometry coefficients, were computed as follows. First we homogenized our measures of area and volume to a single linear dimension

(equivalent to a length) using square and cube root transformations respectively. Second, we log-transformed the homogenized length, area and volume data. Third, we performed a correlation analysis using pairs of transformed variables, and if it was significant we computed a regression line between Y and X variables, the slope of which is the allometry coefficient (growth rate). This coefficient describes how much body part Y grows when body part X grows. When the allometry coefficient equals one, body parts grow at the same rate so body shape does not change with size (isometry). When it differs from one, body parts grow at a different rate so body shape changes with size (allometry). For instance, a coefficient of three means that Y growths at a cubic rate relative to X, i.e. $Y = X^3$. Accordingly, when X gets bigger, Y gets much bigger and body shape changes dramatically. We tested whether the allometry coefficient differed from one (allometry) or not (isometry). Finally, we determined whether queens, workers and 'ISM' exhibit different growth rates by comparing their allometry coefficients for all pairs of body parts. We also compared elevations, which reflect critical size and are related to the intercepts of the regression lines. Allometry coefficients and elevations were computed and compared using (S)MATR 1.0 [21] (http://www.bio.mq.edu.au/ecology/SMATR/). All other statistical analyses were performed with R 2.13 (available at http://cran.r-project.org/).

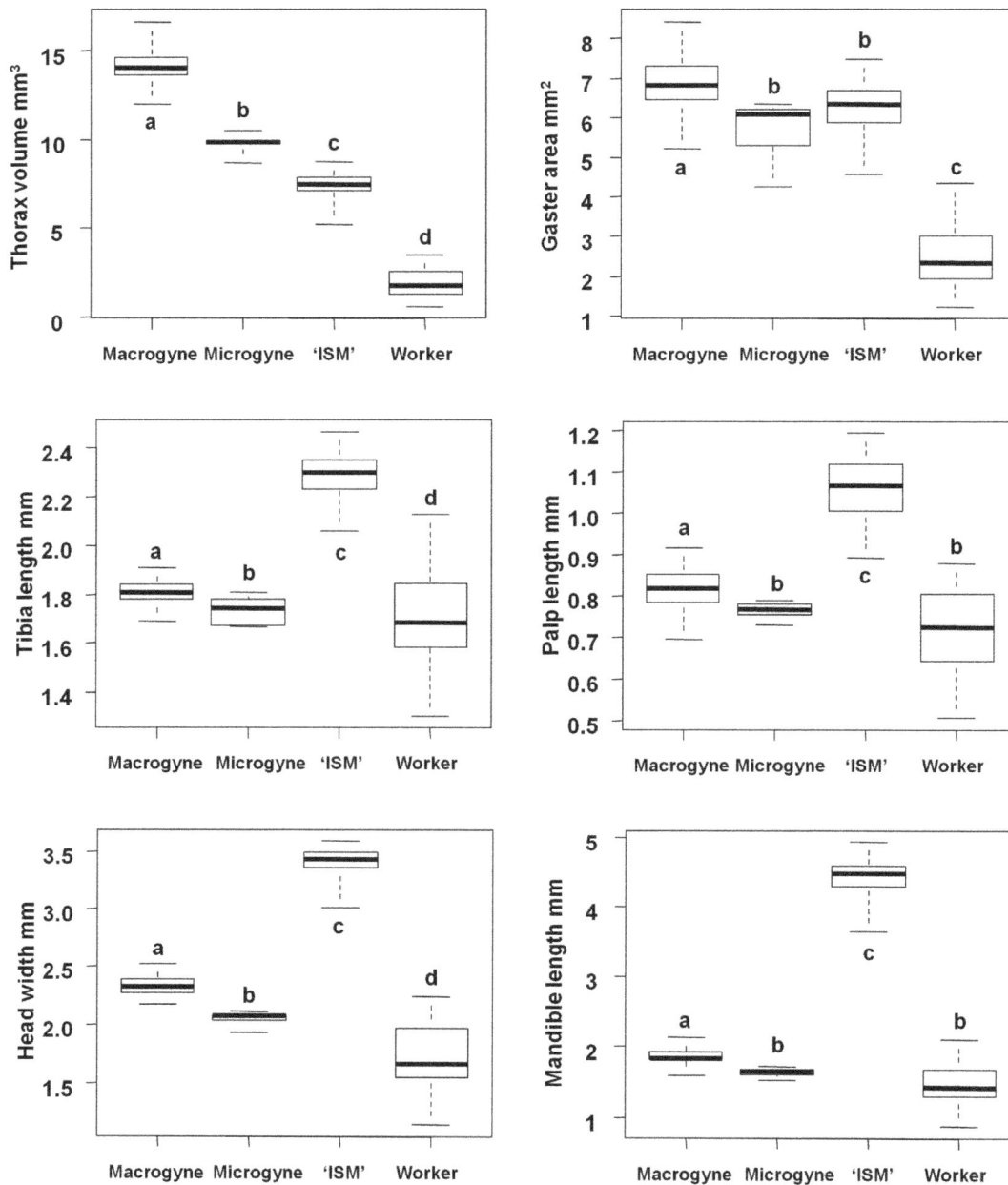

Figure 3. Size of body parts differs between female types in most cases. (52 workers, 47 macrogynes, 9 microgynes and 40 individuals with saber-shaped mandibles 'ISM'). All Kruskal-Wallis rank tests are significant ($P<0.001$), and 38 out of 42 pairwise comparisons using Wilcoxon rank tests with Bonferroni-corrected P-values are significant (different letters indicate significant differences).

Among workers, size distribution was continuous, so we treated them as a single group. Among queens however, size distribution of most body parts was bimodal (Fig. S1). Two queen castes with distinct sizes and growth rules may thus exist, so pulling them together may result in incorrect estimates and erroneous comparisons with 'ISM'. Accordingly, before performing any comparisons among female types, we split queens in two groups: macrogynes and microgynes. 'ISM' were thus compared to workers, macrogynes and microgynes, to test two hypotheses: either they belong to the worker caste, or they are a third caste.

Results

A Distinct Size from Other Castes

Size polymorphism among workers is considerable in *C. bombycina*: the biggest worker's head was 194% wider than the smallest worker's. We compared the size of body parts between female types, and Kruskal-Wallis tests revealed significant differences in size between female types for all body parts ($P<0.001$). 'ISM' had larger tibia, palp, head and mandible than the three other female types (Fig. 3). Their thorax volume and gaster area were intermediate between queens and workers (gaster area did not differ significantly with microgynes). Mandibles

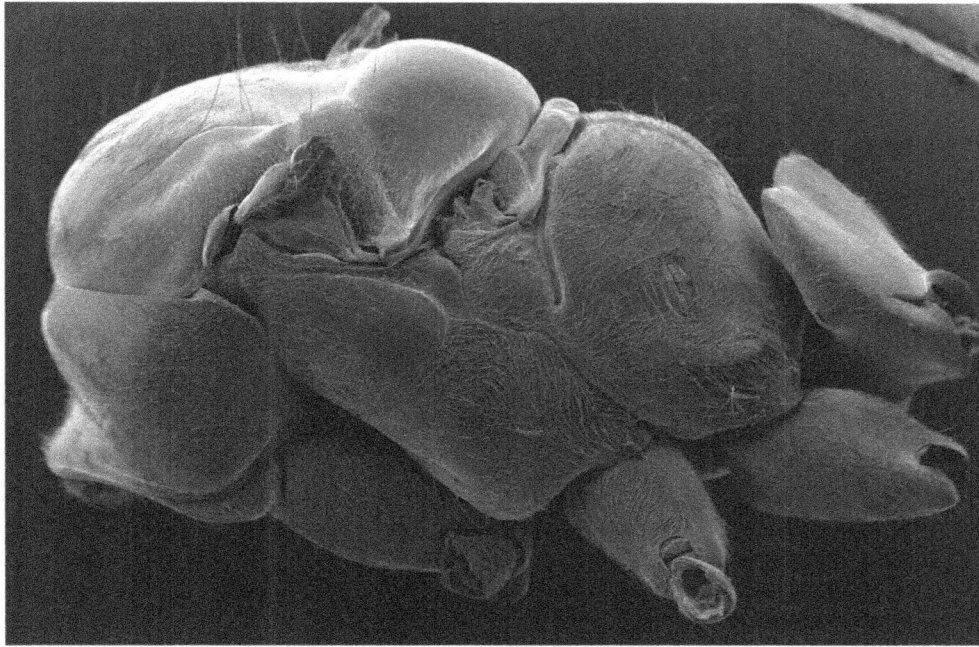

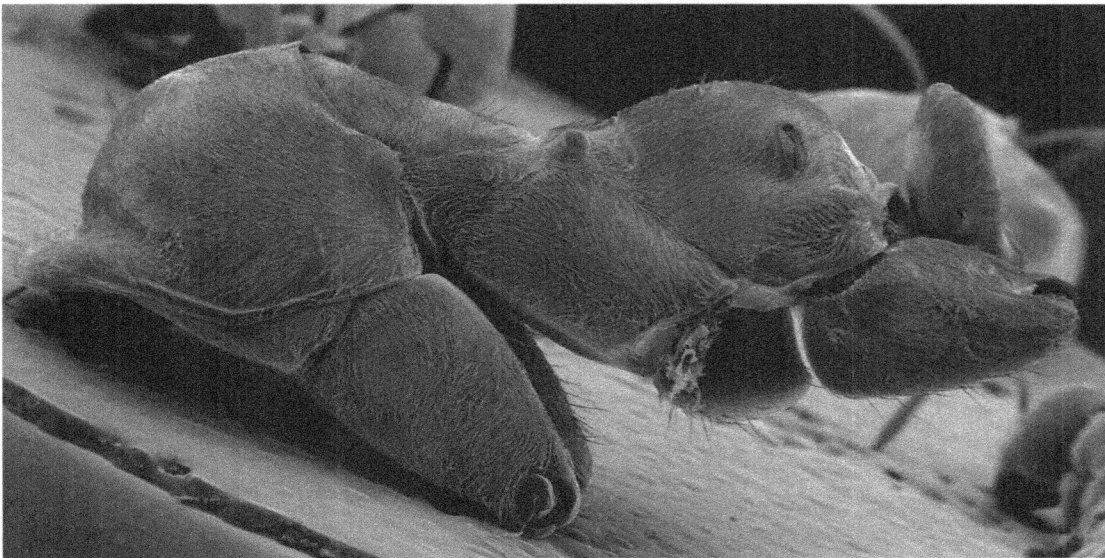

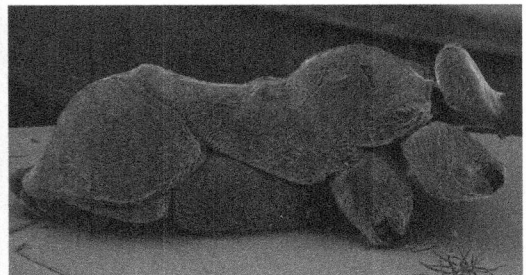

1mm

Figure 4. Scanning electron micrographs of the thorax of macrogyne, individual with saber-shaped mandibles 'ISM', large worker and small worker. (Resp. top, middle, bottom left and bottom right). The thorax of 'ISM' is simplified as in workers: sclerites are fused and mesonotum is reduced.

exhibited the most striking size differences among female types: saber-shaped mandibles were 3.2 times longer than workers' mandibles and 2.4 times longer than macrogynes'.

A Worker-like Thorax but Novel Mandibles

Scanning electron microscopy revealed no noticeable difference in thorax structure between 'ISM' and workers (Fig. 4). The pronotum was prominent, while mesonotum and metanotum were fused. No articulated sclerites or marked grooves were visible. Macrogynes and microgynes differed strongly due to their flight thorax: pronotum was smaller, mesonotum was very developed (it functions to attach wing muscles) and distinct from the metanotum. Sclerites were articulated and separated by grooves. 'ISM' had elongated mandibles with few teeth that strongly differ in shape from the short mandibles with six teeth found in macrogynes and workers (Fig. 5).

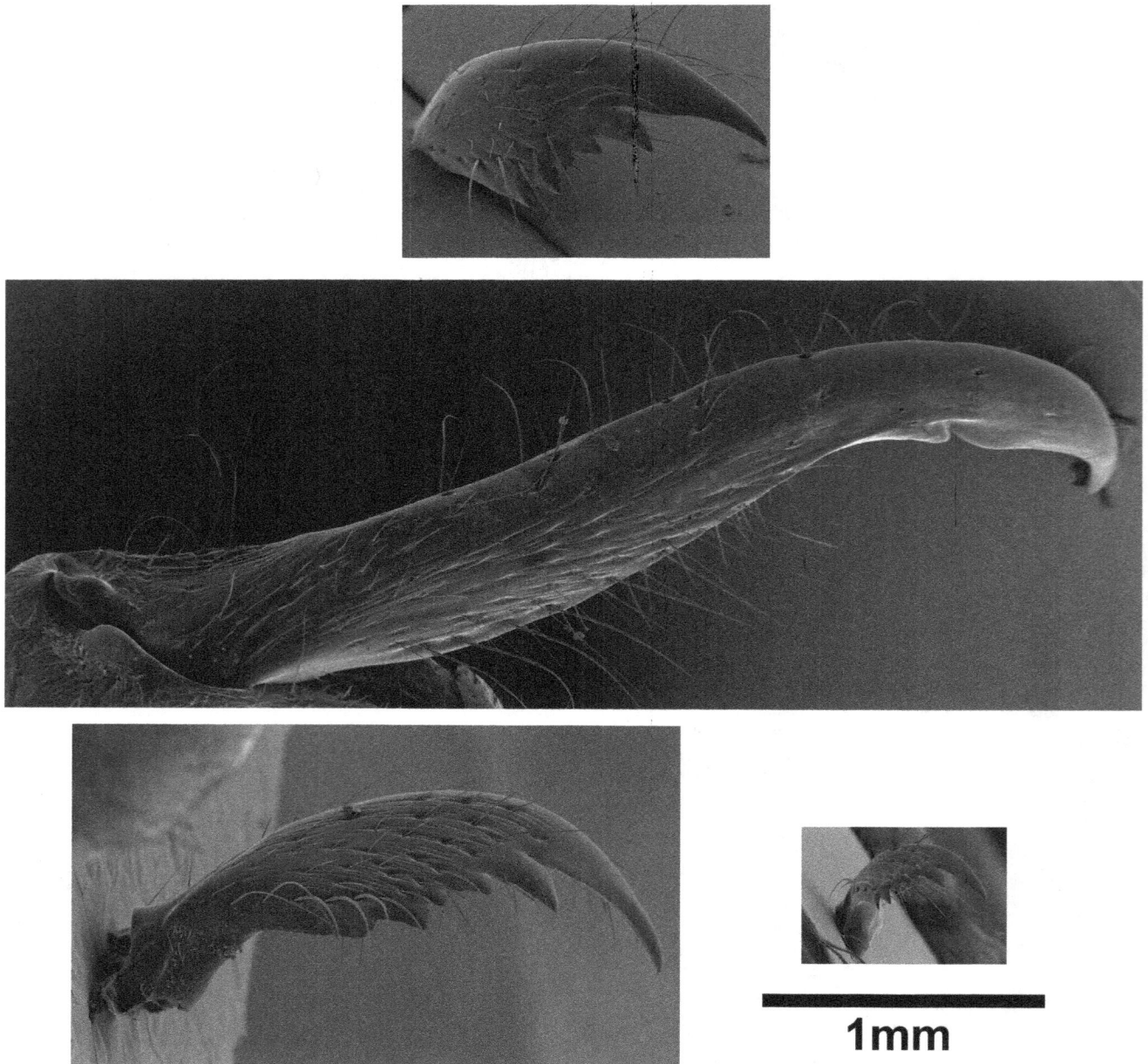

Figure 5. Scanning electron micrographs of the mandibles of macrogyne, individual with saber-shaped mandibles 'ISM', large worker and small worker. (Resp. top, middle, bottom left and bottom right). Macrogynes and workers have similar mandibles, but those of 'ISM' have a completely different shape and size.

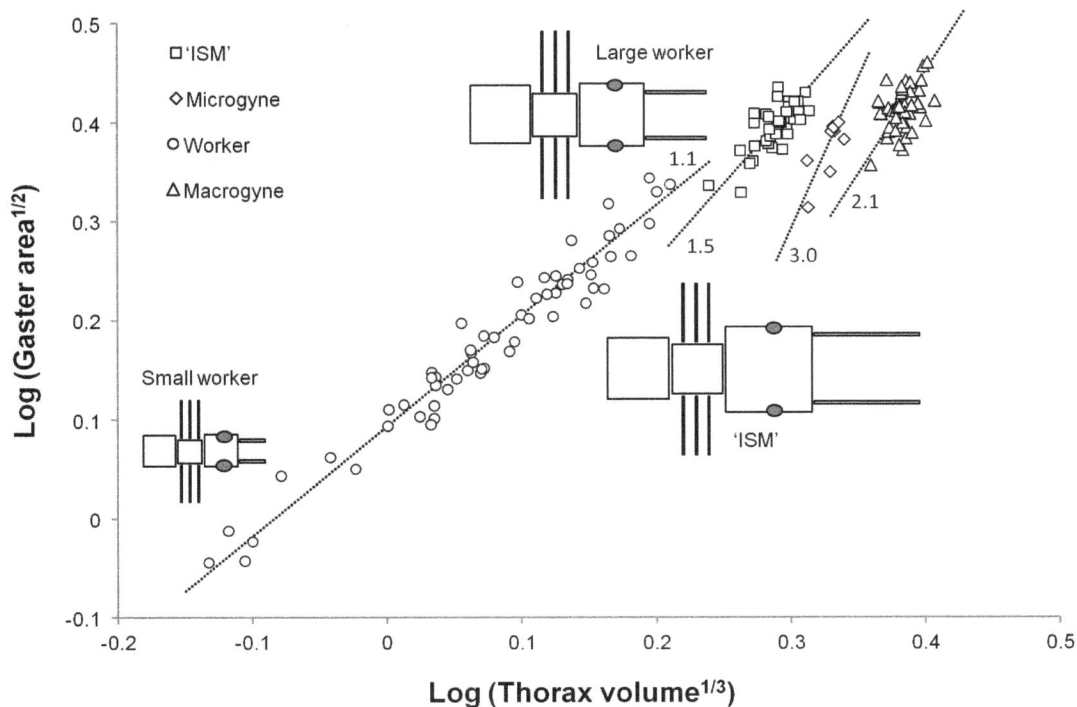

Figure 6. Growth rules of the four female types for one pair of morphological traits. (See Table 1 for some of the other pairs). Circles, squares, diamonds and triangles represent workers, individuals with saber-shaped mandibles 'ISM', microgynes and macrogynes respectively. Dotted lines are major axis regression lines, with their slopes (allometry coefficients) indicated as numbers. These four growth rules differ significantly. However, the allometry coefficients (growth rates) of 'ISM' and microgynes are not significantly different, suggesting that 'ISM' recycle a queen-like developmental program for this pair of traits. Diagrams represent a small worker, a theoretical large worker of tibia length t produced by the allometric growth rules of the worker caste, and an 'ISM' of the same tibia length t produced by the new growth rules. The large worker and the 'ISM' clearly differ. This illustrates that worker growth rules cannot produce a major worker phenotype that would look like an 'ISM' (Fig. 2).

A Mix of Queen and Worker Growth Rates

The sizes of several pairs of body parts in some female types were not significantly correlated. Accordingly, out of the 21 growth rules available, we used the eight growth rules that were significant for at least three female types, in order to compare them fully (Table 1, Fig. 6). The allometry coefficient of 'ISM' was not different from that of workers but different from that of macrogynes and microgynes in one rule out of eight. In contrast, it was not different from that of macrogynes but different from that of workers and microgynes in six rules out of eight. Accordingly, 'ISM' were more similar to macrogynes than to workers or microgynes regarding allometry coefficients (growth rates). However, in all cases where allometry coefficients did not differ between female types, elevation of the regression lines differed, indicating a distinct critical size in 'ISM'.

Discussion

A Soldier Caste, not Major Workers

Our morphometric data show that the large non-reproductive individuals in *C. bombycina* do not follow the growth rules of workers. Their shape cannot be produced by a simple change in critical size along the worker growth curve (Fig. 6). Thus they are a soldier caste according to our definition (Introduction and [18]), and not major workers (Fig. 2). They do not follow the growth rules of queens either. *C. bombycina* is an ideal model to study the evolution of morphological diversity among sterile individuals. Indeed, this species has a highly polymorphic worker caste and a monomorphic soldier caste. This contrasts with what is known in

other species. For instance, *Camponotus festinatus* has a polymorphic worker caste and a polymorphic soldier caste [1], whereas *Pheidole* species and *Atta texana* have a monomorphic worker caste ('minors') and one or two distinct soldier castes [22,23]. Phenotypic diversity within a colony can thus be achieved using multiple developmental options, i.e. by producing novel soldier castes and/or by generating various degrees of intra-caste variability. Our description of unambiguous morphological differences between individuals with saber-shaped mandibles and large/major workers supports the restrictive definition of soldiers advocated by Molet et al. [18], i.e. soldiers are a third caste, neither workers nor queens.

A Mix of Worker and Queen Developmental Programs

New phenotypes require changes in growth rules, and these are caused by changes in gene expression during development. This requires either new mutations or environmental release of cryptic genetic variation [24]. In monomorphic organisms, these new growth rules may lead to the production of lethal or useless phenotypes, and selection over many generations may be required to reach a well-adapted phenotype. In contrast, polyphenic organisms already have distinct sets of growth rules corresponding to each type of phenotype. These growth rules have been selected for and they produce viable phenotypes. Instead of evolving new growth rules, polyphenic organisms may re-utilize existing growth rules and recombine them to produce new phenotypes [18]. In ants, soldiers and ergatoid (permanently wingless) queens may be the product of such a mechanism.

Table 1. Growth rules of the four female types differ, revealing that they are distinct morphological castes.

	Correlation		Allometry coef	Test against isometry			Elevation
	R^2	P	and test	F	P	Conclusion	test
Workers (N = 60)							
Thorax volume vs. Head width	0.98	<0.001	0.85 a	70.2	<0.001	Allometry	a
Thorax volume vs. Mandible length	0.93	<0.001	0.75 a	69.4	<0.001	Allometry	–
Thorax volume vs. Mandible area	0.96	<0.001	0.80 a	78.1	<0.001	Allometry	a
Gaster area vs. Thorax volume	0.95	<0.001	1.11 a	13.2	0.001	Allometry	–
Gaster area vs. Head width	0.95	<0.001	0.94 a	3.5	0.066	Isometry	–
Mandible length vs. Head width	0.94	<0.001	1.14 a	16.5	<0.001	Allometry	–
Mandible length vs. Mandible area	0.98	<0.001	1.06 a	11.4	0.001	Allometry	a
Mandible area vs. Head width	0.97	<0.001	1.07 a	8.5	0.005	Allometry	a
Macrogynes (N = 47)							
Thorax volume vs. Head width	0.39	<0.001	0.68 a	11.9	0.001	Allometry	b
Thorax volume vs. Mandible length	0.12	0.016	0.42 b	50.7	<0.001	Allometry	a
Thorax volume vs. Mandible area	0.32	<0.001	0.65 a	13.2	0.001	Allometry	b
Gaster area vs. Thorax volume	0.18	0.003	2.09 bc	35.4	<0.001	Allometry	a
Gaster area vs. Head width	0.15	0.007	1.41 b	6.6	0.013	Allometry	b
Mandible length vs. Head width	0.31	<0.001	1.63 b	16.7	<0.001	Allometry	b
Mandible length vs. Mandible area	0.50	<0.001	1.56 b	18.9	<0.001	Allometry	–
Mandible area vs. Head width	0.33	<0.001	1.04 ab	0.1	0.73	Isometry	b
Microgynes (N = 9)							
Thorax volume vs. Head width	0.66	0.008	0.80 a	1.0	0.35	Isometry	b
Thorax volume vs. Mandible length	0.11	0.39	–	–	–	–	–
Thorax volume vs. Mandible area	0.07	0.51	–	–	–	–	–
Gaster area vs. Thorax volume	0.57	0.019	3.03 b	29.8	0.001	Allometry	b
Gaster area vs. Head width	0.26	0.16	–	–	–	–	–
Mandible length vs. Head width	0.19	0.24	–	–	–	–	–
Mandible length vs. Mandible area	0.01	0.79	–	–	–	–	–
Mandible area vs. Head width	0.01	0.76	–	–	–	–	–
Individuals with Saber-shaped Mandibles 'ISM' (N = 40)							
Thorax volume vs. Head width	0.58	<0.001	0.87 a	1.8	0.19	Isometry	c
Thorax volume vs. Mandible length	0.14	0.017	0.52 b	21.5	<0.001	Allometry	b
Thorax volume vs. Mandible area	0.34	<0.001	0.61 a	15.1	<0.001	Allometry	c
Gaster area vs. Thorax volume	0.55	<0.001	1.54 c	16.7	<0.001	Allometry	c
Gaster area vs. Head width	0.39	<0.001	1.34 b	5.4	0.026	Allometry	c
Mandible length vs. Head width	0.37	0.023	1.67 b	17.2	<0.001	Allometry	c
Mandible length vs. Mandible area	0.59	<0.001	1.17 a	2.3	0.14	Isometry	b
Mandible area vs. Head width	0.65	<0.001	1.43 b	14.0	0.001	Allometry	c

Column #1: pair of morphological traits for which relative growth rule was assessed (Y vs. X). #2: Pearson correlation test. #3: slope of the regression line (allometry coefficient) and comparison test between female types. #4: comparison of the slope with 1 (isometry). #5: when slopes do not differ between female types, comparison of elevations between female types. Pairs of morphological traits that were not significantly correlated at the same time in three female types were excluded. Thus only eight growth rules are shown out of 21 potential growth rules. Correlations are rarely significant in microgynes due to small sample size.

In *C. bombycina*, we found that the allometry coefficients (growth rates) of soldiers were never different from either worker or queen castes. Instead, they were sometimes similar to workers, and sometimes similar to queens. Thus, soldiers do not exhibit new growth rates, but they recombine growth rates from both worker and queen castes. Hence their production relies on recycling (Fig. 2), and we argue that the ancestral queen-worker polyphenism contributed to the evolution of a novel female phenotype in this species.

In contrast, the elevations of regression lines of soldiers always differed from both worker and queen castes so growth rules were not fully conserved: growth rates were, but critical sizes were not. Moreover, it is not clear how changes in gene expression relative to workers and queens can lead to the modified mandibles of soldiers. 2D morphometrics would be required to assess deformations in mandible shape and discuss the developmental link between these mandible types. Future studies in evo-devo should investigate how much recycling is involved in the production and evolution of novel traits in ants, and whether this recycling could explain the numerous independent evolutions of novel castes across the ant phylogeny.

A Specialization for Defence and Food Storage

Soldiers have saber-shaped mandibles (together with a broad head with powerful muscles) that are similar to those of soldiers in army ants (e.g. *Eciton*; [25]); these are thought to be adapted for defense against vertebrate predators, not arthropods. Accordingly, *C. bombycina* soldiers may be specialized for defense against reptiles. *C. bombycina* is the dominant ant species in the harsh sand dune habitats of North Africa. Their colonies are exceptionally populous for this genus, and the large quantities of brood developing in the deep underground chambers are likely to represent a valuable resource for reptiles. Lizards in the genus *Acanthodactylus* are known to be the main predators of *C. bombycina* [26], and it is possible that they dig ant nests to find brood. This hypothesis is corroborated by the finding of numerous soldiers in deep brood chambers (S. Aron pers. com.). In this species, soldiers are only useful inside the nest because there is no compact resource to defend outside: food items are scattered insect corpses collected up to 100 m away from the nest. Colonies of other *Cataglyphis* species are much smaller and lack a soldier caste [15,27]. Cagniant [27] suggested that soldiers are produced only in older (i.e. bigger) colonies of *C. bombycina*. Unlike Délye [17], we found that soldiers have longer palps than the larger workers, and this could allow them to carry bigger sand pellets [28].

The abdomen of some soldiers and major workers kept in the laboratory with *ad libitum* food was conspicuously distended. Dissections revealed large accumulations of fat bodies. Workers functioning as repletes have been described in other *Cataglyphis* species [27]. *C. bombycina* soldiers have an abdomen with a cross-sectional area that is 2.7 times larger than workers', thus allowing for much more storage in a soldier than in a worker. Accordingly, we suggest that soldiers are also specialized to function as repletes; a similar function (production of trophic eggs) was shown for the soldier caste in *Crematogaster (Orthocrema)* [29]. Many ant species store excess food that is later shared among nestmates [30]. Storage is usually carried out by workers, but some specialized castes have evolved in some species. The latter do not follow the growth rules of workers, and they have queen-like ovaries, hence they fall into our definition of 'soldiers'. *C. bombycina* feeds on dead insects which are an unpredictable resource, so food storage is likely to be adaptive for colonies. It is not clear whether soldiers initially evolved for defense or food storage, however this dual function balances their production cost. Future studies need to determine whether young soldiers function as repletes in deep chambers and older soldiers move to surface chambers to be available for nest defense.

A Second Queen Caste?

Although not the primary focus of our study, we found that *C. bombycina* has dimorphic queens: macrogynes and microgynes.

Microgynes were too few to assess their growth rules and compare them to macrogynes, workers and soldiers. In contrast with soldiers, microgynes in *C. bombycina* may not result from a mosaic development, and an increased sample size is required to test whether or not microgynes are mosaics of macrogynes and workers. Having microgynes in addition to macrogynes can be advantageous for colonies. Macrogynes typically perform independent colony foundation, i.e. they fly away from their natal nest, mate, and start a new colony alone. In contrast, microgynes can stay in their natal nest where they reproduce (polygyny), and/or be involved in dependent colony foundation (fission). In some species, microgynes enter nearby colonies and become social parasites. Accordingly, microgynes are thought to be an adaptation to saturated habitats with high competition [4]. Since they are less costly to produce relative to macrogynes, more of them can be produced with the same amount of resources. The function of microgynes of *C. bombycina* is unknown.

Colonial Life Facilitates the Evolution of New Phenotypes

Our morphometric analysis showed that polyphenic taxa can produce novel phenotypes by recombining growth rules from existing phenotypes. Such mosaics are probably more frequent and viable than new mutants. This original developmental mechanism could enhance the evolvability of polyphenic species, and it may have contributed to the tremendous diversification of ants. In addition, phenotypes that would be suboptimal or lethal in a solitary context can survive in ants because colonies buffer the outside environment. Accordingly, selection for new mosaic phenotypes can be facilitated provided they bring colony-level benefits for defense, food storage or reproduction [18]. The properties of social life as an incubator for evolutionary novelties should be considered in future research.

Acknowledgments

We thank Serge Aron (ULB, Brussels) for providing us with samples of young queens of *C. bombycina*, Roberto Keller for help with thorax morphology, Patrick Landmann for the photograph (Fig. 1) and Romain Péronnet for maintaining colony #1 in the lab. Diana Wheeler provided constructive comments on the manuscript. Fieldwork by C. Peeters occurred during the filming of a documentary for ARTE TV ("Life in Hell", directed by Vincent Amouroux, Mona Lisa Productions).

Author Contributions

Conceived and designed the experiments: MM CP. Performed the experiments: VM. Analyzed the data: VM MM. Contributed reagents/materials/analysis tools: MM CP. Wrote the paper: MM VM CP.

References

1. Wheeler D (1991) The developmental basis of worker caste polymorphism in ants. Am Nat 138: 1218–1238.
2. Hölldobler B, Wilson EO (1990) The Ants. Harvard University Press, Cambridge, MA.
3. Billick I, Carter C (2007) Testing the importance of the distribution of worker sizes to colony performance in the ant species *Formica obscuripes* Forel. Insectes Soc 54: 113–117.
4. Rüppell O, Heinze J (1999) Alternative reproductive tactics in females: the case of size polymorphism in winged ant queens. Insectes Soc 46: 6–17.
5. Lachaud J, Cadena A, Schatz B, Pérez-Lachaud G, Ibarra-Núñez G (1999) Queen dimorphism and reproductive capacity in the ponerine ant, *Ectatomma ruidum* Roger. Oecologia 120: 515–523.
6. Tsuji K (1990) Nutrient storage in the major workers of *Pheidole ryukyuensis* (Hymenoptera: Formicidae). Appl Entomol Zool 25: 283–287.
7. Hasegawa E (1993) Nest defense and early production of the major workers in the dimorphic ant *Colobopsis nipponicus* (Wheeler) (Hymenoptera: Formicidae). Behav Ecol Sociobiol 33: 73–77.
8. Baroni Urbani C (1998) The number of castes in ants, where major is smaller than minor and queens wear the shield of the soldiers. Insectes Soc 45: 315–333.

9. Heinze J, Foitzik S, Oberstadt B, Rüppell O, Hölldobler B (1999) A female caste specialized for the production of unfertilized eggs in the ant *Crematogaster smithi*. Naturwissenschaften 86: 93–95.

10. Gobin B, Ito F (2000) Queens and major workers of *Acanthomyrmex ferox* redistribute nutrients with trophic eggs. Naturwissenschaften 87: 323–326.

11. Wilson EO (2003) *Pheidole* in the New World: A Dominant, Hyperdiverse Ant Genus. Harvard University Press, Cambridge, MA.

12. Powell S (2008) Ecological specialization and the evolution of a specialized caste in *Cephalotes* ants. Funct Ecol 22: 902–911.

13. Peeters C (2012) Convergent evolution of wingless reproductives across all subfamilies of ants, and sporadic loss of winged queens (Hymenoptera: Formicidae). Myrmecological News: 75–91.

14. Peeters C, Ito F (2001) Colony dispersal and the evolution of queen morphology in social Hymenoptera. Annu Rev Entomol 46: 601–630.

15. Lenoir A, Aron S, Cerdá X, Hefetz A (2009) *Cataglyphis* desert ants: a good model for evolutionary biology in Darwin's anniversary year - A review. Isr J Entomol 39: 1–32.

16. Pisarski B (1965) Les fourmis du genre *Cataglyphis* Foerst en Irak (Hymenoptera: Formicidae). Bull L'Academie Pol des Sci II 13: 417–422.

17. Délye G (1957) Observations sur la fourmi saharienne *Cataglyphis bombycina* Rog. Insectes Soc 4: 77–83.

18. Molet M, Wheeler DE, Peeters C (2012) Evolution of novel mosaic castes in ants: modularity, phenotypic plasticity, and colonial buffering. Am Nat 180: 328–341.

19. Leniaud L, Pearcy M, Aron S (2013) Sociogenetic organisation of two desert ants. Insectes Soc in press: DOI 10.1007/s00040-013-0298-2.

20. Molet M, Peeters C, Fisher BL (2007) Winged queens replaced by reproductives smaller than workers in *Mystrium* ants. Naturwissenschaften 94: 280–287.

21. Warton DI, Wright IJ, Falster DS, Westoby M (2006) Bivariate line-fitting methods for allometry. Biol Rev Camb Philos Soc 81: 259–291.

22. Huang MH, Wheeler DE (2011) Colony demographics of rare soldier-polymorphic worker caste systems in *Pheidole* ants (Hymenoptera, Formicidae). Insectes Soc 58: 539–549.

23. Wilson E (1953) The origin and evolution of polymorphism in ants. Q Rev Biol 28: 136–156.

24. Barrett RDH, Schluter D (2008) Adaptation from standing genetic variation. Trends Ecol Evol 23: 38–44.

25. Gotwald H (1978) Trophic Ecology and Adaptation in Tropical Old World Ants of the Subfamily Dorylinae (Hymenoptera: Formicidae). Biotropica 10: 161–169.

26. Wehner R, Marsh A, Wehner S (1992) Desert ants on a thermal tightrope. Nature 357: 586–587.

27. Cagniant H (2009) Le Genre *Cataglyphis* Foerster, 1850 au Maroc (Hyménoptères Formicidae). Orsis 24: 41–71.

28. Bernard F (1951) Adaptation au milieu chez les Fourmis sahariennes. Bull la Société d'Histoire Nat Toulouse 86: 88–96.

29. Peeters C, Lin CC, Quinet Y, Martins Segundo G, Billen J (2013) Evolution of a soldier caste specialized to lay unfertilized eggs in the ant genus *Crematogaster* (subgenus *Orthocrema*).

30. Wheeler D (1994) Nourishment in ants: patterns in individuals and societies. In: Hunt JH, Nalepa CA, editors. Nourishment and Evolution in Insect Societies. Westview Press, Boulder, Colorado. 245–278.

15

Emergence of Small-World Anatomical Networks in Self-Organizing Clustered Neuronal Cultures

Daniel de Santos-Sierra[1]*, **Irene Sendiña-Nadal**[2,1], **Inmaculada Leyva**[2,1], **Juan A. Almendral**[2,1], **Sarit Anava**[3], **Amir Ayali**[3], **David Papo**[1], **Stefano Boccaletti**[4,5]

1 Center for Biomedical Technology, Universidad Politécnica de Madrid, Pozuelo de Alarcón, Madrid, Spain, 2 Complex Systems Group, Universidad Rey Juan Carlos, Móstoles, Madrid, Spain, 3 Department of Zoology, Tel-Aviv University, Tel Aviv, Israel, 4 Istituto dei Sistemi Complessi, Consiglio Nazionale delle Ricerche, Sesto Fiorentino, Florence, Italy, 5 Istituto Nazionale di Fisica Nucleare, Sesto Fiorentino, Florence, Italy

Abstract

In vitro primary cultures of dissociated invertebrate neurons from locust ganglia are used to experimentally investigate the morphological evolution of assemblies of living neurons, as they self-organize from collections of separated cells into elaborated, clustered, networks. At all the different stages of the culture's development, identification of neurons' and neurites' location by means of a dedicated software allows to ultimately extract an adjacency matrix from each image of the culture. In turn, a systematic statistical analysis of a group of topological observables grants us the possibility of quantifying and tracking the progression of the main network's characteristics during the self-organization process of the culture. Our results point to the existence of a particular state corresponding to a *small-world* network configuration, in which several relevant graph's micro- and meso-scale properties emerge. Finally, we identify the main physical processes ruling the culture's morphological transformations, and embed them into a simplified growth model qualitatively reproducing the overall set of experimental observations.

Editor: Matjaž Perc, University of Maribor, Slovenia

Funding: The authors acknowledge financial support from the Spanish Ministerio de Ciencia e Innovación (Spain) under project FIS2009-07072, and from Comunidad de Madrid (Spain) under project MODELICO-CM S2009ESP-1691. DSS is supported by the Comunidad de Madrid through the R+D Activity Program NEUROTEC-CM (2010/BMD-2460). The funders had no role in study design, data collection and analysis, decision to publish, or preparation of the manuscript.

Competing Interests: The authors have declared that no competing interests exist.

* E-mail: daniel.desantos@ctb.upm.es

Introduction

The issue of why and how an assembly of isolated (cultured) neurons self-organizes to form a complex neural network is a fundamental problem [1–3]. Despite their more limited, and yet laboratory-controllable, repertoire of responses [1,4], the understanding of such cultures' organization is, indeed, a basis for the comprehension of the mechanisms involved in their *in vivo* counterparts, and provide a useful framework for the investigation of neuronal network development in real biological systems [3].

Some previous studies highlighted the fact that the structuring of a neuronal cultured network before the attainment of its mature state is not random, being instead governed and characterized by processes eventually leading to configurations which are comparable to many other real complex networks [5]. In particular, networking neurons simultaneously feature a high overall clustering and a relatively short path-length between any pair of them [6]. Such configurations, which in graph theory are termed *small-world* [7], are ubiquitously found in real-world networking systems. Small-world structures have been shown to enhance the system's overall efficiency [8,9], while concurrently warranting a good balance between two apparently antagonistic tendencies for segregation and integration in structuring processes, needed for the network's parallel, and yet synthetic performance [10].

In this paper, we experimentally investigate the self-organization into a network of an *in vitro* culture of neurons during the course of development, and explore the changes of the main topological features characterizing the anatomical connectivity between neurons during the associated network's growth. To that purpose, dissociated and randomly seeded neurons are initially prepared, and the spontaneous and self-organized formation of connections is tracked up to their assembling into a two dimensional clustered network.

Most existing studies in neuronal cultures restricted their attention to functional networks (statistical dependence between nodes activities) and not to the physical connections supporting the functionality of the network [11]. The reason behind this drawback is that the majority of investigations focused on excessively dense cultures, hindering the observation of their fine scale structural connectivity. Although there are studies striving to indirectly infer the underlying anatomical connectivity from the functional network, it has been shown that strong functional correlations may exist with no direct physical connection [12]. Only few studies dealt with the physical wiring circuitry. However, on the one hand, only small networks were considered; on the other hand, how the network state evolves during the course of the maturation process has not been investigated [6].

Here, instead, we focus on intermediate neurons' densities, and provide a full tracking of the most relevant topological features emerging during the culture's evolution. In particular, we show experimentally that *in vitro* neuronal networks tend to develop from a random network state toward a particular networking state, corresponding to a *small-world* configuration, in which several relevant graph's micro- and meso-scale properties emerge. Our approach also unveils the main physical processes underlying the culture's morphological transformation, and allows using such information for devising a proper growth model, qualitatively reproducing the set of our experimental evidence.

Together with confirming several results of previous works on functional connectivity [13], or on morphological structuring at a specific stage of the cultures' evolution [6], we offer a systematic characterization of several topological network's measures from the very initial until the final state of the culture. Such a *longitudinal* study of the network structure highlights as yet unknown self-organization properties of cultured neural networks, such as *i*) a large increase in both local and global network's efficiency associated to the emergence of the small-world configuration, and *ii*) the setting of assortative degree-degree correlation features.

Experimental Set-Up

Neuronal cultures and network growth

In this paper, we report on six cultured networks, which were grown from independent initial sets of dissociated neurons extracted from the frontal ganglion of adult locusts of the *Schistocerca gregaria* species. In all cases, a same protocol was used, involving animals that were daily fed with organic wheat grass and maintained under a 12:12 h light:dark cycle from their fifth nymph growth to their early adult stage of development. At this latter stage, we followed the dissection and culturing protocol thoroughly described in [14]. In brief, the frontal ganglia were dissected from anesthetized animals, and enzymatically treated to soften the sheath. Ganglia were then forced to pass through the tip of a 200 µl pipette to mechanically dissociate the neurons. The resulting suspension of neuronal somata was plated on Concanavalin A pre-coated circular area ($r \sim 5$ mm) of a Petri dish where it was left for 2 h to allow adhesion of neurons at random positions of the substrate. After plating, 2 ml culture medium (Leibovitz L-15) enriched with 5% locust hemolymph was added. Cultures were then maintained in darkness under controlled temperature ($29°$C) and humidity (70%).

The density at which cultures are seeded determines the maturation rate and the spatial organization at the mature state [15,16]. For the purpose of this work, aimed at studying the network evolution into a clustered network, 6 dense cultures of 12 ganglia each ($\sim 1,200$ neurons) were used and monitored during 18 days *in vitro* (DIV). During the entire experiment, the culture medium was not changed.

High-resolution and large scale images of the whole culture were acquired daily using a charge coupled device camera (DS-Fi1, Nikon) mounted on a phase contrast microscope (Eclipse Ti-S, Nikon), with automated control of a motorized XYZ stage (H117 ProScan, Prior Scientific).

A typically observed growth evolution is shown in Fig. 1 (restricted to just a small part of the whole culture) between 3 and 12 DIV. Neurons ranging from 10 to 50 µm in size are initially randomly anchored to a two dimensional substrate, while after 3 DIV (Fig. 1A and B) many cells already start growing neuronal processes (neurites) trying to target neighboring cells. During this growth process, neurites also split and reach other processes forming loops of neurites up to 6 DIV, when the maximum stage of network development takes place (Fig. 1C). At this point, the growth rate decreases and a different mechanism starts shaping the network: tension is generated along the neurites as they stretch between neurons or bifurcation points to form straight segments [17].

The latter process favors neuron migration, giving rise to clusters of neurons, and the fusion of parallel neurites into thicker bundles together with the retraction of those branches which did not target any neuron (see black arrow in Fig. 1C). The resulting network topology shown in Fig. 1D after 12 DIV (and in the enlarged area in Fig. 1E) is characterized by a random distribution of few clusters of aggregated neurons linked by thick nerve-like bundles.

Anatomical graph extraction and complex network statistics

Our experiments consistently show that cultures self-organize from random scattered distributions of bare neurons into spatial networks of interconnected clusters of neurons (compare Fig. 1A and Fig. 1E).

In order to properly quantify the topological and spatial changes of the anatomical neuronal network as cultures approach their mature state, we developed a custom image analysis software in MATLAB to detect the location of neurons, clusters of neurons and neurite paths. The used imaging software has been fully customized for the purpose of the analysis of the present data. The general details of the developed imaging software will be reported elsewhere. The performance of the algorithm is sketched in Fig. 2. The algorithm takes as an input a gray color image of the culture at a particular day (Fig. 2A), upon which it superimposes a layer of new information comprising the contours of the clusters of neurons (red shadows), the traces of the neurites (green lines), and connection points between neurites, as well as those between neurites and clusters (blue dots) (Fig. 2B).

The information contained in the produced layer is then used to map the neuronal network into a graph $\mathcal{G}$ (see Fig. 2C) whose nodes (in blue) are either cluster centroids or connection points, and the links (in green) are straight lines connecting them. Therefore, our graph is made of two types of nodes: neurons or clusters of neurons (v_i) and neurite connection points (u_i). Treating all links as identical, i.e. ignoring edge length and edge directionality, this graph can be described in terms of a symmetric adjacency matrix A whose elements a_{ij} are equal to 1 if nodes i and j are linked, and 0 otherwise.

We focus on the network statistical properties at the level of the v_i nodes, ignoring the dynamics of both neurite connections and branching points. Therefore, we extract from $\mathcal{G}$ the subgraph defining the connectivity among nodes of class v_i in such a way that v_i and v_j are linked either directly or through a connected path of u_i nodes.

The analysis of the networks' evolution requires accounting for the birth and death of links (and, in some cases, nodes) over time. Figure 3 shows the mean values for the number of nodes and the of links at each DIV, calculated for the 6 cultures. During the growth phase, spanning from 0 to 6 DIV, the number of nodes with at least one connection slowly increases with age, while the number of links rises exponentially, reaching a maximum at DIV 6. After this time point, the convergence of parallel neurites and neuronal clusterization induces a more gentle decrease in the number of links, accompanied by a slight reduction in the number of nodes. In order to properly compare networks of different size, we need to refer to a measure which is independent of the network size: the link density, defined as the ratio between the total number of measured links and the number of links characterizing the arrangement of the same number of identified nodes in a complete clique configuration. As illustrated in the inset of Fig. 3A, at any stage of development, the cultured networks are far from being fully connected (only about 2% of all possible connections exist

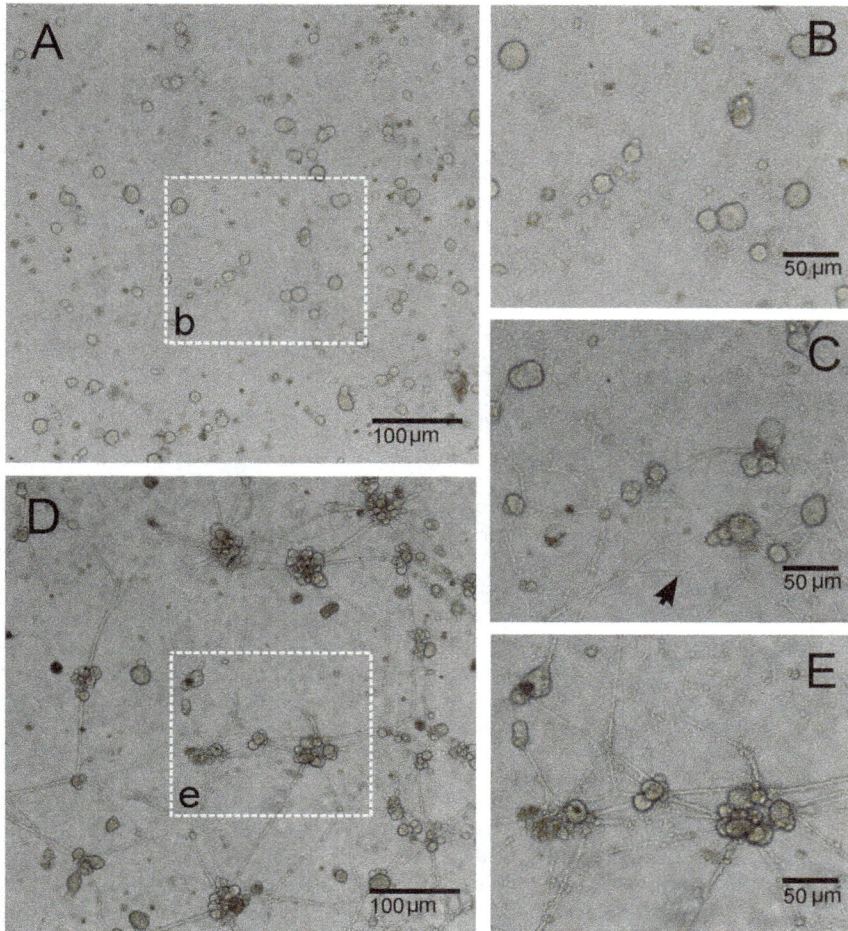

Figure 1. Culture development of locust frontal ganglion neurons into clustered networks. (A) After 3 DIV, completely dissociated neurons had already started growing neuronal processes with continuous branching. The area outlined in (b) is enlarged in B. (C) Same area as in (B) but at 6 DIV. At this stage, neurons and small clusters of neurons are already densely connected and form a complex network. At the same stage, branched neurites (pointed by the black arrow) that failed to contact neighboring neurons start to retract. (D) Migration of neurons due to the tension along neurites leads to the formation of large neuronal clusters and of thicker bundles of neurites. For a better visualization, the area outlined in (e) is enlarged in E.

between nodes), and thus operate in a low-cost regime of sparse anatomical connections.

In such a sparse connectivity regime, we quantify how our networks constrained in 2D space percolate. To do so, we measure the size S_1 of the giant connected component (GCC) and the size S_2 of the second largest component (GCC2) as a function of the

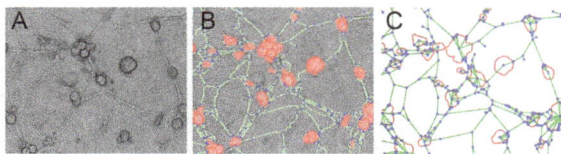

Figure 2. Extraction of the adjacency matrix defining the neural network connectivity. (A) Image cut taken from a 6 DIV culture and (B) the layer on top showing the identification of neurons and clusters of neurons (red), neurites connecting them (green) and neurite branching points (blue). (C) Mapping of the neuronal network into a graph where blue dots represent the nodes and green lines the links of the graph.

age [18,19]. Figure 3B shows that the number of nodes forming such connected components smoothly increases at the same rate along the first days of the network development, up to the DIV 6 when the difference in size between them suddenly and consistently starts to grow. From that point on, the GCC2 starts collapsing and progressively merging into the GCC, and the establishment of an almost fully connected network of clusters characterizes the rest of the culture's life. Figure 3B reports the evolution of the number of nodes belonging to both GCC and GCC2.

A deeper information on the culture evolution can be gathered by monitoring the behavior of a subset of local and network-wide quantities. For that purpose, we calculated several topological properties of the extracted adjacency matrices (using the Matlab Boost Graph Library package and the Brain Connectivity Toolbox [20]), whose definitions are provided in [5,20]. In particular, we analyzed the clustering coefficient (C), the average shortest path length (L), the local (E_{loc}) and global (E_{glob}) efficiency [8], the network assortativity (r) and the cumulative degree distribution ($P_{cum}(k)$), obtained from the degree distribution $P(k)$ as

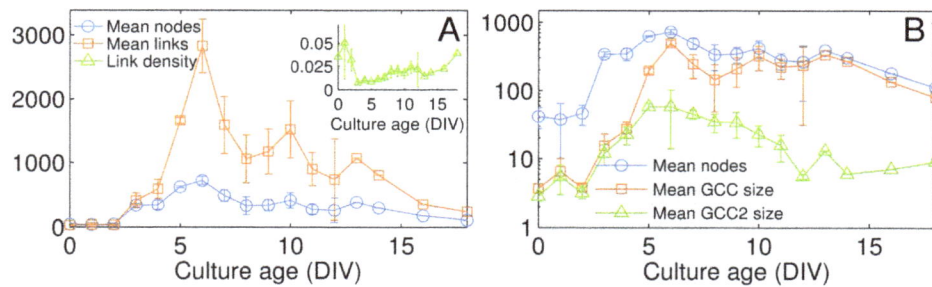

Figure 3. Density of the network as a function of culture age. (A) Mean number of nodes (blue circles), including neurons and clusters of neurons, and links connecting them (red squares), calculated for the 6 cultures vs. age (DIV). Inset: the link density (green triangles) quantifies the actual number of links divided by that of an all-to-all configuration [$N \cdot (N-1)/2$, being N the number of connected nodes at each age]. (B) Log-linear plot of the mean number of nodes having at least one connection (blue circles), of the mean size of the giant connected component (red squares) and of the second largest connected component (green triangles). In all plots, error bars stand for the standard errors of the mean (sem).

$P_{cum}(k) = \sum_{k'=k}^{k'=k_{max}} P(k')$ being k the degree (or number of links) of a node.

In all cases, the calculation of such statistics was restricted to the set of nodes having at least one link, and for the calculation of L to those pairs of nodes belonging to the GCC. Moreover, the experimental values of C and L were also compared with those expected in equivalent random null hypothesis networks, i.e. random networks artificially constructed to have the same number of nodes and links and to display the same degree distribution. Specifically, for each experimental network at a particular age, we generated 20 independent realizations of equivalent random networks, and calculated the corresponding expected network statistics.

Finally, in order to quantify the degree-degree correlation properties, the network assortativity was defined by considering for each node i the average degree of its neighbors k_{nn}, and by computing the linear regression of $\log(\langle k_{nn} \rangle)$ vs. $\log(k_i^p)$. The assortativity coefficient r was then calculated as the Pearson correlation coefficient corresponding to the best fit of $\log(\langle k_{nn} \rangle) \sim p \log(k)$. If $r > 0$ ($r < 0$), the network is set to be assortative (disassortative), while depending upon the obtained value of p, the degree correlation properties are said to be of a linear ($p = 1$), sub-linear ($p < 1$), or super-linear ($p > 1$) nature.

Results

Emergence of small-world structure

The first days of the cultures' development (from DIV 0 to DIV 3) were characterized by networks with very few nodes and links (see Fig. 3A). After DIV 3, the networks showed a very pronounced increase in the number of links and nodes (from DIV 3 to DIV 6) preceding a spatial network reorganization eventually driving the graph into its clustered, mature state.

The associated networks statistics sheds light on the transition from random to non-random properties with a progression of both the clustering coefficient and the average path length (normalized by the GCC size) as a function of age (see Fig. 4A). The first significant result is the simultaneous increase in the clustering coefficient and decrease in the mean path length, a clear fingerprint of the emergence of a small-world network configuration. This configuration becomes prominent at DIV 6 and stays relatively stable through the last two weeks *in vitro*. To properly asses the significance of this finding and isolate the influence of the variable network size and density, we calculated the values of C and L normalized to the corresponding expected values for equivalent random (and lattice) null model networks (see Fig. 4B).

In doing so, we follow the approach that was recently used in similar circumstances for the obtainment of null models [21]. According to Watts and Strogatz's model [7], a small-world network simultaneously exhibits short characteristic path length, like random graphs, and high clustering, like regular lattices. Here, we found a clear change in the trend at DIV 6 where $L_{rand}/L \leq 1$, where the average path length of the cultured network starts to be close to that of a random graph and much smaller than that of a regular graph (L_{reg} is calculated as $L_{reg} = S_1/(2\langle k \rangle)$). At the same time, the clustering coefficient was much higher (between 30–50 times) than that of the corresponding random graphs.

These results are in agreement with previous morphological characterizations of *in vitro* neuronal networks at a single developmental stage [6], where a similar small-world arrangement of connections was evidenced at DIV 6. However, to reinforce the evidence of the emergence of a small-world configuration *during* the graph development (as well as the fact that here the small-world metrics are not influenced by network disconnectedness), we also measured the global and local efficiency, as introduced by Latora and Marchiori in [8]. These latter quantities, indeed, are seen as alternative markers of the small-world phenomenon, in that small-world networks are those propagating information efficiently both at a global and at a local scale. The efficiency curves of the cultured networks are reported in Fig. 4C as a function of age, and compared to the efficiency of the equivalent random graphs. Our results indicate that the connectivity structure of the neuronal networks evolve towards maximizing global efficiency (making it similar to the value of random graphs), while promoting fault tolerance by maximization of local efficiency (which is, instead, larger than the local efficiency of a random graph), and both properties are realized at a relatively low cost in terms of number of links (see again Fig. 3A).

Node degree distribution evolution

Turning now our attention to network statistics at the microscale, we investigated how the node degree distributions evolved during maturation process. At all ages, cultures appeared to belong to the class of single-scale networks, displaying a well defined characteristic mean node degree. Figure 5A shows that the cumulative degree distributions $P_{cum}(k)$ for DIVs 3, 6, 7, and 12 had a fast decay with a non monotonous increase in the average connectivity, with most of the nodes having a similar number of connections and only a few ones with degrees deviating significantly from such a number.

The data were fitted to an exponential scaling law $y(k) \sim exp(-k/b)$ with a level of confidence larger than 95%.

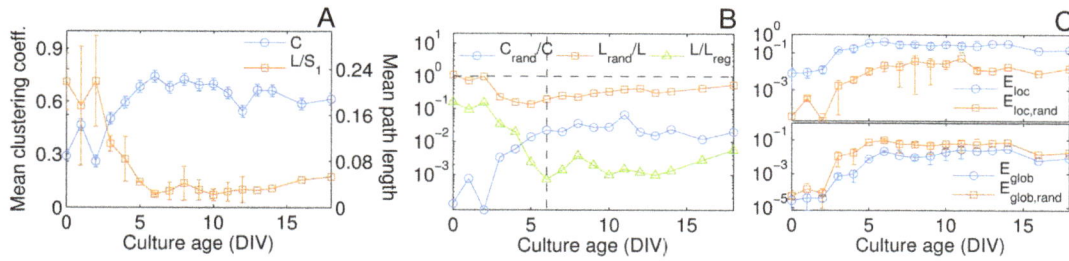

Figure 4. Network clustering and shortest path properties as a function of culture age. (A) Absolute values of the clustering coefficient C (blue circles, left axis) and mean path length L (red squares, right axis) normalized to the size of the largest cluster. (B) Semi-log plot of normalized values of C and L with respect to the expected values for an equivalent random network having the same number of nodes and links and preserving the degree distribution: C_{rand}/C (blue circles) and L_{rand}/L (red squares). The average path length is also compared to the value for a regular lattice as L/L_{reg} (green triangles) with $L_{reg} = S_1/\langle k \rangle$, being $\langle k \rangle$ the average connectivity and S_1 the size of the largest connected component. (C) Local (upper plot) and global (lower plot) efficiency as a function of culture age and compared to their respective values for the random graphs of the null model (see text for an explanation). All quantities are averaged for the set of 6 cultures at each day of measure (DIV). As in the Caption of Fig. 3, error bars represent the standard errors of the mean (sem).

The values of the scaling parameter b were close, within error, to the mean degree $\langle k \rangle$ of the networks at each culture age. It has to be remarked that the distribution of node connections, although always homogeneous, shifted during culture maturation toward much broader distributions, with few highly connected nodes appearing at DIVs 6 and 7. These "hubs" at the peak days of the culture evolution result from a branching process, allowing each single neuron to reach a larger neighborhood. Thus, at variance with scale-free networks [22,23], our cultured and clustered networks are identified as a single-scale homogeneous population of nodes. This is in agreement with reports on many other biological systems like the neuronal network of the worm *Caenorhabditis elegans* [24,25], and suggests the existence of physical costs for the creation of new connections and/or nodes limited capacity [26].

While the number of neighbors (the degree) is a quantity retaining information at the level of a single node, one can go further to inspect degree-degree correlations, i.e. to quantify whether the degrees of two connected nodes are correlated. It is known, indeed, that biological networks feature generally dis-assortative network structures [27], that is connections are more likely to be established between high-degree and low-degree nodes. In our system, we used the assortativity coefficient described in the Experimental set-up section. Figure 5B shows the age evolution of the Pearson coefficient r and of the exponent p that characterizes the scaling behavior of the degree correlation properties

$(\langle k_{nn} \rangle = ak^p)$. At one hand, as r stays positive during the whole development we can generally conclude that our networks are assortative and, on the other hand, the trend of the exponent p indicates that there is a transition from an almost linear (from DIV 0 to DIV 2) to a sub-linear ($p \sim 0.7$) degree-degree correlation regime during the small-world stage.

It is important to remark here that, to the best of our knowledge, this is the first report of assortativity in an *in vitro* cultured neuronal network, and such an evidence actually links to other studies where assortativity was found in simple *in vivo* neuronal systems, like the *C. elegans* neural network structure [28].

Spatial-growth model

A series of previous studies [15,16] singled out tension along neurites and adhesion to the substrate as the two main factors conditioning the neuronal self-organization into a clustered network. Here we go a step ahead, and propose a simple spatial model which not only incorporates migration of neurons but also explicitly considers neurite growth, and the establishment of synaptic connections.

Our model is schematically illustrated in Fig. 6. We start by considering a set of N cells. Each cell is a small disk of radius a randomly distributed in a 2-dimensional circular substrate of area S. The algorithm then evolves the connections and positions of such disks at discrete times, each time step t corresponding to a

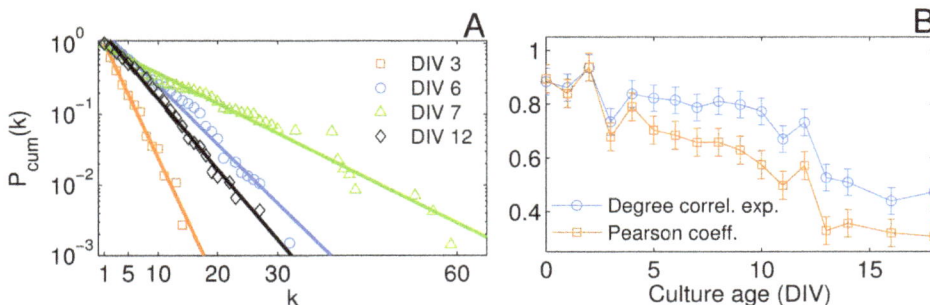

Figure 5. Degree distribution and degree-degree correlation. (A) Cumulative node degree distributions on a semi-log scale for the state of the same culture at DIVs 3, 6, 7, and 12 (see legend for the symbol coding). Solid lines correspond to the best exponential fitting $y(k) \sim \exp(-k/b)$, with $b \simeq \langle k \rangle$ the mean degrees at DIV 3, 6, 7, and 12 respectively. (B) Degree correlation exponent (blue circles) measuring the network assortativity and the corresponding Pearson coefficient (red squares). Both quantities are averaged for the set of 6 cultures at each day of measure (DIV) and error bars represent the sem.

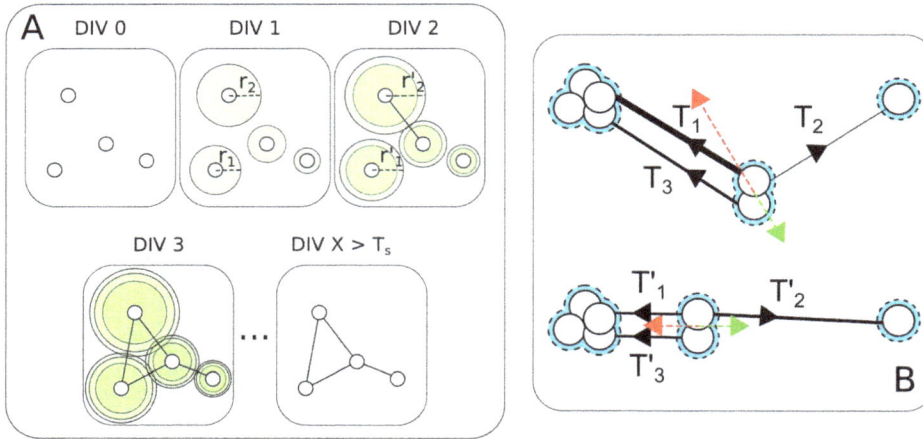

Figure 6. Growth model. (A) Schematic representation of how cells get connected. At DIV 0, 4 cells of radius a are located at random positions. The first iteration of the algorithm, DIV 1, assigns to each cell i a disk of radius r_i (green shade). At the next iteration, DIV 2, the disk's growth rate decreases, r_i, and a link between two cells is established when their disks intersect (DIV 3). This process continues until T_s steps. (B) Force diagram explaining cell migration and clustering. Tension forces T_1, T_2, and T_3 are acting on the central cluster composed of two cells, whose vector sum (red arrow) exceeds the adhesion to the substrate (green arrow). As a result, a new equilibrium state is produced with new tension forces T_1', T_2', and T_3', being the central cluster pulled in the direction of the net force approaching the largest cluster.

DIV of the culture. The complex process of neurite growth and the establishment of synaptic connections is modeled by associating to each cell a time growing disk in such a way that, two cells are linked at a given time step if their outer rings intersect as shown for DIV's 2 and 3 in Fig. 6A. At each time step t, the radius $r_i \geq a$ increases by a quantity δr_i which decays as

$$\delta r_i^t = \frac{V}{t}\left(1 - \frac{1}{K_i}k_i^{t-1}\right)$$

where V is the neurite growth velocity (the same for all cells), K_i a random number in the interval $[1,N]$ and k_i^t the degree of the node (cell) at the time step t. The term k_i^{t-1}/K_i introduces heterogeneity in the cell population, and represents the fraction of links acquired by the cell in the previous steps from the initial randomly assigned endowment K_i. A very large K_i indicates that, potentially, a cell is very active and could connect many other cells. The wiring process is iterated up to a given time step T_s, at which the formation of new connections is stopped.

As for the process of cell migration and clusterization, cells or clusters whose distance is less than $2a$ are then merged into the same new cluster. Furthermore, whenever two cells are connected, an initial tension $\vec{T}_{ij} = 0.1\vec{u}_{ij}$ is created between them, and it is incremented in 0.1 force units at each time step, being $\vec{u}_{ij}$ the unit vector along the direction connecting the two cells. The total force acting on a cell or cluster i is given by $\vec{F}_i = \sum_j \vec{T}_{ij}$ with j running over the cell indexes connected to i, and not belonging to the same cluster. Furthermore, each cell is "anchored" to the substrate by a force $F_a = 10$ force units, and the ith cell can only be detached if there is a net force F_i acting on it larger than F_a. In the case of a cluster of cells, the adhesion force to the substrate is considered to be the sum of the individual adhesions of the cells composing the cluster. Therefore, cells and clusters move in a certain direction in all circumstances in which the net force acting on them overcomes the adhesion force, and an equilibrium point is reached at a new position in which the new net force balances (or is smaller) than the adhesion to the substrate (see Fig. 6B).

In order to validate our model, we ran a large number of simulations for different values of the model parameters N, V and T_s. Remarkably, when comparing the statistical topological features of the simulated networks to those measured from the set of 6 cultures, we found high correlation values exist only in a very narrow window of V and T_s. For instance, the parameter values which better fit the experimental observations for $N = 700$ are $V = 40 \pm 5$ and $T_s = 9 \pm 1$.

Figure 7 shows a typical output of the evolution of a simulated network. The initial number of cells is taken to be of the same order as in the experiments, and we chose as parameters V and T_s those with the highest correlation with experiments. Boundary conditions mimic the real experimental setup by canceling the adhesion force to the substrate outside the culture area. The spatial organization of the network of cells and clusters after 3, 6, and 12 DIV, closely resembles the one observed in the experiments (see Fig. 1).

Despite its relative simplicity, it is remarkable that the model offers a rather good qualitative verification of the trends of all the structural network characteristics measured in the experiments. In particular, Fig. 8 reports a synoptic comparison of the mean number of nodes and links, of C and L, of the mean degree and degree correlation, and of the sizes of the GCC and GCC2, measured in the experiments and those obtained from the

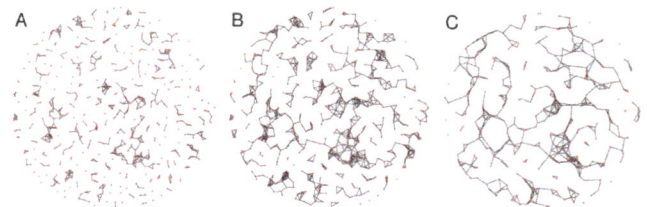

Figure 7. Schematic illustration of the network self-organization at three different instants of the automata generations. (A) DIV 3, (B) DIV 6, and (C) DIV 12. Simulation parameters: $N = 700$, $V = 42$, and $T_s = 10$.

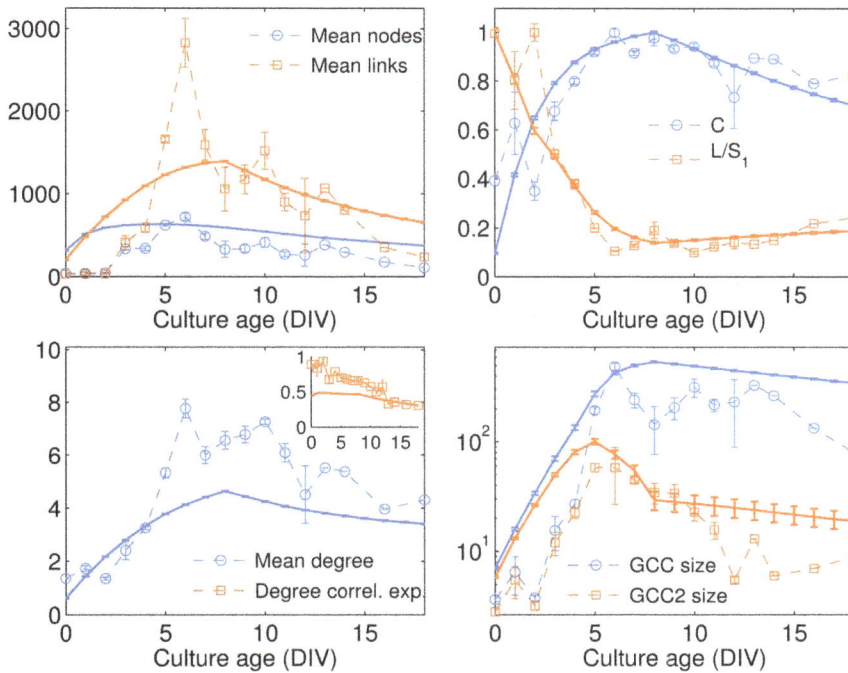

Figure 8. Comparison between model and experiment. Legends in each panel clarifies on the topological quantities measured in experiments (dashed curves), and the corresponding trends of the simulated networks (solid curves). Simulation parameters are the same as in the caption of Fig. 7, and each point is the ensemble average over 50 independent runs of the growth algorithm.

simulations of the model with $N = 700$, $V = 42$, and $T_s = 10$. The main observed difference is found in the mean degree, the reason behind such a slight discrepancy being that the model does not include any neurite branching process, limiting the size of the neighborhood encompassed by the nodes.

Though it would have been unrealistic to expect a perfect quantitative agreement between model and observations, the fact that the model reproduces the main qualitative scenarios of the experiments indicates that it captures the main processes underlying the observed morphological evolution and self-organization of the cultures.

Discussion

In summary, we provided a large scale experimental investigation of the morphological evolution of *in vitro* primary cultures of dissociated invertebrate neurons from locust ganglia. At all stages of the culture's development, we were able to identify neurons' and neurites' location in automated way, and extract the adjacency matrix that fully characterizes the connectivity structure of the networking neurons. A systematic statistical analysis of a group of topological observables has later allowed tracking of the main network characteristics during the self-organization process of the culture, and drawing important conclusions on the nature of the processes involved in the culture' structuring. At early stages of development ($<$DIV 3) characterized by a high neurite growth rate, homogeneous node degree distribution and low clustering resulted in a random topology as expected given the fact that neurons were randomly seeded. Following this immature period, neurite growth rate diminished and tension along neurites started to shift the network to a small-world one with path lengths similar to random configurations but presenting high clustering of connections. This transition from random to small-world con-

curred with the percolation of the culture and the onset of the giant connected network component.

Furthermore, the identification of the main physical processes taking place during the culture's morphological transformations, allowed us to embed them into a simple growth model, qualitatively reproducing the overall scenario observed in the experiments.

Our results extend previous studies where network properties of cultures were investigated at a particular developmental stage and for a lesser number of nodes [6]. These results also systematically characterize several topological network measures along the entire culture's evolution, and unveil many yet unknown self-organization properties, such as *i*) the fact that a small-world configuration spontaneously emerges in connection to a large increase in both local and global network's efficiency, and *ii*) the evidence that cultures tend to organize in a regime of non trivial degree mixing which, in turn, is characterized by assortative degree-degree correlation features. The evolution from an initial random to a small-world topology has also been reported recently in the context of a functional network of a cortical culture [13]. However, although functional connectivity correlates well with anatomical connectivity, there are studies showing that strong functional connections may exist between nodes with no direct physical connection [12]. This suggests that future studies are needed in which both anatomical and functional networks are accessible in order to understand their complex entanglement.

Given the absence of external chemical or electrical stimulations, we conclude that such complex network evolution and morphological structuring is indeed an intrinsic property of neuronal maturation. Our study therefore contributes to the understanding of the complex processes ruling the morphological structuring of cultured neuronal networks as they self-organize from collections of separated cells into clustered graphs, and may help identifying culture development stages in new, specific and targeted, experiments.

Author Contributions

Conceived and designed the experiments: ISN IL SB. Performed the experiments: ISN DSS. Analyzed the data: ISN DSS JA. Contributed reagents/materials/analysis tools: DSS AA SA JA. Wrote the paper: ISN SB DP.

References

1. Marom S, Shahaf G (2002) Development, learning and memory in large random networks of cortical neurons: lessons beyond anatomy. Quarterly reviews of biophysics 35: 63–87.
2. van Pelt J, Vajda I, Wolters PS, Corner MA, Ramakers GJ (2005) Dynamics and plasticity in developing neuronal networks in vitro. Progress in brain research 147: 173–88.
3. Eckman JP, Feinerman O, Gruendlinger L, Moses E, Soriano J, et al. (2007) The physics of living neural networks. Phys Rep 448: 54–76.
4. Ayali A, Fuchs E, Zilberstein Y, Robinson A, Shefi O, et al. (2004) Contextual regularity and complexity of neuronal activity: From stand-alone cultures to task-performing animals. Complexity 9: 25–32.
5. Boccaletti S, Latora V, Moreno Y, Chavez M, Hwang D (2006) Complex networks: Structure and dynamics. Physics Reports 424: 175–308.
6. Shefi O, Golding I, Segev R, Ben-Jacob E, Ayali A (2002) Morphological characterization of in vitro neuronal networks. Phys Rev E 66: 021905.
7. Watts DJ, Strogatz SH (1998) Collective dynamics of small-world networks. Nature 393: 440–442.
8. Latora V, Marchiori M (2003) Economic small-world behavior in weighted networks. The European Physical Journal B - Condensed Matter 32: 249–263.
9. Achard S, Bullmore E (2007) Efficiency and cost of economical brain functional networks. PLoS Computational Biology 3: e17.
10. Rad AA, Sendiña Nadal I, Papo D, Zanin M, Buldú JM, et al. (2012) Topological measure locating the effective crossover between segregation and integration in a modular network. Phys Rev Lett 108: 228701.
11. Feldt S, Bonifazi P, Cossart R (2011) Dissecting functional connectivity of neuronal microcircuits: experimental and theoretical insights. Trends in neurosciences 34: 225–36.
12. Honey C, Sporns O, Cammoun L, Gigandet X, Thiran J, et al. (2009) Predicting human restingstate functional connectivity from structural connectivity. Proceedings of the National Academy of Sciences 106: 2035–2040.
13. Downes JH, Hammond MW, Xydas D, Spencer MC, Becerra VM, et al. (2012) Emergence of a Small-World Functional Network in Cultured Neurons. PLoS Computational Biology 8: e1002522.
14. Anava S, Saad Y, Ayali A (2013) The role of gap junction proteins in the development of neural network functional topology. Insect molecular biology 22: 457–472.
15. Shefi O, Ben-Jacob E, Ayali A (2002) Growth morphology of two-dimensional insect neural networks. Neurocomputing 44-46: 635–643.
16. Segev R, Benveniste M, Shapira Y, Ben-Jacob E (2003) Formation of Electrically Active Clusterized Neural Networks. Physical Review Letters 90: 168101.
17. Anava S, Greenbaum A, Ben-Jacob E, Hanein Y, Ayali A (2009) The regulative role of neurite mechanical tension in network development. Biophysical journal 96: 1661–70.
18. Bollobás B (2001) Random Graphs. Cambridge University Press.
19. Li D, Li G, Kosmidis K, Stanley HE, Bunde A, et al. (2011) Percolation of spatially constraint networks. EPL 93: 68004.
20. Rubinov M, Sporns O (2010) Complex network measures of brain connectivity: Uses and interpretations. NeuroImage 52: 1059–1069.
21. Woiterski L, Claudepierre T, Luxenhofer R, Jordan R, Käs JA (2013) Stages of neuronal network formation. New Journal of Physics 15: 025029.
22. Albert R, Barabási AL (2002) Statistical mechanics of complex networks. Rev Mod Phys 74: 47–97.
23. Barabási AL, Albert R (1999) Emergence of scaling in random networks. Science 286: 509–512.
24. White JG, Southgate E, Thomson JN, Brenner S (1986) The structure of the nervous system of the nematode caenorhabditis elegans. Philosophical Transactions of the Royal Society of London B, Biological Sciences 314: 1–340.
25. Watts DJ, Duncan J (1999) Small Worlds: The Dynamics of Networks Between Order and Randomness. Princeton University Press, Princeton, NJ.
26. Amaral LAN, Scala A, Barthélemy M, Stanley HE (2000) Classes of small-world networks. Proceedings of the National Academy of Sciences 97: 11149–11152.
27. Newman MEJ (2002) Assortative mixing in networks. Phys Rev Lett 89: 208701.
28. Chatterjee N, Sinha S (2008) Understanding the mind of a worm: hierarchical network structure underlying nervous system function in C. elegans. Progress in brain research 168: 145–53.

16

Microsatellite Repeat Instability Fuels Evolution of Embryonic Enhancers in Hawaiian *Drosophila*

Andrew Brittain, Elizabeth Stroebele, Albert Erives*

Department of Biology, University of Iowa, Iowa City, Iowa, United States of America

Abstract

For ~30 million years, the eggs of Hawaiian *Drosophila* were laid in ever-changing environments caused by high rates of island formation. The associated diversification of the size and developmental rate of the syncytial fly embryo would have altered morphogenic gradients, thus necessitating frequent evolutionary compensation of transcriptional responses. We investigate the consequences these radiations had on transcriptional enhancers patterning the embryo to see whether their pattern of molecular evolution is different from non-Hawaiian species. We identify and functionally assay in transgenic *D. melanogaster* the Neurogenic Ectoderm Enhancers from two different Hawaiian *Drosophila* groups: (*i*) the picture wing group, and (*ii*) the modified mouthparts group. We find that the binding sites in this set of well-characterized enhancers are footprinted by diverse microsatellite repeat (MSR) sequences. We further show that Hawaiian embryonic enhancers in general are enriched in MSR relative to both Hawaiian non-embryonic enhancers and non-Hawaiian embryonic enhancers. We propose embryonic enhancers are sensitive to Activator spacing because they often serve as assembly scaffolds for the aggregation of transcription factor activator complexes. Furthermore, as most indels are produced by microsatellite repeat slippage, enhancers from Hawaiian *Drosophila* lineages, which experience dynamic evolutionary pressures, would become grossly enriched in MSR content.

Editor: Arnar Palsson, University of Iceland, Iceland

Funding: AE: National Science Foundation, Award: 1239673. ES: National Institutes of Health, Training Grant: T32GM082729. The funders had no role in study design, data collection and analysis, decision to publish, or preparation of the manuscript.

Competing Interests: The authors have declared that no competing interests exist.

* Email: albert-erives@uiowa.edu

Introduction

Genomic sequences from twelve ecomorphologically diverse *Drosophila* species have been assembled [1] and studied [1–6]. One of these twelve species, *D. grimshawi*, is from the large "picture wing" group, which itself is one of many groups of the remarkably speciose Hawaiian *Drosophila*, corresponding to almost 500 of the ~1500 described *Drosophila* species and others yet to be adequately described [7–16]. The Hawaiian species form a monophyletic group and include recent radiations exemplified by the picture wing group, which diverged from a most recent common ancestor less than one million years ago (~0.5–0.7 Mya), older radiations such as that exemplified by the "modified mouthparts" group, and older still the so-called *Scaptomyza* flies (Fig. 1A). Thus, the *Drosophila* subgenus, known as IDIOMYIA (Hawaiian *Drosophila*+ *Scaptomyza*) illustrates the profound species fecundity of the island forming process that in ~40 million years produced the Hawaiian seamount island chain, which was colonized by *Drosophila* over ~30 million years ago (Fig. 1B).

We consider the consequences of the sustained pattern of frequent species radiations on transcriptional enhancers of the syncytial fly embryo within Hawaiian *Drosophila*. In this evolutionary context, the evolving *Drosophila* egg is being laid in new and ever-changing environments. The associated evolutionary diversification of the syncytial fly embryo (*viz.*, the shape, size, and developmental rate of the embryo as previously shown [17,18]) would have continuously altered embryonic morphogen gradients of each lineage, thus necessitating compensatory evolution of the gradient-sensing responses of target enhancers [4]. We therefore ask whether the pattern of molecular evolution at developmental enhancers that interpret embryonic morphogen gradients in Hawaiian *Drosophila* differs from that in non-Hawaiian *Drosophila*.

To address this question, we considered a group of complex transcriptional enhancers that are important to *Drosophila* morphogenesis: the Neurogenic Ectoderm Enhancers (NEEs) [4,5,19,20]. Unlike protein-coding gene families, which are related by common descent (*i.e.*, homology), the NEEs in a single genome are similar only by molecular convergence (parallelism) and so we define them as a mechanistic "family". Four "canonical" NEEs are present as orthologs across the genus in the unrelated loci *rhomboid* (*rho*), *vein* (*vn*), *brinker* (*brk*), and *ventral nervous system defective* (*vnd*) [4,5,19–21]. The NEEs are responsive to the morphogenic gradient of Dorsal, which patterns the dorsal/ventral axis of the early embryo. While Dorsal is a homolog of the NFkB-enhanceosome forming factor, and is known to work with many different co-activator transcription factors (TFs) along the D/V axis [22], we found that the NEEs contain binding sites for a specific subset of these factors. Binding sites for the activators, Dorsal, Twist, Su(H), and the mesodermal repressor Snail are present in each of the NEEs we have found [4,20], which is

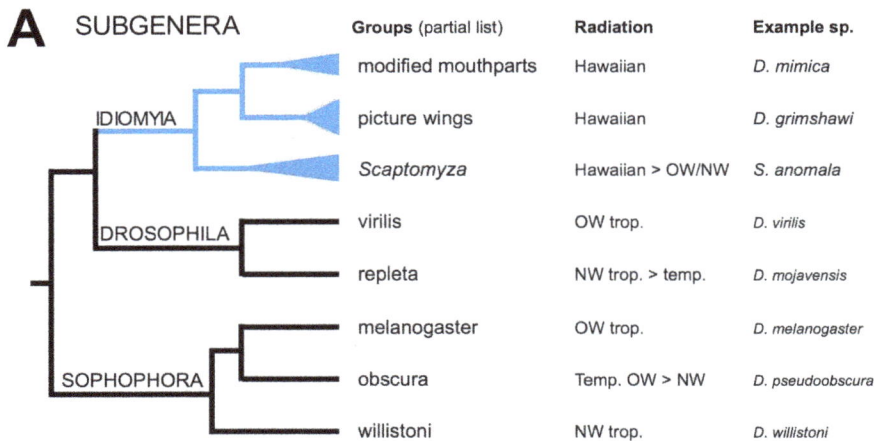

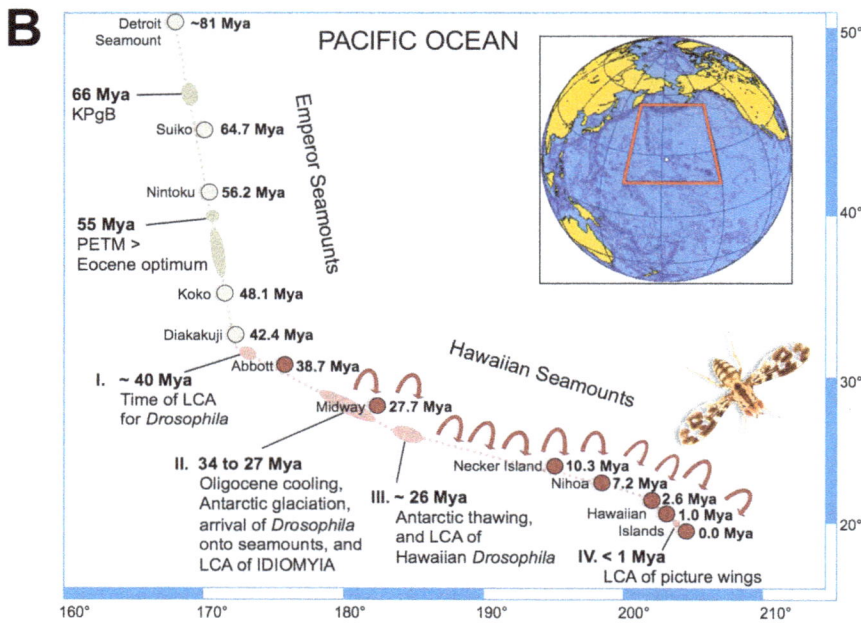

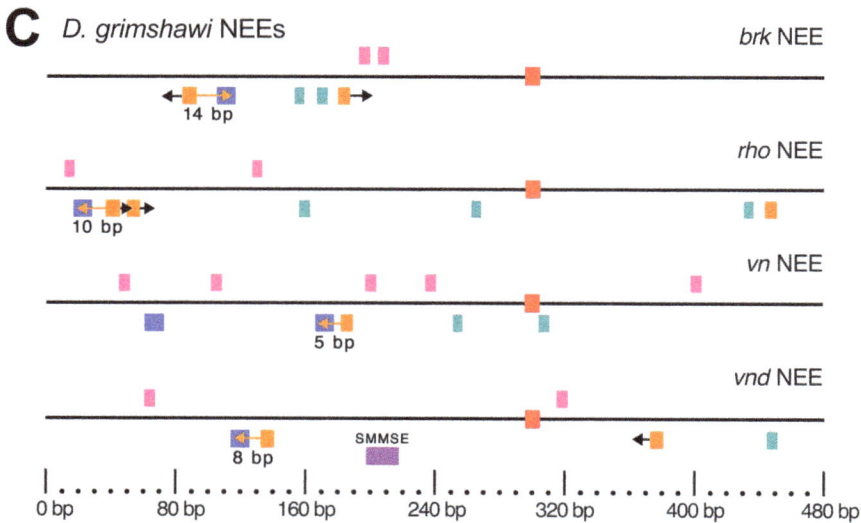

Figure 1. The Neurogenic Ectoderm Enhancers from the Hawaiian species *D. grimshawi*. (**A**) Shown is a phylogenetic tree showing the relationship of the Hawaiian *Drosophila*, which comprise subgenus IDIOMYIA along with the so-called *Scaptomyza* [16]. This subgenus gave rise to multiple clades (highlighted in blue) corresponding to multiple adaptive radiations associated with the last ~30 My history of island formation. Embryonic enhancers from two of these clades, the modified mouthparts group and the picture wing group, are analyzed in this study. (**B**) Shown is a

map of the Hawaiian Seamounts with relevant events in the evolutionary lineage leading to the Hawaiian *Drosophila* groups indicated by the ages of the various islands. The picture wing group (inset shows an example *D. grimshawi* adult fly) is one of the most recent radiations and is closely associated with the newest and easternmost front of the seamounts, *i.e.*, the Hawaiian islands. (**C**) The graphs indicate the site architecture of the four genus-canonical Neurogenic Ectoderm Enhancers (NEEs), which we identified in the *D. grimshawi* genome. The NEEs are a class of early embryonic enhancers downstream of the Dorsal morphogen gradient [4,5,19]. The search for NEEs in the *D. grimshawi* genome was conducted by searching for linked sites for the Twist:Daughterless bHLH heterodimer (Twi:Da) and the rel homology domain-containing TF Dorsal (5′-CACATGT 0–41 bp GGAAABYCC), plus a nearby Su(H) site binding site located up to +/−300 bp away (see Material & Methods). Three horizontal tracks indicate sequences matching different TF binding motifs (tracks are separated to avoid overlap and for ease of visualization when viewed in black & white print or with color-blindness). <u>Above line</u>: pink boxes = Snail (5′-CARRTG) [57]. <u>On line</u>: red boxes = Su(H) (5′-YGTGRGAA). A single Su(H) site is present in each enhancer and is used to anchor each sequence at basepair position 300 bp [the Su(H) motif matches the top strand for all except in the *vn* NEE]. <u>Below line</u>: orange boxes = Twist:Daughterless (Twi:Da, 5′-CACATGT); blue boxes = Dorsal (5′-VGGAAABYCCV); blue and orange boxes connected by arrow = linked Twi:Da–Dorsal sites with text indicating spacer length; and purple boxes = Shnurri:Mad:Medea Silencer Element (SMMSE). The Schnurri/Mad/Medea Silencer Element (SMMSE), which functions to constrain the dorsal border of activity for the NEE at *vnd* [19], matches the *D. melanogaster*/*D. grimshawi* consensus 5′-MYGGCGWCACACTGTCTGS and is highlighted in purple.

consistent with the NEEs representing one specific equivalence class of enhancers; there exist other lateral stripe enhancers and sometimes even lateral stripe "shadow enhancers" [4,23–26] at the same NEE-bearing loci, but they feature binding sites for distinctly different sets of factors other than Dorsal.

Here, we identify and analyze the evolutionary divergence of NEEs in one representative species of the Hawaiian picture wing group, which has a fully sequenced genome (*D. grimshawi*) [1], and in one representative species of the Hawaiian modified mouthparts group (*D. mimica*), for which we cloned, sequenced, and tested their enhancers. We show that relative to *D. virilis*, which is a representative of the continental Old World group of the subgenus DROSOPHILA that gave rise to the Hawaiian *Drosophila*, the intervening DNA sequences between the NEE binding sites have been largely replaced by ***microsatellite repeat*** (**MSR**) sequences. This unique MSR-footprint demarcates the functional binding sites for Dorsal, Twist/Snail, and Su(H). It also demarcates the dedicated Snail binding sites and sites for the general embryonic timing factor Zelda. We also demonstrate that relative enrichment of MSR in Hawaiian *Drosophila* is specific to developmental enhancers of embryogenesis and suggests that diverse enhancers function as enhanceosome scaffolds with sensitive spacing requirements. Because Dorsal, Twist, Su(H), and Zelda are all polyglutamine-rich transcriptional activators, we propose a specific model in which enhancers functioning as scaffolds for polyglutamine-mediated co-factor complexes are both sensitive to *cis*-element spacing and are sites of MSR-enrichment when subjected to evolutionary pressures.

Results

Neurogenic Ectoderm Enhancers (NEEs) from Hawaiian *Drosophila*

To identify the repertoire of NEE functions in a Hawaiian *Drosophila* species, we used the assembled genome from the Hawaiian picture wing fly, *D. grimshawi* (Fig. 1A, and adult pictured in Fig. 1B inset). Specifically, we searched for a Su(H)-binding motif (5′-YGTGRGAA) located within 300 bp of linked Twist and Dorsal binding sites (5′-CACATGT 0–40 bp nGGAAABYCCn, where the Dorsal site could be in any orientation and the n's are included here only to indicate the normal extent of the Dorsal binding site; see methods). We find only the four genus-canonical NEEs at *brk*, *rho*, *vn*, and *vnd*, each having linked Dorsal and Twist binding sites with spacers of length 14 bp, 10 bp, 5 bp, and 8 bp, respectively (Fig. 1C). Previously, we showed the length of the spacer separating these linked Dorsal and Twist sites to be a major determinant of the extent to which the NEE is responsive to the Dorsal morphogen [4,5].

We then designed primers on the basis of the NEE sequences in *D. grimshawi*, and successfully amplified intact fragments, ~500 to

600 bp in length, containing the NEEs from the *rho*, *vn*, and *vnd* loci from a Hawaiian modified mouthparts fly, *D. mimica*, which we have begun culturing in the lab (see Methods). We made standard fusion reporter genes using the −42 *eve*:*lacZ* β-*tub* 3′-UTR reporter construct in a P-element vector and transformed them into *D. melanogaster* to test for function. All three of these enhancers are *bona fide* NEEs by definition of their site composition and organization, and have discernible neuroectodermal enhancer activity in *D. melanogaster* embryos despite ~40 million years of evolution in addition to the accelerated levels of evolution in the Hawaiian system (Fig. 2). We find that the *vnd* NEE still encodes a predicted low threshold response that drives early expression, when the Dorsal nuclear gradient is still increasing ventrally [27], but a later dorsally-repressed expression pattern due in part to a well-conserved Schnurri/Mad/Medea silence element [19] (Fig. 2, E–H).

Extensive Microsatellite Repeat (MSR) Replacement of Hawaiian NEE intersite spacers

Inspection of the Hawaiian NEEs reveals diverse microsatellite repeat patterns besides the known genus-wide enrichment of CA-dinucleotide repeats in NEEs [5]. To visualize precisely this content and to determine the possibility of longer repeats being present, we plotted all direct (tail to head) repeats (two or more) of a unit sequence that is 2–50 bp long (fluorescent green boxes in Fig. 3) (also see Materials & Methods). We find that the *rho* and *vn* NEEs from both Hawaiian species are qualitatively enriched in MSR content relative to both *D. melanogaster* and *D. virilis* (Fig. 3A, B). The enrichment that can be seen in comparison to both non-Hawaiians is made more significant by the fact that *D. virilis* has a much larger genome than *D. melanogaster* while also being more closely related to the Hawaiian lineages [4,5]. The enrichment is not seen in the Hawaiian *vnd* NEEs relative to the non-Hawaiians (Fig. 3C), but this is as expected for the following two reasons. First, the *vnd* NEEs are less variable in activity phylogenetically relative to all other NEEs [4]. Second, the *vnd* NEEs possess a Shnurri:Mad:Medea Silencer Element, which corresponds to a second repressive input from the Dpp morphogen gradient and which ensures its characteristic ventral pattern of expression, critical to its role in patterning the nervous system [19].

To better quantify the MSR enrichment, we also plotted the exact MSR content for all three of these enhancers across all four species (Fig. 4). This shows that the NEEs without Shnurri/Mad/Medea Silencer Elements (SMMSE [19]) from the Hawaiian species have MSR content in the range of 43–57% in a 400 bp window encompassing all of the relevant TF binding sites (labeled "pure NEEs" in Fig. 4). In comparison, the pure NEEs from the non-Hawaiians have much less content in the range of 32–38%, similar to the range for the *vnd* NEEs of all species.

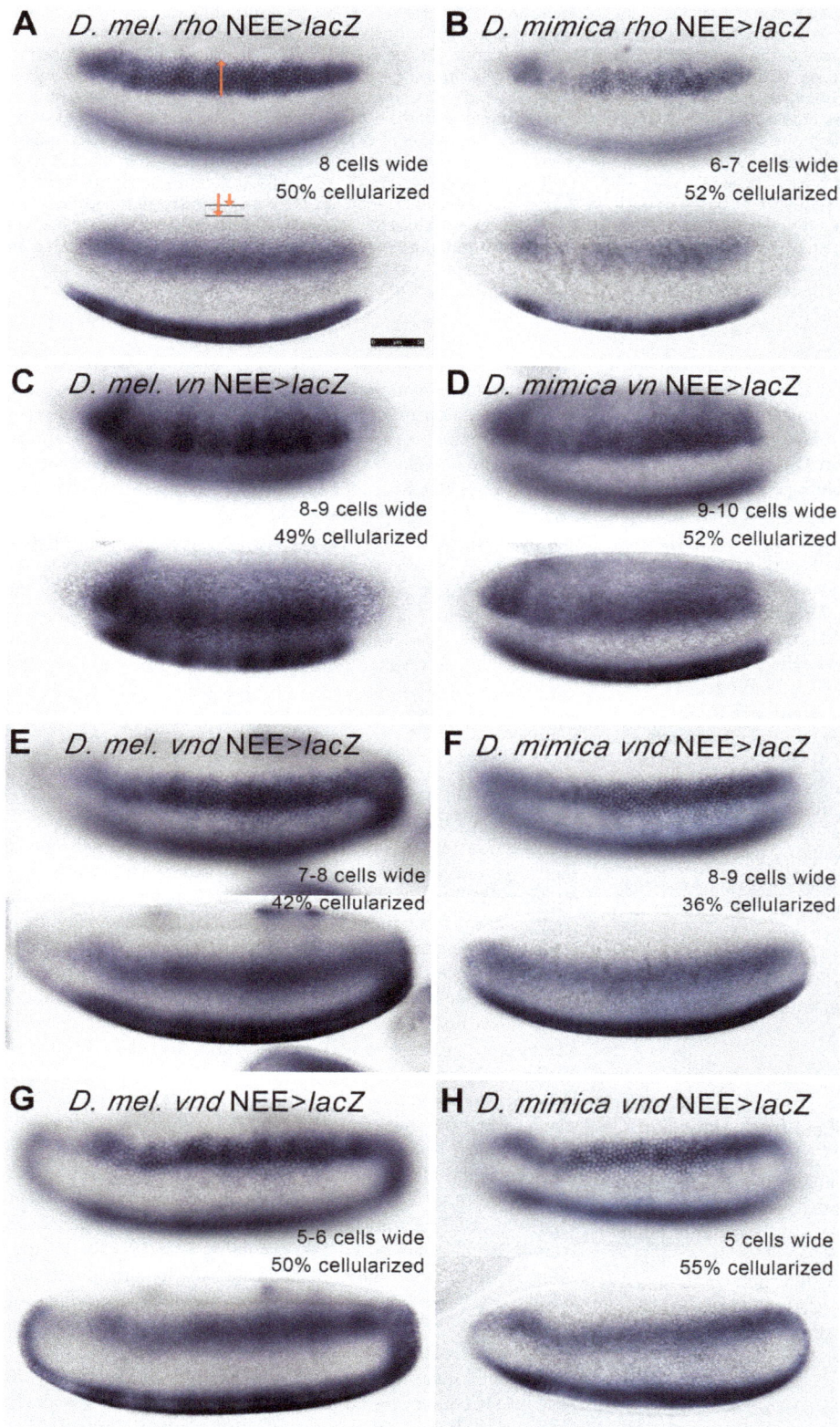

Figure 2. The NEEs from *D. mimica* Can Drive Expression in Neurogenic Ectoderm of *D. melanogaster* Despite Replacement of Inter-Element Spacers with MSR. NEEs from the *D. mimica*, a member of the Hawaiian modified mouthparts clade, were cloned and tested in *D. melanogaster* transgenic reporter assays. Expression of the *lacZ* reporter gene driven from *D. melanogaster* NEEs (**A, C, E, G**) and *D. mimica* NEEs (**B, D, F, H**) as determined by *in situ* hybridization with an anti-*lacZ* probe are shown. In each panel two optical cross-sections are shown. The top image is a surface view allowing determination of the stripe of expression (numbers of cells spanning D/V axis). The bottom image is a cross-section through the dorsal midline allowing determination of the exact stage of embryogenesis (% cellularization as determined at 50% egg length on the dorsal

midline). The expression patterns driven by the *rho* (**A, B**) and *vn* (**C, D**) NEEs are shown for the stages close to 50% cellularization. The expression patterns for the *vnd* NEEs are shown at two time points: an earlier time point at about ~40% cellularization (**E, F**) and a later time point at ~50% cellularization (**G, H**). For both Hawaiian and non-Hawaiian *vnd* NEEs, activity is dorsally repressed by the the Dpp gradient via a conserved binding site for the Shnurri:Mad:Medea complex beginning at about midway through cellularization [19]. All embryos are oriented with anterior pole to the left and dorsal side on top. The 50 micron scale bar shown in (**A**) is the same for all figures.

The striking nature of these enhancers can be summarized as follows. The MSR motifs in the *rho* and *vn* NEEs of Hawaiians (yellow green tracks in Fig. 3) encompass much of the enhancer, and the remaining sequences are either known TF binding sites (for Dorsal, Twist, Su(H), and Snail), or Zelda sites, which are present generally in early embryonic enhancers but have not been specifically pinpointed in the NEEs [28,29]. Zelda is considered a pioneer TF for early embryonic activation of genes, whose expression is patterned along the D/V and A/P axes [29–31].

We list a few examples of Hawaiian MSR enrichment that illustrate the range of repeat patterns and their phylogenetic distributions. There are many examples of both large and small duplicated blocks conserved only in the Hawaiians (Fig. 5, #1), and some of these have since diverged in repeat number (Fig. 5, #2). In some locations, repeats are found in only one of the Hawaiian species, such as an octamer repeat in the *D. grimshawi rho* NEE, which is a fragment of the Snail/Zelda sites (Fig. 5, #3). In other locations, repeats are conserved across the genus but the repeat unit sequence differs indicating a potential region of frequently amplified MSR sourced from different sequences prone to repeat slippage (Fig. 5, #4). Last, there are long repeat sequences present in the Hawaiians that are composed of smaller unit repeats (*i.e.*, repeats of repeats; Fig. 5, #5). In many places, the repeats are evidently diverging based on changes to the repeat unit or appearance of indels that disrupt their repeat pattern. Thus, molecular drive based on MSR slippage is an important mechanism in NEE evolution at the *rho* and *vn* loci of Hawaiian lineages but its subdued presence in the constrained *vnd* NEEs suggests that natural selection continuously acts on this mutagenic source of functional variation [32].

MSR-Enrichment in Embryonic *vs.* Non-embryonic Notch-Target Enhancers in *D. grimshawi*

The extreme MSR-footprint is widespread in the Hawaiian NEEs and here we demonstrate that in general the embryonic enhancers from Hawaiian *Drosophila* are enriched for MSR using two different genome-wide analyses described below. We first asked whether this MSR-enrichment was a general property of Notch-target, Su(H) binding site containing enhancers or only a specific feature of embryonic enhancers targeted by Notch/Su(H). To do this, we undertook an analysis of the entire set of Su(H) binding site repertoire for *D. virilis* and *D. grimshawi*. We first identified all individual Su(H) sites (5′-YGTGRGAA) and/or clusters of sites in each of the two genomes. This dataset was composed of blocks containing up to 270 bp of sequence flanking the Su(H) site or site cluster when possible, but ~3% of sequences had less because of close proximity to the edge of a contig but were not eliminated (385/13,473 and 476/14,904 for *D. grimshawi* and *D. virilis*, respectively). Site clusters were defined as having at least two Su(H)-binding sequences separated by <540 bp (*i.e.*, less than twice the desired flanking distance of 270 bp). Site clusters defined 6.0% of the data for *D. grimshawi* (815/13,473), and 7.1% of the data for *D. virilis* (1,060/14,904).

In order to identify conserved blocks between Hawaiian and non-Hawaiian Su(H)-binding motif containing repertoires, we identified sequence alignment parameters that are suited for the patterns of indel/MSR-mediated divergence that we see in

Drosophila enhancers. We took a heuristic approach to settle on a set of customized regulatory "*rblastn*" parameters that gave the highest alignment scores to NEEs that are homologous to each other across different *Drosophila* species (see Methods). Using this *rblastn* pipeline and an E-value cutoff of <1.0e-15, we identified ~3400 homologous sequences between *D. grimshawi* and *D. virilis*, which includes the canonical NEEs present across the genus at four unrelated loci: *rho*, *vn*, *brk*, and *vnd* [4,5,19–21].

We then took these ~3400 sequences from *D. grimshawi* and split them into two distinct sets (Fig. 6A). The first set of 270 sequences were identified because they had perfect binding sites for the embryonic temporal activator Zelda (5′-CAGGTAR), a pioneer TF for early gene activation [29–31]. As seen in the sequence alignments between the two Hawaiians and other *Drosophila* genomes, Zelda binding sites (Fig. 5 E, F, cyan nucleotides) are readily apparent between diverse MSR signatures (yellow green sequences in Fig. 5). The second set of 1671 sequences was derived by depleting the set of ~3400 conserved blocks of those blocks containing either Zelda binding sequences (5′-CAGGTA, 5′-CAGGCAR, or 5′-TAGGTAR), or more than a single Su(H) binding site (5′-nGTGnGAAn). To ensure that our test and control data sets contained sequences with equivalent levels of conservation, we plotted the distribution of E-values and found that there are still proportionally many highly conserved sequences in both data sets (Fig. 6B).

We find that the Zelda-positive data set contains much more (CA)$_n$- and (CAR)$_n$- MSR content than the control data set using a discriminative MEME [33] analysis (Fig. 6C). (CA)$_n$-MSR is known to be highly enriched in NEEs, and even enriched in *D. virilis* relative to *D. melanogaster*, which has a smaller genome [5]. As CAG-trinucleotide repeats and repeat instabilities are often seen in the protein-coding sequences for many transcriptional activators [34–37], we asked whether the enrichment of this motif was possibly due to the occurrence of nearby protein-coding exons near intronic enhancers. We find that almost all of the (CAR)$_n$-MSR content contributing to the enrichment occurs in non-protein-coding sequence (see File S3 and Table S1).

In sum, these findings suggest first that Notch-target enhancers that are operative in the early embryo are more divergent (and hence potentially faster-evolving) than non-embryonic enhancers containing Su(H)-binding sequences, and second that the divergence is driven in part by a molecular drive mechanism that changes the internal spacing separating TF binding sites. In the Discussion, we propose some molecular phenotypes that could explain why natural selection would act on the prodigious output of this MSR drive mechanism.

MSR-Enrichment in A/P Embryonic Enhancers of Hawaiian *vs.* Non-Hawaiian *Drosophila*

To determine whether unique MSR signatures are also enriched in embryonic, anterior/posterior (A/P) patterned enhancers, we first identified 3975 conserved blocks containing Zelda binding sites in *D. grimshawi* and *D. virilis*, with *rblastn* E-values of less than 1.0e-40 (Fig. 7A). From these we chose the subset of sequences that also contain a binding site for Runt (5′-AACCRCA), which represses the posterior expression domains of Bcd targets in the intermediate regions of the Bcd morphogen

A *rhomboid (rho)* NEE

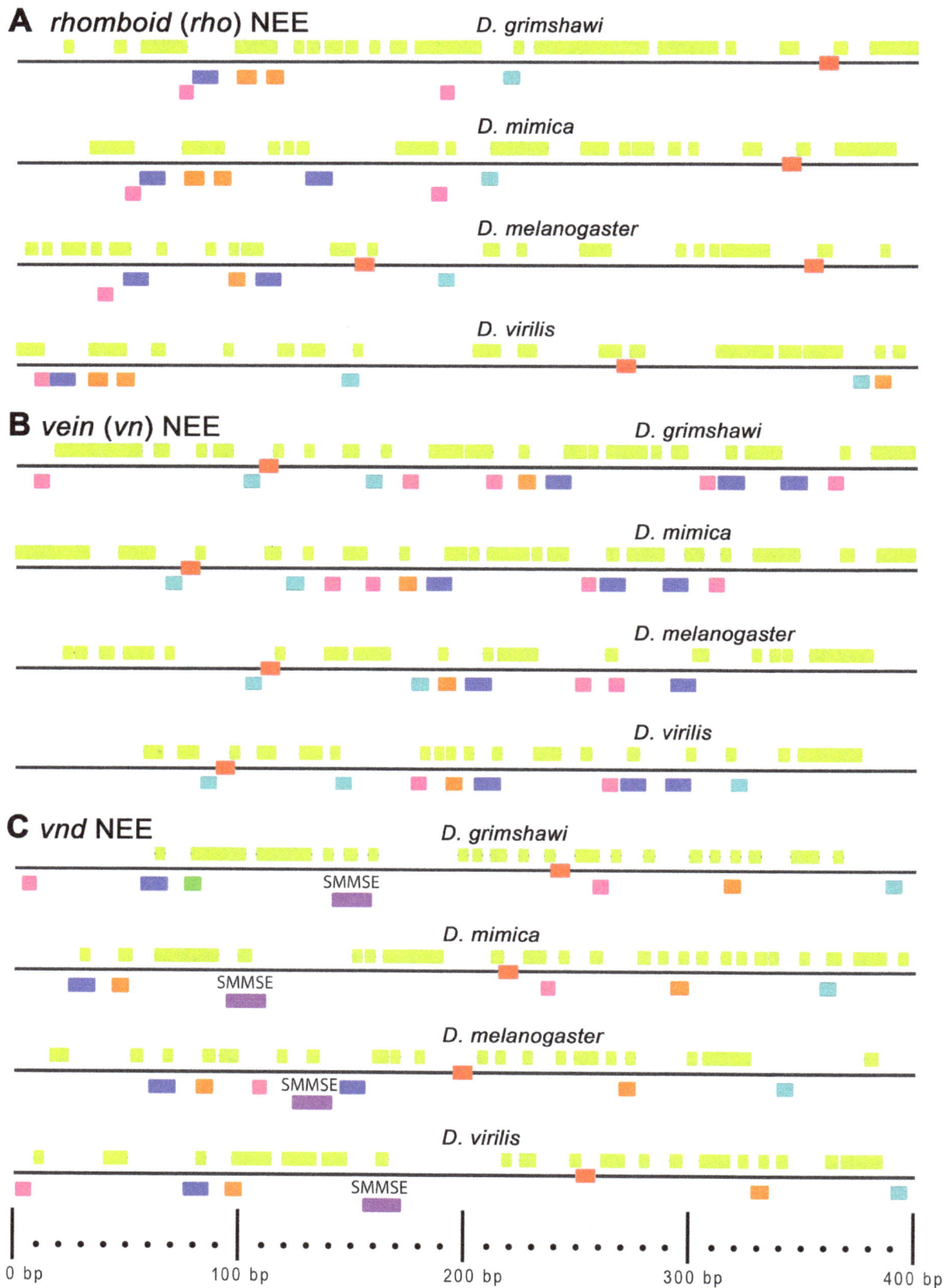

Figure 3. Diverse Direct MSRs Fill the Inter-Element Spacers of NEEs from Hawaiian *Drosophila*. Graph depicts micro-satelleite repeat (MSR) content, also known as simple sequence repeats, for the NEEs of the Hawaiian species, *D. grimshawi* and *D. mimica*, and two non-Hawaiian species, *D. virilis* and *D. melanogaster*, which also represent extremes in genome size (large and small, respectively). Note that *D. melanogaster* is the outgroup species as it is a member of the SOPHOPHORA subgenus. The panels are ordered by orthologous enhancers (*rho*, *vn*, *vnd*) and then by

species within each panel, top to bottom. The colored boxes correspond to the same TF binding motifs depicted in Fig. 1 except that the Dorsal $D\beta$ motif is relaxed at one position to 5'-VGGAAABNCCV (underlined "N") in order to match the site in D. mimica's NEE at vnd. The MSR content is plotted by a UNIX-type regular expression, "(.{2,50})\1" corresponding to two or more direct repeats of a unit sequence that is at least 2 bp or more in length (green yellow highlight above each line). Many such MSR sequences overlap. While difficult to see at first glance the Hawaiian NEEs are much enriched in this type of content. Exactly 400 bp centered on the NEE heterotypic site cluster is shown.

gradient [38]. Unlike the Bcd binding motif, the Runt binding motif is better suited to our question because it is: (**i**) well-defined [38], (**ii**) less variable, (**iii**) not related to binding sites for a large family of TFs such as the homeodomain-containing TFs, and (**iv**) associated with enhancers reading the rate-limiting parts of the Bcd morphogen gradient [38]. Because Runt binding sites were not always found in homologous sequences (either for lack of conservation or due to location of a truncated block near the edge of a contig), we performed a second *rblastn* query to identify only

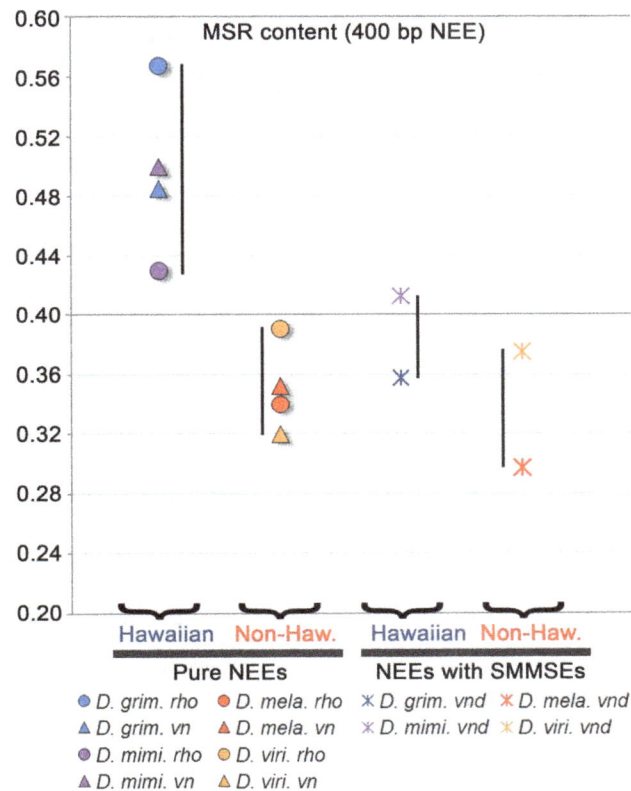

Figure 4. **Hawaiian NEEs without SMM silencer elements are enriched in MSR content relative to Non-Hawaiians.** Graph plots the MSR content in the 400 bp NEE window shown in Figure 3. The points for the "pure" NEEs, which lack Shnurri/Mad/Medea Silence Elements (SMMSE) are plotted separately from the *vnd* NEE, which contains a highly conserved SMMSE [19]. NEEs for Hawaiians and non-Hawaiians are plotted separately for ease of comparison. The *vnd* NEE activity is the least variable in output both ontogenetically and phylogenetically [4]. Its highly constrained early ventral expression is fundamentally important for correct D/V patterning of the nervous system and is thus not likely to be subject to shifting selection for a change in stripe width as we previously demonstrated [19]. Assuming that MSR is a signature of previous selection for changes in threshold responses, these results support the view that MSR content is uniquely enriched in enhancers subject to dynamic evolutionary pressures. Note that the length of the SMMSE that is not also MSR-like is 14 bp (19 bp *minus* 5 bp because of the underlined sequence in the SMMSE motif: 5'-MYGGCGWCACACTGTCTGS.) Thus, the presence of an SMMSE can only account for a reduction of MSR content of no more than 3.5%.

those with reciprocal homologs (Fig. 5A). We then performed a discriminative MEME analysis [33] using the set of homologous Zelda+Runt-containing sequences from D. grimshawi and D. virilis as the positive and negative data sets, respectively.

We find that the binding sites for Runt and Zelda are enriched in D. grimshawi relative to D. virilis (Fig. 7B), and suspect this is likely due to increased homotypic site clustering of these sites [39]. This is consistent with higher rates of binding site turnover, and we hope to investigate this matter in later studies. We also identified diverse MSR motifs, including the $(CA)_n$-dinucleotide and $(CAR)_n$-trinucleotide motifs previously seen in the embryonic Su(H) data set (Fig. 7C). We note that the binding motifs for both Zelda and Runt contain fragments of these sequences (asterisks in Fig. 7B). We also identify a clear $(AG)_n$-dinucleotide MSR motif (Fig. 7C), which best matches the binding site for *Trithorax-like* (*Trl*), which encodes the GAGA-binding factor GAGA that is expressed ubiquitously in the early embryo [40]. Thus, Zelda, Runt, and GAGA may be natural sources of functionally variant spacer alleles, much like the Twist-binding site of the NEEs [19]. This further supports our previously proposed hypothesis [5] that MSR-enrichment of enhanceosome-building enhancers is related to the intrinsic MSR-seeding capabilities of Activator TF binding sites. Last, we identify a T-rich motif, which is likely to also serve as a source of mono-nucleotide runs (Fig. 7D). We find that this motif best matches binding sites for the pair-rule and gap products Slp1 and Hb, consistent with their expression patterns (Fig. 7E). This is also consistent with Zelda's role in early embryonic timing of gene activation, and Runt's role in repressing the posterior borders of expression of Bcd targets in the central region of the embryo.

Discussion

A Model for How Certain Classes of Enhanceosome Scaffold Result in MSR-Enrichment

Previous analyses of *Drosophila* genomic sequences have demonstrated a non-random distribution of microsatellite repeat (MSR) sequences in *Drosophila* genomes [41–43], the presence of compound (*i.e.*, clustered) MSR tracts [44], and an unpredicted excess of long MSR sequences [45]. It was also previously shown that the length of a spacer DNA separating linked Dorsal and Twist activator binding sites sites in the Neurogenic Ectoderm Enhancers (NEEs) can play a functional role [4]. It was then subsequently shown that CA-dinucleotide MSR related to the Twist site is used to source functional variants during evolution [5]. Here, we show that the Hawaiian NEEs offer an extreme case of MSR enrichment in terms of both the amounts and types of MSR content. These observations hint at additional spacer functionalities at other sites within the NEEs, the extent of functionality of which will have to be tested with additional mutagenesis in transgenic reporter assays.

Our results show that MSR-enrichment patterns can be linked to entire classes of regulatory DNAs, which in this case correspond to embryonic enhancer DNAs driven by the A/P and D/V morphogens patterning the syncytial embryo. Specifically, we showed that intervening DNA sequences between the Hawaiian *Drosophila* NEE binding sites have been replaced by microsatellite repeat (MSR) sequences and that these MSR sequences are still

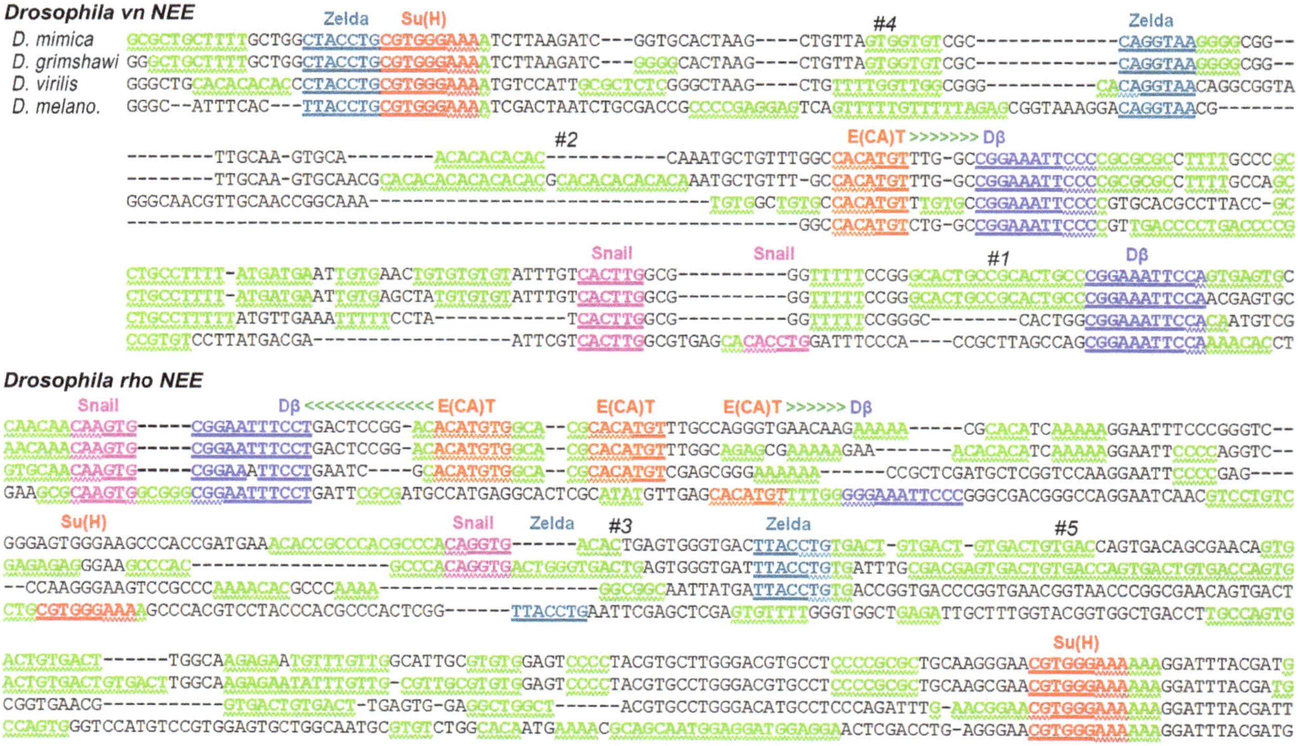

Figure 5. The non-Dpp constrained NEEs at *rho* and *vein* are rapidly diverging via MSR mutagenesis in the Hawaiian lineages. Shown is a sequence alignment for the *vn* (top) and *rho* (bottom) NEEs highlighted to show MSR content (green letters with squiggly underline). The motif coloring follows previous figures. Due to multiple insertions and deletions, these homologous sequences depicted are of different lengths. Numbers indicate the presence of novel MSR sequences present only in Hawaiians (the octapeptide 2x repeat at #1), present only in Hawaiians but still diverging (5 to 13 repeats of CA-microsatellite at #2), sites of MSR repeats of different types at the same position and present only in the DROSOPHILA subgenus (#3) or across the entire genus (#4), and repeats of made of smaller repeats and seen only in Hawaiian species (#5). In addition, other divergent repeat content that is not highlighted by the strict MSR algorithm can also be seen (see Material & Methods).

diverging. We showed that MSR demarcates the majority of the spaces separating the functional binding sites for Dorsal (*i.e.*, the Dβ site), Twist/Snail [*i.e.*, the E(CA)T site], Su(H), as well as the dedicated [non-E(CA)T] Snail binding sites, and sites for the general embryonic timing factor Zelda. Our use of footprinting to describe this effect of MSR enrichment in Hawaiian embryonic enhancers is in line with both phylogenetic footprinting and enzymatic footprinting. In MSR footprinting, phylogenetic footprinting, and enzymatic footprinting, the principle means used to reveal TF binding sites are by highlighting the space separating the sites *via* MSR content, divergence, and digestibility, respectively.

We also showed that the subset of conserved Su(H) site-containing Hawaiian blocks that contain binding sites for Zelda are specifically enriched in MSR motifs relative to non-Zelda containing conserved blocks from the same species, and that there was similar enrichment of MSR motifs in this set of Runt+Zelda containing blocks in the Hawaiian *D. grimshawi* relative to their homologous sequences in *D. virilis*. These results demonstrate that embryonic enhancers from Hawaiian *Drosophila* are enriched in MSR relative to both (*i*) other equally-conserved developmental enhancers from Hawaiian *Drosophila*, and (*ii*) homologous embryonic enhancers from non-Hawaiian *Drosophila*.

We propose that: (*i*) transcription factor activator complexes are sensitive to the spacing of Activator TF binding sites within enhancers that serve as assembly scaffolds for the aggregation of such complexes; and (*ii*) MSR enrichment in embryonic

enhancers is a signature of frequent past selection for as yet unknown complex characteristics (*e.g.*, rate of assembly, complex stability on DNA, complex off-rate, and non-DNA bound complex half-life after assembly). As most indels are produced by microsatellite repeat slippage, enhancers from Hawaiian *Drosophila* lineages that are subjected to frequent evolutionary pressures would become grossly enriched in MSR content. Furthermore, as the embryonic enhancers would be subject to tremendously dynamic evolutionary pressures associated with both life cycle and ecological contexts affecting egg size, egg shape, and embryonic development during adaptive radiations, they would be more enriched in MSR signatures than enhancers operating at later developmental stages. While *Drosophila* lineages in general have exhibited much divergence related to adult pigmentation (*e.g.*, wing spot patterning), adult behavior (*e.g.*, mate choice and courtship song), and potentially other adult systems [46–52], we are unable to identify bioinformatically the enhancer sequences underlying these specific systems selectively without also pulling out many other non-changing adult enhancers. Thus, it is possible that the MSR enrichment we have seen for embryonic enhancers relative to all non-embryonic enhancers could be comparable to a select subset of enhancers underlying dynamically evolving adult phenotypes. Recent studies have indicated the utility of such MSR signatures for enhancer class identification [5,53].

The genetics of microsatellite repeat number has received much attention also because of the role played by CAR-

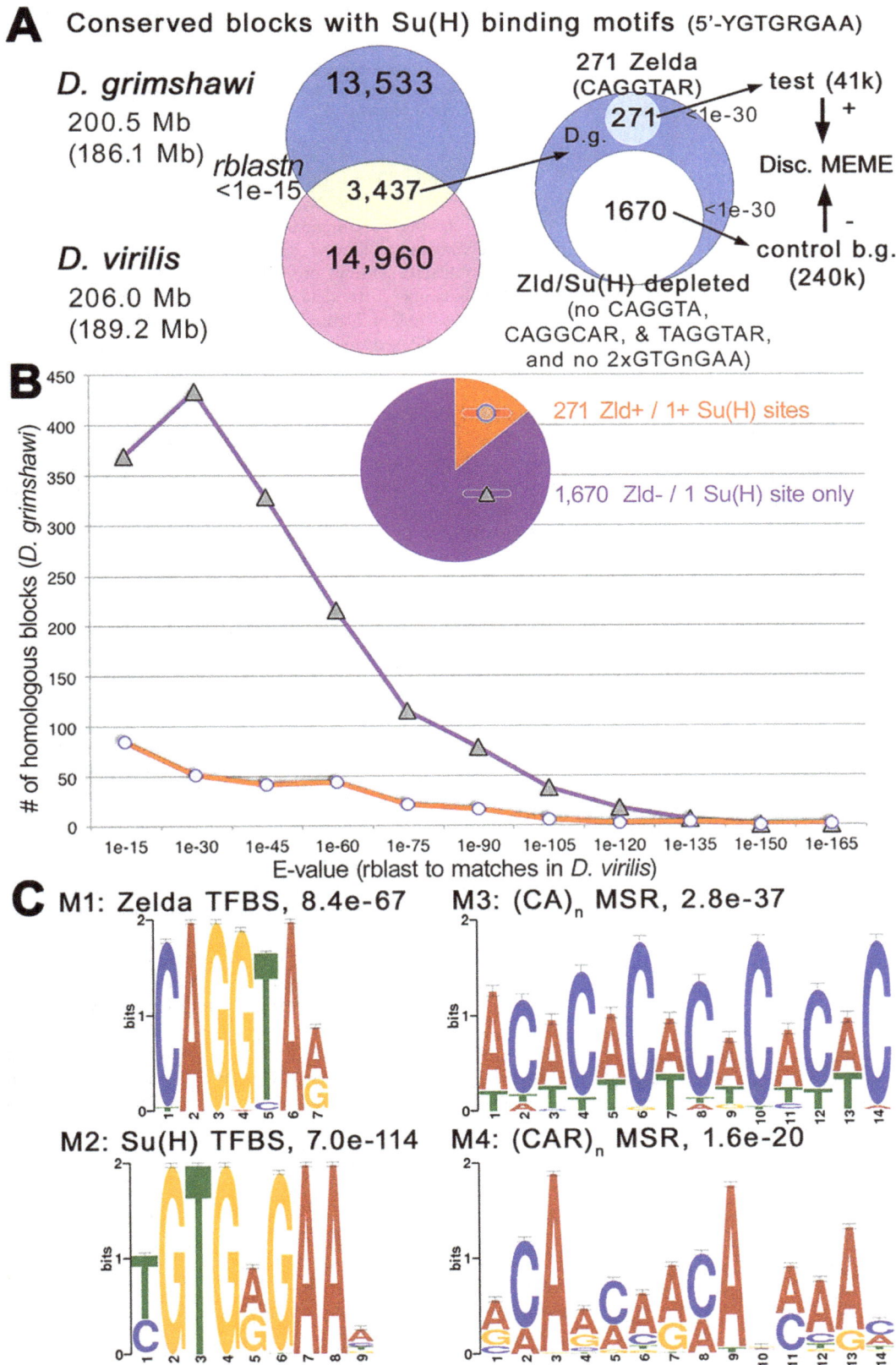

Figure 6. Microsatellite Repeats are Enriched in Conserved Su(H) Site-Containing *D. grimshawi* DNAs of Embryonic Enhancers Relative to Non-Embryonic Enhancers. (**A**) Shown is a flowchart of Venn diagrams showing the identification of 3,437 conserved blocks containing Su(H) sites in *D. grimshawi* and *D. virilis* with regulatory blastn ("*rblastn*") E-value of <1e-15 using parameters calibrated to the NEEs.

Homologous sequences from *D. grimshawi* were separated into test and control data sets for discriminative motif elicitation by maximum expectation (MEME) [33]. The test data set of 271 sequences contains Zelda sites (5'-CAGGTAR). The negative control set of 1,670 sequences is depleted of blocks containing any Zelda site (5'-CAGGTA, 5'-CAGGCAR, and 5'-TAGGTAR) or more than a single Su(H) binding sequence (5'-GTGnGAA). **(B)** The distribution of E-values for *rblastn* hits of the Su(H) strings shows that the test data set of conserved blocks containing Zelda sites are not any less conserved than the control data set lacking these sites. **(C)** MEME analysis identifies CA-dinucleotide and CAR-trinucleotide MSR motifs as being enriched in the Zelda+ dataset relative to other Su(H)-containing conserved blocks.

trinucleotide expansions in many neurodegenerative disorders and this has been extended to *Drosophila* [54]. In a study of length variation and evolution of CAR-trinucleotide microsatellite, or rather their "extreme conservation" in the *Drosophila* gene *mastermind* (*mam*) gene it was suggested that there must be strong selective constraints acting on the spacer lengths [36]. In a test of the null hypothesis that such length divergence arose by chance led to the conclusion that the CAR-MSR content in *mam* evolves both by molecular drive due to frequent repeat slippage and by natural selection on optimal spacer lengths [37]. Sequence data on *de novo* mutations from the HapMap project has also established that MSR-instability is repeat length dependent because similar instabilities are seen across diverse repeat unit sizes and sequences [34]. As we have previously shown the importance of CA-microsatellite repeat slippage emanating from the CA-dinucleotide rich Twist binding site in the NEEs [5], we conclude that MSR repeat variants are generally sourced by selection to adjust functional spacers across both cis-regulatory and protein-coding components of a genome. Thus, routine methods in bioinformatics, such as genomic repeat filtering and genome assembly based on point differences relative to a reference genome (as opposed to *de novo* assembly), may filter out important MSR-based functional variation that differentiates closely related genomes.

Methods

Bioinformatics

UNIX-shell scripts were written using grep, perl, and the BASH command set to identify all Su(H) sites and site clusters in the genome assemblies of *D. grimshawi* (r1.3) and *D. virilis* (r1.2) (see File S1). Site clusters were defined as two or more Su(H) sites located less than twice the desired flanking distance. For Su(H) binding sites (5'-YGTGRGAA) this was defined as 292 bp because (292 bp×2 flanking sequences) +8 bp = 600 bp. For Zelda (5'-CAGGTAR) we defined blocks as +/−300 bp from the Zelda binding site. The special case of not having enough flanking sequence due to proximity to the edge of a contig was also handled and these sequences kept in the data sets. For *blastn* analyses, the UNIX command line version of blast tools was downloaded from NCBI. The parameter set used for *Drosophila* enhancer bioinformatics identified largely by trial and error is the following: "-penalty −4 -reward 5 -word_size 9 -gapopen 8 -gapextend 6 -xdrop_gap_final 90 -best_hit_overhang 0.25 -best_hit_score_edge 0.1". The subset of conserved Su(H) blocks with linked Twist–Dorsal sites (5'-CACATGT 0–41 bp GGAAABYCC) were identified with the UNIX-style regular expression: "CA-CATGT.{0,41}GGAAA[^A][CT]CC". All shell scripts are provided in File S1.

MEME analyses

We performed discriminative motif discovery using Multiple EM for motif elicitation (http://meme.nbcr.net/meme/) and a control data set of negative sequences, and searched for "zero or one" occurrences per sequence [33]. We specified motif limits of 6 to 14 bp, and asked for an optimum number of sites

between 10 and 300 with the upper limit varying depending on the size of the test data set, usually setting it at a maximum of 1.5x the number of sequences. For control data set we chose to use the maximum allowed dataset size of 240,000 characters. For the test data set limit of 60,000 characters we would choose a random sample if the data set was larger. For example, in the analysis depicted in Figure 7, we used 100 random sequences out of the 287 sequences available due to constraints on test data set.

In situ hybridization

Whole-mount anti-sense in situ hybridizations with a digoxigenin UTP-labeled anti-sense RNA probe against *lacZ* were conducted on fixed embryos collected over a four hour egg-laying period held at room temperature. NEE reporters were integrated into the P-element vector between the mini-*white* gene and −42 *eve:lacZ* reporter as previously described [55].

Molecular cloning

Live *D. mimica* were obtained from the UCSD stock center and reared with a protocol similar to that supplied from the stock center. Genomic DNA for PCR amplification was prepared using the Ashburner protocol [56], except that three adult flies instead of a single one were homogenized in a 1.5 mL microcentrifuge tube. Homogenization buffer, lysis buffer, and 8 M K acetate were used as described, followed by phenol-CHCl$_3$ extractions, and EtOH precipitation. A 626 bp fragment of the *rho* NEE from *D. mimica* was cloned using the following oligonucleotide primers based on the *D. grimshawi* reference genome: 5'-AGA TGA AAA TCC GCA ATG CAA CGG (top strand primer), and 5'-AAA CAC AGC AGA AAG TCT CAA GC (bottom strand primer). A 513 bp fragment of the *vn* NEE from *D. mimica* was cloned and sequenced using the following oligonucleotide primers based on the *D. grimshawi* reference genome: 5'-ACA GAA GCT CAG CAT TTG GC (top strand primer), and 5'- GCC AGC GGC AAT TTT ATC TGC (bottom strand primer). A ~500 bp fragment of the *vnd* NEE from *D. mimica* was cloned and sequenced using the following oligonucleotide primers based on the *D. grimshawi* reference genome: 5'-CCA CCG GGT CTC AAA TTC TTT CAC AGT (top strand primer), and 5'-CCA CCG GGT CTC AAA TTC CCA TCA ACA (bottom strand primer). These amplified PCR fragments were cloned into Promega's pGEM-T easy cloning vector. Clones were sequenced, and a few were selected to be cut with EcoR I, gel purified, and ligated into the EcoR I-cut pCaspeR P-element vector carrying the −42 *eve lacZ-tubulin* 3'UTR reporter construct previously reported [4,5]. The cloned enhancers from *D. mimica* have been deposited at GenBank and have accession numbers: KJ814003 (Dmim_rhomboid_NEE), KJ814004 (Dmim_vein_NEE), and KJ814005 (Dmim_vnd_NEE). In addition, the sequences for the NEEs of both Hawaiian *Drosophila* species are included in File S2.

A Conserved blocks with Zelda binding motifs (5'-CAGGTAR)

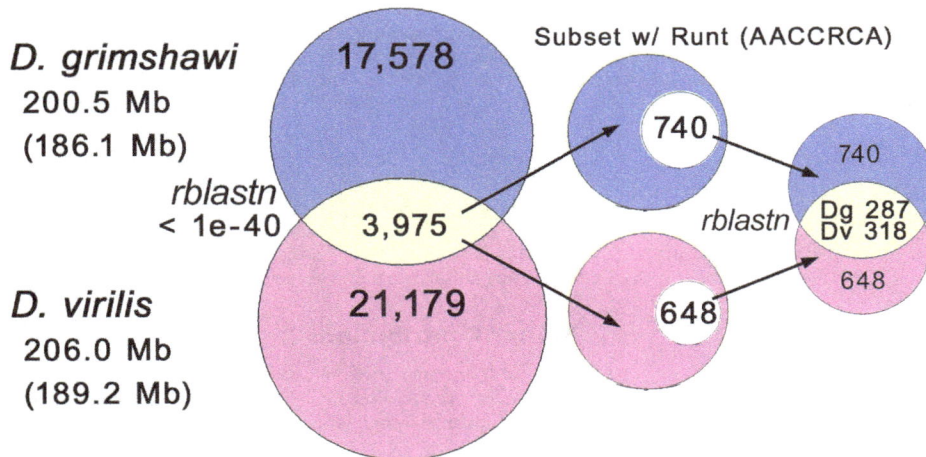

D. grimshawi
200.5 Mb
(186.1 Mb)

17,578

Subset w/ Runt (AACCRCA)

740

740

rblastn
< 1e-40

3,975

Dg 287
Dv 318

rblastn

D. virilis
206.0 Mb
(189.2 Mb)

21,179

648

648

B Conserved blocks with Zld & Runt binding motifs

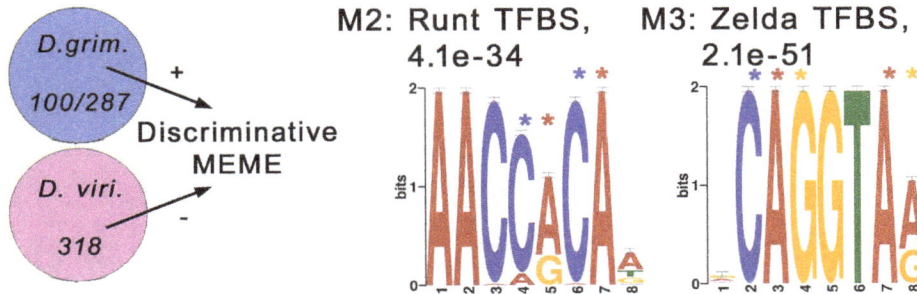

D.grim.
100/287

+

Discriminative
MEME

D. viri.
318

−

M2: Runt TFBS,
4.1e-34

M3: Zelda TFBS,
2.1e-51

C M1: (CAR)$_n$ MSR, 3.2e-81

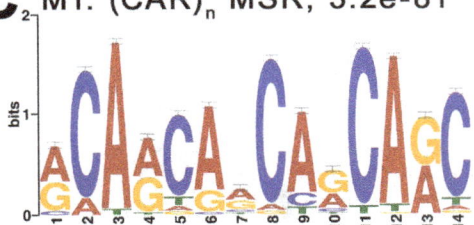

D M5: Slp1 & Hb TFs, 2.3e-28

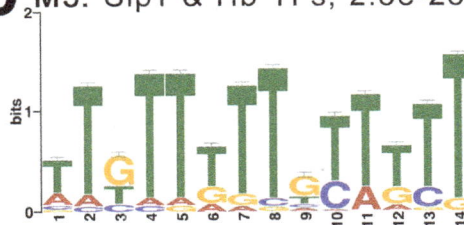

M4: (AG)$_n$ MSR, 3.9e-26

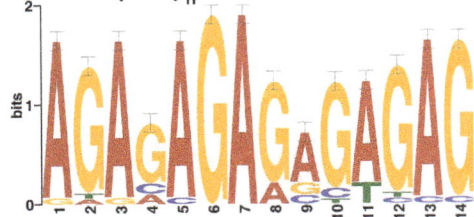

E TOMTOM:
Slp1
(MA0458.1)

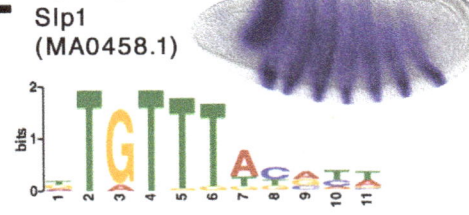

M6: (CA)$_n$ MSR, 4.5e-20

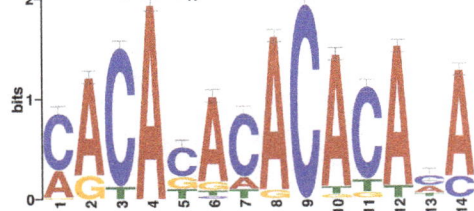

TOMTOM:
Hb
(MA0049.1)

Figure 7. Diverse MSR Motifs are Enriched in the Embryonic A/P Patterning Enhancers of *D. grimshawi* **Relative to their Homologous Sequences in** *D. virilis.* (**A**) Shown is a flowchart of Venn diagrams showing the identification of 3,975 conserved blocks containing Zelda binding sites in *D. grimshawi* and *D. virilis* with an *rblastn* E-value of <1e-40. From these were chosen those that also contain a binding site for Runt (5'-

AACCRCA), which is known to repress the posterior expression domains of targets induced by intermediate levels of the Bcd morphogen gradient. Because the Runt binding sites are not always found in the homologous sequences (either for lack of conservation or due to flanking truncation of the blocks), a second *rblastn* is performed to identify only the ones with reciprocal homologs. (**B**) A discriminative MEME analysis identifies the binding sites for Runt and Zelda relative in *D. grimshawi* vs. *D. virilis*, likely due to increased homotypic site clustering of these sites as well as (**C**) diverse MSR motifs, and (**D**) a motif matching binding sites for the pair-rule and gap products Slp1 and Hb. Numbers within each circle in (**B**) represent number of sequences used in MEME analysis due to constraints on test set (100 random sequences for *D. grimshawi* *versus* control data set (entire data set or 318 sequences for *D. virilis*). Asterisks in (**B**) indicate di- and tri-nucleotide patterns found in the MSRs. (**E**) TOMTOM results showing matches to motif M5, and expression of *slp1* and *hb* in early embryo.

Supporting Information

Table S1 Location of CAR repeat-rich sequences in conserved Su(H) blocks.

File S1 Unix computer scripts.

File S2 FASTA file of *D. mimica* and *D. grimshawi* enhancer sequences.

File S3 Annotated Su(H) blocks from D. grimshawi (Zelda+) with (CAR)$_n$ repeats.

Acknowledgments

We thank Jan Fassler, Gary Gussin, Clinton Rice, and Danielle Beekman for reading and commenting on our manuscript.

Author Contributions

Conceived and designed the experiments: AB AE. Performed the experiments: AB. Analyzed the data: AB ES AE. Contributed to the writing of the manuscript: AE.

References

1. Drosophila 12 Genomes C, Clark AG, Eisen MB, Smith DR, Bergman CM, et al. (2007) Evolution of genes and genomes on the Drosophila phylogeny. Nature 450: 203–218.
2. Huntley MA, Clark AG (2007) Evolutionary analysis of amino acid repeats across the genomes of 12 Drosophila species. Mol Biol Evol 24: 2598–2609.
3. Bhutkar A, Schaeffer SW, Russo SM, Xu M, Smith TF, et al. (2008) Chromosomal rearrangement inferred from comparisons of 12 Drosophila genomes. Genetics 179: 1657–1680.
4. Crocker J, Tamori Y, Erives A (2008) Evolution acts on enhancer organization to fine-tune gradient threshold readouts. PLoS Biol 6: e263.
5. Crocker J, Potter N, Erives A (2010) Dynamic evolution of precise regulatory encodings creates the clustered site signature of enhancers. Nat Commun 1: 99.
6. Roy S, Ernst J, Kharchenko PV, Kheradpour P, Negre N, et al. (2010) Identification of functional elements and regulatory circuits by Drosophila modENCODE. Science 330: 1787–1797.
7. Carson HL, Clayton FE, Stalker HD (1967) Karyotypic stability and speciation in Hawaiian Drosophila. Proc Natl Acad Sci U S A 57: 1280–1285.
8. Carson HL (1973) Ancient chromosomal polymorphism in Hawaiian Drosophila. Nature 241: 200–202.
9. Carson HL (1982) Evolution of Drosophila on the newer Hawaiian volcanoes. Heredity (Edinb) 48: 3–25.
10. Carson HL (1983) Chromosomal sequences and interisland colonizations in hawaiian Drosophila. Genetics 103: 465–482.
11. Ayala FJ, Campbell CD, Selander RK (1996) Molecular population genetics of the alcohol dehydrogenase locus in the Hawaiian drosophilid D. mimica. Mol Biol Evol 13: 1363–1367.
12. Carson HL (1997) The Wilhelmine E. Key 1996 Invitational Lecture. Sexual selection: a driver of genetic change in Hawaiian Drosophila. J Hered 88: 343–352.
13. Carson HL (2002) Female choice in Drosophila: evidence from Hawaii and implications for evolutionary biology. Genetica 116: 383–393.
14. O'Grady PM, Kam MWY, Val FC, Perreira WD (2003) Revision of the Drosophila mimica subgroup, with descriptions of ten new species. Annals of the Entomological Society of America 96: 12–38.
15. Edwards KA, Doescher LT, Kaneshiro KY, Yamamoto D (2007) A database of wing diversity in the Hawaiian Drosophila. PLoS One 2: e487.
16. Powell JR (1997) Progress and prospects in evolutionary biology : the Drosophila model. New York: Oxford University Press. xiv, 562 p.
17. Kambysellis MP, Heed WB (1971) Studies of Oogenesis in Natural Populations of Drosophilidae.1. Relation of Ovarian Development and Ecological Habitats of Hawaiian Species. American Naturalist 105: 31–&.
18. Kambysellis MP, Starmer T, Smathers G, Heed WB (1980) Studies of Oogenesis in Natural-Populations of Drosophilidae.2. Significance of Microclimatic Changes on Oogenesis of Drosophila-Mimica. American Naturalist 115: 67–91.
19. Crocker J, Erives A (2013) A Schnurri/Mad/Medea complex attenuates the dorsal-twist gradient readout at vnd. Dev Biol 378: 64–72.
20. Erives A, Levine M (2004) Coordinate enhancers share common organizational features in the Drosophila genome. Proc Natl Acad Sci U S A 101: 3851–3856.
21. Crocker J, Erives A (2008) A closer look at the eve stripe 2 enhancers of Drosophila and Themira. PLoS Genet 4: e1000276.
22. Stathopoulos A, Levine M (2004) Whole-genome analysis of Drosophila gastrulation. Curr Opin Genet Dev 14: 477–484.
23. Barolo S (2012) Shadow enhancers: frequently asked questions about distributed cis-regulatory information and enhancer redundancy. Bioessays 34: 135–141.
24. Frankel N, Davis GK, Vargas D, Wang S, Payre F, et al. (2010) Phenotypic robustness conferred by apparently redundant transcriptional enhancers. Nature 466: 490–493.
25. Hong JW, Hendrix DA, Levine MS (2008) Shadow enhancers as a source of evolutionary novelty. Science 321: 1314.
26. Perry MW, Boettiger AN, Bothma JP, Levine M (2010) Shadow enhancers foster robustness of Drosophila gastrulation. Curr Biol 20: 1562–1567.
27. Kanodia JS, Rikhy R, Kim Y, Lund VK, DeLotto R, et al. (2009) Dynamics of the Dorsal morphogen gradient. Proc Natl Acad Sci U S A 106: 21707–21712.
28. Harrison MM, Botchan MR, Cline TW (2010) Grainyhead and Zelda compete for binding to the promoters of the earliest-expressed Drosophila genes. Dev Biol 345: 248–255.
29. Liang HL, Nien CY, Liu HY, Metzstein MM, Kirov N, et al. (2008) The zinc-finger protein Zelda is a key activator of the early zygotic genome in Drosophila. Nature 456: 400–403.
30. Harrison MM, Li XY, Kaplan T, Botchan MR, Eisen MB (2011) Zelda binding in the early Drosophila melanogaster embryo marks regions subsequently activated at the maternal-to-zygotic transition. PLoS Genet 7: e1002266.
31. Nien CY, Liang HL, Butcher S, Sun Y, Fu S, et al. (2011) Temporal coordination of gene networks by Zelda in the early Drosophila embryo. PLoS Genet 7: e1002339.
32. Dover G (1982) Molecular drive: a cohesive mode of species evolution. Nature 299: 111–117.
33. Bailey TL, Boden M, Buske FA, Frith M, Grant CE, et al. (2009) MEME SUITE: tools for motif discovery and searching. Nucleic Acids Res 37: W202–208.
34. Ananda G, Walsh E, Jacob KD, Krasilnikova M, Eckert KA, et al. (2013) Distinct mutational behaviors differentiate short tandem repeats from microsatellites in the human genome. Genome Biol Evol 5: 606–620.
35. Ashley CT Jr, Warren ST (1995) Trinucleotide repeat expansion and human disease. Annu Rev Genet 29: 703–728.
36. Newfeld SJ, Schmid AT, Yedvobnick B (1993) Homopolymer length variation in the Drosophila gene mastermind. J Mol Evol 37: 483–495.
37. Newfeld SJ, Tachida H, Yedvobnick B (1994) Drive-selection equilibrium: homopolymer evolution in the Drosophila gene mastermind. J Mol Evol 38: 637–641.
38. Chen H, Xu Z, Mei C, Yu D, Small S (2012) A system of repressor gradients spatially organizes the boundaries of Bicoid-dependent target genes. Cell 149: 618–629.
39. Li L, Zhu Q, He X, Sinha S, Halfon MS (2007) Large-scale analysis of transcriptional cis-regulatory modules reveals both common features and distinct subclasses. Genome Biol 8: R101.
40. Bhat KM, Farkas G, Karch F, Gyurkovics H, Gausz J, et al. (1996) The GAGA factor is required in the early Drosophila embryo not only for transcriptional regulation but also for nuclear division. Development 122: 1113–1124.

41. Harr B, Zangerl B, Brem G, Schlotterer C (1998) Conservation of locus-specific microsatellite variability across species: a comparison of two Drosophila sibling species, D. melanogaster and D. simulans. Mol Biol Evol 15: 176–184.

42. Bachtrog D, Weiss S, Zangerl B, Brem G, Schlotterer C (1999) Distribution of dinucleotide microsatellites in the Drosophila melanogaster genome. Mol Biol Evol 16: 602–610.

43. Bachtrog D, Agis M, Imhof M, Schlotterer C (2000) Microsatellite variability differs between dinucleotide repeat motifs-evidence from Drosophila melanogaster. Mol Biol Evol 17: 1277–1285.

44. Kofler R, Schlotterer C, Luschutzky E, Lelley T (2008) Survey of microsatellite clustering in eight fully sequenced species sheds light on the origin of compound microsatellites. BMC Genomics 9: 612.

45. Dieringer D, Schlotterer C (2003) Two distinct modes of microsatellite mutation processes: evidence from the complete genomic sequences of nine species. Genome Res 13: 2242–2251.

46. Cooley AM, Shefner L, McLaughlin WN, Stewart EE, Wittkopp PJ (2012) The ontogeny of color: developmental origins of divergent pigmentation in Drosophila americana and D. novamexicana. Evol Dev 14: 317–325.

47. Gompel N, Prud'homme B, Wittkopp PJ, Kassner VA, Carroll SB (2005) Chance caught on the wing: cis-regulatory evolution and the origin of pigment patterns in Drosophila. Nature 433: 481–487.

48. Wittkopp PJ, Carroll SB, Kopp A (2003) Evolution in black and white: genetic control of pigment patterns in Drosophila. Trends Genet 19: 495–504.

49. Wittkopp PJ, Williams BL, Selegue JE, Carroll SB (2003) Drosophila pigmentation evolution: divergent genotypes underlying convergent phenotypes. Proc Natl Acad Sci U S A 100: 1808–1813.

50. Shirangi TR, Stern DL, Truman JW (2013) Motor control of Drosophila courtship song. Cell Rep 5: 678–686.

51. Arthur BJ, Sunayama-Morita T, Coen P, Murthy M, Stern DL (2013) Multi-channel acoustic recording and automated analysis of Drosophila courtship songs. BMC Biol 11: 11.

52. Cande J, Andolfatto P, Prud'homme B, Stern DL, Gompel N (2012) Evolution of multiple additive loci caused divergence between Drosophila yakuba and D. santomea in wing rowing during male courtship. PLoS One 7: e43888.

53. Yanez-Cuna JO, Arnold CD, Stampfel G, Boryn LM, Gerlach D, et al. (2014) Dissection of thousands of cell type-specific enhancers identifies dinucleotide repeat motifs as general enhancer features. Genome Res.

54. Jung J, van Jaarsveld MT, Shieh SY, Xu K, Bonini NM (2011) Defining genetic factors that modulate intergenerational CAG repeat instability in Drosophila melanogaster. Genetics 187: 61–71.

55. Erives A, Corbo JC, Levine M (1998) Lineage-specific regulation of the Ciona snail gene in the embryonic mesoderm and neuroectoderm. Dev Biol 194: 213–225.

56. Sullivan W, Ashburner M, Hawley RS (2000) Drosophila protocols. Cold Spring Harbor, N.Y.: Cold Spring Harbor Laboratory Press. xiv, 697 p.

57. Ip YT, Park RE, Kosman D, Bier E, Levine M (1992) The dorsal gradient morphogen regulates stripes of rhomboid expression in the presumptive neuroectoderm of the Drosophila embryo. Genes Dev 6: 1728–1739.

17

The Arrival of the Frequent: How Bias in Genotype-Phenotype Maps Can Steer Populations to Local Optima

Steffen Schaper[¤], Ard A. Louis*

Rudolf Peierls Centre for Theoretical Physics, University of Oxford, Oxford, United Kingdom

Abstract

Genotype-phenotype (GP) maps specify how the random mutations that change genotypes generate variation by altering phenotypes, which, in turn, can trigger selection. Many GP maps share the following general properties: 1) The total number of genotypes N_G is much larger than the number of selectable phenotypes; 2) Neutral exploration changes the variation that is accessible to the population; 3) The distribution of phenotype frequencies $F_p = N_p/N_G$, with N_p the number of genotypes mapping onto phenotype p, is highly biased: the majority of genotypes map to only a small minority of the phenotypes. Here we explore how these properties affect the evolutionary dynamics of haploid Wright-Fisher models that are coupled to a random GP map or to a more complex RNA sequence to secondary structure map. For both maps the probability of a mutation leading to a phenotype p scales to first order as F_p, although for the RNA map there are further correlations as well. By using mean-field theory, supported by computer simulations, we show that the discovery time T_p of a phenotype p similarly scales to first order as $1/F_p$ for a wide range of population sizes and mutation rates in both the monomorphic and polymorphic regimes. These differences in the rate at which variation arises can vary over many orders of magnitude. Phenotypic variation with a larger F_p is therefore be much more likely to arise than variation with a small F_p. We show, using the RNA model, that frequent phenotypes (with larger F_p) can fix in a population even when alternative, but less frequent, phenotypes with much higher fitness are potentially accessible. In other words, if the fittest never 'arrive' on the timescales of evolutionary change, then they can't fix. We call this highly non-ergodic effect the 'arrival of the frequent'.

Editor: Suzannah Rutherford, Fred Hutchinson Cancer Research Center, United States of America

Funding: Funding for this research was received from the Engineering and Physical Sciences Research Council (EPSRC) under grant EP/P504287/1. The funders had no role in study design, data collection and analysis, decision to publish, or preparation of the manuscript.

Competing Interests: The authors have declared that no competing interests exist.

* E-mail: ard.louis@physics.ox.ac.uk

¤ Current address: Aachen Institute for Advanced Study in Computational Engineering Science (AICES), RWTH Aachen University, Aachen, Germany

Introduction

Darwin's account of biological evolution [1] stressed the importance of natural selection: If some individuals are better adapted to their environment than their competitors, their offspring will come to dominate the population. The fittest survive and the less fit go extinct. Yet selection alone is not sufficient to drive evolution because natural selection reduces the very variation that it requires to operate. It was only recognised well after Darwin's day [2], in part through the success of the Modern Synthesis, that the fuel for selection is provided by mutations that make offspring genetically different from their parents. Crucially, mutations change genetically stored information (the *genotype*) while selection operates on the physical expression of this information (the *phenotype*). Understanding the relation between genotypes and phenotypes – the GP map – is therefore crucial to understanding evolutionary dynamics [3].

GP mappings have been studied at different levels of abstraction [4] The most basic systems are concerned with the sequence-to-structure(-to-function) relation of single molecules such as RNA [5] or proteins [6–8], but higher-level systems such as protein complexes [9], gene-regulatory networks [10] and developmental networks [11] have also been studied. Even though these GP maps

arise in quite different contexts, they share several interesting properties:

1) Most basically, the number of possible genotypes N_G is typically much greater than the number of possible phenotypes N_P, so the map is many-to-one. As a consequence, many mutations may conserve the phenotype, leading to mutational robustness. Important prior work has linked such robustness to the concept of neutral spaces, namely the set of all genotypes that map to a particular phenotype, with the additional property that they be linked by neutral mutations [4,5,12].

2) Even though $N_P \ll N_G$, the accessible genetic neighbourhood of a single genotype g that generates a given phenotype p may include significantly fewer alternative phenotypes (potential variation) than is found in the neighbourhood of the (neutral) set $\mathcal{N}_p$ of all $N_p = |\mathcal{N}_p|$ genotypes that map onto phenotype p. Exploration of a neutral space can therefore increase the variety of phenotypes discovered by a population [13,14].

3) Perhaps the most striking commonality of these GP maps is a strong bias in assignment of genotypes to phenotypes: Most phenotypes are realised by a tiny proportion of all genotypes, while most genotypes map into a small fraction of all phenotypes. This property is shared by all the GP maps we

noted before. Typically the number N_p of genotypes per phenotype p and the related phenotype frequencies $F_p = N_p/N_G$ can vary over many orders of magnitude. Such huge variations are likely to have an effect on the course of evolution.

In this paper we study the evolutionary dynamics of a classical Wright-Fisher model, but with explicit microscopic GP maps that capture the three generic properties of such maps introduced above. Motivated by the strong bias in the distribution of the F_p observed for many GP maps, we derive a mean-field like approximation for the average probability ϕ_{pq} that a mutation will change a genotype that generates phenotype q into one that generates phenotype p. This approximation greatly simplifies the dynamics, allowing us to calculate analytic expressions for quantities such as the median time T_p for phenotype p to first appear in the population as a function of population size N, the point mutation rate μ, genome length L and the mutation probabilities ϕ_{pq}.

These approximations are then tested against extensive simulations of two models: firstly, a simple GP map where the genotypes are randomly assigned to phenotypes according to a pre-determined distribution for the frequencies F_p and secondly, the well-known mapping of RNA sequence to secondary structure [4,5,15], which is more complex, but also more biologically realistic. We focus on the case where a population of N individuals has initially equilibrated at a fitness maximum given by phenotype q, and then measure the median time T_p for alternative phenotype p to first arise in the population.

Our analytic expressions agree quantitatively with the simulations in the polymorphic limit where $NL\mu \gg 1$, and also in the opposite monomorphic limit $NL\mu \ll 1$. In between these regimes a single scaling factor must be included. In all regimes the median discovery time $T_p \propto 1/\phi_{pq}$. For the random model $\phi_{pq} \approx F_p$; this scaling also holds for the more complex RNA mapping, although there is significantly more scatter due to local correlations within the neutral spaces and for some phenotypes we find $\phi_{pq} = 0$ even though F_p is large (this can be due to biophysical constraints explained for example in ref. [16]). Despite such higher order effects, the variation of the F_p over many orders translates directly into the T_p. More frequent (higher F_p) phenotypes are therefore discovered more rapidly and more often along evolutionary trajectories. In this way the structure of the GP map can play a key role in determining evolutionary outcomes.

Finally, we employ the RNA GP map to study the case where two phenotypes $p1$ and $p2$ are both more fit than the source phenotype q, but where $F_{p1} \gg F_{p2}$ (or more accurately $\phi_{p1q} \gg \phi_{p2q}$). Direct simulations show that phenotype $p1$, which is more frequent, is much more likely to fix in the population, even if its fitness is much lower than that of $p2$, an effect we call 'the arrival of the frequent'.

Results

Theoretical framework

We study the evolution of a population of N asexual haploid individuals. Each individual i carries a genotype g_i of L letters taken from an alphabet of size K. The individual's phenotype p_i is determined from g_i via the GP map. The population evolves in in discrete, non-overlapping generations according to the classical Wright-Fisher model for haploid individuals: At each generation T, N parents are drawn with replacement with probability proportional to their fitness $1 + s_i$ with the constraint that the

population size (or carrying capacity) N is fixed. Each parent gives rise to one offspring, and the offspring make up the population for the next generation. During reproduction, each base in the genotype of length L mutates to a random alternative base with probability μ. The number of mutations (that is, the Hamming distance) d between parent and offspring is thus distributed binomially according to $h(d) = \binom{L}{d}\mu^d(1-\mu)^{L-d}$. In this way the set $\{g_i\}$ of N genotypes changes at each generation.

The expected number of individuals with phenotype p that arises at generation t can be written as:

$$m_p(t) = \sum_i^N \sum_{d=1}^L h(d)\Phi_p(g_i, s_i, d) \qquad (1)$$

where $\Phi_p(g_i, s_i, d)$ is the probability that a d fold mutation of genotype g_i (selected for reproduction according to fitness $1 + s_i$) generates an individual with phenotype p. It takes into account the mutational connections between the $N_G = K^L$ genotypes that make up the GP map. The probability of not finding p is approximately given by the Poisson distribution as $\exp(-m_p(t))$.

While exact, these dynamic expressions depend implicitly on time through stochastic changes in the set $\{g_i\}$, and are typically very hard to solve. In order to gain intuitive insight, we employ a number of simplifications and approximations, motivated in part by the general properties of GP maps discussed in the introduction. First, we assume that $L\mu \ll 1$ so that for $d > 1$, $h(d) \ll h(1) \approx L\mu$, which means that we can ignore higher order mutations (terms with $d > 1$ in Eq. (1)). For a given source phenotype q (where the fitnesses of all genotypes mapping into q are equal, and so we take $1 + s_q = 1$ for simplicity) we can then calculate the mean probability ϕ_{pq} that a single point mutation will generate another phenotype p:

$$\phi_{pq} = \frac{1}{N_q}\sum_{i=1}^{N_q}\Phi_p(g_i, 0, 1) \qquad (2)$$

where the sum is over the set $\mathcal{N}_q$ of all N_q genotypes that generate phenotype q (see also Figure 1). It is convenient to introduce the robustness of phenotype q as the average probability over all $\mathcal{N}_q$ of neutral mutations: $\rho = \phi_{qq}$. If we consider the case where at generation $t-1$ the whole population is on $\mathcal{N}_q$, then Eq. (1) simplifies in this mean-field (or pre-averaged) approximation to:

$$m_p(t) = NL\mu\phi_{pq} \qquad (3)$$

The polymorphic limit. If $NL\mu \gg 1$ then the population naturally spreads over different genotypes, a regime called the polymorphic limit. Consider the case where $1 + s_p = \delta_{qp}$ so that the population remains on $\mathcal{N}_q$, which is one way to model neutral exploration. In the mean-field approximation the expected number of individuals with phenotype p produced per generation is now independent of time, and given by Eq.(3), as long as double mutations can be ignored. The time $T_p(\alpha)$ when on average the probability of having discovered p is α (so that the median discovery time of p is $T_p(1/2)$) is then given by:

A

B

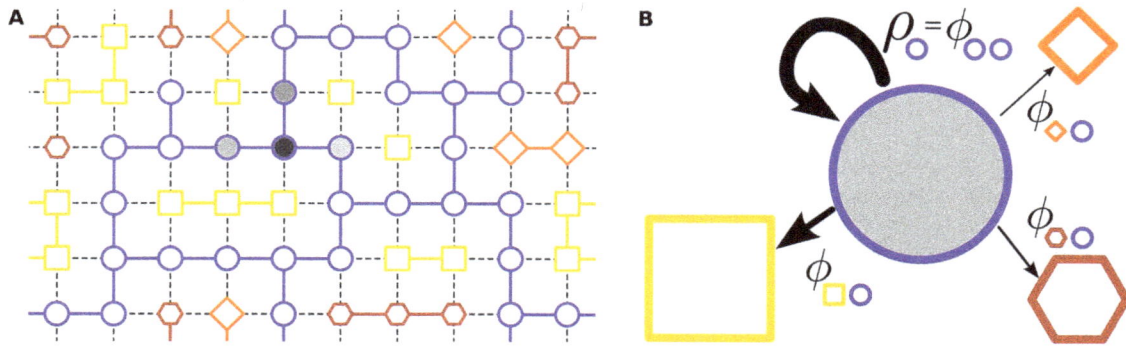

Figure 1. Illustration of the mean field approximation. *A)* An example genotype space: Each point corresponds to a unique genotype; shape and color of the marker indicate the phenotype. Genotypes joined by edges can be interconverted by single mutations. Edges for neutral mutations share the color of the (conserved) phenotype, non-neutral mutations are shown as black dashed lines. The shading of the genotypes illustrates the number of individuals carrying the respective genotype in a hypothetical population. The mutations away from the genotypes occupied by the population determine the accessible phenotypes. *B)* Our meanfield approximation averages over the internal structure of neutral spaces. So neutral spaces are represented by the markers of their phenotypes only, with the size representing the neutral space size (ie. number of genotypes in the space). The uniform shading of the blue neutral space implies that in the meanfield approximation, the population is assumed to continually explore the neighbourhood of its entire neutral space. Mutational outcomes are thus determined from the local frequencies of phenotypes around the neutral space, as measured by the ϕ_{pq} coefficients. This mean field approximation allows us to derive analytic forms that can be compared to simulations of the full GP map.

$$T_p(\alpha) = \frac{-\log(1-\alpha)}{NL\mu\phi_{pq}} \qquad (4)$$

Eqns. (3–4) should provide a good approximation of the full dynamics in the limit that N is large enough that variations between individual genotypes $g_i \in \mathcal{N}_q$ are averaged out, in other words, for the case where the 1-mutant neighbourhood of the population is similar to that of the whole neutral space.

The monomorphic limit. Neutral spaces can be astronomically large [17], much bigger than even the largest viral or bacterial populations. In that case, the local neighborhood of the population may not be fully representative of the neighborhood of the entire space. This scenario can most easily be understood in the monomorphic limit where mutants are rare, $NL\mu \ll 1$, and exploration is dominated by genetic drift. Every neutral mutant has a probability of $1/N$ to go to fixation, allowing the population to move to a new genotype. Thus the timescale of fixations is Kimura's famous result [18] $\tau_f = 1/(L\mu\rho)$, where the robustness ρ is the probability that a mutation is neutral, so that $L\mu\rho$ is the rate of neutral mutations.

Between fixations, the population undergoes periods of genotypic stasis in which only the 1-mutant neighborhood of the current genotype g is explored by (rare) mutations. As there are $(K-1)L$ adjacent genotypes, the timescale of this exploration is $\tau_e = (K-1)L/(NL\mu) = (K-1)/(N\mu)$.

It is instructive to compare the ratio ξ of these two time-scales, defined via

$$\xi = \frac{\tau_f}{\tau_e} = \frac{N}{(K-1)L\rho} \approx \frac{N}{L} \qquad (5)$$

We can use this dimensionless ratio to distinguish between different dynamic regimes. If $\xi \gg 1$, fixation takes much longer than exploration. If we define n_p^g as the number of local neighbours of the genotype g mapping to phenotype p for the

current population, then in this limit, phenotypes with $n_p^g > 0$ are produced continuously (on a time-scale given by τ_e) until the population moves to a different genotype. The dynamics under strong genetic drift therefore induce short-term correlations in the mutant phenotypes. Since $\xi \approx N/L$, we call this regime the large population limit.

In the opposite extreme $\xi \ll 1$, which we call the large genome limit, the population typically moves to a different genotype before all accessible mutants have been explored. In this regime, we do not expect short-term correlations in the mutant phenotypes, simply because every mutant occurs only very rarely.

Actual discovery and neutral fixation times can show strong fluctuations. As our evolutionary process is a Markov process – the next set of mutants depends only on the parents, not on earlier mutants – the first discovery time of a neighbour genotype as well as the arrival time of the neutral mutant "destined" to be fixed, are distributed geometrically (or exponentially in a model with continuous time). Thus the mean of τ_e or τ_f is equal to the respective standard deviation, and any particular evolutionary trajectory can be very different from the average behaviour.

Let τ be the actual time the population stays at the current genotype. In the continuous time approximation, τ is distributed exponentially with mean τ_f. If the genotype g has n_p^g mutations leading to p then the probability that p is found during this time is $1 - \exp(-n_p^g \tau/\tau_e)$. Integrating over the distribution of τ, we have the probability $P(n_p^g)$ that phenotype p is discovered before the next neutral fixation:

$$P(n_p^g) = \int_0^\infty \frac{d\tau}{\tau_f}(1 - e^{-n_p^g \tau/\tau_e})e^{-\tau/\tau_f} = 1 - \frac{1}{1 + n_p^g \xi} \qquad (6)$$

If fixations are the rate-limiting step (ie. $\xi \gg 1$), $P \to 1$ if $n_p^g \neq 0$, as each neighborhood is searched exhaustively before the population moves on. On the other hand, if fixation is faster than exploration ($\xi \ll 1$), the introduction of alternative phenotypes is determined by random fluctuations, as most available mutants are not produced. To leading order, we find $P(n_p^g) \approx n_p^g \xi = Nn_p^g/((K-1)L\rho)$. We note that the inverse dependence on ρ arises from τ_f: More robust neutral spaces are explored faster, but therefore less thoroughly.

The dynamics in the monomorphic regime are thus relatively straightforward. But whether some new phenotype p is discovered still depends on the structure of the neutral space which in turn determines how the available phenotypes change upon a neutral fixation. To describe this structure, we turn again to a mean-field approximation: The mutational neighborhood of each particular genotype $g \in \mathcal{N}_q$ resembles the average over $\mathcal{N}_q$. As the mean number of mutations per genotype leading to p is given by $\bar{n}_{pq} = (K-1)L\phi_{pq}$, the probability that p is accessible after a neutral fixation is $1 - \exp(-\bar{n}_{pq}) \approx \bar{n}_{pq}$ (the approximation is valid provided $n_{pq} \ll 1$, that is p is not accessible from every genotype in the source neutral space; of course, this is just the condition we are interested in, as otherwise neutral exploration would not typically be necessary for phenotype p to arise).

Over a large number of generations $(\tau \gg \tau_f)$, a monomorphic population explores its neutral space uniformly [19]. Assuming that $n_p^g > 1$ can be ignored in practice, we have $T_p(\alpha) = -\tau_f \log(1-\alpha)/(n_{pq}P(1))$. The first discovery time in the large population limit becomes:

$$T_p(\alpha) = \frac{-\tau_f \log(1-\alpha)}{n_{pq}} = \frac{-\log(1-\alpha)}{L^2(K-1)\mu\rho\phi_{pq}} \qquad (7)$$

whereas in the large genome limit we obtain

$$T_p(\alpha) = \frac{-\tau_e \log(1-\alpha)}{n_{pq}} = \frac{-\log(1-\alpha)}{NL\mu\phi_{pq}} \qquad (8)$$

which has the same form as the polymorphic limit, Eq. (4): When the population is too small (compared to the genome length), the exploration of each genotype's mutational neighborhood is typically incomplete. Then, just as in the polymorphic limit, only random fluctuations determine which accessible genotypes are actually realized by the population.

Finally, let us compare our results for large populations in the monomorphic and polymorphic limits. Most importantly, in both cases T_p is inversely proportional to ϕ_{pq}: Rare phenotypes are hard to find. Comparing Equations (4) and (7), the only difference is that N in the polymorphic regime is replaced by $L(K-1)\rho$ in the monomorphic limit. This difference is intuitive: When the population is diverse, every new individual helps exploration and reduces discovery times. But if all individuals have the same genotype, simply having "more of the same" does not make neutral exploration faster. However, repeated mutants may influence the fixation of adaptive phenotypes.

These results suggest that for intermediate $NL\mu$ there should be a smooth transition between these two regimes. To quantify the crossover we introduce a factor γ that multiplies N in Eq.(4); we expect that $\gamma \to 1$ as either $NL\mu$ becomes very large (the polymorphic limit) or $N \ll L$ (the large genome limit), and that $\gamma \to (K-1)L\rho/N$ as $NL\mu \ll 1$ and $N \gg L$ (the large population monomorphic limit).

Simulations in model GP maps

In order to test our mean-field theory we study two kinds of GP maps that both include the generic properties of GP maps that we introduced earlier.

Random GP map. In the random GP map, the total number of phenotypes N_P and the frequencies $\{F_p\}$ can be set arbitrarily (subject to the normalization constraint $\sum_{p=1}^{N_P} F_p = 1$). The $K^L \times F_p$ genotypes mapping into phenotype p are distributed randomly in genotype space. The statistical properties of the map are thus determined by the parameters L, K, and the set $\{F_p\}$.

Studying this map has two motivations: First, ignoring some biophysical detail may help illuminate generic features shared by the systems described in the introduction. Second, a simple model may clarify which deviations from our theory arise from population dynamic effects rather than from detailed (and system-specific) structure in the GP map.

In this simple model, correlations between genotypes are absent, facilitating analysis of the resulting neutral spaces. For example, $\phi_{pq} = F_p$ is a good approximation as long as $N_P \ll N_G$ and $N_q, N_p \gg 1$. Also, there is a percolation threshold $\lambda(K) = 1 - K^{-1/(K-1)}$: thus only phenotypes with $F_q > \lambda(K)$ have completely connected neutral spaces [20].

Here we study a particular random GP map with $L = 12$, and $K = 4$ (as in DNA and RNA) so that there are $N_G = 4^{12} \approx 1.68 \times 10^7$ genotypes. These map onto $N_P = 58$ phenotypes distributed with frequencies $F_p \propto 1.2^{-p}$. The F_p vary over about 5 orders of magnitude, a range similar to the F_p of $L = 12$ RNA (see also Figure S1). To make sure that the largest neutral space percolates, its frequency is set separately as $F_1 = 0.5 > \lambda(4) = 0.37$. For several values of μ, we simulated $N = 1000$ individuals for up to 7×10^{10} generations. The fitness was set as $1 + s_p = \delta_{p,1}$ so that we are effectively modelling neutral exploration on the space $\mathcal{N}_1$, which is convenient for measuring all T_p. We measured first discovery times for the 57 alternative phenotypes over 100 independent simulations to obtain the median time T_p.

Figure 2A depicts these median discovery times T_p for simulations ranging from the polymorphic regime $NL\mu \gg 1$ to the monomorphic limit $NL\mu \ll 1$ (see also Figure S2). We note the following:

1) For all regimes the T_p vary over many orders of magnitude, but they are found in fewer generations for larger μ.

2) Locally frequent phenotypes (i.e. those with high ϕ_{pq}) are much easier to discover. The inset of Figure 2A shows that $\phi_{pq} \approx F_p$, so this conclusion carries over to frequent phenotypes with large F_p.

3) A subset of the phenotypes with $\phi_{pq} > \phi_L \equiv 1/(K-1)L \approx 0.028$ are likely to be in the one-mutation neighbourhood of any genotype. In the monomorphic regime these are then found by exploration of a genome so that T_p is given by Eq. (8), which has the same form as the polymorphic limit, Eq. (4), as can be seen in Figure 2A. Discovery times cross over to the regime where neutral exploration is required when $\phi_{pq} \ll \phi_L$. Such behaviour can be viewed as a finite size effect: N_P typically increases with L. Therefore the largest F_p will likely decrease for larger systems, so that a smaller fraction of phenotypes can be found without neutral exploration.

4) In the fully polymorphic regime where each individual essentially explores independently, any phenotype with $\phi_{pq} > 1/(NL\mu)$ is likely to be part of immediately accessible *standing variation* [21] in the initial population, and is therefore found quickly. Indeed, in Figure 2A for $\mu = 10^{-2}$, where $NL\mu = 120$, these phenotypes are typically found in one or two generations on average. However, for rarer phenotypes, where neutral exploration is important, the T_p are well approximated by Eq. (4). Again, the fraction of phenotypes that are immediately accessible should decrease for larger L.

5) In the intermediate regime $\mu = 10^{-4}$, where $NL\mu = 1.2$, the population spreads over more phenotypes than in the monomorphic regime, but over fewer than in the polymorphic regime. Thus the crossover to the regime where neutral exploration is important occurs at a smaller ϕ_{pq} than for the monomorphic regime. In this intermediate μ regime neither Eq. (4) nor Eq. (7) suffices. Instead, we use the previously introduced factor γ that multiplies N in Eq. (4) to achieve quantitative accuracy. In Appendix S1 and in Figures S3 and S4, we explore the scaling of γ with the parameters N, L, μ, for different dynamic regimes, and for a range of ξ.

In summary then, our theory derived in the previous section accurately describes the median discovery time T_p of this simple random GP map as a function of the parameters N, μ, ϕ_{pq}. We find that $\phi_{pq} \approx F_p$, and thus $T_p \sim 1/F_p$ in all regimes studied. The more frequent the phenotype, the earlier (and more often, see Figure S2) it appears as potentially selectable variation in an evolving population. Given the success of our theory for the random model, we now will test our theory and conclusions for a more complex GP map.

RNA secondary structure mapping. One of the best studied GP mappings has RNA genotypes of length L made up of nucleotides G, C, U and A. The phenotypes are the minimum free-energy secondary structures for the sequences, which can be efficiently calculated [15]. The number of genotypes grows as 4^L, while the number of phenotypes is thought to grow roughly as $N_P \sim 1.8^L$ [4] so that $N_P \ll N_G$. Moreover, sampling and exact enumerations [5,16,22] have shown that the distribution of phenotype frequencies F_p is highly biased, with a small fraction of phenotypes taking up the majority of genotypes. The neutral spaces $\mathcal{N}_q$ are typically broken up into a number of large

components that are connected by single point mutations that allow neutral exploration [16,22]. By exhaustive enumeration of the $L = 20$ RNA mapping (see also Figure S5) we calculate the ϕ_{pq} between several neutral components of the 11,219 distinct secondary structures that the $N_G = 4^{20} \approx 1.1 \times 10^{12}$ genotypes map to.

Figure 2b shows the ϕ_{pq} for the largest component of the phenotype q drawn in the figure. This phenotype is ranked as the 3rd most frequent for $L = 20$ and exhibits behaviour typical of this system. First, the ϕ_{pq} vary over many orders of magnitude. Second, as shown in the inset if $\phi_{pq} \neq 0$, then the *local* ϕ_{pq} are, to first order, proportional to the *global* F_p. Finally, this neutral space connects to just over 75% of the total $N_P = 11,219$ phenotypes in this particular map: Some ϕ_{pq} are zero even though F_p can be quite large. Generally, the number of phenotypes that can be reached from $\mathcal{N}_q$ increases with F_q [13,16].

We performed extensive simulations of the $L = 20$ RNA system. Typical results are shown in Figure 2B. *First*, we note that the median discovery times vary over many orders of magnitude. The most frequent are found in a median time of $T_p \approx 10^3$ generations while after the maximum measured time of 2×10^9 generations, over 42% of the directly accessible phenotypes (with $\phi_{pq} \neq 0$) have still not been found. We estimate that over 10^{13} generations would be needed to discover all accessible phenotypes, giving a ten order of magnitude range in the T_p. *Second*, the local frequency ϕ_{pq} is a good predictor for ranking T_p (see Figure S6 for a comparison of T_p and global frequency F_p). Further, the criterion $\phi_{pq} = 0$ accurately predicts that phenotypes are *not* discovered (see also Figure S6). However, in contrast to the random GP map, the T_p are discovered at a slower rate than predicted by Eq. (7). Instead, we use a single $\gamma < 3L\rho/N$ to renormalise N in Eq. (4). This slower

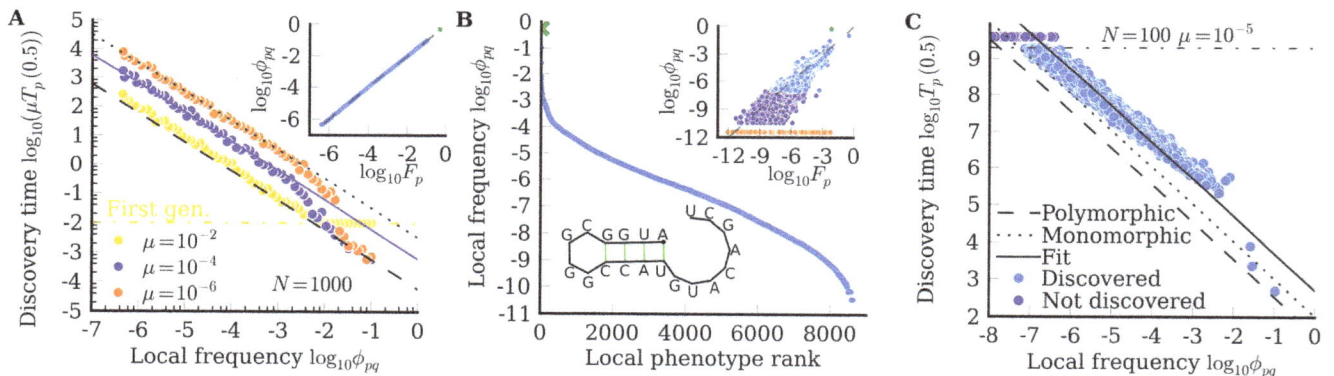

Figure 2. Test of the meanfield model. *A)* Median discovery times T_p for the random GP map averaged over 100 simulations with $N = 1000$ and varying mutation rates. Note that the y-axis is scaled with μ. In the the polymorphic limit ($\mu = 10^{-2}$), Eq. (4) (dashed line) describes discovery times well for $\phi_{pq} < 1/(NL\mu)$. Phenotypes with larger ϕ_{pq} are part of the standing variation typically found in the first generation (yellow dash-dotted line). In the monomorphic limit ($\mu = 10^{-6}$), Eq. (7) (dotted line) quantitatively describes T_p for $\phi_{pq} \ll \phi_L$, whereas Eq. (4) tracks the simulation data with just one fit parameter $\gamma = 0.099$ multiplying N for the intermediate regime with $\mu = 10^{-4}$ (solid line). For $\phi_{pq} \gtrsim \phi_L$ the curves follow Eq. (4), for reasons described in the text. *Inset*: For the random GP map the local phenotype frequency ϕ_{pq} correlates very well with the global frequency F_p. *B)* Local frequency ϕ_{pq} ranked for the 8639 phenotypes that link with single point mutations from the $|\mathcal{N}_q| = 460{,}557{,}583$ genotypes that map to this RNA structure; an example sequence from $\mathcal{N}_q$ is shown in the figure. *Inset*: The local connections ϕ_{pq} are roughly proportional to the global frequency F_p, but there is significant scatter due to the internal correlations of the RNA neutral spaces. Organge points depict the 2580 phenotypes for which $\phi_{pq} = 0$. Light blue points depict the 4933 phenotypes that are discovered in our simulations, and the dark blue points depict the 3705 accessible phenotypes that are not found (q itself is shown in green). *C)* Simulations of T_p (blue dots) versus ϕ_{pq} for the RNA phenotype shown in B), compared to Eq. (4) (solid line) with a factor $\gamma = 0.070$ multiplying N. Here $N = 100$, $\mu = 10^{-5}$ and the simulations were run for 2×10^9 generations. Also shown are the purely polymorphic (dashed) and monomorphic (dotted) predictions. Dark blue dots above 2×10^9 (dot-dashed line) depict some of the 3705 accessible phenotypes that are not found (as can be seen in see the inset of B). We estimate that about 10^{13} generations would be needed to find the phenotypes with the smallest $\phi_{pq} \neq 0$.

discovery rate reflects the internal structure of the RNA: similar genotypes typically have similar mutational neighbourhoods [23], and so the population needs to neutrally explore longer in order to find novelty. Nevertheless, a single γ factor yields a remarkably good fit for all the different phenotypes p (something we find for all source phenotypes q we have so far studied). Finally, we note that the three most frequent phenotypes are found relatively faster because they satisfy $\phi_{pq} \gtrsim \phi_L$. As expected, for this larger system the fraction of phenotypes for which this holds is lower than for the random GP map with smaller L.

Overall, the evolutionary dynamics of this rather complex RNA system resembles that of the much simpler random GP map. Most importantly, the discovery times vary over many orders of magnitude. More precisely, as long as $\phi_{pq} \neq 0$, $T_p \propto 1/\phi_{pq}$ for both the monomorphic and polymorphic regimes: Phenotypic bias leads to a simple, systematic ordering in the discovery of novel phenotypes.

The arrival of the frequent

The many orders of magnitude difference in the arrival rate of variation between phenotypes should have many important implications for evolutionary dynamics. Consider for example the situation where the population has equilibrated to a phenotype q, which was the fitness peak, when subsequently the environment changes so that a different phenotype p has a higher fitness $1 + s$. In order to fix, the alternative phenotype must first be found. If the time-scale T_E on which the environment changes again is much longer than T_p then it likely that the population will discover and fix p. However, if $T_E \ll T_p$, then a new phenotype p' may become more fit before p has time to fix. T_p can vary over many orders of magnitude, so many potentially highly adaptive phenotypes may satisfy $T_p > T_E$ and thus never be found.

Consider also the situation where two phenotypes $p1$ and $p2$ are both more fit than q after an environmental change. If $s_2 > s_1 \gtrsim 1/2N$, then in a standard population genetics picture, we would expect $p2$ to fix rather than $p1$ as long as $T_{p2} \lesssim T_E$. However, this argument ignores the rate at which variation arises. If, for example, $\phi_{p1q} \gg \phi_{p2q}$, then $p1$ may fix well before $p2$ is discovered and fixes.

To illustrate this effect, we study the $L = 12$ RNA system depicted in Figure 3, where the source neutral space has $N_q = 1932$ genotypes, while the two target phenotypes have $\phi_{p1q} = 0.067$ and $\phi_{p2q} = 0.0015$, so $\phi_{p2q}/\phi_{p1q} \approx 0.022$, a relatively modest ratio compared to the what could be found from e.g. Fig. 2. For this particular system $\phi_{p1p2} = 0$: there are no direct single mutation connections between the two target phenotypes – $p1$ and $p2$ are distinct peaks of the fitness landscape.

We simulated a population of $N = 1000$ individuals with fixed $s_1 = 0.002 > 1/2N$, but with varying ratios $s_2/s_1 \geq 1$. The population begins on phenotype q and evolves until $p1$ or $p2$ fixes.

Results are shown in Figure 4. As the mutation rate increases and the system moves from the monomorphic to polymorphic regime, the probability that $p2$ is discovered at least once increases (and is largely independent of fitness). Nevertheless, phenotype $p1$ is discovered much earlier and also much more often because $\phi_{p1q} \gg \phi_{p2q}$. Furthermore, in the monomorphic regime where $\xi \gg 1$ the population remains on a single genotype g much longer than it takes to explore all the neighbours. Thus if $p1$ is accessible from g, then $p1$ is likely arise repeatedly in relatively quick succession (in "bursts"). This effect, which arises naturally in our microscopic model [24], can significantly enhance the probability of fixation over that predicted by origin-fixation models [25] which ignore the discreteness of the source neutral space.

Overall, our simulations show how the more frequent phenotype $p1$ can fix at the expense of the more fit phenotype $p2$. Given the many orders of magnitude difference possible between the T_p, such an "arrival of the frequent" effect may prevent the arrival of the fittest: If a highly beneficial phenotype is never discovered, a much less adaptive but easily accessible phenotype may go to fixation instead.

Finally, phenotype $p2$ is significantly less mutationally robust than $p1$ (more frequent phenotypes are typically more robust [13,16]), and so once discovered, produces deleterious mutants at a higher rate, making it harder for $p2$ to fix at higher mutations rates, a phenomenon known as "survival of the flattest" [26], observed here for the lower ratios s_2/s_1 at higher μ. Thus both the "arrival of the frequent" and the "survival of the flattest" mitigate against the fixation of phenotypes with lower frequency F_p, even if their fitness is much higher.

We note that differences in neutral network size have traditionally also been taken into account in terms of free fitness [27], which – in analogy with free energy in statistical physics [28] – incorporates an entropy-like component to account for mutational effects such as genetic drift and mutational robustness. This picture provides a theoretical foundation for the "survival of the flattest" [26] effect we observe at high mutation rates in Figure 4. However, the "arrival of the frequent" effect is fundamentally different because it does not rely on mutation-selection balance and quasi-equilibrium or steady-state assumptions like free-fitness theory does. Rather, it reflects the strongly non-equilibrium effect that p_2 is rarely or never found. In the example above, the difference in discovery times between p_1 and p_2 is rather modest, and so at large enough mutation rates p_2 is found fairly regularly and free-fitness could be used to analyse results in that regime. But as can be seen for instance in Figure 2 for $L = 20$ RNA, differences in discovery times can vary over many more orders of magnitude than is the case for our particular example, so that in practice highly adaptive yet rare phenotypes may not be discovered at all, even on very long timescales.

Discussion

Mutations provide the fuel for natural selection. Based on this principle, we have presented a detailed model of evolutionary dynamics that focuses on a microscopic description of the outcome of mutations. The phenotypic effect of mutations is mediated by the genotype-phenotype (GP) map which is therefore a crucial ingredient. As outlined in the introduction, several generic features are shared by many different example maps, independent of model details. Here we mainly focussed on the fact that these mapping are highly *biased*: Some phenotypes are realised by orders of magnitude more genotypes than most other phenotypes.

Our calculations for a simplified random mapping and for the more complex RNA secondary structure model predict that the large bias observed in the GP maps translates into a similar order of magnitude variation in the median discovery times T_p for a range of population genetic parameters. For both maps the local frequencies ϕ_{pq} (which predict discovery times) are a good predictor for the discovery times T_p. For the random GP map $\phi_{pq} \approx F_p$. For RNA this relationship provides a rough first order estimate, but the local frequencies can also deviate strongly, especially when $\phi_{pq} = 0$, which can occur even when the global frequency F_p is large. For both maps the strong bias in the GP map leads to a systematic *ordering* of the median discovery times of alternative phenotypes, an effect that we postulate may hold for other GP maps as well.

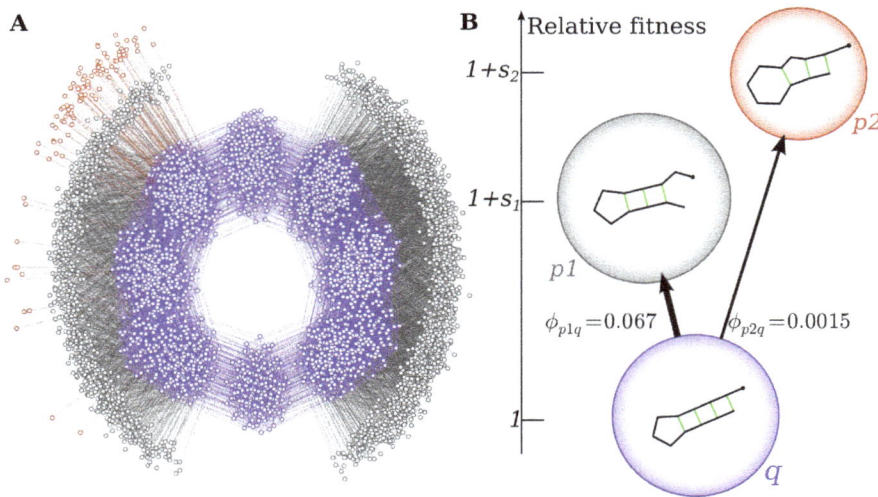

Figure 3. Interconnections of neutral spaces in RNA influence evolutionary trajectories. *A)* $L = 12$ RNA neutral component for phenotype q with $N_q = 1932$ genotypes (drawn in blue). Lines depict single mutations to itself, or to two alternative phenotypes $p1$ (grey) and $p2$ (red). The genotypes were ordered using the Fruchterman-Reingold algorithm [30]. *B)* Illustration of the fitness landscape.

In light of the simplicity of our mean-field approximation, its success in predicting the first-discovery time T_p (cf. Figure 2) is rather striking. In the random GP map, the excellent agreement probably arises because all genotypes in the source neutral space are similar in the sense that they have the same probability distribution to have a certain mutational neighbourhood. There are static fluctuations because the number of neighbours is less than the number of states with $\phi_{pq} \neq 0$. But while these fluctuations have an effect on processes like fixation, they average out over the many runs used to find the mean or median T_p. By contrast, in the RNA GP map mutational neighbourhoods of adjacent genotypes are often correlated [13,23] so that a single neutral mutation does not completely re-shuffle the accessible phenotypes (as the mean-field assumption would assume). This effect explains why the value of the exploration parameter γ we obtain by fitting is below the value suggested by our mean-field model, and also why we still observe around 1 order of magnitude variation in T_p for very

similar values of ϕ_{pq} (see Figure 2). Despite such correlations (which we postulate may occur in other realistic GP maps), rare phenotypes (low ϕ_{pq}) remain hard to find; the strong phenotypic bias in the RNA GP map provides a good a posteriori justification for our mean-field calculations: The many orders of magnitude range in ϕ_{pq} dominates the scale of the phenotype discovery times.

The large differences we observe in the rate with which potential variation appears should have many consequences for evolutionary dynamics. There is of course a long history of invoking processes that impose directionality on the pathways available for evolutionary exploration (see ref. [29] for a recent discussion). Here, by solving microscopic population genetic models, we show in detail just how strong these orienting processes can be. Other authors have also pointed out how evolution may favour phenotypes with large neutral networks for RNA, see e.g. refs. [5,22]. Similar points have been made for protein models [12]. Consider, for example, our $L = 20$ RNA system. Despite its rather modest size, we find 10 orders of magnitude difference between the discovery times of frequent and rare phenotypes. These differences should be even more pronounced for larger L. In nature, selectable RNA phenotypes are of course characterised by more than just their secondary structure, and evolutionary processes don't always work at constant L. Nevertheless, it is hard to see how such enormous variations in T_p would not persist in some form in much more sophisticated treatments of biological RNA. Similar arguments can be made for the other GP maps we listed above. More generally we emphasise that including the GP map in population genetic calculations may be of importance to a wide range of evolutionary questions.

We explicitly showed how phenotypes with a high local frequency can fix at the expense of locally rare phenotypes, even if the latter have much higher fitness. Taken together, these arguments suggest that the vast majority of possible phenotypes may never be found, and thus never fix, even though they may globally be the most fit: Evolutionary search is deeply non-ergodic. When Hugo de Vries was advocating for the importance of mutations in evolution, he famously said "Natural selection may explain the survival of the fittest, but it cannot explain the arrival of the fittest" [2]. Here we argue that the fittest may never arrive. Instead evolutionary dynamics can be dominated by the "arrival of the frequent".

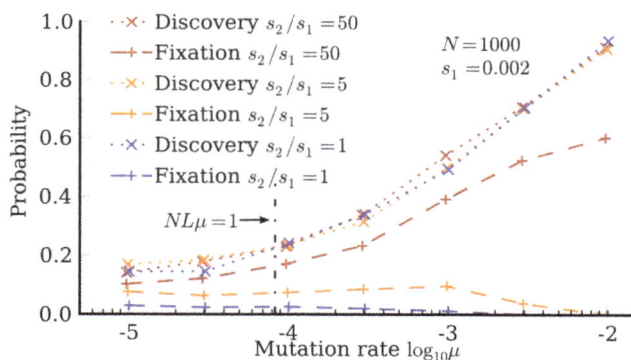

Figure 4. The arrival of the frequent. Probability that phenotype $p2$ is discovered (dotted lines) or is fixed (dashed lines) as a function of mutation rate μ for different relative selection coefficients s_2/s_1 for $Ns_1 = 2$. The probability that $p2$ is discovered is independent of relative fitness (within statistical simulation errors). Phenotype $p1$ is much more likely to fix than phenotype $p2$, even when the latter is much more fit, due to an "arrival of the frequent" phenomenon.

Methods

Simulations

In the dynamic simulations, all N individuals of the population are initially assigned to a single random genotype in the source neutral space. Then the population evolves for $10N$ generations to reach a steady-state dispersal on the neutral space before measurements are started.

RNA

Secondary structures for RNA were predicted from sequence using the Vienna package [15], version 1.8.5 with all parameters set to their default values.

Supporting Information

Appendix S1

Figure S1 Static properties of the random GP map. *A*) Global phenotypes frequencies. In addition to the distribution of frequencies F_p used in our simulations (orange), the diagram also shows the frequencies of RNA secondary structures at $L = 12$, obtained by exhaustive enumeration using the Vienna package, Version 1.8.5 with all parameters set to their default values [15]. *B*) Comparison of global frequencies F_p and local frequencies ϕ_{pq} for the source neutral space q with rank 1. The robustness of phenotype q ($\rho \equiv \phi_{qq}$) is marked in green; alternative phenotypes ($p \neq q$) are shown in light blue. The dashed line marks the equality of global and local frequency $F_p = \phi_{pq}$. The relative size of deviations becomes more severe as F_p becomes small: The less genotypes map into p, the less will frozen fluctuations in the GP map average out.

Figure S2 Total number of mutants per phenotype in different dynamic settings. The diagram shows the total number of mutants $M_p = \sum_{t=1}^{T} m_p(t)$ carrying phenotype p that were produced during a total of $T = 10^4/(N\mu)$ generations of simulation under the random GP map. Dots show the average over 100 simulations, error bars show the standard deviation. The dashed lines correspond to the mean-field theory $M_p = NL\mu\phi_{pq}T$ that follows directly from Eq. (3). In panels B and D, the populations are in the highly polymorphic regime ($NL\mu \gg 1$) and hence evolve towards greater robustness [19] so that the total number of non-neutral mutants is reduced.

Figure S3 Scaling of $N\gamma$ with population dynamic parameters. The diagram shows the dependence of γ on: *A*) mutation rate μ, *B*) population size N and *C*) number of mutants per generation $NL\mu$. Note that the y-axis has been scaled by population size N.

Figure S4 Scaling of γ with population dynamic parameters. The diagram shows the dependence of γ on: *A*) mutation rate μ, *B*) population size N and *C*) number of mutants per generation $NL\mu$. In contrast to Figure S3, the y-axis shows γ without any scaling factors.

Figure S5 Phenotypic bias for RNA secondary structures of length $L = 20$. *A*) Global phenotype frequencies F_p for all $N_P = 11,219$ secondary structures. It required about 1 CPU-year on typical present-day hardware to fold all $4^{20} \approx 10^{12}$ sequences once using the fold-routine of the Vienna package [15], version 1.8.5 with all default parameters. *B–D*) Local phenotype frequencies ϕ_{pq} around 3 neutral spaces. An example sequence and its secondary structure is given in each panel; starting from this sequence, the ϕ_{pq} can be obtained exactly by tracing out all possible neutral mutations and counting how often each phenotype is produced. *Insets*: Comparison of global and local frequencies. Accessible phenotypes ($\phi_{pq} > 0$) are drawn in blue, inaccessible phenotypes ($\phi_{pq} = 0$) are shown in orange and the phenotype corresponding to the neutral space itself is shown in green ($\phi_{qq} \equiv \rho$). The dashed line marks the equality of local and global frequencies $F_p = \phi_{pq}$ and the dotted line indicates the minimal (non-zero) local frequency $\phi_{min,q} = 1/(3LN_q)$, corresponding to only a single mutation away from one of the N_q genotypes in the neutral space. Inaccessible phenotypes with very small global frequencies are omitted for clarity. Note that all these phenotypes are relatively rare ones when compared to Fig. 2b.

Figure S6 Predictions based on global frequency. The diagram shows the same median discovery times of alternative RNA secondary structures that are displayed in Figure 2c, but here as a function of the phenotypes' global frequencies F_p rather than their local frequencies ϕ_{pq}. The different colors indicate: Accessible phenotypes that are typically discovered within the simulation time ($T_p(1/2) \leq 2 \times 10^9$), $\phi_{pq} > 0$, light blue); accessible phenotypes that are typically not discovered ($T_p(1/2) > 2 \times 10^9$, $\phi_{pq} > 0$, dark blue); inaccessible phenotypes that are typically discovered ($T_p(1/2) \leq 2 \times 10^9$, $\phi_{pq} = 0$, orange); inaccessible phenotypes that are typically not discovered ($T_p(1/2) > 2 \times 10^9$, $\phi_{pq} = 0$, red). The lines correspond to the prediction for T_p based on global rather than local frequencies: $T_p(1/2) = \log 2/(NL\mu F_p)$ (cf. Eq. (4)), dashed) and $T_p(1/2) = \log 2/(3L^2\mu\rho F_p)$ (cf. Eq. (7)). In contrast to the predictions based on the local frequencies ϕ_{pq} in Figure 2c, we note the following: 1) Several phenotypes arise even earlier than predicted by the analogue of the polymorphic limit (points below dashed line). 2) Many phenotypes are not discovered even though other phenotypes of comparable (and even much lower) frequency do arise during the simulation. 3) 4 of the most frequent, but locally inaccessible phenotypes are discovered on a time-scale when double mutations become relevant (orange dots; since $N = 100$ and $\mu = 10^{-5}$, double mutants occur on the timescale $t_2 \approx 1/(N(L\mu)^2 = 2.5 \times 10^5$, so if double mutations were to lead to globally random phenotypes, we expect phenotypes with $\phi_{pq} = 0$ to be discovered around $T_p \approx t_2 \log 2/F_p$.)

Author Contributions

Conceived and designed the experiments: AAL SS. Performed the experiments: AAL SS. Analyzed the data: AAL SS. Contributed reagents/materials/analysis tools: AAL SS. Wrote the paper: AAL SS.

References

1. Darwin CR (1859) On the Origin of Species. Murray.
2. de Vries H (1904) Species and Varieties, Their Origin by Mutation. The Open Court Publishing Company.
3. Alberch P (1991) From genes to phenotype: dynamical systems and evolvability. Genetica 84: 5–11.
4. Wagner A (2005) Robustness and Evolvability in Living Systems. Princeton University Press, Princeton, NJ.

5. Schuster P, Fontana W, Stadler PF, Hofacker IL (1994) From sequences to shapes and back: A case study in RNA secondary structures. Proc Roy Soc B 255: 279–284.

6. Li H, Helling R, Tang C, Wingreen N (1996) Emergence of preferred structures in a simple model of protein folding. Science 273: 666–669.

7. England JL, Shakhnovich EI (2003) Structural determinant of protein designability. Physical review letters 90: 218101.

8. Ferrada E, Wagner A (2010) Evolutionary innovations and the organization of protein functions in genotype space. PLoS ONE 5: e14172.

9. Ahnert SE, Johnston IG, Fink TMA, Doye JPK, Louis AA (2010) Self-assembly, modularity, and physical complexity. Phys Rev E 82: 026117.

10. Raman K, Wagner A (2011) Evolvability and robustness in a complex signalling circuit. Mol BioSyst 7: 1081–1092.

11. Borenstein E, Krakauer DC (2008) An end to endless forms: Epistasis, phenotype distribution bias, and nonuniform evolution. PLoS Comp Biol 4: e1000202.

12. Ferrada E, Wagner A (2012) A comparison of genotype-phenotype maps for rna and proteins. Biophysical Journal 102: 1916–1925.

13. Wagner A (2008) Robustness and evolvability: a paradox resolved. Proc Roy Soc B 275: 91–100.

14. Wagner A (2008) Neutralism and selectionism: a network-based reconciliation. Nat Rev Genet 9: 965–974.

15. Hofacker IL, Fontana W, Stadler PF, Bonhoeffer LS, Tacker M, et al. (1994) Fast folding and comparison of RNA secondary structures. Monatsh Chemie 125: 167–188.

16. Schaper S, Johnston IG, Louis AA (2012) Epistasis can lead to fragmented neutral spaces and contingency in evolution. Proc Roy Soc B 279: 1777–1783.

17. Jörg T, Martin O, Wagner A (2008) Neutral network sizes of biological RNA molecules can be computed and are not atypically small. BMC Bioinf 9: 464.

18. Kimura M (1985) The Neutral Theory of Molecular Evolution. Cambrige University Press, Cambridge, UK.

19. van Nimwegen E, Crutchfield JP, Huynen M (1999) Neutral Evolution of Mutational Robustness. Proc Nat Acad Sci USA 96: 9716–9720.

20. Reidys CM (1997) Random induced subgraphs of generalized n-cubes. Adv Appl Math 19: 360–377.

21. Barrett RD, Schluter D (2008) Adaptation from standing genetic variation. Trends in Ecology & Evolution 23: 38–44.

22. Cowperthwaite MC, Economo EP, Harcombe WR, Miller EL, Meyers LA (2008) The Ascent of the Abundant: How Mutational Networks Constrain Evolution. PLoS Comp Biol 4: e1000110.

23. Huynen MA (1996) Exploring phenotype space through neutral evolution. J Mol Evol 43: 165–169.

24. Schaper S (2013) On the significance of neutral spaces in adaptive evolution. Ph.D. thesis, University of Oxford.

25. Yampolsky LY, Stoltzfus A (2001) Bias in the introduction of variation as an orienting factor in evolution. Evolution & Development 3: 73–83.

26. Wilke CO, Wang JL, Ofria C, Lenski RE, Adami C (2001) Evolution of digital organisms at high mutation rates leads to survival of the attest. Nature 412: 331–333.

27. Iwasa Y (1988) Free fitness that always increases in evolution. Journal of Theoretical Biology 135: 265–281.

28. Sella G, Hirsh AE (2005) The application of statistical physics to evolutionary biology. Proceedings of the National Academy of Sciences of the United States of America 102: 9541–9546.

29. Lynch M (2007) The frailty of adaptive hypotheses for the origins of organismal complexity. Proc Nat Acad Sci USA 104: 8597.

30. Fruchterman TMJ, Reingold EM (1991) Graph drawing by force-directed placement. Software: Practice and experience 21: 1129–1164.

18

Fitness Costs of Thermal Reaction Norms for Wing Melanisation in the Large White Butterfly (*Pieris brassicae*)

Audrey Chaput-Bardy[1,2*], **Simon Ducatez**[1,3], **Delphine Legrand**[1,4], **Michel Baguette**[1,4]

1 Muséum National d'Histoire Naturelle, UMR 7205 Institut Systématique Evolution Biodiversité, Paris, France, 2 INRA, Equipe Ecotoxicologie et Qualité des Milieux Aquatiques, UMR 985 Ecologie et Santé des Ecosystèmes, INRA-Agrocampus, Rennes, France, 3 Department of Biology, McGill University, Montreal, Quebec, Canada, 4 Station d'Ecologie Expérimentale du CNRS à Moulis, CNRS USR 2936, Moulis, France

Abstract

The large white butterfly, *Pieris brassicae*, shows a seasonal polyphenism of wing melanisation, spring individuals being darker than summer individuals. This phenotypic plasticity is supposed to be an adaptive response for thermoregulation in natural populations. However, the variation in individuals' response, the cause of this variation (genetic, non genetic but inheritable or environmental) and its relationship with fitness remain poorly known. We tested the relationships between thermal reaction norm of wing melanisation and adult lifespan as well as female fecundity. Butterflies were reared in cold (18°C), moderate (22°C), and hot (26°C) temperatures over three generations to investigate variation in adult pigmentation and the effects of maternal thermal environment on offspring reaction norms. We found a low heritability in wing melanisation ($h^2 = 0.18$). Rearing families had contrasted thermal reaction norms. Adult lifespan of males and females from highly plastic families was shorter in individuals exposed to hot developmental temperature. Also, females from plastic families exhibited lower fecundity. We did not find any effect of maternal or grand-maternal developmental temperature on fitness. This study provides new evidence on the influence of phenotypic plasticity on life history-traits' evolution, a crucial issue in the context of global change.

Editor: Casper Breuker, Oxford Brookes University, United Kingdom

Funding: MB and DL are funded by the Agence Nationale de la Recherche, via the programs (1) open call DIAME, (2) 6th extinction MOBIGEN, and (3) open call INDHET. MB and DL are part of the Laboratoire d'Excellence TULIP (ANR-10-LABX-41). SD was funded by a post-doctoral fellowship from the Fondation Fyssen. The funders had no role in study design, data collection and analysis, decision to publish, or preparation of the manuscript.

Competing Interests: The authors have declared that no competing interests exist.

* E-mail: chaput_bardy_audrey@hotmail.com

Introduction

Knowing the mechanisms allowing organisms to cope with rapid variations of their environmental conditions is a research priority in the current era of anthropogenic global changes and species extinction [1], [2]. Adaptation is the only alternative to extinction for those organisms that are unable to move fast enough to track the shift of their climatic envelopes [3]. Adaptation occurs through two main molecular mechanisms, polymorphism of gene expression (phenotypic plasticity) and polymorphism of gene sequences (allelic fixation) [4]. Phenotypic plasticity consists in changes in phenotypic expression of a genotype in response to environmental factors [5], [6] and has been shown to have significant evolutionary consequences [7], [8]. Plasticity is adaptive if the phenotypes produced in two different environments result in higher average fitness across both environments than either fixed phenotype would [9], [10]. Adaptive phenotypic plasticity thus allows organisms to maintain fitness in rapidly changing environments [11–15].

In spite of this benefit, phenotypic plasticity is not ubiquitous at all [16]. Plasticity of organisms is limited in the range of environments they can respond to [17]. There are indeed some constraints such as limits or costs of plasticity that are not straightforward to show [18]. Plasticity costs are defined as any fitness reduction incurred by a plastic individual compared with a non-plastic individual that expresses the same trait value [19]. Studies attempting to quantify these costs have found either no or limited support to fitness decrease in plastic organisms [16], [18]. After a thorough review of studies that measure the costs of plasticity, Auld et al. [16] concluded that a potentially common correlation between environment-specific trait values and the magnitude of trait plasticities (i.e. multi-collinearity) could result in imprecise and/or biased estimates of costs. However, plasticity costs are expected to be common because developmental responses to environmental changes show evidence of imperfect adaptation [12–18]. Phenotypic plasticity creates resource allocation trade-offs during development, i.e. increased investment into one trait may decrease investments in others [17]. Given the occurrence of phenotypic plasticity and the impact that a cost of plasticity could have on phenotypic evolution, identifying costs of plasticity is thus an important issue to evolutionary ecologists [18], [20].

Phenotypic plasticity is commonly estimated as the response of an organism to an environmental gradient. Importantly, this response can be affected by environmental conditions experienced by the parents of this organism [12], [21]. Integrating sources of non genetic inheritability such as the parental environmental effects on phenotypic variation [22], [23] might thus bring

Figure 1. Dorsal wings of female (a) and male (b) *Pieris brassicae.* The colouration of ventral wings is the same for both sexes; a female is shown here (c). Total black areas of dorsal and ventral forewings were measured.

information of major interest to better understand the evolution of phenotypic plasticity. Especially, we could expect non genetic transgenerational effects to affect the way organisms cope with phenotypic plasticity costs [24].

Here we tackle these issues by investigating the cost of plasticity in wing melanisation in response to temperature variation. We use the large white butterfly, *Pieris brassicae* L., as model species. This butterfly shows seasonal polyphenism, the spring morph being usually darker than the summer one because of larger melanised areas on the fore and the hind wings [25]. Studies in other Pierid butterflies have reported that wing melanisation have a high to moderate heritability [26], [27]. This variation in wing melanisation heritability may be due to genotype-environment interactions and then we can expect that those differences in mean wing melanisation can be adaptive. The synthesis of melanin, which is a complex nitrogen-rich polymer with a heavy molecular weight, involves costs in butterflies, especially since the allocation of nitrogen-rich pigment material is constrained within the closed metabolic system of a developing pupa, as shown both by various indirect and direct evidences [28–32]. Melanin and components of the melanin synthesis pathway ensure a wide range of functions in insects, from thermoregulation to immune defence and tegument coloration but also in traits as diverse as wound healing, cuticle sclerotisation (hardening) and egg tanning [29], [33–35]. Given the complexity of the melanin synthesis pathway and the various

roles played by melanin in insect homeostasis, we expect significant trade-offs between the melanin production per se, and fitness-related traits [29].

In *Pieris brassicae*, the experimental exposition of full-sibling caterpillars to contrasted temperatures during their development is supposed to induce the expression of the spring or the summer morphs [25]. Such plasticity of wing melanisation in response to photoperiod and/or temperature is supposed to be adaptive in Pierid butterflies with respect to thermoregulation [26], [29], [36]. Increased melanisation may increase the direct absorption of solar radiation, thus favoring a faster heating rate or giving the ability to reach higher body temperatures. The large white butterfly typically basks by opening its wings at angles of 5–75° to the incident sunlight. This behavior allows dorsal wings to act as mirrors that reflect the solar radiation onto the thorax and/or the more melanised part of the wings [35], [37], [38]. When butterflies use reflectance basking, melanisation of the central and distal parts of the dorsal forewing is thus expected to influence body temperature [35], [37], [38]. Decreased melanisation of spots and distal parts of forewings is expected to favor the efficiency of reflectance basking [35], [37], [38]. In a related species to *P. brassicae*, *Pieris rapae*, Stoehr and Goux found a plastic response of the distal part of dorsal forewings to temperature variation [36]. In addition, dorsal forewing patterns are expected to play an important role in sexual selection and mate choice in *Pieris* butterflies. Especially, UV reflectance is known to affect mate detection and choice in *P. rapae* [39]. As UV reflectance is strongly affected by both wing melanin patterns and external light conditions [40], we may also expect variation in melanisation across time and space according to the expected light conditions [40]. Overall, dorsal wing melanisation in *P. brassicae* may thus be affected by various, potentially opposed, selection pressures. Consequently, plasticity in wing melanisation may depend on combined effects of temperature and light, and can be affected by selection on thermoregulation efficiency and sexual selection.

In this study, we address the following questions:

1. What is the relative importance of the additive genetic variation as opposed to the environmental variation in the wing melanisation? We used 'animal model' methods to estimate the heritability of wing melanisation and to partition the relative roles of these two sources of variation.

2. Does the level of plasticity in wing melanisation vary among full-sib families? Phenotypic plasticity was quantified from variance between environments and from the slope of the reaction norm, which is the phenotypic trait value expressed as a function of environmental conditions [13].

3. Is there a fitness cost associated with the plasticity in wing melanisation (after taking into account direct costs associated with mean melanisation), and is it affected by non-genetic parental effects? We assumed plasticity costs would appear when the strength of the reaction norm negatively affects fitness independently from the character values expressed within single environments [19].

Materials and Methods

Ethics Statement

No permits were necessary for the sampling of individuals and the experimental work on the large white butterfly, and this project did not involve endangered or protected species.

Rearing Experiment

Experiments were performed at the Muséum National d'Histoire Naturelle in Brunoy, France, using butterflies from a recently

Table 1. Model selection for 'animal models' of melanin variation to estimate forewing melanisation heritability.

Random effect	Fixed effect	DIC
–	–	7444.30
animal	–	7413.91
animal+RT	–	7381.75
animal+RT+MT	–	7389.82
animal +RT+MT+GMT	–	7402.16
animal+RT	FA	7030.22
animal+RT	sex	6352.47
animal+RT	**FA+sex**	**5759.95**

Model selection is based on deviance information criterion (DIC). Forewing melanisation heritability was estimated from the most parsimonious model in bold.
RT is the rearing temperature.
MT is the mother rearing temperature.
GMT is the grand-mother rearing temperature.
FA is the forewing area.

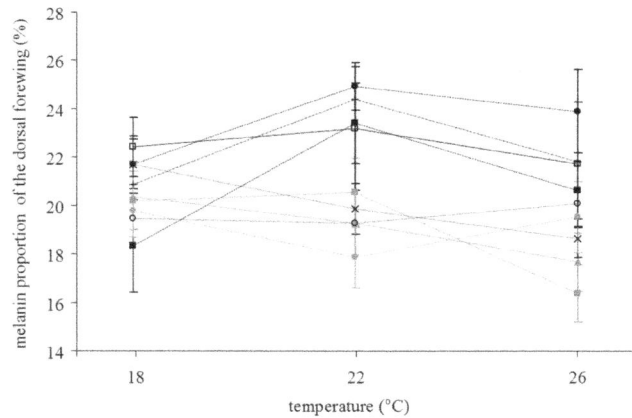

Figure 2. Thermal reaction norms (mean ± standard error) for melanin proportion of the dorsal forewing in 9 full-sib families of *Pieris brassicae*.

Table 2. Effect of rearing temperature on melanin proportion in 9 butterfly families (GLM results).

Family 1	Estimate	SE	z	p
Intercept	−1.624	0.0202	−80.2	<0.0001
Temperature	0.021	0.0009	22.19	<0.0001
Family 2				
Intercept	−1.7402	0.0299	−58.22	<0.0001
Temperature	0.0184	0.0013	13.67	<0.0001
Family 3				
Intercept	−0.9731	0.0251	−38.82	<0.0001
Temperature	−0.0214	0.0012	−18.35	<0.0001
Family 4				
Intercept	−0.8607	0.0184	−46.66	<0.0001
Temperature	−0.0236	0.0008	−28.55	<0.0001
Family 5				
Intercept	−0.8563	0.0222	−38.5	<0.0001
Temperature	−0.0274	0.0011	−25.84	<0.0001
Family 6	Estimate	SE	z	p
Intercept	−1.5451	0.0156	−99.35	<0.0001
Temperature	0.0062	0.0007	9.36	<0.0001
Family 7				
Intercept	−1.2953	0.0231	−56.07	<0.0001
Temperature	−0.0069	0.0011	−6.24	<0.0001
Family 8				
Intercept	−1.2003	0.0274	−43.8	<0.0001
Temperature	−0.0021	0.0014	−1.49	0.136
Family 9				
Intercept	−1.5584	0.023	−67.88	<0.0001
Temperature	0.0128	0.0012	10.57	<0.0001

'Estimate' corresponds to the slope of the thermal reaction norm and SE stands for the standard error.

established breeding. The stock originated from eggs collected in

Table 3. Effects of forewing melanisation plasticity ('plasticity'), melanin proportion ('melanin'), sex, mating and rearing temperature on lifespan in *Pieris brassicae*: results of model averaging on linear mixed models with family as a random effect.

(a) plasticity = variance between environments

Parameter	Estimate	SE	Adjusted SE	z	p	Relative importance
Intercept	2.92E+01	1.96E+01	1.96E+01	1.491	0.1361	1
mate	−1.09E+01	9.32E+00	9.33E+00	1.168	0.2428	1
melanin	−1.03E+00	7.47E-01	7.49E-01	1.373	0.1699	0.83
sex	**−1.97E+01**	**8.81E+00**	**8.83E+00**	**2.233**	**0.0255**	**1**
plasticity	**9.47E-06**	**4.01E-06**	**4.60E-06**	**2.058**	**0.0396**	**1**
rearing temperature	−6.40E-01	9.42E-01	9.44E-01	0.678	0.4976	1
mate:sex	**6.95E+00**	**3.59E+00**	**3.60E+00**	**1.934**	**0.0531**	**1**
mate:plasticity	4.17E-06	2.48E-06	2.48E-06	1.677	0.0935	0.72
melanin:temperature	6.05E-02	3.24E-02	3.24E-02	1.865	0.0622	0.77
sex:plasticity	**2.05E-06**	**9.85E-07**	**9.87E-07**	**2.075**	**0.0379**	**0.73**
sex:temperature	7.90E-01	4.07E-01	4.07E-01	1.938	0.05257	1
plasticity:temperature	**−4.83E-07**	**1.48E-07**	**1.48E-07**	**3.258**	**0.0011**	**1**
melanin:plasticity	−1.65E-07	8.37E-08	8.38E-08	1.969	0.0489	0.27
melanin:sex	−2.54E-01	2.46E-01	2.46E-01	1.033	0.3015	0.16
mate:melanin	5.08E-01	4.99E-01	5.00E-01	1.016	0.3096	0.19
mate:temperature	3.52E-01	3.55E-01	3.56E-01	0.989	0.3228	0.19

(b) plasticity = slope of the reaction norm

Parameter	Estimate	SE	Adjusted SE	z	p	Relative importance
Intercept	7.66806	15.34079	15.35699	0.499	0.6175	
mate	−14.59819	13.81057	13.82346	1.056	0.2909	1
melanin	−0.07471	0.63756	0.63822	0.117	0.9068	0.81
sex	−12.59627	8.20778	8.21625	1.533	0.1252	1
plasticity	720.96612	356.4199	403.42805	1.787	0.0739	1
rearing temperature	0.41018	0.69892	0.6996	0.586	0.5577	1
mate:sex	7.87137	4.42812	4.43427	1.775	0.0759	0.88
mate:plasticity	268.41227	154.4337	154.73693	1.735	0.0828	0.65
melanin:temperature	0.02767	0.03843	0.03847	0.719	0.4719	0.35
sex:plasticity	**191.65618**	**99.01677**	**99.20913**	**1.932**	**0.0534**	**0.7**
sex:temperature	0.50927	0.31558	0.31601	1.612	0.1071	0.8
plasticity:temperature	**−34.87847**	**10.44949**	**10.47155**	**3.331**	**0.0009**	**1**
melanin:plasticity	−14.21438	10.79973	10.8175	1.314	0.1888	0.43
melanin:sex	−0.19636	0.24174	0.24225	0.811	0.4176	0.26
mate:melanin	0.45264	0.59386	0.59464	0.761	0.4465	0.36
mate:temperature	0.3734	0.39313	0.39384	0.948	0.3431	0.37

Plasticity is estimated by the variance of forewing melanisation between temperatures (a) and the absolute slope of the thermal reaction norm (b). Females were taken as references for the calculation of coefficients. Significant variables ($p < 0.05$) with a relative importance > 0.60 are bolded. SE stands for the standard error.

Brunoy (Essonne, France, 48°42'00"N, 2°30'00"E), Moulis (Ariège, France, 42°57'40"N, 01°05'27"E) and Mesnil-Eglise (Belgium, 50°10'00"N, 45°58'00"E) during the summer 2007.

Twenty mating pairs were isolated from the tenth generation of captive breeding, 1507 eggs were collected from these mating pairs and served as basis for the first generation of our experiment (F1). The rearing experiment consisted of three temperature treatments over three generations to estimate melanin heritability and to test the existence of transgenerational effects of parental and grandparental developmental temperature on offspring plasticity. Full-

sibling eggs were randomly divided over three climate rooms set to constant temperatures of 18°C (cold treatment), 22°C (moderate treatment), and 26°C (hot treatment). Optimum temperature for development seems to be between 20°C and 26°C [41]. Moreover, a pilot experiment with F9 individuals showed that larvae reared above 26°C and below 18°C experienced high mortality (>90%). Larvae were separated in familial groups of thirty individuals. They were reared on cabbage, *Brassica oleracea* provided *ad libitum*, in 15*9*9 cm boxes under a constant light cycle (Light : Dark 14: 10 h) that induces direct development without pupal diapause.

Temperature, light brightness and hygrometry were controlled daily within each climate rooms (POL-EKO ST500), using the same thermometer, hygrometer and light meter, and the only significant difference we observed between the rooms was the temperature (which is the manipulated variable here). Rearing boxes with caterpillars were moved from their position in a climate room every day to avoid a position effect in the room. After emergence, imagoes were placed in 60*60*60 breeding cages, at $25°C \pm 1°C$ with unrelated individuals, where their survival was recorded daily. Mating pairs were isolated in 20*20*20 cm laying cages to produce the next generation (F2). Every day, the eggs laid by each female on cabbage leaves were counted before splitting them over the three temperature treatments. Eggs from the same clutches were full siblings as females could mate only once. The number of offspring at each developmental stage was also recorded. We used the same protocol to produce the next generation (F3). The developmental temperature of F1 and F2 generations was manipulated in order to test the importance of maternal and grand-maternal effects on F3 generation, while the developmental temperature of F3 generation was manipulated to test potential costs of phenotypic plasticity.

Picture Acquisition and Processing

We developed a system to take standardised digital photography of living imagoes after their emergence. Each individual was photographed on the day of its emergence in order to limit wing attrition effect on colouration and area of wing. Butterflies were anaesthetized with nitrogen monoxide using an Inject+Matic Sleeper TAS and held between two transparent plastic pieces. The butterfly was then placed into a light tent on a standard white background. The tent diffused homogeneous and constant light from two lamps on both sides. Grey reflectance standards were included in pictures at the start of a photography session [42]. A decimeter arranged under the butterfly allowed us to control the scale of each photograph. The digital camera was placed on a tripod and in a dark room at a distance of 30 cm from the butterfly. We used a Nikon D300 digital camera equipped with a 105 mm macro lens. The shutter lag time was 10 seconds to impede vibrations. We manually set the white balance according to the light intensity and we maintained a constant integration time and a constant lens aperture [42].

The left and the right forewings were extracted from the digital picture (4288×2488 pixels) with the Gimp shareware (http://gimp.org/), and wing variables were measured with the ImageJ shareware (http://rsb.info.nih.gov/ij/). The total areas of the right and left forewing of each individual were measured. Melanised areas had a grey reflectance score above the threshold of 120 (where 0 = white, 255 = black). This threshold reliably separated melanised from unmelanised scales on the wings according to the grey reflectance standards [36]. Based on a subsample of 30 individuals, wing measures were highly repeatable (99% of the variation in repeated measurements, including repeated photographing, is due to between individual variations).

Inheritance of Wing Melanisation

A total of 829 butterflies (403 males and 426 females) on 3 000 harvested eggs over three generations were photographed. Adult size was estimated by averaging the area of left and right forewings (i.e. 'size' variable). Total black area was also averaged over both forewings for each side (i.e. ventral and dorsal, see Figure 1). As ventral melanised area was correlated with dorsal melanised area (Pearson's correlation test, $r^2 = 0.87$, $t = 51.21$, $df = 827$, $p < 0.0001$), we used dorsal melanised area for subsequent analyses.

We estimated wing melanisation heritability from the pedigree data (829 butterflies over 3 generations) fitting animal models with the R package MCMCglmm [43]. The animal model is a mixed-effect model that allows assessing the genetic and non genetic components of phenotypic variation. Individuals' pedigree is included as an explanatory random variable to estimate genetic variance. This approach allowed to partition the additive genetic variance (animal effect), the variance due to maternal effects, and the residual environmental variance of melanisation. The forewing melanised area constituted the response variable. Forewing area and sex were treated as fixed effects and animal effect (pedigree), rearing temperature, mother temperature and grand-mother temperature were treated as random effects. The MCMC chains were run for 1 300 000 iterations with a burn-in period of 300 000 to ensure satisfactory convergence [44]. Parameters, estimated standard errors and confidence intervals were performed by sampling 1 000 times the posterior parameter distribution. The deviance information criterion (DIC) was used for model selection. The difference between DIC-values of two competing models should not exceed 10 [45]. Adding a random effect 'rearing box' nested in the animal effect did not improve the models, and we thus excluded this effect from the models.

Wing Melanisation Plasticity

For analyses of wing melanisation plasticity we used the offspring of 9 females (9 families, 585 individuals) from the F3 generation. Thereby we measured wing melanisation plasticity at the family level. As estimators of phenotypic plasticity, we considered: (i) the variance of melanin proportion (i.e. the ratio between dorsal melanised area and total wing area) between temperatures, and (ii) the absolute slope of the linear regression between the melanin proportion and temperature in each family. We used the absolute slope of the reaction norm, as we were interested in the magnitude of the variation, and the direction of the variation was indicated by a supplementary binary variable (i.e. the 'slope sign').

We tested the effect of wing melanisation plasticity for dorsal forewings on adult lifespan, using linear mixed effect models (LME) with sex, melanin proportion, rearing temperature, rearing cage, rearing temperature of the mother and the grand-mother, mating status and first order interactions as supplementary fixed effects, and family as a random effect. The mating status, hereafter called 'mate' indicates whether an individual mated or not. This variable is known to negatively affect adult lifespan due to an energetic cost [46], [47]. The melanin proportion was added to models in order to verify that lifespan was influenced by melanin

Table 4. Effect of forewing melanisation plasticity (i.e. within family variance in forewings melanisation between temperatures) on lifespan in females and males at a rearing temperature of 18, 22 and 26°C.

Sex	Rearing temperature	Estimate	SE	t	p
females	18	−1.30E-06	1.03E-06	−1.261	0.2090
	22	−1.54E-07	1.36E-06	−0.113	0.9100
	26	−3.05E-06	1.24E-06	−2.457	0.0159
males	18	1.74E-06	9.32E-07	1.864	0.0647
	22	1.65E-06	1.57E-06	1.048	0.2990
	26	−3.58E-06	1.37E-06	−2.621	0.0104

SE stands for the standard error.

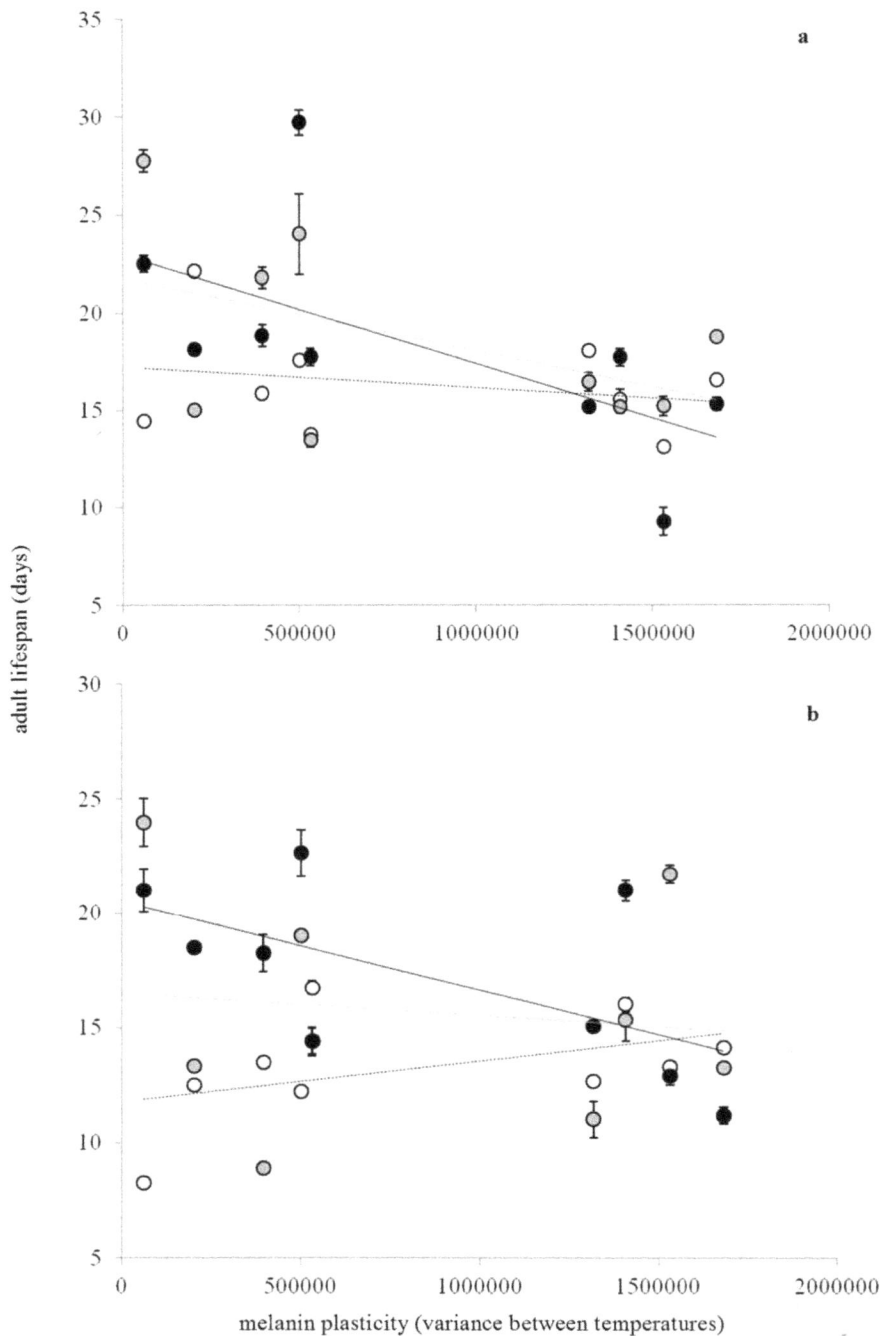

Figure 3. Effect of the interaction between wing melanisation plasticity (i.e. variance between environments) and rearing temperature on mean adult lifespan (± standard error) in females (a) and males (b). The continuous black line and closed circles correspond to 26°C, the continuous grey line and grey circles to 22°C and the dotted black line and open circles to 18°C.

plasticity and not by individual melanisation only. We performed a model selection approach using the Akaike Information Criterion (AIC, [48]) and a model averaging procedure. Akaike weights (Wi) that represent the relative probability for a model i to be the best among considered models, were calculated for the subset of models having (AICbest −AICi) ≤2. In a model averaging procedure, averaged parameters and their corresponding unconditional standard errors were calculated from the smallest subset of

AIC-ranked models for which $\sum Wi$ was ≥0.95. The relative importance of each variable within the averaged model was estimated by adding Wi-$values$ of those models within the 95% confidence set containing that variable [48–50].

We tested the effect of wing melanisation plasticity on female fecundity (30 females from F3 generation) using generalized linear mixed models (GLMM) with the number of hatching eggs as response variable, and wing melanisation plasticity, melanin

Table 5. Effects of forewing melanisation plasticity and female lifespan on fecundity.

(a) plasticity = variance between environments

Parameter	Estimate	SE	t	p
Intercept	4.42E+00	4.24E-01	10.433	<0.0001
lifespan	5.53E-02	2.46E-02	2.249	0.0329
plasticity	−9.29E-07	4.93E-07	−1.884	0.0704

(b) plasticity = slope of the reaction norm

Parameter	Estimate	SE	t	p
Intercept	4.65115	0.43775	10.625	<0.0001
lifespan	0.05701	0.02422	2.353	0.0261
plasticity	−65.86106	27.25042	−2.417	0.0227

Here is presented the most parsimonious model. Plasticity is estimated by the variance of forewing melanisation between temperatures (a) and the absolute slope of the thermal reaction norm (b). We used a GLMM and applied a deviance analysis for model selection (see Methods). SE stands for the standard error.

proportion, rearing temperature, rearing temperatures of the mother and the grand-mother, female lifespan and their interactions as fixed effects, and family as a random effect. The number of hatching eggs was analyzed using a quasi-Poisson error and a logarithmic function link (a Poisson GLM correcting for overdispersion [51]). As in quasi-Poisson models, the AIC is not defined, we performed analysis of deviance to compare two nested models (full and nested models [52]). The best model was obtained by stepwise deletion of non significant terms (p>0.05). No transformations of response variables were needed to meet the assumptions of normality and homoscedasticity. Statistical analyses were performed using R, version 2.11.1 (The R Foundation for Statistical Computing, Vienna).

Results

Heritability of Wing Melanisation

The most parsimonious model included forewing area and sex as fixed effects and animal effect and rearing temperature as random effects (Table 1). The animal model provided evidence for equivalent effects of additive genetic variation (CVa = 40.03; 95% Highest Posterior Density Interval (HDPI) = 20.00–62.26), developmental temperature variation (CV = 35.19; 95% HDPI = 10.13–527.08) and environmental (residual) variation (CVe = 48.91; 95% HDPI = 36.54–60.98) on wing melanisation, which resulted in a rather low heritability estimate for melanin (h^2 = 0.18; HDPI = 0.01–0.41). It is noteworthy that maternal and grand-maternal temperatures were not included in the best model.

Thermal Reaction Norm

The 9 families showed contrasted reaction norms (i.e. the relationship between melanin and temperature, Figure 2). Mean reaction norms exhibited positive, flat and negative slopes (Figure 2). Testing the effect of interaction between temperature and family on wing melanisation indicated that variations between environments were significantly different across families (Likelihood Ratio Chi-Square = 1190, DF = 8, p<0.0001). Wing melanisation increased significantly with temperature in 4 families (0.006< slope of the reaction norm <0.021), decreased signifi-

cantly in 4 families (−0.007< slope of the reaction norm<−0.027), and did not significantly vary in one family (slope = − 0.002, p = 0.136, Table 2).

Effect of Plasticity on Adult Lifespan

We did not find a parental or grand-parental effect of rearing temperature on adult lifespan, as mother and grand-mother rearing temperatures were not included as random factors in the best models. Wing melanisation plasticity (the slope of the reaction norm and the variance between environments), melanin proportion, rearing temperature, sex, mating status and interactions accounted for adult lifespan (Table 3). Only the magnitude of the thermal reaction norm (the absolute slope of the reaction norm) accounted for wing melanisation plasticity, as the direction of the reaction norm (the slope sign) did not affect lifespan.

In order to better understand the effects of interactions between melanin plasticity, sex and rearing temperature on lifespan (Table 3), we made post-hoc analyses by sex within each temperature. These post-hoc analyses were linear models with lifespan as response variable and plasticity (i.e. variance between environments) as fixed effect (Table 4 and Figure 3). Lifespan was differently affected by temperature according to the sex. The lifespan of males tended to be longer in high wing melanisation plasticity families at 18°C and shorter at 26°C (Table 4 and Figure 3). The lifespan of females was also shorter at 26°C when plasticity increased (Table 4 and Figure 3).

Effect of Plasticity on Females' Fecundity

Females' fecundity was also negatively influenced by wing melanisation plasticity although this effect was weaker considering variance of melanin between temperatures instead of the slope of the thermal reaction norm (Table 5). There was no developmental and parental effect of temperature on fecundity. Females from plastic families produced fewer eggs. Females that lived longer produced more eggs than those with a shorter lifespan (Table 5).

Discussion

Roff [53] compared the heritability of suites of life history, behavioural, physiological and morphological traits in a large panel of organisms and showed that life-history traits are always less heritable than other traits. We report here a rather low heritability of wing melanisation in *P. brassicae* (h^2 = 0.18), which is in the range of heritabilities of life-history traits rather than of morphological or physiological traits. This result is in contradiction with studies on other Pierids that showed a high to moderate heritability of melanin [26], [27]. We suggest that this low heritability of melanin deposition on the wings reflects more developmental plasticity associated with the multiple functions of melanin in the homeostasis of insects [29], [33–35] rather than the transmission of pure morphological or physiological attributes. The effect of environment on wing melanisation revealed here by the animal model is also in line with adaptive plasticity reported in Pierid butterflies with respect to thermoregulation [26], [29], [36]. This adaptive plasticity does not seem to be influenced by the developmental temperature experienced by the previous generations. This result confirms findings of other cross-generation studies in which effects of parental temperature on offspring life-history traits were weak or absent [54], [55].

Wing melanisation is classically supposed to decrease with increasing temperature [56], [57]. Here we found that some butterfly families were darker at warmer temperatures. Besides, the reaction norms of these 9 families were markedly different. Our results seem to support a quantitative genetic variation in thermal reaction norms for wing melanisation, rather than a simple, general mechanistic relationship linking temperature to melanin deposition. Families of a same population can diverge not only in the average amount of plasticity expressed but also in their patterns of inter-individual variation in plasticity. The maintenance of this variation may be explained by the existence of different life history and ecological strategies within populations, or by differences in individuals' exposure to selective pressures in the wild. Such a reversal in the temperature-phenotype rule, together with between-family variation in thermal reaction norms were also observed for the body size in natural populations of *Pieris rapae* [58]. Adaptive plasticity consists in producing an optimum phenotype corresponding to the prevalent local environmental conditions [59]. We would expect individuals that experienced particular environmental conditions during their larval development to be better adapted to similar conditions during their adult stage. However, we showed here that adult lifespan was not higher when larvae developed at 26°C, which is close to adult rearing temperature (25°C ±1°C). Similarly to ours, studies on thermal reaction norms based on the breeding of larvae at contrasted temperatures [29], [36], [58] investigated the adaptability of adult phenotypes at only one temperature regime. Alongside with the recommendation of Gilchrist and Huey [55], we suggest that it would be highly informative to study the relationship between life-history traits and temperature on adults manipulating adult breeding temperatures according to the ranges of developmental temperatures.

We have highlighted the coexistence of high and low wing melanisation plasticity families in our experimental population (Figure 2). Among males, high plasticity seemed to confer a lifespan benefit at low temperature but not at moderate temperature, and became costly at high temperature. Among females, high wing melanisation plasticity incurred a lifespan cost at high temperature and a fecundity cost regardless the developmental temperature. Overall, in our experimental setting, wing melanisation plasticity was costly for females, whereas males seemed to benefit from plasticity at low temperature. The evolutionary dynamics of this potential trade-off certainly deserves further research; we suggest again that this exciting issue could be

fully addressed only by comparing fitness of adults bred under the same ranges of temperatures than those experienced by larvae during their growth.

Here, adult butterflies experienced a homogeneous thermal environment across 13 generations in rearing conditions. Such stable conditions probably maintained constant the costs of wing melanisation plasticity. Differences in fitness between high and low wing melanisation plasticity families tend to confirm that homogeneous environmental conditions could select against plasticity or favour canalisation [18]. Plastic individuals must indeed invest resources in maintaining the molecular/physiological 'machinery' needed to detect, monitor and respond to various environmental conditions [16] and therefore may be counter-selected under stable conditions.

Temperature is one of the environmental cues that influences the expression of many phenotypic traits [13]. We reported here on fitness costs of thermal reaction norms for wing melanisation in a butterfly, which potentially entail selection against the more plastic individuals. Making quantitative predictions on the extent to which the homogenisation of the environmental conditions [60] associated with the current era of global changes will affect phenotypic plasticity is thus an important challenge.

Acknowledgments

We thank Marielle Peroz and Sandrine Chertouk for rearing maintenance. We also acknowledge Charlotte Joreau, Lucas Auer and Christine Li who helped us in pictures' acquisition and Gabriel Nève (University of Provence), Vincent Debat (MNHN) and Marc Théry (CNRS) for their advice in picture analysis. Pierre-Yves Henry (MNHN), Céline Teplitsky (MNHN) and Thomas Tully provided us with statistical advice. Finally, we thank Lhorine François (University of Bordeaux) for her English corrections.

Author Contributions

Conceived and designed the experiments: ACB MB. Performed the experiments: ACB SD. Analyzed the data: ACB SD. Contributed reagents/materials/analysis tools: ACB SD MB. Wrote the paper: ACB SD DL MB.

References

1. Hoffmann A, Sgro C (2011) Climate change and evolutionary adaptation. Nature 470: 497–485.
2. Bell G, Gonzalez A (2011) Adaptation and evolutionary rescue in metapopulations experiencing environmental deterioration. Science 332: 1327–1330.
3. Berg MP, Kiers ET, Driessen G, Vand Der Heijden M, Kooi BW, et al. (2010) Adapt or disperse: understanding species persistence in a changing world. Glob Chang Biol 16: 587–598.
4. Pigliucci M, Murrannand CJ, Schlichting CD (2006) Phenotypic plasticity and evolution by genetic assimilation. J Exp Biol 209: 2362–2367.
5. Schlichting CD (1986) The evolution of phenotypic plasticity in plants. Annu Rev Ecol Syst 17: 667–693.
6. Bradshaw AD (1965) Evolutionary significance of phenotypic plasticity in plants. Adv Genet 13: 115–155.
7. Schlichting CD (2004) The role of phenotypic plasticity in diversification. In: DeWitt TJ, Schneider SM, editors. Phenotypic plasticity: Functional and Conceptual Approaches. Oxford: Oxford University Press. 191–200.
8. Murren CJ, Denning W, Pigliucci M (2005) Relationships between vegetative and life history traits and fitness in a novel field environment: impacts of herbivores. Evol Ecol 19: 583–601.
9. van Kleunen M, Fisher M (2005) Constraints on the evolution of adaptive phenotypic plasticity in plants. New Phytol 166: 49–60.
10. Davidson AM, Jennions M, Nicotra AB (2011) Do invasive species show higher phenotypic plasticity than native species and, if so, is it adaptive? A meta-analysis. Ecol Lett 14: 419–431.
11. Schlichting CD, Pigliucci M (1998) Phenotypic evolution: a reaction norm perspective. Sunderland, MA: Sinauer. 387 p.
12. West-Eberhard MJ (2003) Developmental plasticity and evolution. New york: Oxford University Press. 794 p.
13. Via S, Gomulkiewicz R, de Jong G, Schneider SM, Schlichting CD, et al. (1995) Adaptive phenotypic plasticity: consensus and controversy. Trends Ecol Evol 10: 212–217.
14. Moran NA (1992) The evolutionary maintenance of alternative phenotypes. Am Nat 139: 971–989.
15. Pigliucci M (2001) Phenotypic plasticity: beyond nature and nurture. Baltimore, Maryland: John Hopkins University Press. 344 p.
16. Auld JR, Agrawal AA, Relyea RA (2010) Re-evaluating the costs and limits of adaptive phenotypic plasticity. Proc R Soc Biol Sci Ser B 277: 503–511.
17. Moczek AP (2010) Phenotypic plasticity and diversity in insects. Philos Trans R Soc B Biol Sci 365: 593–603.
18. Van Buskirk J, Steiner UK (2009) The fitness costs of developmental canalization and plasticity. J Evol Biol 22: 852–860.
19. DeWitt TJ, Sih A, Wilson DS (1998) Costs and limits of phenotypic plasticity. Trends Ecol Evol 13: 77–81.
20. Pigliucci M (2005) Evolution of phenotypic plasticity: where are we going now? Trends Ecol Evol 20: 481–486.
21. Mousseau TA, Fox CW (1998) The adaptive significance of maternal effects. Trends Ecol Evol 13: 403–407.
22. Bonduriansky R, Day T (2009) Non genetic inheritance and its evolutionary implications. Annu Rev Ecol Evol Syst 40: 103–125.
23. Danchin E, Wagner RH (2010) Inclusive Heritability: combining genetic and nongenetic information to study animal behavior and culture. Oikos 119: 210–218.
24. Ducatez S, Baguette M, Stevens VM, Legrand D, Fréville H (2012) Complex interactions between paternal and maternal effects: parental experience and age at reproduction affect fecundity and offspring performance in a butterfly. Evolution 66: 3558–3569.

25. Feltwell J (1982) Large white butterfly, the biology, biochemistry and physiology of *Pieris brassicae* (Linnaeus). The Hague, the Netherlands: Dr. W. Junk Publisher. 535 p.

26. Kingsolver JG, Wiernasz DC (1991) Seasonal polyphenism in wing-melanin patterns and thermoregulatory adaptation in *Pieris* butterflies. Am Nat 137: 816–830.

27. Ellers J, Boggs CL (2002) The evolution of wing color in Colias butterflies: heritability, sex linkage, and population divergence. Evolution 56: 836–840.

28. Stoehr AM (2006) Costly melanin ornaments: the importance of taxon? Funct Ecol 20: 276–281.

29. Stoehr AM (2010) Responses of disparate phenotypically-plastic, melanin-based traits to common cues: limits to the benefits of adaptive plasticity. Evol Ecol 24: 287–298.

30. Talloen W, Van Dyck H, Lens L (2004) The cost of melanization: butterfly wing coloration under environmental stress. Evolution 58: 360–366.

31. Lee KP, Simpson SJ, Wilson K (2008) Dietary protein-quality influences melanization and immune function in an insect. Funct Ecol 22.

32. Ma W, Chen L, Wang M, Li X (2008) Trade-offs between melanisation and life-history traits in *Helicoperva armigera*. Ecol Entomol 33: 37–44.

33. Nappi AJ, Christensen BM (2005) Melanogenesis and associated cytotoxic reactions: applications to insect innate immunity. Insect Biochem Mol Biol 35: 443–459.

34. Sugumaran M (2002) Comparative biochemistry of eumelanogenesis and the protective roles of phenoloxidase and melanin in insects. Pigm Cell Res 15: 2–9.

35. Kingsolver JG (1985) Thermoregulatory significance of wing melanization in *Pieris* butterflies (Lepidoptera: Pieridae): physics, posture, and pattern. Oecologia 66: 546–553.

36. Stoehr AM, Goux H (2008) Seasonal phenotypic plasticity of wing melanisation in the cabbage white butterfly, *Pieris rapae* L. (Lepidoptera: Pieridae). Ecol Entomol 33: 137–143.

37. Kingsolver JG (1985) Thermal ecology of Pieris butterflies: A new mechanism of behavioral thermoregulation. Oecologia 66: 540–545.

38. Kingsolver JG (1987) Evolution and coadaptation of thermoregulatory behavior and wing pigmentation pattern in pierid butterflies. Evolution 41: 472–490.

39. Obara Y, Majerus MEN (2000) Initial mate recognition in the British cabbage butterfly, *Pieris rapae rapae*. Zoolog Sci 17: 725–730.

40. Obara Y, Koshitaka H, Kentaro A (2008) Better mate in the shade: enhancement of male mating behaviour in the cabbage butterfly, *Pieris rapae crucivora*, in a UV-rich environment. J Exp Biol 211: 3698–3702.

41. Maercks H (1934) Untersuchungen zur ökilogie des kohlweisslings (*Pieris brassicae* L.), die temperaturreaktionen und das feuchtigkeitoptimum. Z Morph Ökol Tiere 28: 692–721.

42. Stevens M, Parraga CA, Cuthill IC, Partridge JC, Troscianko TS (2007) Using digital photography to study animal coloration. Biol J Linn Soc 90: 211–237.

43. Hadfield JD (2010) MCMC methods for multiple-response generalized linear mixed models: MCMCglmm R package. J Stat Softw 33: 1–22.

44. Wilson AJ, Réale D, Cements MN, Morrissey MM, Postma E, et al. (2010) An ecologist's guide to the animal model. J Anim Ecol 79: 13–26.

45. Spiegelhalter DJ, Best NG, Carlin BP, van der Linde A (2002) Bayesian measures of model complexity and fit. J R Stat Soc Ser B 64: 583–639.

46. Fowler K, Partridge L (1989) A cost of mating in female fruitflies. Nature 338: 760–761.

47. Partridge L, Harvey PH (1988) The Ecological context of life history evolution Science 241: 1449–1455.

48. Burnham KP, Anderson DR (2002) Model selection and multimodel inference: A practical information-theoretic approach. New York: 2nd ed. Springer-Verlag. 488 p.

49. Burnham KP, Anderson DR (2004) Multimodel inference: Understanding AIC and BIC in model selection. Sociol Methods Res 33: 261–304.

50. Grueber CE, Nakagawa S, Laws RJ, Jamieson IG (2011) Multimodel inference in ecology and evolution: challenges and solutions. J Evol Biol 24: 699–711.

51. Richards SA (2008) Dealing with overdispersed count data in applied ecology. J Appl Ecol 45: 218–227.

52. Zuur AF, Ieno EN, Walker NJ, Saveliev AA, Smith GM (2009) Mixed effects models and extensions in Ecology with R. New York: Springer. 574 p.

53. Roff DA (2002) Life History Evolution. Sunderland, MA: Sinauer Associates. 465 p.

54. Crill WD, Huey RB, Gilchrist GW (1996) Within- and between-generation effects of temperature on the morphology and physiology of *Drosophila melanogaster*. Evolution 50: 1205–1218.

55. Gilchrist GW, Huey RB (2001) Parental and developmental temperature effects on the thermal dependence of fitness in *Drosophila melanogaster*. Evolution 55: 209–214.

56. Kingsolver JG (1983) Thermoregulation and flight in Colias butterflies: elevational patterns and mechanistic limitations. Ecology 64: 534–545.

57. Van Dyck H, Matthysen E (1998) Thermoregulatory differences between phenotypes in the speckled wood butterfly: hot perchers and cold patrollers? Oecologia 114: 326–334.

58. Kingsolver JG, Massie KR, Ragland GJ, Smith MH (2007) Rapid population divergence in thermal reaction norms for an invading species: breaking the temperature–size rule. J Evol Biol 20: 892–900.

59. Moczek AP (2011) Evolutionary biology: the origins of novelty. Nature 473: 34–35.

60. Western D (2001) Human-modified ecosystems and future evolution. Proc Natl Acad Sci U S A 98: 5458–5465.

19

Naturally Occurring Deletions of Hunchback Binding Sites in the *Even-Skipped* Stripe 3+7 Enhancer

Arnar Palsson[1,2,3,4]*, **Natalia Wesolowska**[4,5], **Sigrún Reynisdóttir**[1,2], **Michael Z. Ludwig**[4], **Martin Kreitman**[4]

1 Faculty of Life and Environmental Sciences, University of Iceland, Reykjavik, Iceland, **2** Institute of Biology, University of Iceland, Reykjavik, Iceland, **3** Biomedical Center, University of Iceland, Reykjavik, Iceland, **4** Department of Ecology and Evolution, University of Chicago, Chicago, Illinois, United States of America, **5** Cell Biology and Biophysics Unit, European Molecular Biology Laboratory (EMBL), Heidelberg, Germany

Abstract

Changes in regulatory DNA contribute to phenotypic differences within and between taxa. Comparative studies show that many transcription factor binding sites (TFBS) are conserved between species whereas functional studies reveal that some mutations segregating within species alter TFBS function. Consistently, in this analysis of 13 regulatory elements in *Drosophila melanogaster* populations, single base and insertion/deletion polymorphism are rare in characterized regulatory elements. Experimentally defined TFBS are nearly devoid of segregating mutations and, as has been shown before, are quite conserved. For instance 8 of 11 Hunchback binding sites in the stripe 3+7 enhancer of *even-skipped* are conserved between *D. melanogaster* and *Drosophila virilis*. Oddly, we found a 72 bp deletion that removes one of these binding sites (Hb8), segregating within *D. melanogaster*. Furthermore, a 45 bp deletion polymorphism in the spacer between the stripe 3+7 and stripe 2 enhancers, removes another predicted Hunchback site. These two deletions are separated by ~250 bp, sit on distinct haplotypes, and segregate at appreciable frequency. The Hb8Δ is at 5 to 35% frequency in the new world, but also shows cosmopolitan distribution. There is depletion of sequence variation on the Hb8Δ-carrying haplotype. Quantitative genetic tests indicate that Hb8Δ affects developmental time, but not viability of offspring. The Eve expression pattern differs between inbred lines, but the stripe 3 and 7 boundaries seem unaffected by Hb8Δ. The data reveal segregating variation in regulatory elements, which may reflect evolutionary turnover of characterized TFBS due to drift or co-evolution.

Editor: Alan M. Moses, University of Toronto, Canada

Funding: This project was funded by the National Institutes of Health (Grant R01 GM61001 to M.K. and M.Z.L.), Icelandic research fund (IRF 070260021 to A.P.), Marie Curie international reintegration grant (MIRG-CT-2007-046510 to A.P.) and the University of Iceland research fund to A.P. The funders had no role in study design, data collection and analysis, decision to publish, or preparation of the manuscript.

* E-mail: apalsson@hi.is

Introduction

Evolution of Transcriptional Regulatory Sequences

The molecular basis for phenotypic divergence and standing variation is often attributed to differences in the regulation of transcription[1–3]. The mechanistic principles of regulatory DNA and factor structure and function such as; multiple transcription factor binding sites (TFBS), TFBS motif degeneracy, cooperativity and number of trans factors [3,4] and interactions between transcription factors (TFs), enhancers and promoters [5,6] impose unique rules on their evolution. Regulatory DNA has no single "active-site", since most regions consist of multiple transcription factor binding sites. Evolutionary analyses of experimentally verified TFBS demonstrate examples of conservation, but also reveal evolutionary turnover of TFBS, were some sites are lost and others gained [7–9].

It has been postulated that selection mainly acts on the transcriptional output of a gene (timing, location and amount) and does not preserve individual TFBS [10,11]. That is, changes in TFBS and even losses are permitted, if the transcriptional output is preserved. Such models of stabilizing selection acting on transcriptional output can account for both loss of functional binding sites and evolutionary fine-tuning of regulatory elements

[12]. They also suggest that positive selection may sometimes play a role, acting on compensatory mutations in *cis* or *trans*. Several studies [13–16] have investigated the evolutionary origin of TFBS, including co-evolution within regulatory sequences. From first principles one would predict both co-evolution in *cis* (promoters, regulatory modules, more distantly located signals like insulators) and co-evolution of sequence elements with the *trans* environment (abundance of transcription factor, mediator or holenzyme components). The model of *trans* co-evolution is corroborated by studies of between-species hybrids [17], which *e.g.* reveal mis-expression of genes in hybrids of *D. melanogaster* and *D. simulans*, two closely related species, most likely due to species-specific *cis-trans* compensatory evolution. Also, genome-wide changes in *cis* elements of co-expressed genes in two distantly related yeast species document the co-evolution of the TF repertoire of an organism and the regulatory elements of coordinately expressed genes [18,19]. Numerical models show how mutation and drift can generate binding sites, and predictably that selection can speed up fixation of new TFBS [20]. Crucially, functional polymorphism (both single nucleotide polymorphism: SNPs or insertion/deletion polymorphism: indels) in human enhancers, are shaped by positive selection [21].

Insertion and Deletion Polymorphism in Regulatory DNA

Population genetics studies have largely neglected indels, perhaps because they represent a minority of segregating variation in most genomes [22]. Deletion polymorphism in the intergenic region of *Adh* in *Drosophila pseudoobscura* does not conform to neutral evolution, but exhibits signatures of purifying selection, *i.e.* deletion (but not insertion) polymorphism was removed from introns over time [23]. On a larger scale, Comeron and Kreitman [24] revealed a bias in the insertion and deletion frequency distribution in *D. melanogaster* populations. While deletion events were more common and on average longer, insertions were at significantly higher frequency. This may reflect both mutational bias (because the mechanisms causing deletions are different from those causing insertions) and a difference in selection pressures, with purifying selection keeping a large fraction of deletions at low frequency in the population [24,25]. Ometto *et al.* [25], on the other hand, also concluded that weak positive selection might increase the population frequency of some insertions, which is supported by a genome-wide study in *D. melanogaster* [26]. Population genetic analyses of Bicoid response genes in *D. melanogaster* revealed single nucleotide polymorphism (SNPs) in 13 of 85 predicted Bicoid binding sites [27]. Most notable was the high frequency of SNP and indel polymorphism in the *Orthodenticle* (*Otd*) early head enhancer. These polymorphisms clustered on two haplotypes, both at intermediate frequency. Transgene tests showed that the *Otd* haplotypes differ in transcriptional output [27]. Similarly, studies of the *Endo16* promoter and other sea-urchin enhancers [28–30] show that many TFBS are affected by segregating indel variation. In particular, in *Endo16* two rare insertions affect the same part of the promoter. One of these generated a functional repressor module [29].

Enhancers of Eve as a Model of Regulatory Evolution

Early embryonic development in *D. melanogaster* is regulated by numerous genes through a complex network of activation and repression, resulting in segmental boundaries along the embryo length [31–35]. The accurate temporal and spatial expression of these genes is mainly achieved by integration of multiple TFs and their binding to regulatory sequences. Some regulatory functions (required for a given expression pattern) are aggregated in distinct modules like the *eve* stripe 2 enhancer (*s2e*) and the stripe 3 and 7 enhancer (*s3+7e*). These experimentally verified "minimal" enhancers [36,37] suffice to generate 4–7 cell-wide Eve stripes in early development. Not all regulatory sequences contain modular enhancers, and often spacer sequences (separating regulatory modules) have function, meaning that the length of these sequences matters for proper function of flanking *cis*-modules [10,38].

The cumulative effect of nucleotide changes in *s2e* between species is a turnover of functional motifs within enhancers [8,11,39]. Notably, the *s2e* from *D. erecta* is less effective than *s2e* from the more distantly related *D. pseudoobscura* at complementing a deletion of the *s2e* in *D. melanogaster* [11]. Quantitative analysis of the amount of Eve in stripe 2 illustrated the functional deficiency of the *D. erecta s2e* in the *D. melanogaster* genetic background. This means that, for a given enhancer, the spatial and temporal features of the expression pattern are highly conserved, but the quantity of gene product probably less so. The expression level of developmentally-specific gene products may exhibit changes over evolutionary time, possibly reflecting "developmental system drift" [40,41].

The aim of the current study was to gauge the level of polymorphism in the well-characterized regulatory regions in *D. melanogaster*, with particular focus on insertion and deletion polymorphism. Consistent with other studies and evolutionary theory, SNP and indel-polymorphism are rare in TFBS. However we find two peculiarly large and common deletions in and close to the *eve* stripe 3 and 7 enhancer. Both deletions remove binding sites for Hunchback, prompting analysis of the genetics and phylogeography of one of those polymorphisms and its potential phenotypic effects. The data provide insights into the nature of variation segregating in *cis*-regulatory elements.

Materials and Methods

Flies and Populations

Several populations of flies where studied. The population genetic surveys were done on collections of inbred lines derived from North Carolina, collected in 2000 and 2005 [27,42], and a Costa Rican sample from Peter Andolfatto, made isogenic for the second chromosome by three generations of crosses. Walter Eanes provided DNA from thirteen US East coast populations [43]; a total of 380 individuals used to test for clinal variation in the *eve* region. Jean-Claude Walser provided a sample of 46 cosmopolitan populations [44], in which DNA from 100 lines in each population was pooled.

PCR, DNA Sequencing and Genotyping

Primers were designed with primer 3 version 0.3 (frodo.wi.mit.edu [45]) for 13 well-characterized early developmental enhancers or promoters and several other non-coding regions (see Table S1). The regions studied were several parts of the *eve* locus (the *late element*, *s2e*, *s3+7e*, and the promoter, along with two spacer sequences), *Kruppel* promoter and *CD1*, *salm* wing blade enhancer, *ems* abdominal enhancer, *en* regulatory region and promoters of *Antp*, *Ubx-bxd*, *tll*, *Act57B*, *RpL29/CG30390* and *RpL30*. The sequence variation in those regions was assessed by PCR followed directly by DNA sequencing. PCR was done as before [46] with Takara Taq and MJ Tetrad machines on 96 well plates. Products where purified by Qiagen purification columns or Exo-sap. DNA sequencing was done on purified PCR products, with the forward and reverse primer using Applied biosystems reagents. The ethanol purified reaction products where run in the University of Chicago sequencing facility or the ABI sequencing machine at the Institute of Biology, University of Iceland.

The deletion of the Hb8 site in *s3+7e* (see below) and the wild type allele were genotyped with PCR using allele-specific primers (Table S1). We ran separate reactions for both alleles on individuals from the East coast sample and on bulk DNA samples from the cosmopolitan sample. This was used to infer geographic distribution of specific variants, but does of course not yield information about frequency. All sequences were submitted as Popset data to NCBI (accession numbers: KJ465109–KJ465866), except two alignments that were shorter than 200 bp (provided in fasta format as Supporting information S1 and S2).

Population Genetic Analysis

Metrics of population genetics (S, π, θ, Haplotype number) were calculated for SNPs and indels with Tassel vs. 2.1 (www.maizegenetics.net [47]), either for individual regions or as a sliding window for the haplotype analysis. Tassel was also used to calculate LD, and R (www.r-project.org version 12.3 [48]) for testing of contingency tables. DNAsp vs. 4.1 (www.ub.edu/dnasp [49]) was also used to test for deviations of Tajima's D and Fu and Li's estimators. Furthermore Hudson's haplotype test [50] (utilizing the ms program and the psub option) was used to test for positive selection in four *eve* regions.

Phylogenetic Shadowing

A 2 kb region surrounding the stripe 3+7 minimum enhancer was blasted against the 12 finished genomes (insects.eugenes.org/species), and the orthologous regions extracted (except *D. willistoni* which did not return a significant blast hit). The Drosophila species (abbreviated) and contig names and locations are listed; *D. melanogaster* (*D. mel*), release 4, *D. simulans* (*D.sim*) chromosome 2R, bases 4491595 to 4494659, *D. sechellia* (*D. sech*) scaffold 359, bases 7623 to 10695, *D. erecta* (*D. ere*) scaffold 4929, bases 8504394 to 8507885, *D. yakuba* (*D. yak*) chromosome 2L, bases 18628840 to 18632292, *D. ananassae* (*D. ana*) scaffold 13266, bases 15371395 to 15373454, *D. pseudoobscura* (*D. pse*) chromosome 3, bases 10879010 to 10881069, *D. persimilis* (*D. per*) scaffold 4, bases 6230662 to 6232721, *D. virilis* (*D. vir*) scaffold 12875, bases 1335449 to 1337479, *D. grimshawi* (*D. gri*) scaffold 15245, bases 9663295 to 9665324, *D. mojavensis* (*D. moj*) scaffold 6496, bases 4426987 to 4430428. The sequences were aligned with MAVID (baboon.math.berkeley.edu/mavid [51]). Divergence in these sequences is considerable, requiring manual curating in Genedoc (www.psc.edu/biomed/genedoc [52]), with special devotion to characterized TFBS from redfly.ccr.buffalo.edu [53] and ORegAnno [54]. In addition two additional Hb sites (Hb15 and Hb16) found by Stanojovic *et al.* [55] and two Stat binding sites discovered by Yan *et al.* [32] were included. We found that the *D. melanogaster* Stat binding sites differ from the genomic sequence, probably due to sequencing error (Stat-1 was reported to start with an **A** and stat-2 was reported as **G**TTCCCCGAA**A**, highlighted bases differ).

We also used (jaspar.genereg.net [56]) to predict Hb binding sites (score above 6) in the ~8000 bp upstream of *eve*, in *D. melanogaster*, *D. sechellia*, *D. yakuba* and *D. pseudoobscura*. Based on multiple alignments from Mavid, and Multiz alignments from the Santa Cruz genome browser (downloaded in December 2013), we mapped predicted Hb binding sites in orthologous and more rapidly evolving regions.

Testing the Effects of a Segregating Deletion on Adult Phenotypes

A set of 20–60 healthy inbred lines from NC [46] were used for the two experiments conducted to test the effects of a 72 bp deletion within *s3+7e* (called *Hb8Δ*, see below) on viability and developmental time. The first was a set of controlled crosses to lines deficient for *eve*, and the second was phenotyping of 60 genotyped inbred lines. All fly-rearing took place on cornmeal food at constant temperature, 25°C.

We first crossed the inbred lines to four stocks with characterized *eve* mutations. Ten inbred lines, homozygous for each allele (*Hb8Δ* or *wt*) were crossed to each *eve* mutant. The Bloomington stock numbers and genotypes are; BL-4084: *eve[5]/SM6a*, BL-5344: *eve [1]/CyO; P{ry[+t7.2] =ftz/lacC}*, BL-1719: *Df(2R)X3/CyO, Adh[nB]* and 1702: *Df(2R)X1, Mef2[X1]/CyO, Adh[nB]*. Three virgins of a mutant stock were crossed with 3 males from each of the 20 inbred lines, and allowed to lay eggs for 2–3 days. The offspring were counted and sexed, between 10 and 11 am, from day 10 to 18. The experiment was fully balanced and repeated three times, several weeks apart. The parents of all lines used in the crosses had been grown for 2 generations under controlled density (parents discarded between days 2 and 5 depending on visual assessment of egg number). We recorded both the total number of offspring (viability), and developmental time, summarized as the average time to eclosion for a given combination of, mutation, cross, genotype, sex and replicate.

For the association tests, 60 inbred lines where studied. The *Hb8Δ/wt* polymorphism was genotyped in three individuals of each line in the generation that was phenotyped. The rearing and measuring procedure was identical to the first experiment, except no crosses were required and only replicates were measured (two weeks apart).

Embryo Collections, Fixing and Staining

The embryos were collected, fixed and stained with standard protocols, as we have done before [8]. Four inbred lines with (NC25 and NC128) and without (NC006, NC017) the *Hb8Δ* laid eggs for 4–5 hrs at 22°C. Briefly, we collected embryos from each of the four lines, and they were fixed. Multiple embryo collections were pooled before staining with Eve primary antibody and a secondary antibody. The histochemical LacZ staining reaction was run for 12 minutes. The stained embryos were stored in 70% glycerol at 4°C, and photographed within a week.

Photography and Measurements

Each embryo in the appropriate developmental stage range was photographed three times at 20X magnification with water immersion on a Zeiss microscope. First a DIC sagittal section yielding maximum length of embryo and then two sections (DIC sagittal and bright-field) captured the stripes. Tiff photographs were saved and the X and Y coordinates of stripe boundaries assessed in ImageJ (rsb.info.nih.gov/ij/ [57]). First, a straight line was superimposed on the sagittal image, and the X-Y coordinates of anterior and posterior of the embryo recorded. Second, the same guideline was superimposed on the other two images and X-Y landmarks of the anterior and posterior boundary of each stripe were visually assessed and recorded. Third, the rotation of the embryo along the Dorsal/Ventral axis was scored. Finally, the stage of development was also visually assessed from *eve* pair-rule expression, in increments of 0.5 on the scale from 1 to 5, around cellularization [11]. The same investigator (AP) did all measurements.

Summarizing the Expression Pattern

The raw landmark data indicating the length of the embryo and placement of stripes were processed in two ways. The relative positions of stripe boundaries were estimated by calculating distance of landmarks from the anterior and posterior end using standard geometric formula. First, the length of the embryos was estimated. Second, the relative distance from one embryo tip to the anterior and posterior boundary of each stripe was calculated.

Statistical Analysis of Adult and Embryonic Phenotypes

SAS version 8.2 [58] was used for analyses of phenotypes. The viability and developmental time analyses were conducted with mixed model ANOVA (proc MIXED). The model for the testcross was:

$$Y = M + C + MXC + G + MXG + CXG + S + O + L(CXG) + error$$

Denoting the fixed effects of the mutation (M), that is the 4 different *eve* deficiencies or point mutations, the cross (C) designating the balancer (CyO) or the "loss of function" (LoF) *eve* mutation, the genotype (G) term which evaluates the effects of *Hb8Δ*, sex (S) and appropriate interaction terms. The effects of Line (L) and replicate vials (R) are considered random factors. Furthermore, the total number of offspring (O) was included as a covariate. As a large factorial model with 4 fixed terms runs the risk of being overly parametrized, higher order terms were

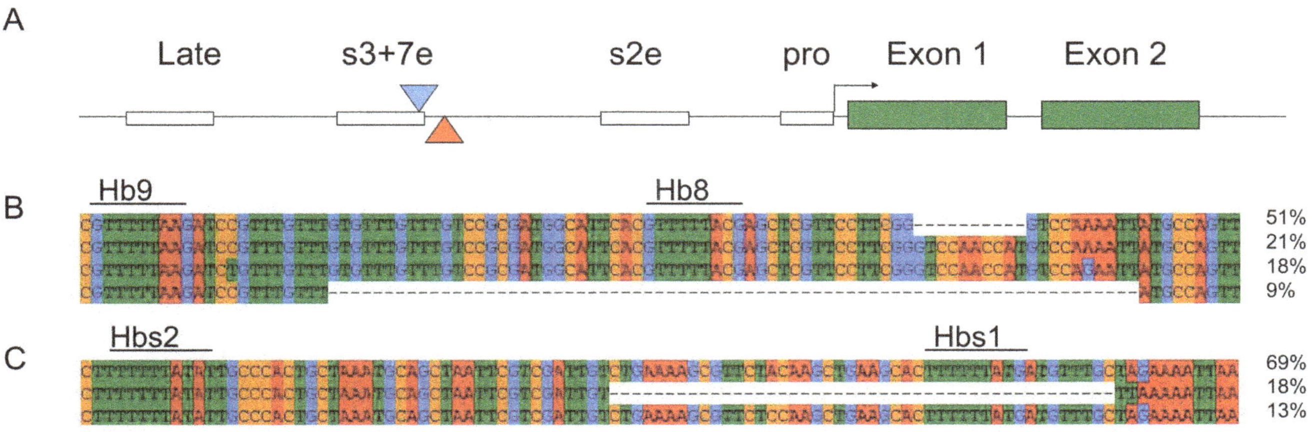

Figure 1. Two large deletions remove conserved Hunchback binding sites in the *eve* stripe 3+7 enhancer. A) The structure of the upstream region of *eve*, open boxes represent the late element, *s3+7e*, *s2e* and promoter regions, and green boxes the two exons. The deletions are shown by blue (*Hb8Δ*) and red (*Hbs1Δ*) triangles. B) Detailed structure of the *Hb8Δ* and frequency of the four alleles at this position in a Costa Rican population. C) Structure of *Hbs1Δ* and frequency of alleles in the same population. D) The conservation of a subset of TFBS in the *s3+7e* and the Hbs1 site. Full species names are provided in Materials and Methods and data for other *s3+7e* binding sites in Table S3.

evaluated and dropped if they were not significant at the 0.05 level. The association tests of the inbred lines data were simpler, with only terms denoting genotype, sex and total number of offspring, and not described here.

The relative location of histochemically detected Eve stripes was studied similarly. In order to remove the effects of orientation, a reduced model was fit, and the residuals were used in the subsequent analysis. The positioning of stripes was analyzed with a mixed model ANOVA. The dependent variables of interest are the relative positioning of stripe boundaries, with the anterior boundary of stripe 3 (S3A) and the posterior position of stripe 7 (S7P) being particular candidates given prior evidence on Hunchback distribution in the embryo [59]. The ANOVA model had the general form:

$$Y = G + T + GXT + L(G) + error$$

Where G, indicating genotype (the presence or absence of Hb8), is a fixed main effect. The covariate T (for developmental time) captures the developmental progression and L is a random term for different inbred lines. The relative stripe position matrix (anterior/posterior boundary of all 7 stripes) was also summarized with Principle component (Proc PRINCOMP) on the correlation

matrix. Only the first component, with eigenvalue 7.42, was analyzed for dependence on Hb8 genotype.

Results

Polymorphism in Regulatory DNA Includes Large Deletions of TFBS

First we surveyed the molecular variation, *i.e.* nature, frequency and distribution of polymorphisms, in 13 well studied *Drosophila* regulatory elements and several less well defined elements and spacer sequences. Few indel polymorphisms are found in the regulatory regions, 8 of the regions have no indels (Table 1). Purifying selection seems to affect both SNP and indel polymorphism, as there is a significant correlation between θ for SNPs and indels ($r = 0.48$, $p = 0.03$, Figure S1A). The size and frequency of indels in characterized *cis*-elements was contrasted to those in noncoding regions surrounding two developmental genes, *hairy* and *EGFR* [46,60]. As was previously observed [46] most indels are short, and rarely do large indels (more than 10 bp) reach appreciable frequency (Figure S1B). The notable exception is a 72 bp deletion in the stripe 3 and 7 enhancer (s3+7e) of *eve* (Figure 1A and B). Interestingly this deletion removes a DNase I characterized Hunchback (Hb) binding site [55], and is henceforth called *Hb8Δ*. Bioinformatic analyses in Jaspar show that this site has a PWM score of 8.5, suggesting the notion that this a

Table 1. Single base and indel polymorphism in *D. melanogaster* regulatory elements.

Gene	region	Sites	SNPs				Indels			
			S	π	θ	Tajima's D	S	π	θ	Tajima's D
Antp	promoter	512	9	0.0037	0.0055	−1.24	2	0.0006	0.0010	−1.20
salm	wing blade enhancer	432	12	0.0088	0.0090	−0.05	0			
en	regulatory element	554	6	0.0031	0.0034	−0.36	0			
Ubx	Bxd promoter	346	1	0.0008	0.0009	−0.29	0			
ems	Abdominal enhancer	604	10	0.0043	0.0052	−0.68	0			
Hb	promoter spacer#	184	0				0			
tll	promoter	399	9	0.0070	0.0071	−0.05	0			
Act57B	promoter	489	12	0.0082	0.0077	0.24	2	0.0007	0.0013	−1.28
RpL29/CG30390	promoter	482	19	0.0108	0.0124	−0.53	4	0.0009	0.0026	−2.19
RpL30	promoter	563	9	0.0053	0.0048	0.35	0			
Kr	promoter	492	6	0.0053	0.0037	1.47	0			
Kr	CD1(a)*	444	20	0.0138	0.0121	0.53	1	0.0002	0.0006	−1.16
Kr	CD1(b)*	334	22	0.0124	0.0181	−1.17	2	0.0006	0.0016	−1.39
eve	promoter	373	6	0.0043	0.0050	−0.44	0			
eve	promoter spacer#	545	6	0.0036	0.0029	0.79	2	0.0003	0.0010	−1.40
eve	s2e	593	11	0.0051	0.0041	0.67	1	0.0001	0.0004	−0.87
eve	s3+7e	283	3	0.0024	0.0025	−0.05	2	0.0012	0.0016	−0.49
eve	s3+7e spacer#	432	12	0.0088	0.0065	1.07	4	0.0009	0.0022	−1.41
eve	Late element	386	8	0.0040	0.0046	−0.36	0			

*Kr CD1 region was sequenced in two parts – and is presented as such due to incomplete genotyping.
#The region just upstream of the *eve* and *hb* promoters are called ''promoter spacer'', and similarly the region proximal of *s3+7e*.

Table 2. Polymorphism in four regulatory elements of *eve* among inbred lines from North Carolina.

Region	Length	Sample*	S	π	Dxy	Haplotypes	Hd
Late	327	All	4	0.0034	0.0036	7	0.8
		wt	4	0.0034		6	0.748
		Hb8Δ	3	0.0028		5	0.663
s3+7e	262	All	6	0.0103	0.0131	7	0.805
		wt	6	0.0098		7	0.8
		Hb8Δ	0	0		1	0
s2e	547	All	11	0.0050	0.0057	18	0.859
		wt	11	0.0048		13	0.862
		Hb8Δ	8	0.0021		6	0.447
Pro	565	All	12	0.0052	0.0054	15	0.864
		wt	11	0.0056		12	0.863
		Hb8Δ	7	0.0020		5	0.442

*Sample size: All (N = 63), *wt* (N = 43), *Hb8Δ* (N = 20).
S: segregating sites.
Dxy: Average number of nucleotide substitutions per site between *wt* and *Hb8Δ* samples.
Hd: Haplotype diversity.
Pro: Promoter.

transcription factor binding site presence/absence polymorphism. Oddly enough, less than 250 bp away (in the spacer separating *s3+7e* and *s2e*), another segregating large deletion also removes a putative Hunchback binding site (Figure 1C). This site (here called Hbs1) is predicted with high PWM score, 11.2. That is the fourth highest score of 60 predicted Hb sites in the 8 kb region upstream of *eve* in *D. melanogaster* (Figure S2A and Table S2). Most of the 21 DNaseI characterized Hb sites in *s3+7e* and *s2e* have lower scores than Hbs1. This 45 bp deletion in the spacer is referred to as *Hbs1Δ*. This putative Hb binding site has probably been unnoticed for two reasons. It sits outside the fragments tested for enhancer function, presumably because of restriction site locations [10,37]. Also, the *D. melanogaster* reference genome sequence contains the deletion. To iterate, the 45 deleted bases do not appear in the standard versions of the *D. melanogaster* genome and are only visible in genomic alignments with close *Drosophila* relatives or population genetic sequence data. The two deletions sit on distinct haplotypes, and are never found in the same inbred lines. They are both at appreciable frequency, in a sample of 55 Costa Rican chromosomes the *Hb8Δ* and *Hbs1Δ* are at 9% and 17% frequency respectively (Figure 1B and C). This leads to the question, are these deletions harmful, neutral or beneficial?

Phylogenetic Footprinting of s3+7e shows the Hb Sites are Conserved

Comparative genomic alignments of the *s3+7e* and the adjacent regions with 12 publicly- available *Drosophila* genomes [61] were used to assess the functional importance of these two predicted Hb binding sites, and other characterized Hb, Kni and Stat sites [32,55,59]. Similarly to the *eve s2e*, TFBS in s3+7e are highly conserved (Table S3); 3 of 13 Hb sites are identical from *D. melanogaster* to *D. mojavensis* and 9 have none or only one mutation between *D. melanogaster* and *D. persimilis*. The Hb8 site is found in all of the 12 species, except *D. ananassae* (most probably due to a gap in the genomic sequence), but has experienced several substitutions (Figure 1D). The PWM score for Hb8 is 8.2 in *D. melanogaster* and *D. simulans*, but 9.9 in *D. yakuba* and *D. pseudoobscura* (Figure S2 and table S2). On the other hand, the predicted Hbs1

site (with a PWM score of 11.2) is completely conserved between *D. melanogaster* and *D. yakuba*, but was not found in distantly related species. Those data suggest considerable evolutionary constraints on those sequences, arguing that they could indeed be functional Hb binding sites. But in the absence of functional tests they must regarded as putative Hb binding sites.

Additionally, the *Hb8Δ* also removes half of a putative Sloppy Paired 1 (Slp1) binding site. The putative Slp1 site is less conserved then the characterized Slp1 site in *s2e* [62] (Table S4), but no SNPs within either of these two (characterized and putative) Slp1 binding sites in *eve*, in 104 sequenced alleles, suggests selective constraint within *D. melanogaster* at least. The genome comparisons confirm that both Hb binding sites in *eve* affected by these two deletions have been protected by purifying selection. This prompts the question, why do these deletions of conserved TFBS occur at such high frequencies in populations? Here we focus mainly on studying the population genetics of *Hb8Δ* and assess its potential impact on development and fitness.

Polymorphism on the *Hb8Δ* and *wt* Haplotypes

How can a deletion removing a conserved binding site be at such high frequency in the population? One possibility is that the deletion of Hb8 is buffered by compensatory mutations (sitting on the same haplotype). To assess this, and to evaluate the polymorphism in the region, two strategies were deployed. One was deeper sequencing of four *eve* regions (the promoter, *s2e*, *s37e* and the *late element*) in inbred strains from North Carolina, and the other, a contrast of sequence diversity in alleles with or without the *Hb8Δ* in ~8 kb around *s3+7e*.

The *Hb8Δ* is at 32% frequency in the NC population (N = 63), and there is less variation on the *Hb8Δ* haplotypes compared to the *wt* haplotypes (Table 2). For instance π (which captures the number of substitutions and their frequency) is 25% to 100% lower on the *Hb8Δ* haplotypes. This is most extreme in the *s3+7e*, and notably weaker in flanking regions. This tendency was captured by other population genetic summary statistics (S, Haplotypes, haplotype diversity and Dxy – a measure of differences in nucleotide substitution rate between samples).

A

B

C

eve region

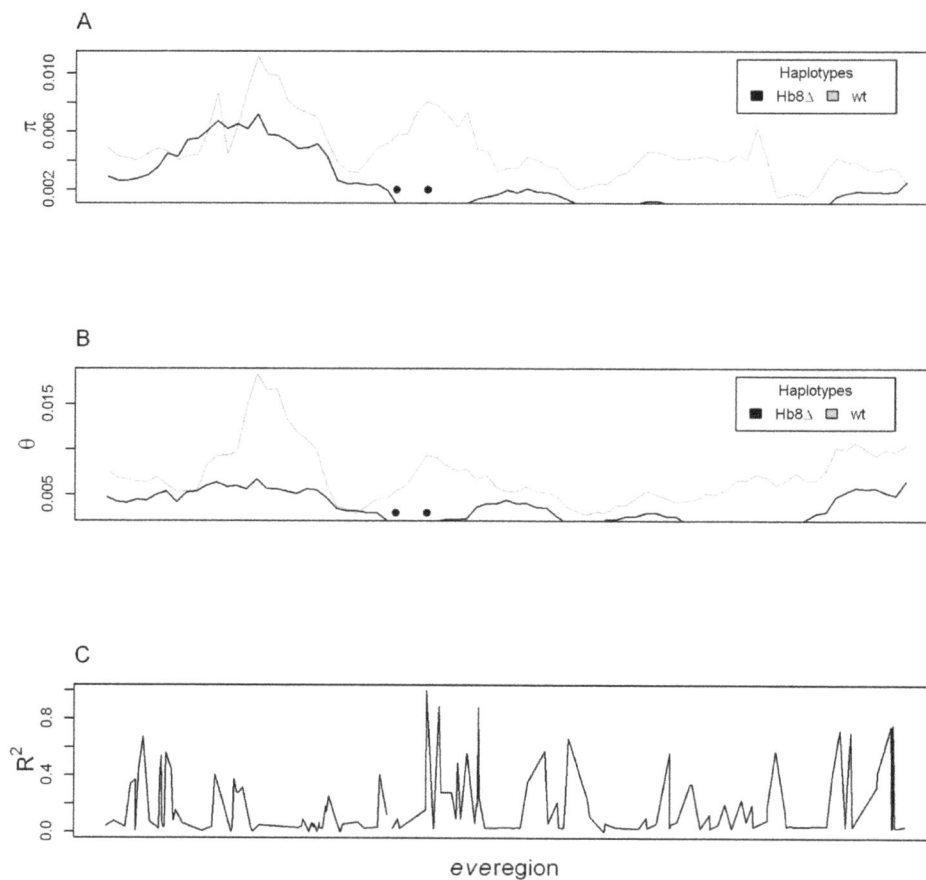

Figure 2. Polymorphism in the ~8200 bp *eve* region. Visualized are positions 5,860,182–5,868,302 on 2R, with the *Hb8Δ* at position 3292 and *Hbs1Δ* at 3602 (black dots). Contrast of polymorphism in the *Hb8Δ* (black) and *wt* haplotypes (gray), with π in A) and θ in B), in 800 bp windows, sliding 100 bp. C) LD between the *Hb8Δ* and other variant in the region, estimated with r^2.

Furthermore, no unique mutations are found on the eve-*Hb8Δ* haplotypes; the variation observed on the *Hb8Δ* haplotypes is all presumed to be due to recombination. These observations suggest positive selection favors the *Hb8Δ* or linked variants. However none of the standard population genetics tests (Tajima's D or Fu and Li's statistics) indicate positive selection (data not shown); neither did the Hudson *et al.* (1993) haplotype test (p>0.73 for each of the four regions).

We next compared more extensively the sequence variation on the *Hb8Δ* and *wt* chromosomes and screened for variants that might possibly compensate for the loss of this Hb binding site. We estimated the polymorphism on two distinct haplotypes carrying either the *wt* or deletion polymorphism, by sequencing 16 (*Hb8Δ*) and 18 (*wt*) chromosomes of each type. The 8200 bp region we selected spans the *eve* neighborhood, from the 3′UTR of CG12134 to the end of the transcript. There is reduced polymorphism (π and θ) on the *Hb8Δ* haplotypes compared to *wt* haplotypes (Figure 2A and B), which is consistent with selection for the *Hb8Δ* bearing haplotype. Another indicator of long haplotypes is high LD between *Hb8Δ* and polymorphic sites in the region (Figure 2C). Several sites more than 3 kb away from Hb8 are in high LD ($r^2 >$ 0.7) with the deletion. Additionally, most polymorphism in the region shows perfect coupling or repulsion LD to *Hb8Δ* (data not shown). (The *Hbs1Δ* was only found in 3 (*wt*) lines. Omission of those 3 lines did not affect the outcome of the polymorphism

analyses - data not shown). Furthermore, no variants are unique to the *Hb8Δ* haplotype. Finally, no potential compensatory mutations that strengthened or generated other Hb sites were observed. The data do reveal less diversity on the *Hb8Δ* haplotype, compared to the *wt* haplotype. Note however, standard tests of natural selection can not be deployed on these data because the sampling was not random from a population; lines were picked for sequencing to get similar representation of *wt* and *Hb8Δ* chromosomes.

Geographic Distribution of the *Hb8Δ*

What is the geographic distribution of *Hb8Δ* and does it correlate with geographic attributes? To study the geographic distribution, bulk DNA samples from 51 cosmopolitan samples, from Europe, Africa, Asia and South America [44] were genotyped with allele specific primers. There was evidence of *Hb8Δ* in 43 of the 51 populations (Table S5), consistent with an evolutionarily old and broadly distributed polymorphism. The cosmopolitan distribution of the *Hb8Δ* is unlikely if it was strongly deleterious.

Does this binding site deletion show any relationship with geographic attributes? To assess this we genotyped *Hb8Δ* in 13 east coast samples, from Maine to Florida [43]. The frequency ranged from 5% to 35% (Table 3) but there was not a significant relation between latitude and frequency of *Hb8Δ* (b = −0.006, p = 0.1). For comparison the *s2e* was also sequenced in the same

Table 3. Frequency of $Hb8\Delta$ and $s2e$ polymorphism along the east coast of North America.

Populations		$Hb8\Delta$		$s2e$				
Location, State	Latitude	Freq.	F_{ST}	N	Sites	π	θ	F_{ST}
Homestead, FL	25° 2′	0.32		24	5	0.0020	0.0024	
Merrit Island, FL	28° 3′	0.16	0.051	26	7	0.0024	0.0033	0
Jacksonville, FL	30° 2′	0.19	0.000	29	7	0.0022	0.0032	0
Eutawville, SC	33° 2′	0.20	0.000	26	5	0.0021	0.0024	0
Smithfield, NC	35° 3′	0.14	0.000	16	4	0.0024	0.0022	0
Richmond, VA	37° 3′	0.05	0.033	33	5	0.0018	0.0022	0
Churchville, MD	39° 3′	0.17	0.052	23	5	0.0018	0.0024	0.051
Middlefield, CT	41° 3′	0.09	0.017	37	5	0.0019	0.0022	0.007
Concord, MA	42° 0′	0.19	0.030	41	5	0.0014	0.0021	0.009
Whiting, VT	43° 6′	0.17	0.000	30	2	0.0015	0.0010	0
All		0.17	0.055(0.04)*	285	11	0.0020	0.0032	0.029(0.02)*

The $s2e$ amplicon was 555 bp.
Sample size for $Hb8\Delta$ was 380.
*Average F_{ST} (standard deviation). None of these pairwise F_{ST} are significant after Bonferroni correction.

individuals. Again, no unique SNPs are found on the $Hb8\Delta$ haplotype. Thus, nothing in in this broader N-American sample suggests complementary mutations in $s2e$. Curiously however, there is a significant reduction in $s2e$ polymorphism with latitude (p = 0.02 for π and θ). This does not explain the prevalence of $Hb8\Delta$, but suggests geography (or history) affects variation in the regulatory regions of some developmental genes.

Testing for Effects of $Hb8\Delta$ on Viability and Developmental Time

Test crosses and analysis of inbred lines were used to gauge the putative impact of $Hb8\Delta$ on the number of offspring hatching and developmental time. Here developmental time is assessed as the time to eclosion (see methods).

Consistently with earlier studies [63,64] hemizygosity at eve reduces viability (Table 4) by about 20% in all crosses except to eve^5 (DF vs. Cy in Figure S3). However offspring number was not affected by the deletion of Hb8 binding site (Genotype term in Table 4). Number of hatching offspring differs between the four eve mutant stocks (Table 4) most likely due to varying genetic backgrounds. We also asked about factors influencing developmental time. The ANOVA's indicate difference among eve alleles, and potential effects of hemizygosity at the locus (Table 4). Most notably, $Hb8\Delta$ seems to reduce developmental time (Table 4) – while hemizygosity at eve increases it. In three of the four crosses did $Hb8\Delta$ individuals develop significantly faster than the wt flies (Figure 3). The $Hb8\Delta$ flies eclose on average 3.5 hours earlier, but again no effects are seen in eve^5. This effect was also seen if the effect is estimated for sexes separately. In 13 of the 16 Mutation-Cross-Sex combinations $Hb8\Delta$ developed faster than flies with wt $s3+7e$, which is significant in a sign-test (binomial, p = 0.02). Note the $Hb8\Delta$ is tested in heterozygous form, thus in these crosses it appears to have dominant effects on developmental time.

We also examined the effects of $Hb8\Delta$ with association tests in 60 inbred lines. As before, $Hb8\Delta$ had no effect on offspring number. Peculiarly, the data do not confirm the association between $Hb8\Delta$ and developmental time (lower part of Table 4). The estimated developmental time is in the same range for both experiments suggesting they are not systematically different.

Together these data suggest an effect of $Hb8\Delta$ on developmental time, but further tests are needed to confirm or refute this.

Histochemical Staining of Eve Expression

Proximal phenotypes, like protein level at a specific time and location in the embryo, might be associated with functional variation in regulatory elements. To test this we stained for Eve in stage 14A embryos of four inbred lines, two $Hb8\Delta$ and two wt. Mixed model ANOVA shows that the relative positioning of the Eve stripe boundaries differs between the four inbred lines studied (Table S6). Both developmental stage and embryo orientation affect the anterior and posterior boundaries of stripes. Those sources of error were accounted for by i) working with the residuals after fitting the embryo orientation and ii) using developmental stage as a covariate. The average developmental stage does not differ between lines (p = 0.8), suggesting that rate of early development does not contribute to the line differences.

Hb repression establishes the anterior boundary of stripe 3 and posterior boundary of stripe 7 [62]. Thus, a priori, those features are most likely to be affected by $Hb8\Delta$. However, the mixed model ANOVA does not indicate effects of the $Hb8\Delta$ on these stripe 3 and 7 boundaries (Figure 4). It is possible that this Hb site has broader function. The only putative signal in the data was with stripe 5; according to least square means stripe 5 is found more anteriorly in $Hb8\Delta$. But this is not formally significant after Bonferroni correction for all 14 tests. A complementary analysis of principle components (PC) of the relative stripe positions does not implicate $Hb8\Delta$ in stripe positioning. The two largest principle components capture variation in (PC1) the central stripes and (PC2) the anterior – posterior axis of the embryo. The contribution of $Hb8\Delta$ to principle component 1 is not formally significant ($F_{1,10} = 4.25$, $p = 0.07$). These results do not suggest that $Hb8\Delta$ affects Eve pattern in the early development.

Discussion

Sequence comparisons of close and more distantly related species show how TFBS emerge, change and get lost [8,65]. Is this turnover of functional sequences due to relaxed purifying selection, or does positive selection play a role [66–68]? There is substantial

Table 4. ANOVAs testing for the effect of *Hb8Δ* (genotype) on viability and developmental time.

Exp[a]	Term/ Var.Comp	Viability			Term/ Var.Comp	Developmental time		
		df	F/ Est(SE)[b]	P		df	F/ Est(SE)[b]	P
Test Cross	Mutation	3,493	55.15	9.4E−31	Mutation	3,486	5.52	9.8E-04
	Cross	1,36	20.06	7.3E-05	Cross	1,36	0.46	5.0E-01
	M X C	3,493	6.81	1.7E-04	M X C	3,486	2.64	4.9E-02
	Genotype	1,36	0.09	0.77	Genotype	1,36	12.62	1.1E-03
	M X G	3,493	6.39	3.0E-04	M X G	3,486	1.46	0.22
	C X G	1,36	0	1.00	C X G	1,36	0.28	0.60
	Sex	1,493	1.04	0.31	Sex	1,486	6.67	0.01
	$V_{Line(CG)}$[c]		10.8(4.1)	3.9E-03	Offspring	1,486	4.53	0.03
	V_{error}[c]		80.9(5.2)	9.3E-56	$V_{Line(CG)}$[c]		25.5(15.1)	0.05
					V_{error}[c]		538.7(34.4)	1.9E-55
Inbred lines	Genotype	2,53	0.26	0.77	Genotype	2,53	1.71	0.19
	Sex	1,136	3.93	0.05	Sex	1,132	3.4	0.07
	G*S	2,136	0.26	0.77	G*S	2,132	0.73	0.48
	$V_{Line(G)}$[c]		109.2(26.4)	1.8E-05	Offspring	1,132	0.02	0.90
	$V_{Rep(L)}$[c]		27.2(8.6)	8.0E-04	$V_{Line(G)}$[c]		231.6(63.4)	1.3E-04
	V_{error}[c]		53.1(6.4)	8.2E-17	$V_{Rep(L)}$[c]		122.4(29)	1.2E-05
					V_{error}[c]		107.1(13.6)	2.0E-15

Mutation tests for differences among *eve* allele stocks, Cross the balancer vs loss-of-function *eve* allele, and genotype the *wt* vs. *Hb8Δ*.
[a]Experiment: a test cross of 20 lines with defined genotype to four *eve* mutants and genotype tests on the 60 inbred lines.
[b]For fixed terms the F-statistic is reported and for the random terms the estimated variance components (e.g. $V_{Line(C\ G)}$) with standard error.
[c]The significance of the variance components was determined by the z-function. The variance component for Developmental time was multiplied by 1000 for representation.

variation in gene expression among individuals and the bulk of expression QTLs map in *cis* [69–71]. The exact nature of those *cis* variants is rarely known, but a systematic review by Rockman and Wray [72] shows that SNPs, indels and length polymorphism in repeats can abolish TF binding and affect expression of neighboring genes.

Hunchback Site Polymorphisms are not Deleterious

Here we report that two large deletions segregating at moderate frequency remove predicted Hunchback binding sites in, and next to, the stripe 3 and 7 enhancer of *eve*. Both sites have high PWM scores and are evolutionarily conserved. One of them (Hb8) was characterized molecularly [55]. Three observations suggest that Hbs1, removed by a 45 bp deletion, is a true Hb binding site. It has among the highest PWM score of Hb sites in the *eve* region. It is evolutionarily conserved between *D. melanogaster* and *D. erecta* and resides less than 250 bp away from the Hb8 site. Stanojevic *et al.* [55] footprinted 4 Hb sites in the spacer between *s2e* and *s3+7e*, and recent thermodynamic models and quantitative measurements of TF abundances indicate that the spacer between *s2e* and *s3+7e* contains functional Hb motifs [73]. However functional assays are required to confirm that Hb binds to these two sites *in vivo* and modulates *eve* expression.

Our initial hypothesis was that these deletions of Hb binding sites are deleterious, as the loss or modulation of a single TFBS can have measurable effects [72,74,75]. This is refuted by several facts: 1) both mutations are at appreciable frequency, 2) individuals homozygous for each of those deletions survive as inbred stocks, 3) *Hb8Δ* has cosmopolitan distribution and 4) *Hb8Δ* does not seem to reduce viability and, if anything, it speeds up developmental time.

The genetic assays had sufficient statistical power to detect the effects of *eve* hemizygosity on offspring number (consistent with reported partial haplo-insufficiency at the locus [63,64]) and less so developmental time. Thus we conclude that the *Hb8Δ* is not strongly deleterious. The alternate scenarios are that the two deletions are either (nearly) neutral or favored by positive selection.

The most parsimonious explanation is that *Hb8Δ* is neutral and drifts in the population. This scenario is supported by haplotype tests, which do not point to the involvement of positive selection. However, the fact that the two deletions destroy binding sites for the same TF in the same enhancer is rather puzzling. Thus, it is tempting to hypothesize that the two Hb binding site deletions are favored by selection. Curiously, no other Hb sites in the *s3+7e* or *s2e* are affected, no substitutions are seen in more than 100 sequenced lines.

Variation in Early Development

Several studies have documented substantial variation in early Drosophila gene expression, with expression arrays [76], RNA seq [77] and *in-situs* [78]. As the deletions are found in *s3+7e*, it is most probable that they could affect Eve stripes 3 or 7. Hb is abundant in the anterior of the embryo, and drops adjacent to the anterior boundary of *eve* stripe 3. Hb is also produced in a narrower domain in the posterior, close to the posterior boundary of *eve* stripe 7 [62]. Hb demarcates the boundaries of those stripes (and stripes 4 and 6). Thus deletions of Hb sites would be expected to lead to an anterior shift of stripe 3 and posterior shift of stripe 7, because this regulatory module would be less sensitive to Hb repression (the absence of its full complement of binding sites). Our analysis of Eve expression in four inbred lines does not reveal effects of *Hb8Δ*

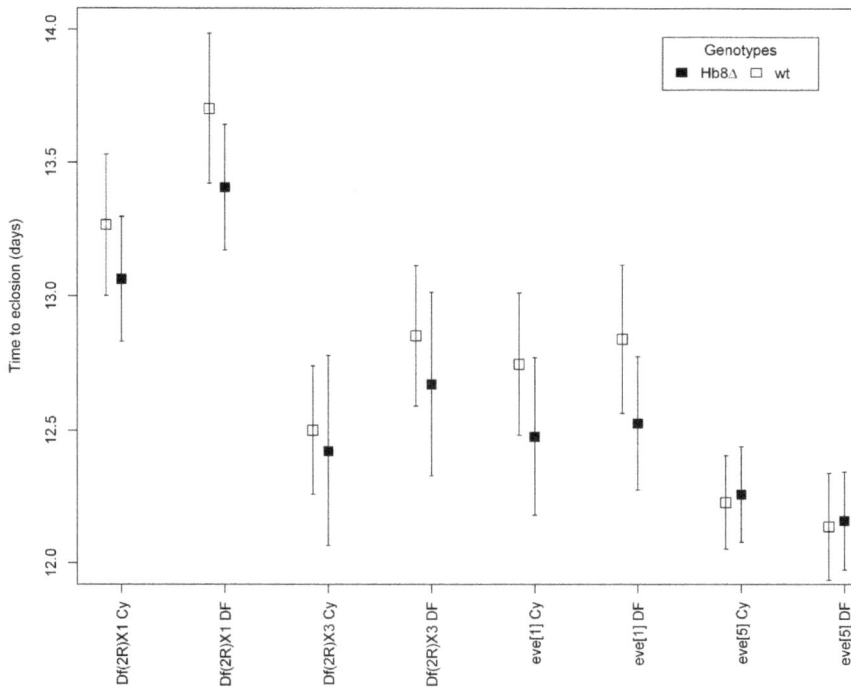

Figure 3. Effects of *Hb8Δ* on developmental time. Represented are least square mean estimates for combination of eve mutation (alleles and deficiency chromosomes), and balancer (Cy) or mutation carrying chromosome (DF). Error bars represent 95% confidence intervals. Developmental time was estimated as the time to eclosion, see methods.

on Eve stripe placement. Genetic and maternal factors affect the placement of expression boundaries; physical or environmental attributes like egg size do as well [78–80]. Note, lack of evidence does not prove the alternative. These results do not prove that the *Hb8Δ* does not affect Eve expression. The ideal test of the functionality of *Hb8Δ* and *Hbs1Δ* requires transgenic constructs in a common genetic background or homologous recombination into the *eve* locus of a particular line. It is unclear how such alterations would affect proximal or distal features of development. The quantitative tests suggest *Hb8Δ* acts dominantly, and speeds up development by ~ 3 hours. This seems unrealistic as the Eve pair rule pattern only takes ~50 minutes to mature [81], thus it is impossible that these effects (if real) are due to Hb and *eve* interaction during early development. But curiously both *eve* and *hb* also play a role in the developing neuronal system [82,83] but the functional interaction of Hb and *eve* in those tissues is largely unexplored. In the absence of functional or genetic confirmation we argue for cautious interpretation of the observed association of *Hb8Δ* and developmental time in the test-crosses. Finally, it is also possible that these deletions affect proximal developmental events, but that those effects are a minute or acceptable noise in the system.

Can Co-evolution Explain the High Frequency Hb TFBS Deletions?

Co-evolution can occur via neutral changes (e.g. in the network neighborhood [84]) or via positive selection favoring compensatory changes in the genome. Here two co-evolution models that may account for these two Hb binding site deletions in *eve* are entertained. Those are i) *cis*-changes within *eve* or, ii) *trans*-changes in the function or abundance of activators and/or repressors.

First, the relatively high frequency of those two deletions could reflect co-evolution within *eve*. Hunchback acts both as a transcriptional activator and repressor during development [85–87]. Hb positively influences expression via the *eve* stripe 2 enhancer, but is part of two-tier repressor system that demarcates the boundaries of stripes 3, 4, 6 and 7 [62]. Stripes 3 and 7 are known to be activated by D-stat [32], an ubiquitously available activator (other agents may also play a role). The high frequency of Hb binding site deletions could be a co-evolutionary response to increased activation of stripes 3 and 7 expression, for instance via altered Dstat binding. This is unlikely as the two D-stat sites in *s3+7e* have not diverged between *D. melanogaster* and *D. erecta* (Table S2) and no polymorphism is found in those sites within *D. melanogaster*. Binding sites for other agents activating *eve* stripes 3 and 7 may have changed; TFBS that could reside elsewhere in regulatory regions around *eve*. The *eve* regulatory region is 85–95% identical between *D. melanogaster* and *D. simulans*. We scanned the *eve* region of both species with Jaspar [56], and found hundreds of TFBS differing between the species (data not shown). Nonetheless, no changes in Hb or Dstat sites were found. It is also possible that miRNA docking sites or other regulatory elements in *eve* have changed, thus leading to selection for higher frequency of those two Hb site deletions.

Alternatively, changes in structure or function of *trans*-factors, like Hb itself, may have led to the increased frequency of those two Hb binding site deletions. It is improbable that a protein change is responsible, as the differences between the *D. melanogaster* and *D. simulans* Hb proteins are all on the *D. simulans* branch (unpublished results, Dagmar Yr Arnardottir and Arnar Palsson). We find it more plausible that the spatial or temporal amount of *trans*-factors has changed, for instance a lower amount of Dstat. The most intuitive scenario is, quantitative, temporal or even

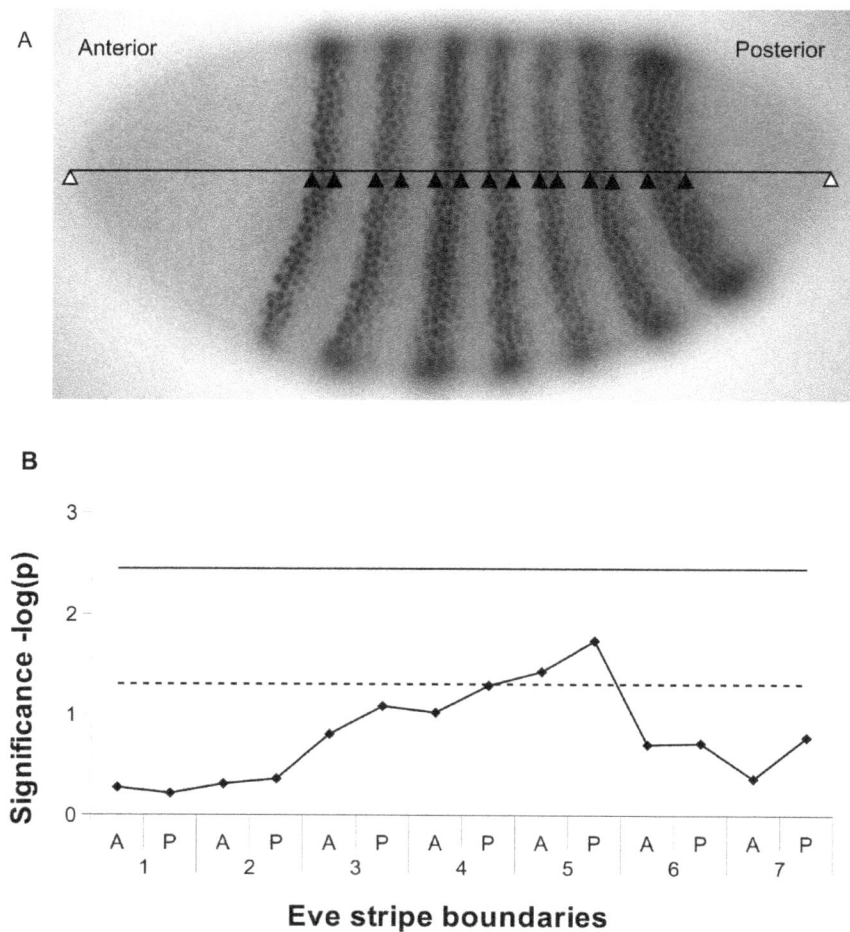

Figure 4. Testing for effects of *Hb8Δ* on Eve stripes. A) Measurement of *eve* stripe positioning. A surface image is used for measurement of stripe boundaries. A line was superimposed on the embryo and stripe boundaries visually assessed and recorded as X-Y coordinates (black triangles). Coordinates for embryo ends (white triangles) are measured from sagittal slices (not shown). B) Significance (negative log of p for genotype; *Hb8Δ* vs. *wt*) along the embryo. Shown are lines corresponding to the -log ($p = 0.05$) cutoff (dashed line) and the Bonferroni correction for 14 tests - log($p = 0.0035$) (solid line).

spatial changes in Hb expression in the embryo – which may have prompted co-evolution in regulatory elements sensitive to quantitative changes in Hb amount in development. The *eve s3+7e* might be such a critical Hb-target element. This is of course speculation, but in this scenario, one would expect that other Hb target enhancers, which produce expression overlapping the spatial and temporal patterns of *eve s3+7e* might also have experienced altered selection pressure. Thus, other Hb such target genes could also exhibit point mutations or deletions of conserved and presumably functional Hb binding sites. Note, we are not arguing positive selection is necessarily responsible; changes in Hb dose could lead to relaxation of selection for a subset of Hb target genes, and thus previously detrimental mutations in these genes could drift to higher frequency.

Conclusions

The genetic network governing early *Drosophila* development has been used to discover many of the basic principles of developmental genetics, regulatory DNA function and regulatory evolution [6,10,88,89]. Recent technical and analytical improvements have enabled quantitative analyses of enhancer function and logic

[87,88,90–92] and dosage compensation [77,93]. Developmental networks must cope with variation due to chance, the internal and external environment, and in the relevant genetic components. Studies point to the involvement of positive selection in the gain and loss of TFBS in *Drosophila* [66,94] and co-evolution within enhancers [39,95]. Furthermore, non-clocklike evolution of the *s2e* from four *Drosophila* species [11], indicates co-evolution of TF abundance and functional elements in *cis*-regulatory modules. The fact that two large deletions removing TFBS for Hb are found in close proximity in a regulatory element, might be an example of such co-evolution. However we favor the cautious explanation that these high frequency deletions reflect developmental system drift [40,41], i.e. permitted deviations in parameters of the *Drosophila* developmental regulatory network.

Supporting Information

Figure S1 Constraints on SNPs and indels in regulatory DNA. A) The relationship between single base and indel polymorphism (summarized with *θ*) in 19 enhancers and promoters in *D. melanogaster*. Many of the characterized enhancers have no indels, and sit therefore at Y = 0. B) Size and frequency of indels in characterized regulatory DNA and proximate promoters (dark circles) vs. indels in non-coding regions (open circles) around two developmental genes (*hairy* and *EGFR*).

Figure S2 Comparative genomics of predicted Hb binding sites in *eve*. The strength (height of bar) and location of Hb binding sites predicted with JASPAR in the ~8 kb region upstream of *eve* transcription start site, in four *Drosophila* species, A) *D. melanogaster*, B) *D. sechellia*, C) *D. yakuba* and D) *D. pseudoobscura*. The three characterized regulatory elements (the late element, stripes 3+7 enhancer and stripe 2 enhancer) are graphed as gray boxes in A), and the two predicted Hb sites (Hb8 to the left and Hbs1 on the right) affected by the deletions in *D. melanogaster* are indicated by black circles. Coordinates are according to a manually edited Multiz alignment of 12 *Drosophila* species.

Figure S3 Effects of *Hb8Δ* alleles on viability (above) and developmental time (below). Represented are least square mean estimates for combination of eve mutation (alleles and deficiency chromosomes), balancer (Cy) or mutation carrying chromosome (DF) and sex. Developmental time was estimated as the time to eclosion.

Table S1 Oligonucleotide primers used for PCR amplification, DNA sequencing and/or genotyping. Chimeric primers were used to PCR and sequence the *eve* locus, with a 5′ tag corresponding to the M13 universal sequencing primers (lowercase).

Table S2 Predicted Hb binding sites in the regulatory region upstream of *eve*, in 5 *Drosophila* species and the source alignments. Sheet one lists the Jaspar predicted Hb sites in *D. melanogaster* (*D.mel*), *D. simulans* (*D.sim*), *D. sechellia* (*D.sec*), *D. pseudoobscura* (*D.pse*) and *D. yakuba* (*D.yak*). Coordinates are according to a manually edited Multiz alignment of 12 *Drosophila* species. Hb8 is at 4495 and Hbs1 is at 4871. See materials and methods for details. Sheet two contains multiple alignments of the *eve* region.

Table S3 Conservation of binding sites in the eve stripe 3+7 enhancer. Transcription factor binding site numbering of sites follows Stanjovic et al 1989, Small et al 1996 and Yan et al 1996. Hb binding site 16 is on the opposite strand. Full species names and accession numbers are listed in material and methods. (*) indicate bases shared by two overlapping binding sites. (N/A) sites not identified in these species. Full species names and accession numbers are listed in material and methods. (*) indicate bases shared by two overlapping binding sites. (N/A) sites not identified

in these species. The order reflects approximately phylogenetic relationship available on http://insects. eugenes.org/species. There is length variation in T stretch between Kni5 and Hb11c; extra 1 and 2 bases in *D. sim* and *D. gri* respectively. As these are monomorphic stretches the core binding sites are presumably not affected.

Table S4 Little evolutionary conservation of a putative sloppy-paired site in the *eve* stripe 3+7 enhancer. Full species names and accession numbers are listed in material and methods. Orthology of the sloppy-paired binding site region was determined by colinearity of binding sites in the stripe 3+7 region, were Hb8 and Hb9 flank the sloppy-paired binding site. Fewer than 50 bp separated Hb8 and Hb9 in all species. The exception is *D. ananassae*, were Hb8 was not detected.

Table S5 The presence of the *Hb8Δ* in a world wide sample of populations. A deletion specific primer, annealing to regions joined by the mutation was used in a PCR on pooled DNA (100 individuals) from each of the 51 populations. Pop: Population.

Table S6 Mixed model ANOVA on *eve* stripe positioning.

Supporting information S1 Alignment of population sequencing of a part of *evenskipped* stripes 3+7 enhancer from North Carolina, in fasta format.

Supporting information S2 Alignment of population sequencing of a part of the *hunchback* regulatory region from North Carolina, in fasta format.

Acknowledgments

Thanks to Walter Eanes, Jean Claude Walser, Peter Andolfato and Ian Dworkin for flies or DNA from flies. Thank to Casey Bergman for help with for primer design. Thanks to Einar Arnason and his staff for help with sequencing. We thank two reviewers for good comments on the manuscript.

Author Contributions

Conceived and designed the experiments: AP MK MZL. Performed the experiments: AP NW SR MZL. Analyzed the data: AP. Contributed reagents/materials/analysis tools: AP MZL. Wrote the paper: AP NW MK. Fly work: AP. Molecular work: AP NW SR. Embryo work: AP MZL.

References

1. Raff RA (1996) The Shape of Life: Genes, Development, and the Evolution of Animal Form. University Of Chicago Press. Available: http://www.amazon.com / The-Shape-Life-Development-Evolution/dp/0226702669. Accessed 21 August 2013.
2. Gould S (1977) Ontogeny and Phylogeny. Boston: Belknap Press of Harvard University Press.
3. Wray GA, Hahn MW, Abouheif E, Balhoff JP, Pizer M, et al. (2003) The evolution of transcriptional regulation in eukaryotes. Molecular biology and evolution 20: 1377–1419. Available: http://www.ncbi.nlm.nih.gov/ pubmed/ 12777501. Accessed 8 August 2013.
4. Carroll SB (2008) Evo-devo and an expanding evolutionary synthesis: a genetic theory of morphological evolution. Cell 134: 25–36. Available: http://www.ncbi.nlm.nih.gov/pubmed /18614008. Accessed 26 May 2013.
5. Payankaulam S, Li LM, Arnosti DN (2010) Transcriptional repression: conserved and evolved features. Current biology?: CB 20: R764–71. Available: http://www.pubmedcentral.nih.gov / articlerender.fcgi?artid = 3033598&tool = pmcentrez&rendertype = abstract. Accessed 10 March 2013.
6. Arnosti DN (2003) Analysis and function of transcriptional regulatory elements: insights from Drosophila. Annual review of entomology 48: 579–602. Available: http://www.ncbi.nlm.nih.gov/ pubmed/12359740. Accessed 19 August 2013.
7. Dermitzakis ET, Clark AG (2002) Evolution of transcription factor binding sites in Mammalian gene regulatory regions: conservation and turnover. Molecular biology and evolution 19: 1114–1121. Available: http://www.ncbi.nlm.nih.gov/ pubmed /12082130. Accessed 19 May 2013.
8. Ludwig MZ, Patel NH, Kreitman M (1998) Functional analysis of eve stripe 2 enhancer evolution in Drosophila: rules governing conservation and change. Development (Cambridge, England) 125: 949–958. Available: http://www.ncbi. nlm.nih.gov/pubmed /9449677. Accessed 19 March 2013.
9. Ludwig MZ, Kreitman M (1995) Evolutionary dynamics of the Enhancer region of even-skipped in Drosophila. Molecular biology and evolution 12: 1002–1011.
10. Ludwig MZ (2002) Functional evolution of noncoding DNA. Current opinion in genetics & development 12: 634–639. Available: http://www.ncbi.nlm.nih.gov/ pubmed/12433575. Accessed 19 May 2013.

11. Ludwig MZ, Palsson A, Alekseeva E, Bergman CM, Nathan J, et al. (2005) Functional evolution of a cis-regulatory module. PLoS biology 3: e93. Available: http://www.pubmedcentral.nih.gov/ articlerender.fcgi?artid = 1064851&tool = pmcentrez&rendertype = abstract. Accessed 19 March 2013.

12. Crocker J, Tamori Y, Erives A (2008) Evolution acts on enhancer organization to fine-tune gradient threshold readouts. PLoS biology 6: e263. Available: http://www.pubmedcentral.nih.gov/ articlerender.fcgi?artid = 2577699&tool = pmcentrez&rendertype = abstract. Accessed 6 August 2013.

13. Carter AJR, Wagner GP (2002) Evolution of functionally conserved enhancers can be accelerated in large populations: a population-genetic model. Proceedings Biological sciences/The Royal Society 269: 953–960. Available: http://www.pubmedcentral.nih.gov/ articlerender.fcgi?artid = 1690979&tool = pmcentrez&rendertype = abstract. Accessed 12 March 2013.

14. Gerland U, Hwa T (2002) On the selection and evolution of regulatory DNA motifs. Journal of molecular evolution 55: 386–400. Available: http://www.ncbi.nlm.nih.gov/ pubmed/12355260. Accessed 11 March 2013.

15. Mustonen V, Lässig M (2005) Evolutionary population genetics of promoters: predicting binding sites and functional phylogenies. Proceedings of the National Academy of Sciences of the United States of America 102: 15936–15941. Available: http://www.pubmedcentral.nih.gov/ articlerender.fcgi?artid = 1276062&tool = pmcentrez&rendertype = abstract.

16. Sinha S, Siggia ED (2005) Sequence turnover and tandem repeats in cis-regulatory modules in drosophila. Molecular biology and evolution 22: 874–885. Available: http://www.ncbi.nlm.nih.gov/ pubmed/15659554. Accessed 21 August 2013.

17. Landry CR, Wittkopp PJ, Taubes CH, Ranz JM, Clark AG, et al. (2005) Compensatory cis-trans evolution and the dysregulation of gene expression in interspecific hybrids of Drosophila. Genetics 171: 1813–1822. Available: http://www.pubmedcentral.nih.gov/ articlerender.fcgi?artid = 1456106&tool = pmcentrez&rendertype = abstract. Accessed 21 August 2013.

18. Tanay A, Regev A, Shamir R (2005) Conservation and evolvability in regulatory networks: the evolution of ribosomal regulation in yeast. Proceedings of the National Academy of Sciences of the United States of America 102: 7203–7208. Available: http://www.pubmedcentral.nih.gov /articlerender.fcgi?artid = 1091753&tool = pmcentrez&rendertype = abstract. Accessed 8 May 2013.

19. Ihmels J, Bergmann S, Gerami-Nejad M, Yanai I, McClellan M, et al. (2005) Rewiring of the yeast transcriptional network through the evolution of motif usage. Science (New York, NY) 309: 938–940. Available: http://www.ncbi.nlm.nih.gov/pubmed/16081737. Accessed 8 March 2013.

20. Stone JR, Wray GA (2001) Rapid evolution of cis-regulatory sequences via local point mutations. Molecular biology and evolution 18: 1764–1770. Available: http://www.ncbi.nlm.nih.gov/pubmed/11504856. Accessed 21 August 2013.

21. Rockman M V, Hahn MW, Soranzo N, Zimprich F, Goldstein DB, et al. (2005) Ancient and recent positive selection transformed opioid cis-regulation in humans. PLoS biology 3: e387. Available: http://www.pubmedcentral.nih.gov/articlerender. fcgi?artid = 1283535&tool = pmcentrez&rendertype = abstract. Accessed 12 March 2013.

22. Kim J, He X, Sinha S (2009) Evolution of regulatory sequences in 12 Drosophila species. PLoS genetics 5: e1000330. Available: http://www.pubmedcentral.nih. gov/articlerender.fcgi?artid = 2607023&tool = pmcentrez&rendertype = abstract. Accessed 21 August 2013.

23. Schaeffer SW (2002) Molecular population genetics of sequence length diversity in the Adh region of Drosophila pseudoobscura. Genetical research 80: 163–175. Available: http://www.ncbi.nlm.nih.gov/pubmed/12688655. Accessed 21 August 2013.

24. Comeron JM, Kreitman M (2000) The correlation between intron length and recombination in drosophila. Dynamic equilibrium between mutational and selective forces. Genetics 156: 1175–1190. Available: http://www.pubmedcentral.nih. gov/articlerender.fcgi?artid = 1461334&tool = pmcentrez&rendertype = abstract. Accessed 21 August 2013.

25. Ometto L, Stephan W, De Lorenzo D (2005) Insertion/deletion and nucleotide polymorphism data reveal constraints in Drosophila melanogaster introns and intergenic regions. Genetics 169: 1521–1527. Available: http://www.pubmedcentral. nih.gov/articlerender.fcgi?artid = 1449560&tool = pmcentrez&rendertype = abstract. Accessed 21 August 2013.

26. Leushkin E V, Bazykin G a, Kondrashov AS (2013) Strong mutational bias toward deletions in the Drosophila melanogaster genome is compensated by selection. Available: http://www.pubmedcentral.nih.gov/articlerender. fcgi?artid = 3622295&tool = pmcentrez&rendertype = abstract. Accessed 14 August 2013.

27. Goering LM, Hunt PK, Heighington C, Busick C, Pennings PS, et al. (2009) Association of orthodenticle with natural variation for early embryonic patterning in Drosophila melanogaster. Journal of experimental zoology Part B, Molecular and developmental evolution 312: 841–854. Available: http://www.pubmedcentral. nih.gov/articlerender.fcgi?artid = 2784951&tool = pmcentrez&rendertype = abstract. Accessed 21 August 2013.

28. Romano LA, Wray GA (2003) Conservation of Endo16 expression in sea urchins despite evolutionary divergence in both cis and trans-acting components of transcriptional regulation. Development (Cambridge, England) 130: 4187–4199. Available: http://www.ncbi.nlm.nih.gov/pubmed/12874137. Accessed 19 May 2013.

29. Balhoff JP, Wray GA (2005) Evolutionary analysis of the well characterized endo16 promoter reveals substantial variation within functional sites. Proceedings of the National Academy of Sciences of the United States of America 102: 8591–8596. Available: http://www.pubmedcentral.nih.gov/articlerender. fcgi?artid = 1150811&tool = pmcentrez&rendertype = abstract. Accessed 19 May 2013.

30. Garfield D, Haygood R, Nielsen WJ, Wray GA (2012) Population genetics of cis-regulatory sequences that operate during embryonic development in the sea urchin Strongylocentrotus purpuratus. Evolution & development 14: 152–167. Available: http://www.ncbi.nlm.nih.gov/pubmed/23017024. Accessed 21 August 2013.

31. Nüsslein-Volhard C, Wieschaus E (1980) Mutations affecting segment number and polarity in Drosophila. Nature 287: 795–801. Available: http://www.ncbi. nlm.nih.gov/pubmed/6776413. Accessed 17 March 2013.

32. Yan R, Small S, Desplan C, Dearolf CR, Darnell JE (1996) Identification of a Stat gene that functions in Drosophila development. Cell 84: 421–430. Available: http://www.ncbi.nlm.nih.gov/pubmed/8608596.

33. Schroeder MD, Pearce M, Fak J, Fan H, Unnerstall U, et al. (2004) Transcriptional control in the segmentation gene network of Drosophila. PLoS biology 2: E271. Available: http://www.pubmedcentral.nih.gov/articlerender. fcgi?artid = 514885&tool = pmcentrez&rendertype = abstract. Accessed 6 August 2013.

34. Jaeger J, Blagov M, Kosman D, Kozlov KN, Manu, etal. (2004) Dynamical analysis of regulatory interactions in the gap gene system of Drosophila melanogaster. Genetics 167: 1721–1737. Available: http://www.pubmedcentral.nih.gov/articlerender. fcgi?artid = 1471003&tool = pmcentrez&rendertype = abstract. Accessed 22 August 2013.

35. Jaeger J, Surkova S, Blagov M, Janssens H, Kosman D, et al. (2004) Dynamic control of positional information in the early Drosophila embryo. Nature 430: 368–371. Available: http://www.ncbi.nlm.nih.gov/pubmed/15254541. Accessed 22 August 2013.

36. Small S, Kraut R, Hoey T, Warrior R, Levine M (1991) Transcriptional regulation of a pair-rule stripe in Drosophila. Genes & development 5: 827–839. Available: http://www.ncbi.nlm.nih.gov/pubmed/2026328. Accessed 22 August 2013.

37. Small S, Arnosti DN, Levine M (1993) Spacing ensures autonomous expression of different stripe enhancers in the even-skipped promoter. Development (Cambridge, England) 119: 762–772. Available: http://www.ncbi.nlm.nih.gov/ pubmed/8187640. Accessed 22 August 2013.

38. Hiromi Y, Kuroiwa A, Gehring WJ (1985) Control elements of the Drosophila segmentation gene fushi tarazu. Cell 43: 603–613. Available: http://www.ncbi. nlm.nih.gov/pubmed/3935327. Accessed 22 August 2013.

39. Ludwig MZ, Bergman C, Patel NH, Kreitman M (2000) Evidence for stabilizing selection in a eukaryotic enhancer element. Nature 403: 564–567. Available: http://www.ncbi.nlm.nih.gov/pubmed/10676967. Accessed 12 March 2013.

40. True JR, Haag ES (2001) Developmental system drift and flexibility in evolutionary trajectories. Evolution & development 3: 109–119. Available: http://www.ncbi.nlm.nih.gov/pubmed/11341673. Accessed 22 August 2013.

41. Gibson G (2000) Evolution: hox genes and the cellared wine principle. Current biology?: CB 10: R452–5. Available: http://www.ncbi.nlm.nih.gov/pubmed/ 10873798. Accessed 22 August 2013.

42. Palsson A, Gibson G (2004) Association between nucleotide variation in Egfr and wing shape in Drosophila melanogaster. Genetics 167: 1187–1198. Available: http://www.pubmedcentral.nih.gov/articlerender.fcgi?artid = 1470961&tool = pmcentrez&rendertype = abstract. Accessed 2 June 2013.

43. Verrelli BC, Eanes WF (2001) Clinal variation for amino acid polymorphisms at the Pgm locus in Drosophila melanogaster. Genetics 157: 1649–1663. Available: http://www.pubmedcentral.nih.gov/articlerender.fcgi?artid = 1461594&tool = pmcentrez&rendertype = abstract. Accessed 22 August 2013.

44. Walser J-C, Chen B, Feder ME (2006) Heat-shock promoters: targets for evolution by P transposable elements in Drosophila. PLoS genetics 2: e165. Available: http://www.pubmedcentral.nih.gov/articlerender. fcgi?artid = 1592238&tool = pmcentrez&rendertype = abstract. Accessed 14 August 2013.

45. Rozen S, Skaletsky H (2000) Primer3 on the WWW for general users and for biologist programmers. Methods in molecular biology (Clifton, NJ) 132: 365–386. Available: http://www.ncbi.nlm.nih.gov/pubmed/10547847. Accessed 14 August 2013.

46. Palsson A, Rouse A, Riley-Berger R, Dworkin I, Gibson G (2004) Nucleotide variation in the Egfr locus of Drosophila melanogaster. Genetics 167: 1199–1212. Available: http://www.pubmedcentral.nih.gov/articlerender.fcgi?artid = 1470963&tool = pmcentrez&rendertype = abstract. Accessed 20 August 2013.

47. Bradbury PJ, Zhang Z, Kroon DE, Casstevens TM, Ramdoss Y, et al. (2007) TASSEL: software for association mapping of complex traits in diverse samples. Bioinformatics (Oxford, England) 23: 2633–2635. Available: http://www.ncbi. nlm.nih.gov/pubmed/17586829. Accessed 23 May 2013.

48. R Development Core Team (2011) R: A language and environment for statistical computing. R Foundation for Statistical Computing. Available: http:// www.r-project.org/.

49. Rozas J, Sánchez-DelBarrio JC, Messeguer X, Rozas R (2003) DnaSP, DNA polymorphism analyses by the coalescent and other methods. Bioinformatics (Oxford, England) 19: 2496–2497. Available: http://www.ncbi.nlm.nih.gov/ pubmed/14668244. Accessed 22 August 2013.

50. Hudson RR, Bailey K, Skarecky D, Kwiatowski J, Ayala FJ (1994) EVIDENCE FOR POSITIVE SELECTION IN THE SUPEROXIDE-DISMUTASE (SOD) REGION OF DROSOPHILA-MELANOGASTER. Genetics 136: 1329–1340. Available: http://www.pubmedcentral.nih.gov/articlerender.

fcgi?artid = 1205914&tool = pmcentrez&rendertype = abstract. Accessed 21 May 2013.

51. Bray N, Dubchak I, Pachter L (2003) AVID: A global alignment program. Genome research 13: 97–102. Available: http://www.pubmedcentral.nih.gov/articlerender.fcgi?artid = 430967&tool = pmcentrez&rendertype = abstract. Accessed 15 August 2013.

52. Nicholas KB, Nicholas HB Jr, Deerfield DWI (1997) GeneDoc: Analysis and Visualization of Genetic Variation. EMBNEW NEWS. Available: Http://www.psc.edu/biomed/genedoc.

53. Halfon MS, Gallo SM, Bergman CM (2008) REDfly 2.0: an integrated database of cis-regulatory modules and transcription factor binding sites in Drosophila. Nucleic acids research 36: D594–8. Available: http://www.pubmedcentral.nih.gov/articlerender.fcgi?artid = 2238825&tool = pmcentrez&rendertype = abstract. Accessed 22 August 2013.

54. Griffith OL, Montgomery SB, Bernier B, Chu B, Kasaian K, et al. (2008) ORegAnno: an open-access community-driven resource for regulatory annotation. Nucleic acids research 36: D107–13. Available: http://www.pubmedcentral.nih.gov/articlerender.fcgi?artid = 2239002&tool = pmcentrez&rendertype = abstract. Accessed 18 August 2013.

55. Stanojević D, Hoey T, Levine M (1989) Sequence-specific DNA-binding activities of the gap proteins encoded by hunchback and Krüppel in Drosophila. Nature 341: 331–335. Available: http://www.ncbi.nlm.nih.gov/pubmed/2507923. Accessed 22 August 2013.

56. Sandelin A, Alkema W, Engström P, Wasserman WW, Lenhard B (2004) JASPAR: an open-access database for eukaryotic transcription factor binding profiles. Nucleic acids research 32: D91–4. Available: http://www.pubmedcentral.nih.gov/articlerender.fcgi?artid = 308747&tool = pmcentrez&rendertype = abstract. Accessed 21 January 2014.

57. Rasband WS (1997) Image J. Available: http://imagej.nih.gov/ij/.

58. Institute S (2001) SAS.

59. Small S, Blair a, Levine M (1996) Regulation of two pair-rule stripes by a single enhancer in the Drosophila embryo. Developmental biology 175: 314–324. Available: http://www.ncbi.nlm.nih.gov/pubmed/8626035.

60. Robin C, Lyman RF, Long AD, Langley CH, Mackay TFC (2002) hairy: A quantitative trait locus for drosophila sensory bristle number. Genetics 162: 155–164. Available: http://www.pubmedcentral.nih.gov/articlerender.fcgi?artid = 1462234&tool = pmcentrez&rendertype = abstract. Accessed 22 August 2013.

61. Clark AG, Eisen MB, Smith DR, Bergman CM, Oliver B, et al. (2007) Evolution of genes and genomes on the Drosophila phylogeny. Nature 450: 203–218. Available: http://www.ncbi.nlm.nih.gov/pubmed/17994087. Accessed 8 August 2013.

62. Andrioli LPM, Vasisht V, Theodosopoulou E, Oberstein A, Small S (2002) Anterior repression of a Drosophila stripe enhancer requires three position-specific mechanisms. Development (Cambridge, England) 129: 4931–4940. Available: http://www.ncbi.nlm.nih.gov/pubmed/12397102. Accessed 22 August 2013.

63. Fujioka M, Yusibova GL, Patel NH, Brown SJ, Jaynes JB (2002) The repressor activity of Even-skipped is highly conserved, and is sufficient to activate engrailed and to regulate both the spacing and stability of parasegment boundaries. Development (Cambridge, England) 129: 4411–4421. Available: http://www.pubmedcentral.nih.gov/articlerender.fcgi?artid = 2709299&tool = pmcentrez&rendertype = abstract. Accessed 22 August 2013.

64. Nüsslein-Volhard C, Kluding H, Jürgens G (1985) Genes affecting the segmental subdivision of the Drosophila embryo. Cold Spring Harbor symposia on quantitative biology 50: 145–154. Available: http://www.ncbi.nlm.nih.gov/pubmed/3868475. Accessed 22 August 2013.

65. Moses AM, Pollard DA, Nix DA, Iyer VN, Li X-Y, et al. (2006) Large-scale turnover of functional transcription factor binding sites in Drosophila. PLoS computational biology 2: e130. Available: http://www.pubmedcentral.nih.gov/articlerender.fcgi?artid = 1599766&tool = pmcentrez&rendertype = abstract. Accessed 4 March 2013.

66. He BZ, Holloway AK, Maerkl SJ, Kreitman M (2011) Does positive selection drive transcription factor binding site turnover? A test with Drosophila cis-regulatory modules. PLoS genetics 7: e1002053. Available: http://www.pubmedcentral.nih.gov/articlerender.fcgi?artid = 3084208&tool = pmcentrez&rendertype = abstract. Accessed 21 May 2013.

67. Stranger BE, Forrest MS, Dunning M, Ingle CE, Beazley C, et al. (2007) Relative impact of nucleotide and copy number variation on gene expression phenotypes. Science (New York, NY) 315: 848–853. Available: http://www.pubmedcentral.nih.gov/articlerender.fcgi?artid = 2665772&tool = pmcentrez&rendertype = abstract. Accessed 27 February 2013.

68. Chan YF, Marks ME, Jones FC, Villarreal G, Shapiro MD, et al. (2010) Adaptive evolution of pelvic reduction in sticklebacks by recurrent deletion of a Pitx1 enhancer. Science (New York, NY) 327: 302–305. Available: http://apps.webofknowledge.com/full_record.do?product = WOS&search_mode = GeneralSearch&qid = 10&SID = Y196NGp9aKnAOdl3mGB&page = 1&doc = 1. Accessed 30 May 2013.

69. Kirst M, Myburg AA, De León JPG, Kirst ME, Scott J, et al. (2004) Coordinated genetic regulation of growth and lignin revealed by quantitative trait locus analysis of cDNA microarray data in an interspecific backcross of eucalyptus. Plant physiology 135: 2368–2378. Available: http://www.pubmedcentral.nih.gov/articlerender.fcgi?artid = 520804&tool = pmcentrez&rendertype = abstract. Accessed 22 August 2013.

70. Brem RB, Kruglyak L (2005) The landscape of genetic complexity across 5,700 gene expression traits in yeast. Proceedings of the National Academy of Sciences of the United States of America 102: 1572–1577. Available: http://www.pubmedcentral.nih.gov/articlerender.fcgi?artid = 547855&tool = pmcentrez&rendertype = abstract. Accessed 22 August 2013.

71. Emilsson V, Thorleifsson G, Zhang B, Leonardson AS, Zink F, et al. (2008) Genetics of gene expression and its effect on disease. Nature 452: 423–428. Available: http://www.ncbi.nlm.nih.gov/pubmed/18344981. Accessed 27 February 2013.

72. Rockman M V, Wray GA (2002) Abundant raw material for cis-regulatory evolution in humans. Molecular biology and evolution 19: 1991–2004. Available: http://www.ncbi.nlm.nih.gov/pubmed/12411608. Accessed 30 April 2013.

73. Kim A-R, Martinez C, Ionides J, Ramos AF, Ludwig MZ, et al. (2013) Rearrangements of 2.5 kilobases of noncoding DNA from the Drosophila even-skipped locus define predictive rules of genomic cis-regulatory logic. PLoS genetics 9: e1003243. Available: http://dx.plos.org/10.1371/journal.pgen.1003243. Accessed 16 September 2013.

74. Arnosti DN, Barolo S, Levine M, Small S (1996) The eve stripe 2 enhancer employs multiple modes of transcriptional synergy. Development (Cambridge, England) 122: 205–214. Available: http://www.ncbi.nlm.nih.gov/pubmed/8565831.

75. Shimell MJ, Peterson AJ, Burr J, Simon JA, O'Connor MB (2000) Functional analysis of repressor binding sites in the iab-2 regulatory region of the abdominal-A homeotic gene. Developmental biology 218: 38–52. Available: http://www.ncbi.nlm.nih.gov/pubmed/10644409. Accessed 23 August 2013.

76. Kalinka AT, Varga KM, Gerrard DT, Preibisch S, Corcoran DL, et al. (2010) Gene expression divergence recapitulates the developmental hourglass model. Nature 468: 811–814. Available: http://www.ncbi.nlm.nih.gov/pubmed/21150996. Accessed 27 May 2013.

77. Lott SE, Villalta JE, Schroth GP, Luo S, Tonkin LA, et al. (2011) Noncanonical compensation of zygotic X transcription in early Drosophila melanogaster development revealed through single-embryo RNA-seq. PLoS biology 9: e1000590. Available: http://www.pubmedcentral.nih.gov/articlerender.fcgi?artid = 3035605&tool = pmcentrez&rendertype = abstract. Accessed 6 March 2013.

78. Lott SE, Kreitman M, Palsson A, Alekseeva E, Ludwig MZ (2007) Canalization of segmentation and its evolution in Drosophila. Proceedings of the National Academy of Sciences of the United States of America 104: 10926–10931. Available: http://www.pubmedcentral.nih.gov/articlerender.fcgi?artid = 1891814&tool = pmcentrez&rendertype = abstract. Accessed 13 August 2013.

79. Houchmandzadeh B, Wieschaus E, Leibler S (2002) Establishment of developmental precision and proportions in the early Drosophila embryo. Nature 415: 798–802. Available: http://www.ncbi.nlm.nih.gov/pubmed/11845210. Accessed 23 August 2013.

80. Gregor T, Bialek W, de Ruyter van Steveninck RR, Tank DW, Wieschaus EF (2005) Diffusion and scaling during early embryonic pattern formation. Proceedings of the National Academy of Sciences of the United States of America 102: 18403–18407. Available: http://www.pubmedcentral.nih.gov/articlerender.fcgi?artid = 1311912&tool = pmcentrez&rendertype = abstract. Accessed 23 August 2013.

81. Surkova S, Kosman D, Kozlov K, Manu, Myasnikova E, et al. (2008) Characterization of the Drosophila segment determination morphome. Developmental Biology 313: 844–862. Available: http://www.sciencedirect.com/science/article/pii/S0012160607014662. Accessed 16 September 2013.

82. Cleary MD, Doe CQ (2006) Regulation of neuroblast competence: multiple temporal identity factors specify distinct neuronal fates within a single early competence window. Genes & development 20: 429–434. Available: http://www.pubmedcentral.nih.gov/articlerender.fcgi?artid = 1369045&tool = pmcentrez&rendertype = abstract. Accessed 10 August 2013.

83. Doe CQ, Smouse D, Goodman CS (1988) Control of neuronal fate by the Drosophila segmentation gene even-skipped. Nature 333: 376–378. Available: http://www.ncbi.nlm.nih.gov/pubmed/3374572. Accessed 28 August 2013.

84. Wagner A (2008) Neutralism and selectionism: a network-based reconciliation. Nature reviews Genetics 9: 965–974. Available: http://www.ncbi.nlm.nih.gov/pubmed/18957969. Accessed 16 August 2013.

85. Hülskamp M, Pfeifle C, Tautz D (1990) A morphogenetic gradient of hunchback protein organizes the expression of the gap genes Krüppel and knirps in the early Drosophila embryo. Nature 346: 577–580. Available: http://www.ncbi.nlm.nih.gov/pubmed/2377231. Accessed 13 August 2013.

86. Zuo P, Stanojević D, Colgan J, Han K, Levine M, et al. (1991) Activation and repression of transcription by the gap proteins hunchback and Krüppel in cultured Drosophila cells. Genes & development 5: 254–264. Available: http://www.ncbi.nlm.nih.gov/pubmed/1671661. Accessed 29 August 2013.

87. Papatsenko D, Levine MS (2008) Dual regulation by the Hunchback gradient in the Drosophila embryo. Proceedings of the National Academy of Sciences of the United States of America 105: 2901–2906. Available: http://www.pubmedcentral.nih.gov/articlerender.fcgi?artid = 2268557&tool = pmcentrez&rendertype = abstract.

88. Arnosti DN (2011) Transcriptional repressors: shutting off gene expression at the source affects developmental dynamics. Current biology?: CB 21: R859–60. Available: http://www.ncbi.nlm.nih.gov/pubmed/22032193. Accessed 27 April 2013.

89. Stern DL, Orgogozo V (2008) The loci of evolution: how predictable is genetic evolution? Evolution; international journal of organic evolution 62: 2155–2177.

Available: http://www.pubmedcentral.nih.gov/articlerender. fcgi?artid = 2613234&tool = pmcentrez&rendertype = abstract. Accessed 9 August 2013.

90. Jaeger J, Reinitz J (2006) On the dynamic nature of positional information. BioEssays 28: 1102–1111. Available: http://www.ncbi.nlm.nih.gov/pubmed/ 17041900. Accessed 9 April 2013.

91. Wunderlich Z, DePace AH (2011) Modeling transcriptional networks in Drosophila development at multiple scales. Current opinion in genetics & development 21: 711–718. Available: http://www.ncbi.nlm.nih.gov/pubmed/ 21889888. Accessed 18 August 2013.

92. Segal E, Raveh-Sadka T, Schroeder M, Unnerstall U, Gaul U (2008) Predicting expression patterns from regulatory sequence in Drosophila segmentation. Nature 451: 535–540. Available: http://www.ncbi.nlm.nih.gov/pubmed/ 18172436. Accessed 6 August 2013.

93. Manu, Ludwig MZ, Kreitman M (2013) Sex-specific pattern formation during early Drosophila development. Genetics 194: 163–173. Available: http://www. ncbi.nlm.nih.gov/pubmed/23410834.

94. Ni X, Zhang YE, Nègre N, Chen S, Long M, et al. (2012) Adaptive evolution and the birth of CTCF binding sites in the Drosophila genome. PLoS biology 10: e1001420. Available: http://www.pubmedcentral.nih.gov/articlerender. fcgi?artid = 3491045&tool = pmcentrez&rendertype = abstract. Accessed 13 August 2013.

95. Barrière A, Gordon KL, Ruvinsky I (2012) Coevolution within and between regulatory loci can preserve promoter function despite evolutionary rate acceleration. PLoS genetics 8: e1002961. Available: http://www.pubmedcentral.nih. gov/articlerender.fcgi?artid = 3447958&tool = pmcentrez&rendertype = abstract. Accessed 21 August 2013.

20

Morphology and Efficiency of a Specialized Foraging Behavior, Sediment Sifting, in Neotropical Cichlid Fishes

Hernán López-Fernández[1,2¤a*ɔ], **Jessica Arbour**[2ɔ], **Stuart Willis**[1¤b], **Crystal Watkins**[1], **Rodney L. Honeycutt**[1¤c], **Kirk O. Winemiller**[1]

1 Program in Ecology and Evolutionary Biology, and Department of Wildlife and Fisheries Sciences, Texas A&M University, College Station, Texas, United States of America,
2 Department of Ecology and Evolutionary Biology, University of Toronto, Toronto, Ontario, Canada

Abstract

Understanding of relationships between morphology and ecological performance can help to reveal how natural selection drives biological diversification. We investigate relationships between feeding behavior, foraging performance and morphology within a diverse group of teleost fishes, and examine the extent to which associations can be explained by evolutionary relatedness. Morphological adaptation associated with sediment sifting was examined using a phylogenetic linear discriminant analysis on a set of ecomorphological traits from 27 species of Neotropical cichlids. For most sifting taxa, feeding behavior could be effectively predicted by a linear discriminant function of ecomorphology across multiple clades of sediment sifters, and this pattern could not be explained by shared evolutionary history alone. Additionally, we tested foraging efficiency in seven Neotropical cichlid species, five of which are specialized benthic feeders with differing head morphology. Efficiency was evaluated based on the degree to which invertebrate prey could be retrieved at different depths of sediment. Feeding performance was compared both with respect to feeding mode and species using a phylogenetic ANCOVA, with substrate depth as a covariate. Benthic foraging performance was constant across sediment depths in non-sifters but declined with depth in sifters. The non-sifting *Hypsophrys* used sweeping motions of the body and fins to excavate large pits to uncover prey; this tactic was more efficient for consuming deeply buried invertebrates than observed among sediment sifters. Findings indicate that similar feeding performance among sediment-sifting cichlids extracting invertebrate prey from shallow sediment layers reflects constraints associated with functional morphology and, to a lesser extent, phylogeny.

Editor: Dennis M. Higgs, University of Windsor, Canada

Funding: Funding for this project was provided by the NSF Undergraduate Mentoring in Environmental Biology Program (grant #0203992), NSF DEB grant #0516831 to KOW, RLH and HLF, and by an NSERC Discovery Grant to HLF. The funders had no role in study design, data collection and analysis, decision to publish, or preparation of the manuscript.

Competing Interests: The authors have declared that no competing interests exist.

* E-mail: hernanl@rom.on.ca

ɔ These authors contributed equally to this work.

¤a Current address: Department of Natural History, Royal Ontario Museum, Toronto, Ontario, Canada
¤b Current address: School of Biological Sciences, University of Nebraska-Lincoln, Lincoln, Nebraska, United States of America
¤c Current address: Natural Science Division, Seaver College, Pepperdine University, Malibu, California, United States of America

Introduction

Adaptive divergence of morphology and behavior has long interested biologists because it provides evidence of biological diversification in response to natural selection. In particular, food intake has an obvious and direct effect on fitness, and as a consequence, foraging behavior has received considerable attention. Modern teleost fishes are particularly good models for comparative research on foraging ecology because the mechanics of their functional morphology are relatively well understood (e.g. [1–5]). Studies of fish feeding generally focus on functional morphology and biomechanics of prey capture in the water column (e.g. [6–8]), but comparatively little attention has been given to taxa specialized for benthic invertebrate feeding [9,10].

Consumption of benthic infauna (i.e. prey buried beneath loose sediments, such as sand, silt and particulate detritus) by teleosts usually involves two steps: a) ingestion of a mouthful of sediment using a suction or scooping action, and b) separation of prey items from sediments within the oropharyngeal chamber by processes referred to as sifting [11] or winnowing [9]. The first step involves bringing sediment and buried food items into the mouth cavity. The second step, winnowing of food items from the ingested sediment, involves a series of contractions and expansions of the orobranchial chamber via adduction/abduction of the gill cover and hyoid apparatus. Such action causes cyclical hydraulic currents that move the food/sediment mix back and forth inside the orobranchial chamber. In each cycle, the pharyngeal jaws are used to rake the mix, directing food items into the esophagus and debris towards the gill openings or mouth for expulsion [9].

An ability to extract food particles buried within loose sediments is common among unrelated lineages of teleost fishes that grub or root for buried items [12]. For example, substrate grubbers (rooting with the snout within loose sediments to locate and ingest single food items) include the common carp (*Cyprinus carpio*), callichthyid and doradid catfishes of the Neotropics, and loaches (Cobitidae) of Asia. Digging and sifting (winnowing) behavior is observed among many, if not most of the diverse percomorph fishes; however, there are examples of convergent morphological and behavioral specialists that feed almost exclusively by sifting sediments using the two steps described earlier. These specialized sifters include marine mojarras (Gerreidae), goatfishes (Mullidae), surfperches (Embiotocidae), and certain gobies such as *Awaous* spp. (Gobiidae). Among the Cichlidae, sediment sifting is widespread, with specialized sifters found in African rivers (e.g., *Chromidotilapia* spp., *Tylochromis* spp., *Sargochromis codringtoni*), African lakes (e.g., *Callochromis* spp., *Grammatotria* spp. and *Xenotilapia* spp. in Lake Tanganyika; *Lethrinops* spp. and *Taeniolethrinops* spp. in Lake Malawi) and Neotropical rivers [13,14]. Among Neotropical cichlids, the South American tribe Geophagini contains two clades with independently derived specialized sediment-sifting genera. The "*Geophagus* clade" includes *Geophagus* sensu lato, *Gymnogeophagus*, *Mikrogeophagus*, and *Biotodoma*, and the "*Satanoperca* clade" includes *Acarichthys*, *Satanoperca*, and *Guianacara*. The Central American heroine genera *Thorichthys* and *Astatheros* are also independently evolved sediment sifters with similar external morphology to that of South American geophagines [13,15,16]. Morphological, behavioral and dietary convergence among sediment-sifting Neotropical cichlid clades is widespread. Both clades of sediment-sifting Geophagini and the Heroini *Astatheros* and *Thorichthys* occupy common areas of morphospace (e.g. [13,15]), share stereotypical sifting-winnowing behaviors [13,16], and have similar diets with high proportions of benthic items [13,14,17]. Among geophagine clades, convergence is also evident in oral jaw biomechanical attributes interpreted as optimized for suction feeding [18]. Additionally, most sediment-sifting taxa within Geophagini have an "epibranchial lobe", an anteroventral expansion of the first epibranchial bone (e.g. [19–21]) that has been found to be correlated with benthic and epibenthic diets [14].

Although many specialized sediment-sifting cichlids appear to have convergent head morphologies (e.g. long snouts, subterminal mouths, [12–15]), little is known about the correlation between these morphological attributes and foraging efficiency for benthic prey embedded within sediments as compared to non-sifting taxa. These convergent morphological and behavioral traits may, for example, enable sifters to dig deeper into loose sediments (e.g., longer snouts, eyes positioned high on the head) or winnow with greater efficiency (e.g., large oropharyngeal chamber volume, morphology of gill rakers used in sifting). Alternatively, morphological specialization may not affect foraging depth, but be associated with increased sediment-sifting efficiency by fine-tuning biomechanical attributes associated with winnowing [18] or improving access to shallow-buried prey. We are not aware of studies that have used experimentally manipulated foraging conditions to address foraging behavior and efficiency of sediment-sifting fishes. In this paper, we examine the link between feeding behavior, foraging performance and morphological adaptation to 1) test whether cichlid species sharing a specialized feeding behavior exhibit convergent morphology that is not simply an artifact of evolutionary relatedness (i.e., is adaptive), 2) test whether the morphology and behavior associated with substrate-sifting relates to more efficient performance in terms of foraging for benthic prey than seen in non-sifting taxa lacking these traits.

Methods

Ethics statement

This study was performed in accordance with the recommendations in the Guidelines for the Use of Fishes in Research of the American Fisheries Society. The protocol was approved by the Institutional Animal Care and Use Committee of Texas A&M

University (AUP# 2005-117). Every effort was made to minimize stress to the fishes used in feeding trials. Morphometrics analyses were performed on specimens on loan from and with permission of the ichthyology collections at the Royal Ontario Museum (ROM), Toronto, Canada, the Museo de Ciencias Naturales de Guanare (MCNG), Guanare, Venezuela, and the Museu de Ciencias da Pontificia Universidade Catolica do Rio Grande do Sul (MCP), Porto Alegre, Brazil.

Ecomorphological correlates of feeding

To compare variation in functional attributes associated with sifting and non-sifting foraging tactics, we measured eleven morphological traits of the head of 128 specimens from 27 Neotropical cichlid species including those in our feeding experiments (Table 1, and see below), using specimens requested on loan from the ichthyology collections at the Royal Ontario Museum (ROM), Toronto, Canada, the Museo de Ciencias Naturales de Guanare (MCNG), Guanare, Venezuela, and the Museu de Ciencias da Pontificia Universidade Catolica do Rio Grande do Sul (MCP), Porto Alegre, Brazil. These species included 13 sediment-sifting species, likely representing two or more origins of sediment-sifting. *Thorichthys ellioti* belongs to a genus of sediment-sifters nested well within the Central American heroines, a clade that includes piscivores, detritivores, rheophilic invertivores, algae eaters, frugivores and generalist feeders (e.g. [13,17]), while all other sifters examined are South American geophagines. Even within the tribe Geophagini, sifting may have originated more than once; all *Satanoperca* species are more closely related to the non-sifting *Crenicichla* species and to *Guianacara stergiosi* than to any other geophagine sifter, and may be separated from the most recent common ancestor of all geophagine sifters by more than 50 Ma [15,18]. The morphological dataset also included 14 non-sifting species including piscivores (ex: *Cichla temensis*), detritivores (ex: *Mesonauta egregius*), benthivores (ex: *Dicrossus filamentosus*), generalist feeders (ex: *Guianacara stergiosi* and *Amatitlania siquia*) and a filamentous algae specialist (ex: *Hypsophrys nematopus*) [14]. We measured between 2 and 5 individuals of each species, a sample size previously shown to accurately represent interspecific morphological variation in Neotropical cichlids (e.g. [13–15,22]).

In addition to recording SL (distance between the tip of the upper lip with mouth completely closed to the midpoint of the caudal peduncle where the caudal fin rays insert into the hypural plates), various head measurements were taken with vernier calipers to the nearest millimeter. Measurements of pharyngeal attributes were performed after dissection of the pharyngeal basket using an ocular micrometer attached to a dissecting stereomicroscope to the nearest tenth of a millimeter. Measurements taken are as follows (abbreviations in parentheses refer to illustrations of measurements in Fig. S1): *head length* (HL) measured from the tip of the upper lip with the mouth completely closed to the caudal edge of the operculum; *head height* (HH) as the vertical distance through the center of the eye between the dorsal and ventral edges of the head; *gape width* (GW) as the horizontal internal distance between the tips of the premaxilla with the mouth fully open and protruded; *eye position* (EP) as the vertical distance between the center of the eye and the ventral edge of the head; *eye diameter* (ED) as the longest horizontal distance between the anterior and posterior edges of the eye; *snout length* (SnL) as the distance from the center of the eye to the center of the upper lip (i.e. the symphysis of the premaxilla) with mouth closed; *ceratobranchial length* (CbL) measured as the straight distance between the joint of the basibranchial with the first ceratobranchial arch and the joint between the first ceratobranchial and the epibranchial; *ceratobran-*

Table 1. Species examined in a linear discriminant function analysis of ecomorphology of sediment-sifting and non-sifting cichlids.

Sediment-sifters	Non-sifters
Satanoperca mapiritensis (G)	Guianacara stergiosi (G)
Satanoperca daemon* (G)	Crenicichla sp. "orinoco lugubris" (G)
Geophagus' brasiliensis (G)	Crenicichla sp. "orinoco wallaci" (G)
Geophagus abalios (G)	Crenicichla sveni (G)
Geophagus dicrozoster (G)	Crenicichla geayi (G)
Geophagus brachybranchus*(G)	Dicrossus filamentosus (G)
Geophagus' steindachneri* (G)	Hoplarchus psittacus (H)
Gymnogeophagus rhabdotus (G)	Amatitlania siquia* (H)
Gymnogeophagus balzanii (G)	Hypsophrys nematopus* (H)
Biotodoma wavrini (G)	Mesonauta egregius (H)
Mikrogeophagus ramirezi (G)	Cichlasoma orinocense (Cs)
Mikrogeophagus altispinosus* (G)	Cichla temensis (Ci)
Thorichthys ellioti* (H)	Cichla orinocensis (Ci)
	Astronotus sp. (A)

Letters in parenthesis identify the Neotropical cichlid tribes included in the morphological analysis: Geophagini (G), Heroini (H), Cichlasomatini (Cs), Cichlini (Ci) and Astronotini (A). Species used in feeding experiments are highlighted with an asterisk.

C_n. Although, realistically, members of different populations or sub-species may not share equal evolutionary history, we feel this is a reasonable assumption given the evolutionary time-scales being considered in these analyses and the fact that all tested fishes are full to half siblings (see [15]). We also found that the mean of the residuals calculated using C_n were identical to the residuals calculated using C and the mean species character values and therefore C_n and C produce consistent results at least at the species level.

v = total length of tree

$c_{i,j}$ = shared evolutionary history (expected covariance) of species i and j

$$C = \begin{bmatrix} v & c_{1,2} & c_{1,3} \\ c_{2,1} & v & c_{2,3} \\ c_{3,1} & c_{3,2} & v \end{bmatrix}$$

$$C_n = \begin{bmatrix} v & v & c_{1,2} & c_{1,2} & c_{1,3} & c_{1,3} \\ v & v & c_{1,2} & c_{1,2} & c_{1,3} & c_{1,3} \\ c_{2,1} & c_{2,1} & v & v & c_{2,3} & c_{2,3} \\ c_{2,1} & c_{2,1} & v & v & c_{2,3} & c_{2,3} \\ c_{3,1} & c_{3,1} & c_{3,2} & c_{3,2} & v & v \\ c_{3,1} & c_{3,1} & c_{3,2} & c_{3,2} & v & v \end{bmatrix}$$

The residuals of the regression of each variable on SL were used in a linear discriminant analysis (LDA) comparing the sifter and non-sifter classes. Assignment of individuals of each species to each of the two classes was based on Winemiller et al. [13], Hulsey and de León [16], and López-Fernández et al. [14,15]. We used a procedure similar to that of a phylogenetic ANOVA [25] to determine whether the results of the LDA could have occurred under a random-walk, Brownian motion process or whether an adaptive process is more likely. Phylogenetic correction was based on the Neotropical cichlid maximum clade credibility (MCC) chronogram provided by López-Fernández et al. [15] after pruning it to include the species used in this study (Fig. 1). Following phylogenetic ANOVA [25], a null distribution of F values for the LDA were generated from data produced from 1000 BM simulations based on this tree; observed F-values were compared against this simulated distribution. The p-value summarizes the frequency of BM simulations that produced a higher F statistic than the observed data. To account for intraspecific variation, we sampled 24 new observations for each species based on its simulated mean value and its observed standard deviation in feeding performance ("rnorm" from R package "stats"). We also calculated how frequently the discriminant function correctly classified sifters vs. non-sifters from the BM simulated datasets and compared this to the observed results. See File S2 for phylogenetic LDA R script (function "phyl.lda").

chial gill-raker space (CbGRsp) as the average distance between gill rakers on the first ceratobranchial arch from five measurements; *epibranchial lobe length* (EBL) the longest distance between the base of the epibranchial lobe (if present) and its tip, excluding gill rakers; *lower pharyngeal jaw width* (LPJW) measured as the maximum external distance between the horns; and *lower pharyngeal jaw length* (LPJL) as the maximum distance from the imaginary midline between the caudal edge of the horns and the anterior-most tip of the plate.

With the exception of epibranchial lobe length (which was expressed as a proportion of head length to accommodate values of 0), a phylogenetically-corrected least-squares linear regression was performed to account for variation in morphological traits resulting from body size variation. All species were analyzed in a single regression of each morphological variable against SL, and the residuals of these phylogenetically-corrected regressions were used as size-corrected character values. A phylogenetically-corrected least squares regression includes a transformation by a variance-covariance matrix derived from phylogenetic branch lengths [23]. This transformation accounts for the fact that species trait values are not independent of one another as a result of shared evolutionary history, which would otherwise violate an assumption of regression analysis [23]. We used a modified version of the "phyl.resid" function from the "phytools" R package [23] to allow for multiple individuals of the sample species which is described below (see File S1 for R script, function "phyl.resid.intra"). Morphological variables, including body size (SL), were log-transformed prior to regression analysis, to account for skew associated with body size dependent traits.

We adjusted the C matrix (evolutionary variance-covariance matrix) which summarizes the shared evolutionary history between species pairs [24] such that an individual within a species shares equal evolutionary history with all other members of that species. An example of the C matrix and modified C matrix is given below for 3 species, each of which has 2 individuals in matrix

Species included in live experimental trials

Of the 27 species examined in the morphological analysis, we selected five representative species of sediment-sifting Neotropical cichlids and two non-sifting species to examine foraging efficiency. Four of these species (*Geophagus* cf. *brachybranchus*, '*Geophagus*' *steindachneri*, *Mikrogeophagus altispinosus* and *Satanoperca daemon*) belong to two potentially convergent clades within the South American tribe Geophagini [26] while the fifth species, *Thorichthys ellioti*, is part of a specialized sediment-sifting genus in the Central

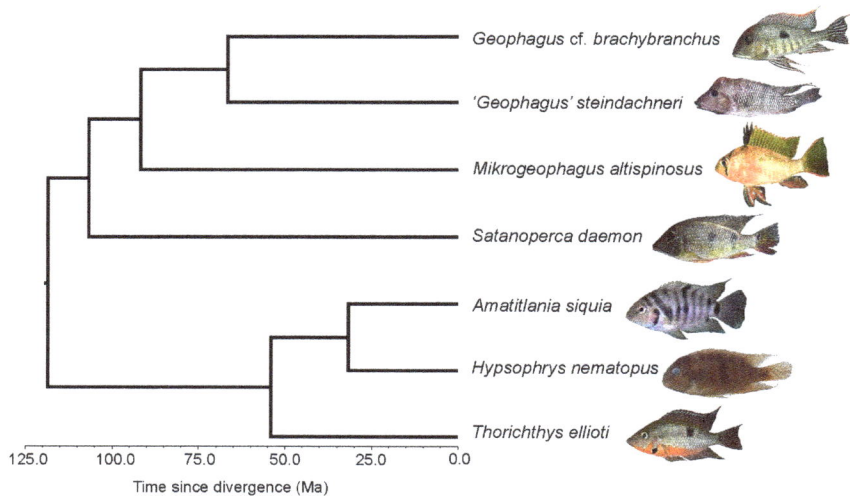

Figure 1. Species of Neotropical cichlids used in foraging experiments. A *Geophagus* cf. *brachybranchus*, B '*Geophagus*' *steindachneri*, C *Mikrogeophagus altispinosus*, D *Satanoperca daemon*, E *Thorichthys ellioti* (picture shown is of the congeneric *T.* cf *meeki*), F *Amatitlania siquia*, G *Hypsophrys nematopus*. Phylogeny and times of divergence follow López-Fernández et al. [15].

American tribe Heroini, a group that lacks an epibranchial lobe but displays morphological and dietary attributes convergent with those of sediment sifters in the tribe Geophagini [13,15,16]. All of these sediment-sifting species inhabit river and stream habitats with sand, mud, particulate organic matter or a combination of these sediments (e.g. [13,15–17]). Two species with different morphology from that of sediment-sifters, and therefore not expected to perform well when feeding on benthic invertebrates, were the Central American heroine cichlids *Amatitlania siquia* (a morphologically generalized omnivore) and *Hypsophrys nematopus*, a filamentous benthic algae specialist [13,15].

Experimental setup

We used an experimental protocol to estimate efficiency of fishes feeding on invertebrates buried beneath layers of sand at variable depths. Fish used in the experiments belonged to cohorts produced in our laboratory from parental stocks obtained from the pet-trade and raised together in the same aquarium room where experiments were performed. Water in all aquariums was prepared with de-ionized water remineralized with a salt mixture (2 parts $CaCO_3$, 2 parts $MgCl_2$, 1 part $CaCl_2$, and 1 part $MgSO_4$ by volume, for a final conductivity of <50 uS; ~1 tbsp/210 L). Commercially available frozen chironomid larvae (Hikari brand bloodworms) were used in all experiments. Chironomid larvae (Diptera) are an important dietary component of many benthivorous Neotropical cichlids [13,14]. Frozen chironomid larvae were thawed by gently rinsing them with warm tap water and then floating them in a 30% solution of sucrose. This procedure allowed for undamaged larvae to be recovered with a fine mesh net as they floated on the top of the solution, while damaged exoskeletons and other debris sank to the bottom [27]. Once recovered, whole larvae were rinsed with tap water to eliminate the sucrose, and gently blotted with a paper towel until moist but without water visible on the material. An electronic balance was used to partition larvae into 5-g portions. These portions were either used immediately or frozen for later use. To ensure that no weight was lost during freezing, thawed portions were weighed again prior to use in trials.

For each trial, weighed chironomid larvae were evenly spread across the bottom of a "20-gallon-long" (75.7 L, bottom area = 2,250 cm^2), all-glass aquarium. Clean pool-filter sand of uniform grain-diameter was either left bare (0 cm substrate depth) or carefully spread over the chironomid larvae at a uniform depth (1, 2 or 3 cm). Freshly prepared water (see above) was added to the tank without disturbing the sand (a plastic tray was temporarily placed over the sand during filling and gently removed afterwards). During trials, an airstone provided aeration. Both holding and experimental tanks were maintained in the aquarium room at a temperature between 26–28°C.

Before any data collection, we performed a series of trial experiments to determine the suitable amount of food and trial duration that would allow discrimination of performance among individuals and species. In experimental trials, fish were offered a known amount of food and allowed to forage for a fixed period of time. After each trial, the difference between the initial amount of food and the amount remaining in the experimental tank was used as an indicator of feeding efficiency (see below). By combining different initial amounts of food and different foraging periods, we determined that an initial amount of 5 g of food (approximately ~0.002 g/cm^2 or an average of ~730 individual chironomids) and a trial duration of 3 h consistently yielded measurable amounts of uneaten food. Preliminary experiments were run with aquarium-reared *Geophagus* cf. *brachybranchus* ($N = 6$, 40–65 mm standard length, SL) and *Mikrogeophagus altispinosus* ($N = 6$, 35–50 mm SL). We then performed a series of control tests without any fish ($N = 10$) to determine the mean and variance of food weight loss associated with handling and other aspects of the experimental procedure.

Feeding experiments were started within 20 min after the experimental tanks had been set up. A single fish that had not eaten for 24 h was introduced into each experimental tank and permitted to forage for 3 h, during which time the aquarium room was not disturbed. At the end of each trial, the fish was removed, measured for standard length (SL), and placed in a stock tank that identified individuals that had been tested. To avoid bias associated with individual subjects, each fish was used in a single feeding trial. After each trial, the sand and uneaten chironomid larvae were removed from the tank, and placed in a container with a 30% sucrose solution. The sand was gently stirred until all chironomid larvae had been recovered after floating to the surface of the solution (see above). Chironomid larvae were then rinsed, blotted dry and weighed as described above. Six experimental replicates at four substrate depths (no sand, 1, 2 and 3 cm) were performed for each of the seven cichlid species, so that a total of 24 individuals of each species were tested: *Geophagus* cf. *brachybranchus* (45–69 mm SL), *Mikrogeophagus altispinosus* (35–50 mm SL), '*Geophagus' steindachneri* (45–63 mm SL), *Satanoperca daemon* (53–62 mm SL), *Thorichthys ellioti* (35–66 mm SL), *Amatitlania siquia* (40–66 mm SL), and *Hypsophrys nematopus* (45–60 mm SL).

At the conclusion of the experiment, it was obvious that one species, *Hypsophrys nematopus*, used a different foraging tactic to extract chironomid larvae buried under sand. The other six cichlid species repeatedly thrust their jaws into the loose sediments to obtain sand mixed with food, and then winnowed the food from the sand within the confines of the oropharyngeal chamber. This action typically was performed at frequent intervals (every 3–10 sec) at positions throughout the aquarium. In contrast, *Hypsophrys* used its mouth as well as sweeping movements of its body and fins to excavate large pits in the sand from which it consumed exposed food items one at a time without ingesting sand. Therefore, we designed a second experiment to test the hypothesis that this foraging tactic increases feeding efficiency when buried food is patchily distributed rather than evenly dispersed. The protocol of the second experiment was the same as the first, except that the 5 g of chironomid larvae were placed on the bottom of the tank in two equal clumps before the bottom of the tank was carefully covered with a layer of sand. We tested the clumped food pattern at 0, 1, 2 and 3 cm depths of sand using a different individual *Hypsophrys* for each trial. Six replicate trials at each depth were run using individuals from the same cohort (each used only once) for each of the 4 treatments.

Measure of feeding efficiency

Feeding efficiency was quantified as the difference between the initial wet weight and the recovered wet weight of chironomid larvae consumed by each experimental fish. To account for differences in body size among individual fish, the amount of larvae consumed was standardized per unit of consumer body length (ln[SL in mm]). Statistical significance of feeding behavior (sifter, non-sifter) or species and their interaction was evaluated as predictors of feeding performance using phylogenetically corrected analysis of covariance (ANCOVA) with sand-depth as the covariate. Phylogenetic ANCOVA was performed following Garland et al. [25], which tests whether the results of an ANCOVA could have been generated under a process of Brownian motion (BM) evolution (i.e., a neutral, random walk). We used a modified version of the function "phylANOVA" (R package "phytools", [23]) and the function "ancova" (R package "HH") to carry out these analyses. The R code for these modified functions can be found in File S3 (function "phylANOVA.intra"). In the case of *Hyposphrys*, a factorial (2×4) analysis of variance based on untransformed data was used to test for statistical differences between dispersion patterns (e.g. clumped versus evenly distributed chironomids) and sand depths.

Results

Ecomorphology

Analysis of two classes of feeding behavior resulted in one discriminant function of morphology (i.e. number of classes - 1) that strongly separated specialized sifters from non-sifters. In general, sifters tended to have larger eyes placed more dorsally, wider gapes and pharyngeal jaws and deeper heads than non-sifting species. The presence of the epibranchial lobe was also characteristic of sifters in species from both clades of Geophagini, compared with non-sifting geophagines and all heroines, both of which lack the lobe (Table 2). The discriminant function of the observed data was able to accurately predict feeding behavior from the residuals of the morphological characters for 93% of individuals examined (Fig. 2). Those specimens that were identified as belonging to the wrong class (sifter or non-sifter) were either *Thorichthys ellioti* (5/5), *Biotodoma wavrini* (1/5) or *Mikrogeophagus ramirezi* (3/4), and all of these were sediment sifters misclassified as non-sifters. Linear discriminant analysis was better able to explain observed variation in morphological traits of the two classes compared to BM expectations, based on the null (simulated) distribution of F-values (p<0.001). Furthermore, the linear discriminant functions of the BM simulated datasets were equally as accurate or more accurate at identifying sifters and non-sifters in only 2.7% of simulations. Morphological convergence among sediment-sifters within Geophagini (e.g. *Satanoperca* and *Geophagus*) and at least the heroine genus *Thorichthys* is, therefore, unlikely to have arisen by chance under a BM evolutionary process.

Feeding efficiency

Phylogeny-corrected ANCOVA showed a significant difference in mean feeding performance between sediment-sifting and non-sifting taxa ($F_{1,164} = 48.46$, p<0.0001), and feeding performance varied significantly with sand depth as a covariate ($F_{1,164} = 33.38$, p<0.0001). There was a significant interaction between feeding behavior and sand depth ($F_{1,164} = 9.81$, p<0.01), with sediment sifters having a significantly more negative relationship between feeding performance and depth (Fig. 3). We observed that the difference in mean feeding performance (between sifters and non-sifters) could have occurred by chance under a Brownian motion process (p = 0.069), but both the effect of depth on feeding performance and the interaction between feeding behavior and sand depth differed significantly from that generated under a random walk, BM evolutionary process (both p = 0.001). The difference in the relationship between feeding performance and depth in sediment-sifters (decrease in performance with depth) versus non-sifters (performance roughly equal across depths) is therefore unlikely to have occurred simply as an artifact of shared evolutionary history among the taxa examined.

Species exhibited significantly different mean feeding performance under a phylogeny-corrected ANCOVA ($F_{6,154} = 17.5$, p<0.0001). Feeding performance varied significantly with sand depth ($F_{1,154} = 42.2$, p<0.0001; S2), and there was a significant interaction between species and sand depth on feeding performance ($F_{6,154} = 3.62$, p<0.01). However, the difference in mean feeding performance between species could have occurred under BM evolution (p = 0.218). While sand depth still represented a significant covariate compared to BM evolution expectations (p = 0.001), the interaction of sand depth and species on feeding performance was marginally non-significant (p = 0.067) compared to BM expectations. Therefore, changes in feeding performance with depth were more strongly associated with feeding behavior (which was significantly different from BM expectations) than with taxonomy (which was not significantly different from BM expectations).

Mean foraging efficiency declined with increasing sand depth for sediment-sifting species (Fig. 3, S2). *Amatitlania siquia* and *Hypsophrys nematopus* revealed small differences in mean foraging efficiency in relation to sand depth, with no overall trend, and the standard deviation of mean foraging efficiency increased with sand depth for *Hypsophrys* (Fig. 4). This unusual pattern for *Hypsophrys* was associated with a foraging strategy that was unique among

Table 2. Coefficients of the linear discriminant function of ecomorphology for each variable examined.

	Coefficients	Non-sifter means	Sediment-sifter means
Head length	1.34	−0.00855(0.0403)	0.000114(0.0273)
Head height	3.80	−0.0539(0.183)	0.0831(0.0381)
Gape width	5.51	−0.010465(0.0863)	0.0165(0.0847)
Eye position	−5.18	−0.0488(0.171)	0.0739(0.0465)
Eye diameter	7.27	−0.0283(0.0794)	0.0391(0.0385)
Snout length	0.358	−0.0265(0.0801)	0.0576(0.0455)
Ceratobranchial length	−8.77	−0.0234(0.0460)	0.0103(0.0509)
Ceratobranchial inter gill raker spacing	0.313	0.0193(0.0566)	−0.0482(0.117)
Epibranchial lobe length	8.21	0(0)	0.433(0.166)
Lower pharyngeal jaw width	−7.20	0.0296(0.0890)	−0.0373(0.0476)
Lower pharyngeal jaw length	3.88	−0.0151(0.0572)	0.00109(0.0462)

Mean values for each variable for sediment sifters and non-sifters (standard deviations in parentheses).

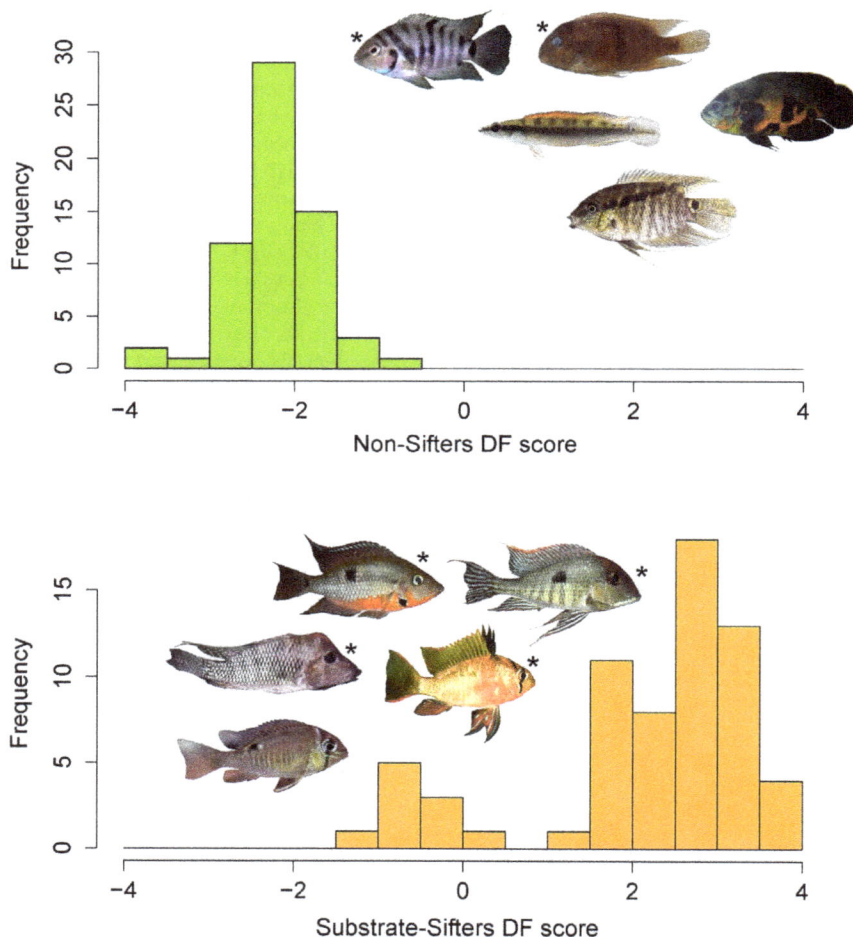

Figure 2. Linear discriminant function analysis (LDA) of morphological attributes in 27 species of Neotropical cichlids. LDA produced an axis of variation that effectively separated non-sifters (top panel) from specialized sediment-sifting species (bottom panel) by their morphological attributes. Among sediment-sifters, the model distribution to the left represents individuals "misclassified" by the LDA analysis as non-sifters, including *Thorichthys ellioti* (5/5), *Mikrogeophagus ramirezi* (4/4) and *Biotodoma wavrini* (1/5). Images marked with an "*" depict genera used in feeding efficiency experiments.

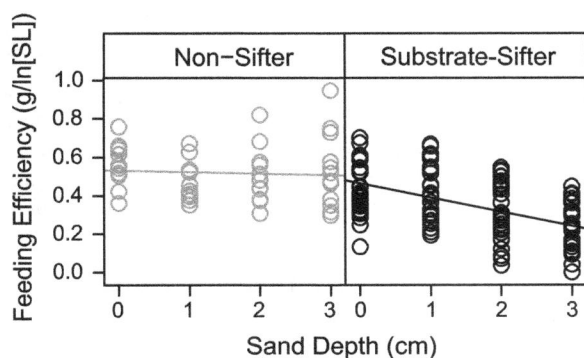

Figure 3. Mean consumption of chironomid larvae buried at 0, 1, 2, or 3 cm depth by non-sifting (2 species) and sediment-sifting (5) Neotropical cichlids. Consumption by each species is illustrated in S2.

species tested. *Hypsophrys* dug large pits using its mouth to move sand by grasping or suctioning, followed by ejection of the particles; and also by performing sweeping movements with its body and fins to excavate large pits in the sand. Chironomid larvae exposed within a pit were consumed individually without any obvious ingestion of sand. A second experiment tested the effect of clumped versus even-dispersed chironomid larvae on the foraging efficiency of *Hypsophrys*: there was a significant effect of dispersion pattern ($F_{1,40} = 4.89$, $p < 0.033$) and sand depth ($F_{3,40} = 3.07$, $p < 0.039$), and their interaction was non-significant ($F_{3,40} = 1.61$, $p < 0.202$). The mean foraging efficiency was nearly the same for clumped and evenly dispersed chironomid larvae at 0 and 1 cm sand depth, but mean foraging efficiency was greater for evenly-dispersed food items at 2 and 3 cm (Fig. 4, circles).

Discussion

Based on a linear discriminant analysis of 27 species of Neotropical cichlids, morphological convergence among sifters (vs. non-sifters) was greater than expected by chance under a

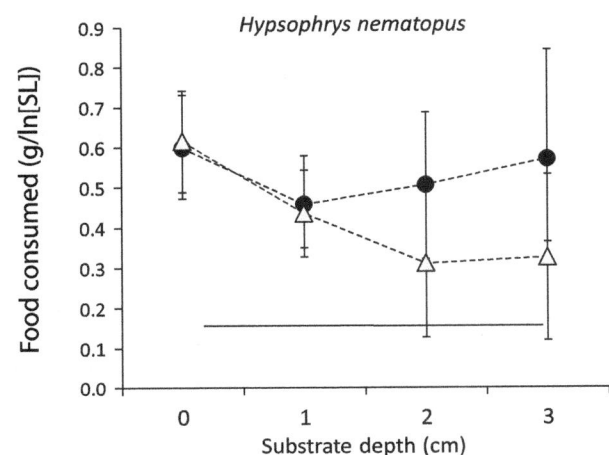

Figure 4. Mean consumption of chironomid larvae buried at 0, 1, 2, or 3 cm depth by the heroine Neotropical cichlid *Hypsophrys nematopus*. Even (filled circles) versus clumped (empty triangles) distributions. The horizontal line indicates weight loss of chironomid larvae in control tanks.

Brownian motion evolutionary process (we obtained a higher correct classification frequency in only 2.7% of BM simulations). The discriminant function was not 100% effective at predicting feeding mode from ecomorphology, which may relate to an inherent property of morphological and mechanical diversity. The principle of "many-to-one mapping" of form and function allows taxa to converge in functional output with different morphological adaptations, and can weaken the relationship between morphological adaptation and ecological performance [28–31]. It is possible that because many functional systems are incorporated into sediment sifting (ex: suction feeding ability, hyoid depression, pharyngeal jaw movement, oral jaw protrusion), this principle may have resulted in functionally equivalent morphological variation with respect to sediment-sifting performance. Under "many-to-one-mapping", ancestral trait values form the starting point for potentially differing morphological evolutionary trajectories that nevertheless result in functionally equivalent endpoints [29,30,32]. *Thorichthys ellioti* specimens may be functionally but not morphologically convergent with geophagine substrate sifters simply as a result of different evolutionary starting points (Fig. 2). More functionally informative traits (e.g. lever biomechanics, jaw protrusion) may have a greater potential to demonstrate morphological convergence among feeding strategies in future studies. Given the possibility of functional redundancy, the strength of convergence observed in ecomorphological traits among sifters versus non-sifters was somewhat surprising, and supports a role for adaptive constraint on morphological diversification associated with this specialized feeding behavior in Neotropical cichlids.

The feeding efficiency experiments did not reveal a foraging advantage for sediment-sifting Neotropical cichlids when feeding on small benthic invertebrates buried in increasingly deep sand. Instead, the specialized sediment-sifting geophagines and the heroin *Thorichthys* all showed a sharp decline in their ability to capture buried prey as substrate depth increased (Figs. 3, S2). *Hypsophrys nematopus*, a Central American heroine with a relatively small head and small compact jaws ill-suited for scooping and sifting sediments, displayed high foraging efficiency at all sand depths. *Amatitlania siquia*, another Central American heroine, has a generalized cichlid morphology and also revealed relatively high foraging efficiency (S2). By revealing that specialized sifters are not more proficient in extracting chironomid larvae buried in sand, our findings suggest that the distinct morphological attributes of sediment-sifting cichlids do not provide an advantage for digging deeper into loose sediments in search for prey. Rather, the negative relationship between feeding performance and sand depth suggests that sediment-sifting taxa forage most efficiently for prey embedded in sediments at shallow depths.

Sediment-sifting among cichlids is a specialized behavior that apparently evolved independently among phylogenetically disparate taxa possessing similar but not necessarily identical morphological traits [14,15,18]. Sifting allows fish to process large amounts of sediment efficiently. Our results indicate that deeply buried food items are less accessible for the sifters we tested. Interestingly, this specialized morphology and behavior for sorting food from sediments does not appear to result in a high degree of dietary specialization, at least with regards to prey types. López-Fernández et al. [14] reported that sediment-sifting geophagine cichlids feed on diverse benthic/epibenthic invertebrates and detritus. Bastos et al. [33] found that gastropods and vascular plant fragments were the most common items among stomach contents of 'Geophagus' brasiliensis, and Winemiller et al. [13] found the diet of *Geophagus* spp. to be composed predominantly of insects, seeds/fruit and detritus. All these resources are available to fishes at the interface between the water column and the shallow horizons of sandy or muddy substrates. Sediment sifting could, nonetheless, facilitate resource partitioning in terms of differential efficiencies for sediment types within different habitats and microhabitats (and see [17,34,35]).

Among Neotropical cichlids, sediment sifters have comparatively high species richness. Typical lowland South American communities can include a large number of coexisting sediment-sifting taxa. For example, communities in the Cinaruco River (Orinoco Basin) and Casiquiare River (Amazon Basin) in Venezuela harbor coexisting species of *Geophagus*, *Satanoperca*, *Biotodoma*, *Apistogramma* and *Biotoecus* [34,36,37], all of which have large components of benthic or epibenthic invertebrates in their diet [14]. Although not as diverse as South American Geophagini, Central American Heroini contains several sediment-sifting species. Soria-Barreto and Rodíles-Hernández [35] reported two species of sediment-sifting *Thorichthys* syntopic within the Usumacinta River Basin in Mexico. In the Bladen River of Belize, *Thorichthys meeki* coexists with sediment-sifting *Astatheros robertsoni* [17]. In natural habitats, cichlids forage on a variety of substrates, including sand, silt, and fine and coarse particulate organic matter. The ability to thrust the jaws deep into the substrate may not be as important as being able to separate small prey from sediments of different types and sizes. In most habitats, meiofauna density is probably greatest at shallow substrate depths, and selection favoring deeper thrusts may not be strong for benthivorous cichlids. Dietary segregation among sifters could be facilitated by interspecific differences in biomechanical attributes [18]. For example, species with relatively small mouths and short snouts, such as *Biotodoma wavrini* and *Mikrogeophagus ramirezi*, may be better able to pick and then sift benthic invertebrates from the surface of sediments. Species with larger gapes, such as *Satanoperca* spp., *Geophagus* spp. and *Retroculus lapidifer* [14], may be more efficient winnowers of invertebrates embedded within sediments. We did not examine the role of prey size, and the chironomid larvae used in our experiments may have been too large to reveal a foraging advantage for *Thorichthys* and the geophagine species. To test the hypothesis that geophagines and morphologically and behaviorally convergent heroine cichlids, such as *Thorichthys* and *Astatheros* species, are more efficient foragers for tiny invertebrate components of the infauna, future experiments should manipulate prey size and sediment particle size.

An unexpected finding from our experiments was the divergent foraging tactic displayed by *Hypsophrys nematopus*. This species was included in the study because it has a morphology that is poorly suited for effective scooping and sifting of sediments, and as a result, was expected to provide a sort of null case for comparison with sediment-sifting species. However, *Hypsophrys* was able to consume buried chironomid larvae efficiently by moving large amounts of sand with the mouth as well as by sweeping motions of the body and fins. Species of the genus *Hypsophrys* inhabit streams and rivers with moderate to fast flow velocities, and excavate holes for nesting and brood guarding (Coleman, 1999). While not as specialized for excavation of holes for nesting, *Amatitlania siquia*, the popular convict cichlid of the aquarium hobby, is well known for its habit of moving large amounts of loose sediment to construct nests. Thus, these Central American species appear to have behavioral repertoires adaptive for nesting as well as locating invertebrate prey buried in sand or other loose sediments.

Functional morphology of feeding in cichlids and other fishes has been studied extensively, but most investigations have focused on use of the oral jaws to capture and manipulate elusive prey. It should be recognized that a significant portion of the family Cichlidae, as well as the global diversity of fishes, consists of species that sift food items from sediments via winnowing within the orobranchial chamber. Our experiments revealed aspects of morphology that may influence feeding efficiency among sediment-sifting cichlids and may influence feeding efficiency in other substrate sifting fishes. Our results show a direct impact of feeding behavior specialization on ecological performance and a corresponding convergence in morphological traits, both of which could not be explained by random-walk evolutionary processes. These results also included some unexpected correlations between morphology and feeding that further illustrate the complexity of relationships between morphology, behavior, and ecology. Given the commonness of sediment sifting within the Cichlidae, further research that integrates functional morphology and ecological performance for this foraging mode should enhance our understanding of evolutionary diversification in this hyperdiverse fish family. Studies in cichlids may also contribute to understand one of the most widespread behaviors in teleost fishes.

Supporting Information

Figure S1   Illustration of morphometric measurements used in this paper. A. Body and head measurements. B. Lower pharyngeal jaw measurements. C. First gill arch measurements. All measurements as linear distances between points. Abbreviations follow those given in "Methods" section. See text for descriptions.

Figure S2   Mean consumption of chironomid larvae buried at 0, 1, 2, or 3 cm depth by seven species of Neotropical cichlid fishes.

File S1   R code for function "phyl.resid.intra" which carries out a phylogenetic size correction for more than one individual per species/tip. See methods and supplementary file for details.

File S2   R code for function "phyl.lda" which carries a linear discriminant analysis and compares the results to a set of Brownian Motion simulated values. The function can include more than one individual per species/tip. See methods and supplementary file for details.

File S3   R code for function "phylANOVA.intra" which performs and ANOVA/ANCOVA analysis by comparing the results to a set of Brownian Motion simulated values. The function can include more than one individual per species/tip. See methods and supplementary file for details.

Acknowledgments

We thank Donald C. Taphorn (MCNG) and Roberto E. Reis (MCP) for the loan of specimens in their care. Valuable assistance with laboratory trials and fish husbandry was provided by Cassidy Crane, Alexandra McEvoy, and Lisa Berryhill.

Author Contributions

Conceived and designed the experiments: HLF SCW KOW. Performed the experiments: HLF SCW CW KOW. Analyzed the data: HLF JHA KOW SCW. Contributed reagents/materials/analysis tools: HLF JHA KOW RLH. Wrote the paper: HLF JHA SCW KOW RLH. Designed R code used in analyses: JHA.

References

1. Liem K (1980) Adaptive significance of intra- and interspecific differences in the feeding repertoires of cichlid fishes. American Zoologist 20: 295–314.
2. Lauder GV (1982) Patterns of evolution in the feeding mechanism of Actynopterygian fishes. American Zoologist 22: 275–285.
3. Barel CDN (1983) Towards a constructional morphology of cichlid fishes (Teleostei, Perciformes). Netherlands Journal of Zoology 33: 357–424.
4. Wainwright PC (1996) Ecological explanation through functional morphology: the feeding biology of sunfishes. Ecology 77: 1336–1343.
5. Westneat MW (2004) Evolution of levers and linkages in the feeding mechanisms of fishes. Integrative and Comparative Biology 44: 378–389.
6. Waltzek T, Wainwright PC (2003) Functional morphology of extreme jaw protrusion in Neotropical cichlids. Journal of Morphology Supplement 257: 96–106.
7. Carroll AM, Wainwright PC, Huskey S, Collar DC, Turingan RG (2004) Morphology predicts suction feeding performance in centrarchid fishes. Journal of Experimental Biology 207: 3873–3881.
8. Higham TE, Hulsey CD, Rican O, Carroll AM (2007) Feeding with speed: prey capture evolution in cichilds. Journal of Evolutionary Biology 20: 70–78.
9. Drucker EG, Jensen JS (1991) Functional analysis of a specialized prey processing behavior: winnowing by surfperches (Teleostei: Embiotocidae). Journal of Morphology 210: 267–287.
10. Ferry-Graham L, Wainwright PC, Westneat MW, Bellwood DR (2002) Mechanisms of benthic prey capture in wrasses (Labridae). Marine Biology 5: 819–830.
11. Heiligenberg W (1965) A quantitative analysis of digging movements and their relationship to aggressive behaviour in cichlids. Animal Behavior XIII: 163–170.
12. Sazima I (1986) Similarities in feeding behaviour between some marine and freshwater fishes in two tropical communities. Journal of Fish Biology 29: 53–65.
13. Winemiller KO, Kelso-Winemiller LC, Brenkert AL (1995) Ecomorphological diversification and convergence in fluvial cichlid fishes. Environmental Biology of Fishes 44: 235–261.
14. López-Fernández H, Winemiller K, Montaña C, Honeycutt R (2012) Diet-morphology correlations in the radiation of South American geophagine cichlids (Perciformes: Cichlidae: Cichlinae). PLoS ONE 7: e33997.
15. López-Fernández H, Arbour JH, Winemiler KO, Honeycutt RL (2013) Testing for ancient adaptive radiations in neotropical cichlid fishes. Evolution 67: 1321–1337.
16. Hulsey C, García de León F (2005) Cichlid jaw mechanics: linking morphology to feeding specialization. Functional Ecology 19: 487–494.
17. Cochran-Biederman J, Winemiller K (2010) Relationships among habitat, ecomorphology and diets of cichlids in the Bladen River, Belize. Environmental Biology of Fishes 88: 143–152.
18. Arbour JH, López-Fernández H (2013) Ecological variation in South American geophagine cichlids arose during an early burst of adaptive morphological and functional evolution. Proceedings of the Royal Society B-Biological Sciences 280: 20130849.
19. Kullander SO (1986) Cichlid fishes of the Amazon river drainage of Peru. Stockholm: Swedish Museum of Natural History. 431 p.
20. Cichocki F (1976) Cladistic history of cichlid fishes and reproductive strategies of the American genera Acarichthys, Biotodoma and Geophagus [PhD Dissertation]. Ann Arbor: The University of Michigan. 710 p.
21. López-Fernández H, Honeycutt RL, Stiassny MLJ, Winemiller KO (2005) Morphology, molecules, and character congruence in the phylogeny of South American geophagine cichlids (Perciformes: Cichlidae). Zoologica Scripta 34: 627–651.
22. Winemiller KO (1991) Ecomorphological diversification in lowland freshwater fish assemblages from five biotic regions. Ecological Monographs 61: 343–365.
23. Revell LJ (2012) phytools: an R package for phylogenetic comparative biology (and other things). Methods in Ecology and Evolution 3: 217–223.
24. Revell LJ (2009) Size-Correction and Principal Components for Interspecific Comparative Studies. Evolution 63: 3258–3268.
25. Garland T, Dickerman A, Janis C, Jones J (1993) Phylogenetic analysis of covariance by computer simulation. Systematic Biology 42: 265–292.
26. López-Fernández H, Winemiller K, Honeycutt RL (2010) Multilocus phylogeny and rapid radiations in Neotropical cichlid fishes (Perciformes: Cichlidae: Cichlinae). Molecular Phylogenetics and Evolution 55: 1070–1086.
27. Barmuta L (1984) A method for separating benthic arthropods from detritus. Hydrobiologia 112: 105–108.
28. Hulsey C, Wainwright PC (2002) Projecting mechanics into morphospace: disparity in the feeding system of labrid fishes. Proceedings of the Royal Society of London 269: 317–326.
29. Alfaro ME, Bolnick DI, Wainwright PC (2004) Evolutionary dynamics of complex biomechanical systems: an example using the Four-Bar Mechanism. Evolution 58: 495–503.
30. Wainwright PC, Alfaro ME, Bolnick DI, Hulsey CD (2005) Many-to-one mapping of form to function: a general principle in organismal design? Integrative and Comparative Biology 45: 256–262.
31. Collar DC, Wainwright PC (2006) Discordance between morphological and mechanical diversity in the feeding mechanism of centrarchid fishes. . Evolution 60: 2575–2584.
32. Wainwright PC (2007) Functional versus morphological diversity in macroevolution. Annual Review of Ecology, Evolution, and Systematics 38: 381–401.
33. Bastos R, Condini M, Junior A, Garcia A (2011) Diet and food consumption of the pearl cichlid Geophagus brasiliensis (Teleostei: Cichlidae): relationships with gender and sexual maturity. Neotropical Ichthyology 9: 825–830.
34. Montaña CG, Winemiller KO (2010) Local-scale habitat influences morphological diversity of species assemblages of cichlid fishes in a tropical floodplain river. Ecology of Freshwater Fish 19: 216–227.
35. Soria-Barreto M, Rodiles-Hernández R (2008) Spatial distribution of cichlids in Tzendales river, biosphere reserve Montes Azules, Chiapas, Mexico. Environmental Biology of Fishes 83: 459–469.
36. Willis SC, Winemiller KO, López-Fernández H (2005) Habitat structural complexity and morphological diversity of fish assemblages in a Neotropical floodplain river. Oecologia 142: 284–295.
37. Winemiller KO, López-Fernández H, Taphorn DC, Nico LG, Duque AB (2008) Fish assemblages of the Casiquiare River, a corridor and zoogeographical filter for dispersal between the Orinoco and Amazon basins. Journal of Biogeography 35: 1551–1563.

21

Gα~o~ and Gα~q~ Regulate the Expression of *daf-7*, a TGFβ-like Gene, in *Caenorhabditis elegans*

Edith M. Myers*

Department of Biological and Allied Health Sciences, Fairleigh Dickinson University, College at Florham, Madison, New Jersey, United States of America

Abstract

Caenorhabditis elegans enter an alternate developmental stage called dauer in unfavorable conditions such as starvation, overcrowding, or high temperature. Several evolutionarily conserved signaling pathways control dauer formation. DAF-7/ TGFβ and serotonin, important ligands in these signaling pathways, affect not only dauer formation, but also the expression of one another. The heterotrimeric G proteins GOA-1 (Gα~o~) and EGL-30 (Gα~q~) mediate serotonin signaling as well as serotonin biosynthesis in *C. elegans*. It is not known whether GOA-1 or EGL-30 also affect dauer formation and/or *daf-7* expression, which are both modulated in part by serotonin. The purpose of this study is to better understand the relationship between proteins important for neuronal signaling and developmental plasticity in both *C. elegans* and humans. Using promoter-GFP transgenic worms, it was determined that both *goa-1* and *egl-30* regulate *daf-7* expression during larval development. In addition, the normal *daf-7* response to high temperature or starvation was altered in *goa-1* and *egl-30* mutants. Despite the effect of *goa-1* and *egl-30* mutations on *daf-7* expression in various environmental conditions, there was no effect of the mutations on dauer formation. This paper provides evidence that while *goa-1* and *egl-30* are important for normal *daf-7* expression, mutations in these genes are not sufficient to disrupt dauer formation.

Editor: Anne C. Hart, Brown University, United States of America

Funding: Work was supported by an National Institutes of Health (NIH) Postdoctoral NRSA to EMM (while in the Koelle lab) as well as start up funds and a Becton College Grant in Aid (Fairleigh Dickinson University). The funders had no role in study design, data collection and analysis, decision to publish, or preparation of the manuscript.

Competing Interests: The author has declared that no competing interests exist.

* E-mail: emyers@fdu.edu

Introduction

Under unfavorable environmental conditions, developing *Caenorhabditis elegans* enter an alternative stage called dauer. In dauer, growth and feeding arrest. Dauer worms also have sealed orifices and form thickened cuticles. The metabolic and morphological changes that accompany dauer increase the likelihood of the animals' survival under harsh conditions. The dauer stage is reversible, and larvae resume development when environmental conditions improve (reviewed in [1]).

Dauer formation is controlled in part by the DAF-7/TGFβ-like signaling pathway ([2] and reviewed in [1]). DAF-7 is expressed in the ASI sensory neurons and is required during larval development to inhibit dauer formation [3,4]. Environmental cues such as starvation and high temperature that trigger dauer formation also downregulate *daf-7* expression [3]. While several genes are required for normal *daf-7* expression [5–7], the signaling pathways that control *daf-7* expression and its sensitivity to environmental signals are still not well understood.

One of the genes required for both *daf-7* expression and dauer formation encodes tryptophan hydroxylase, TPH-1 [6]. TPH-1 is the rate-limiting enzyme required for serotonin biosynthesis. Serotonin signals through the heterotrimeric G proteins GOA-1 and EGL-30 to control several *C. elegans* behaviors [8–11]. GOA-1 and EGL-30 share a high degree of homology with human Gα~o~ and Gα~q~ [12]. In the human nervous system, Gα~o~ and Gα~q~ act downstream of many neurotransmitters, including serotonin. In *C.* *elegans*, GOA-1 and EGL-30 also act upstream of *tph-1* to regulate its expression [13]. It is possible then, that *goa-1* and *egl-30* are important for regulating *daf-7* expression and dauer formation, and may do so by regulating either serotonin signaling or biosynthesis. These experiments explore, in a tractable model organism, a new relationship between evolutionarily conserved pathways and proteins important for neuronal signaling and developmental plasticity.

Results and Discussion

Gα~o~ and Gα~q~ do not Affect Morphology of *daf-7*-expressing Cells

daf-7 is expressed in two head sensory neurons called the ASIs [3,4]. Structural changes in ASI cilia accompany dauer formation [14]. Two neuronal heterotrimeric G proteins, GPA-2 and GPA-3, affect dauer formation by affecting the sensory cilia of ASIs [15–17]. By altering the sensory cilia, GPA-2 and GPA-3 presumably affect the way *C. elegans* sense the environmental signals that regulate the dauer developmental switch. It is possible that additional G proteins such as GOA-1 and EGL-30 (which, unlike GPA-2 and GPA-3, have homologues in humans) could also primarily affect dauer formation or *daf-7* expression through altering the morphology of ASI. To first determine whether ASI morphology was affected by mutations in either *goa-1* or *egl-30*, gross neuronal structure was visualized using DiD. Worms were incubated in DiD, a lipophilic dye, that is only taken up by those sensory neurons making direct contact with the environment. Therefore, if ASI morphology were altered in either *goa-1* or *egl-30* mutants, the neurons would not fill with dye. ASIs in wild type, *goa-1 (n1134)* partial loss-of-function, *goa-1(sa734)* null, *egl-30 (n686* and *ad805)* partial loss-of-function (*lf*) mutants as well as *egl-30 (tg26* and *js126)* gain-of-function (*gf*) mutants filled with DiD (Figure 1 and data not shown). These worms also exhibited normal gross morphology of ASIs, suggesting that signaling through either *goa-1* or *egl-30* is not required for ASI development.

Gα~o~ and Gα~q~ are Required for *daf-7* expression but not Dauer Formation

daf-7 expression peaks during the first and second stages of larval development (L1 and L2; [3]) just before the dauer decision is made. When animals were raised in a favorable environment (with food at 20°C), all wild-type L1 larvae exhibited strong *daf-7* expression. *daf-7* expression was markedly reduced in larvae with (*lf*) alleles of *goa-1* or *egl-30* (Figure 2). In addition, *daf-7* expression was reduced in larvae with *egl-30 gf* alleles.

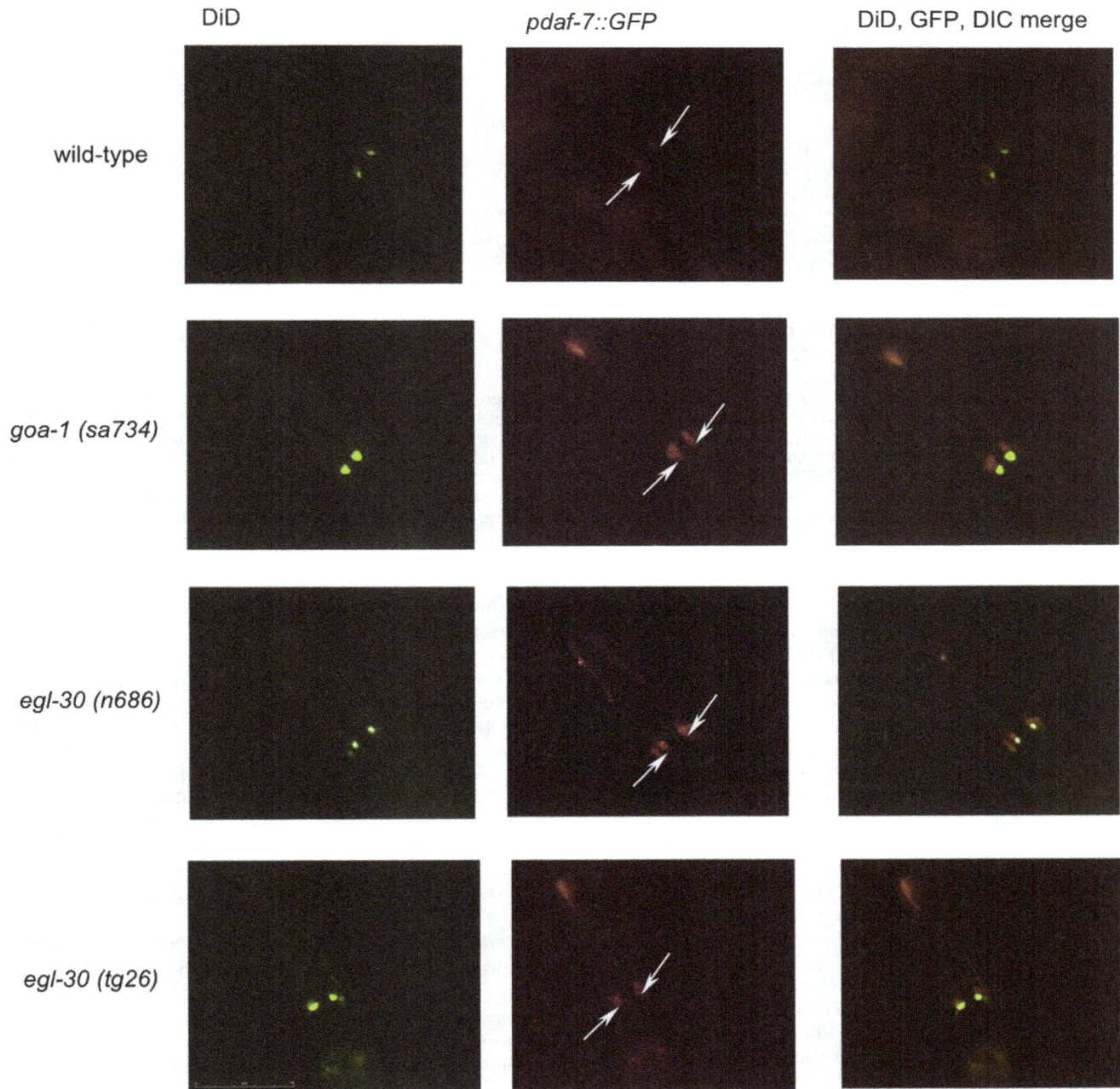

Figure 1. Mutations in *goa-1* and *egl-30* do not affect ASI development. Representative photomicrographs showing DiD filling (first column) and *pdaf-7*::GFP (second column) in wild type, *egl-30*, and *goa-1* mutant backgrounds. DiD filling was not significantly different between ASI neurons (indicated by white arrows) in wild type and mutant worms. ASI projections, visualized with *pdaf-7*::GFP, do not appear significantly altered in mutant worms. The last column shows a merge between DiD, GFP, and DIC images. *pdaf-7*::GFP strains used express GFP under the putative *daf-7* promoter, and only express GFP in the ASI neurons [5].

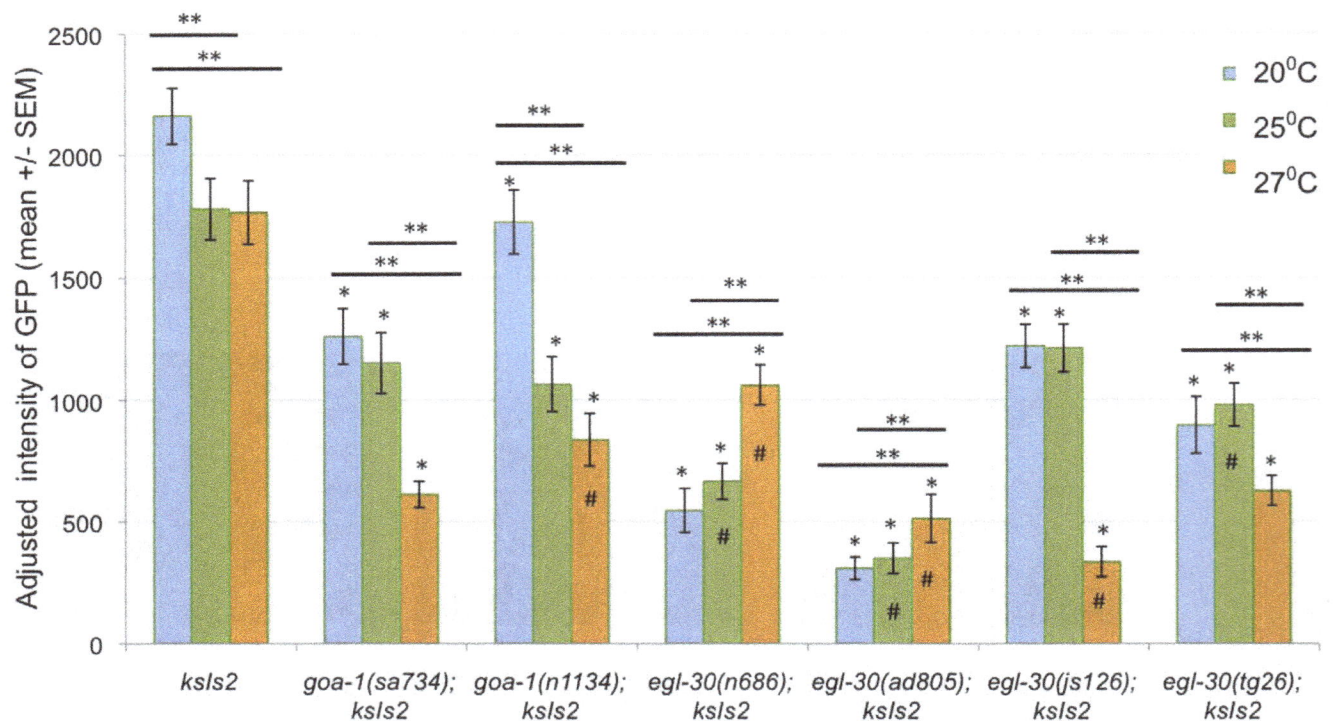

Figure 2. *goa-1* and *egl-30* regulate *daf-7* expression at multiple temperatures. *pdaf-7*::GFP levels were measured in late L1 larvae raised at 20°C, 25°C or 27°C. *goa-1* (*sa734 null* and *n1134lf*), *egl-30* (*n686lf* and *ad805lf*), and *egl-30* (*js126gf* and *tg26gf*) mutant larvae exhibited significantly less *daf-7* expression when compared to wild type larvae (*ksIs2*) raised at 20°C. *daf-7* expression was also significantly lower in all mutant larvae when compared to wild type larvae at both 25°C and 27°C. The decrease in *daf-7* in *goa-1(n1134lf)* and *egl-30(js126gf)* mutants was significantly greater than in wild type larvae raised at 27°C. *pdaf-7*::GFP expression actually increased as temperature increased to 25°C in *egl-30(n686lf)*, *egl-30(ad805lf)* and *egl-30(tg26gf)* mutants, and increased again at 27°C in *egl-30(n686lf)* and *egl-30(ad805lf)* mutants. * = significant difference from wild type (*ksIs2*) worms at the same temperature (Student's t-test, p<0.05). ** = significant difference between treatments in the same genotype (Student's t-test, p<0.05). # = significant interaction between genotype and change in temperature, as compared to wild type, *ksIs2* larvae (ANOVA, p<0.05).

These data were unexpected for several reasons. First, one would expect that loss-of-function mutations and gain-of-function mutations in the same gene might have opposite effects on their target, in this case *daf-7*. However, both the (*lf*) and (*gf*) mutations in *egl-30* decreased *daf-7* expression.

The second unexpected result was that *daf-7* expression was reduced in both *goa-1* and *egl-30* (*lf*) mutants. This was surprising because signaling through GOA-1 is thought to antagonize signaling through EGL-30. These two G proteins are thought to act antagonistically because they have opposite effects on many *C. elegans* behaviors [9,18,19]. In addition, *goa-1* and *egl-30* have opposite effects on *tph-1* expression; *goa-1* represses *tph-1* expression while *egl-30* promotes *tph-1* expression [13]. Since *tph-1* promotes *daf-7* expression [6], one would expect that *goa-1* (*lf*) mutations would cause an increase in *daf-7* expression while *egl-30* (*lf*) mutations might case a decrease in *daf-7* expression. The data suggest that *goa-1* and *egl-30* are both required to maintain *daf-7* expression, and that any perturbation to either signaling pathway results in decreased *daf-7* expression. Because *goa-1* and *egl-30* are expressed in many cells throughout the worm, it is possible that *goa-1* and *egl-30* act in distinct subsets of cells that could have opposite effects on *daf-7* expression. For instance, *goa-1* could be required to activate a set of neurons that promotes *daf-7* expression, while *egl-30* could act to inhibit the activity of a set of neurons that inhibits *daf-7* expression. In a scenario such as this, (*lf*) mutations in both *goa-1* and *egl-30* would result in decreased *daf-7* expression. The decrease in *daf-7* expression seen in *goa-1* and

egl-30 mutants at 20°C was not sufficient, however, to elicit dauer formation (Figure 3).

How precisely EGL-30 and GOA-1 regulate *daf-7* expression may be difficult to elucidate because of a complex feedback loop that exists between *daf-7* and *tph-1*. While TPH-1 upregulates *daf-7* expression [6], DAF-7 downregulates *tph-1* expression [20]. *tph-1* expression is also elevated in dauer larvae [21] when *daf-7* expression is low. It is unlikely that GOA-1 or EGL-30 act downstream of *daf-7* to regulate *tph-1* expression (and then *daf-7* expression) because *egl-30* and *goa-1* are not necessary for the increase in *tph-1* seen in dauer larvae [21]. GOA-1 and EGL-30 may instead be acting downstream or independently of *tph-1* to regulate *daf-7* expression.

Gα$_o$ and Gα$_q$ Alter Temperature-induced Changes in *daf-7* Expression but not Dauer Formation

High temperatures can induce dauer formation in some mutants that do not readily form dauers at moderately high temperatures [22,23]. It was possible that while *goa-1* and/or *egl-30* mutants did not form dauers at favorable temperatures (20°C, Figure 3), they would enter dauer at high temperatures. Moderately high (25°C) or high (27°C) temperatures were both insufficient to induce dauer formation in any of the G protein mutants tested (Figure 3). While most non-dauer worms developed into full adults, *goa-1(sa734)* null and *egl-30(tg26gf)* mutants did not. These non dauer worms appeared to arrest as larvae; either L1 or partial dauers (determined by SDS sensitivity). The larval arrest in

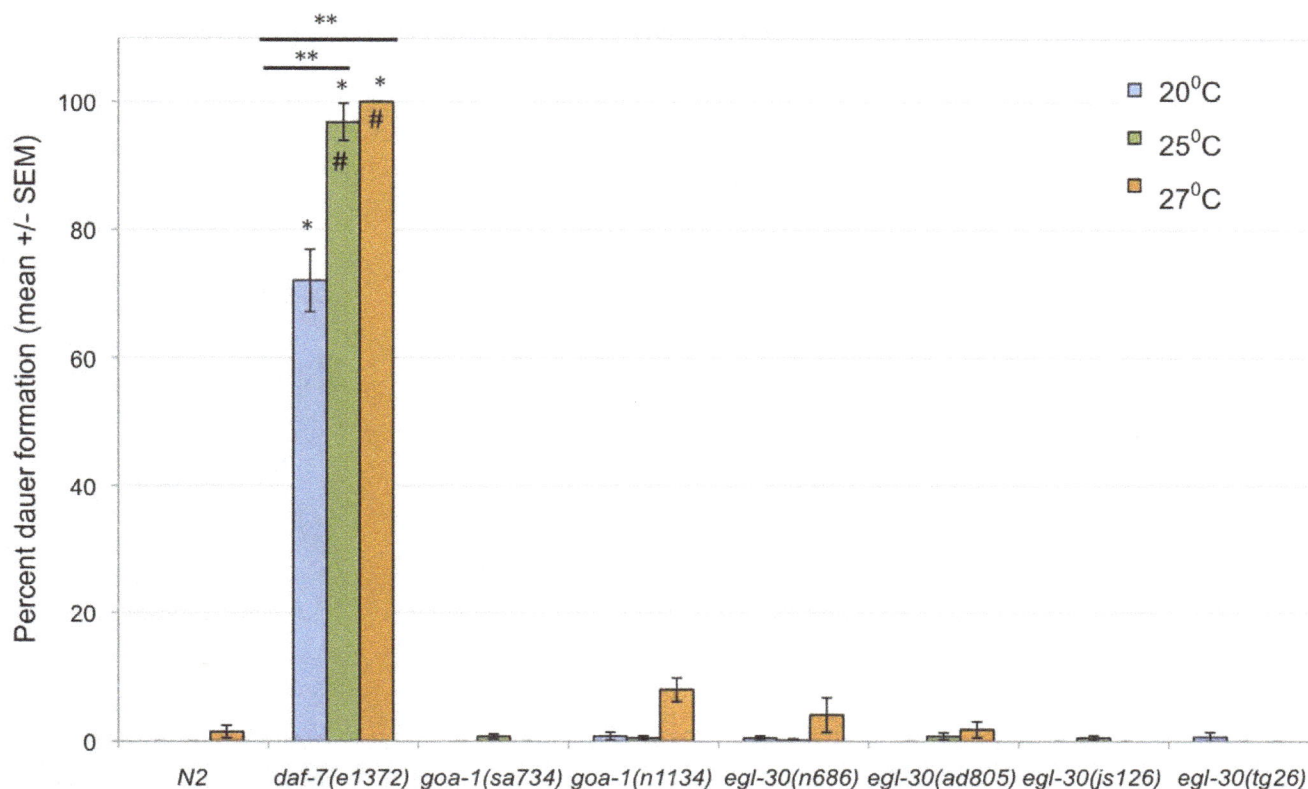

Figure 3. Neither *goa-1* nor *egl-30* inhibit dauer entry at high temperatures. Dauer formation was assayed in larvae raised at 25°C or 27°C. None of the mutant strains tested exhibited a significant increase in dauer formation at higher temperatures. It was not possible to assay dauer formation in *goa-1(sa734)* null or *egl-30 (tg26gf)* mutants at 27°C, because these mutants formed partial dauers or arrested at the L1 stage. Data for these strains at 27°C were therefore not included in the figure. *daf-7(e1372)* mutants are constitutive dauers at 25°C and 27°C, and served as a positive control. * = significant difference from wild type (N2) worms at the same temperature (Student's t-test, $p<0.05$). ** = significant difference between treatments in the same genotype (Student's t-test, $p<0.05$). # = significant interaction between genotype and change in temperature (from 20°C), as compared to wild type larvae (ANOVA, $p<0.05$).

these mutants occurred prior to the time in development when the dauer decision is made, and suggests that normal development at 27°C was disrupted by the *sa734* and *tg26* alleles.

Despite the absence of dauer formation seen at 25°C or 27°C, *daf-7* expression was significantly altered in all mutant strains tested (Figure 2). In almost all strains, there was a significant difference in the way temperature affected *daf-7* expression. These data suggest that GOA-1, and EGL-30 in particular, relay some sensory cues important for *daf-7* expression. EGL-30 appears to be important for downregulating *daf-7* in response to high temperatures. In fact, the effect of temperature on *daf-7* expression is reversed in the *egl-30(lf)* mutants and exaggerated in the *egl-30 (js126gf)* mutant. Other signaling pathways likely contribute to the behavioral/developmental response to high temperature since the decrease in *daf-7* expression in *egl-30 (gf)* mutants was not sufficient to induce dauer formation at high temperatures.

Gα$_o$ and Gα$_q$ Alter the Response to Limiting Amounts of Food

In addition to temperature, food availability affects the course of *C. elegans* development. When larvae hatch from eggs in the absence of food, their development arrests at the L1 stage (prior to the time in development when the dauer decision is made) and *daf-7* expression is reduced in the arrested L1s [3]. Signaling through GOA-1 and EGL-30 modulates food sensitivity in adult *C. elegans*

[24–26], so it is possible *goa-1* and *egl-30* are required for mediating the effect of food in larvae. As in wild-type L1s, starvation caused a significant reduction in *daf-7* in *goa-1* (*lf*) mutants (Figure 4). As expected, an equivalent reduction was seen in the *egl-30(tg26gf)* mutant and an increase was seen in the *egl-30(n686lf)* mutant. While these mutants still appeared to be in the arrested L1 stage, the data suggest that *egl-30* mediates the *daf-7* response to starvation. These results and those from studies in adult worms [25] suggest that EGL-30 plays the same role in the response to starvation throughout development.

Reduced *daf-7* expression was not seen in worms expressing the other *egl-30* (*gf*) mutant allele (*js126*), however. The difference between the *egl-30* (*gf*) phenotypes may be caused by differences in the way each mutation affects the EGL-30 protein. The *tg26* (*gf*) allele is thought to contain a mutation that alters guanine nucleotide binding [27]. The *js126* (*gf*) allele is thought to contain a mutation that alters GTPase activity [28]. Both *tg26* and *js126* mutants have (*gf*) phenotypes with respect to other EGL-30-dependent behaviors such as egg laying and movement [28,29], however it is not clear whether it is reasonable to predict that both alleles would affect all ELG-30-dependent processes in the same way.

When larvae hatch in the presence of limiting amounts of food, they progress through the L1 stage and then enter dauer [30]. If *egl-30* and *goa-1* are important for relaying the food cues important for normal development, one would expect that mutations in

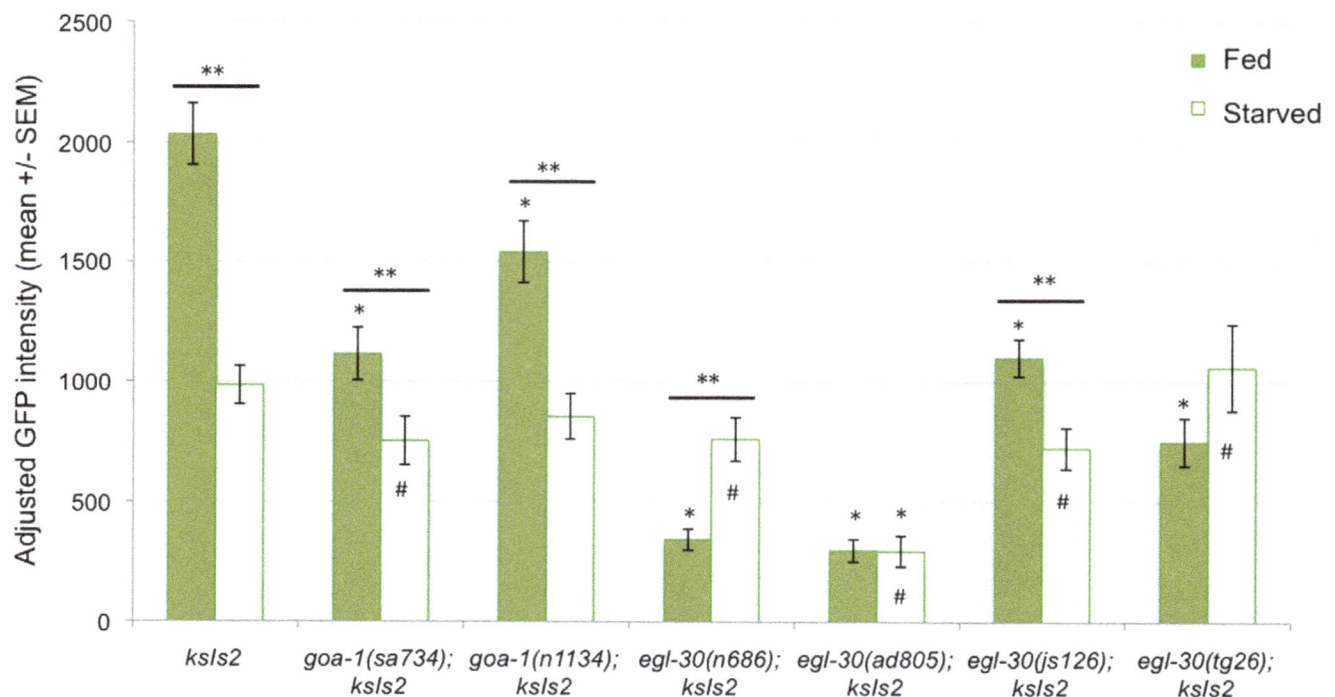

Figure 4. *goa-1* and *egl-30* regulate *daf-7* expression in response to starvation. *pdaf-7*::GFP levels were measured in well-fed late L1 larvae raised at 20°C or starved L1s. *daf-7* expression was lower with starvation in all genotypes except *egl-30(n686lf)*, *egl-30 (ad805lf)* and *egl-30(tg26gf)*, in which *pdaf-7*::GFP expression was unchanged or increased with starvation. The decrease in *daf-7* in *goa-1(sa734)* null, and *egl-30 (js126gf)* mutants was significantly greater than in wild type larvae. * = significant difference from wild type (*ksIs2*) worms at the same temperature (Student's t-test, p<0.05). ** = significant difference between treatments in the same genotype (Student's t-test, p<0.05). # = significant interaction between genotype and change in temperature, as compared to wild type, *ksIs2* larvae (ANOVA, p<0.05).

either *goa-1* or *egl-30* would disrupt dauer formation caused by low food levels. When *goa-1* and *egl-30* mutant larvae were exposed to limiting amounts of food, they did arrest at an early stage of development (Figure 5). However, based on size, most larvae appeared to be arrested at the L1 or partial dauer stages and not as full dauers. Most of the small larvae exhibited pharyngeal pumping and were sensitive to SDS, indicating that they did not arrest as full dauers [31]. These data suggest that *goa-1* and *egl-30* mutant worms are still sensitive to alterations in food availability, because they did not fail to arrest development in the presence of limiting food.

Overall, the experiments in this study showed that *goa-1* and *egl-30* regulate *daf-7* expression in early development. While *goa-1* and *egl-30* mutations significantly decreased *daf-7* expression, they did not affect dauer formation. These results suggest that other signaling pathways act in concert with GOA-1 and EGL-30 to decrease *daf-7* expression to levels sufficient to induce dauer formation.

Materials and Methods

Worm Strains

C. elegans worm strains were maintained on NGM plates with *Escherichia coli* OP50 as the food source [32]. Strains were provided by the Caenorhabditis Genetic Center (CGC) and were derived from the wild-type N2 Bristol strain. Strains used were as follows: N2, JT734 *goa-1(sa734)*, KO96 *goa-1(n1134lf)*, MT1434 *egl-30(n686sd)*, NM1380 *egl-30(js126gf)*, KY26 *egl-30(tg26gf)*, DA823 *egl-30 (ad805)*, and CB1372 *daf-7(e1372ts)*. Strains containing G

protein mutations were crossed into the FK181 *ksIs2 [pdaf-7::GFP, rol-6(su1006)]* strain for *pdaf-7*::GFP analysis.

Microscopy

For all assays, the developmental stages of larvae were carefully synchronized. For temperature assays, gravid adults laid eggs on NGM plates for 4 hours. Adults were removed and eggs were grown on NGM plates for 18–24 hours. For starvation assays, gravid adults were bleached to isolate eggs. Eggs were grown on NGM plates for 18–24 hours (fed) or in M9 medium for 48 hours (starved). Larvae were transferred to a 4% agarose pad on a microscope slide, immobilized with 10 mM levamisole, and viewed using a Leica DM5500 microscope. ASI images were captured with a fixed exposure time using a Hammamatsu Orca ER camera and Leica Microsystems Image capture software. GFP intensity was quantified using NIH Image J software version 1.44o. The intensity of GFP in each ASI cell body was quantified. The intensity of a similarly sized background selection was subtracted from the ASI GFP intensity to get the adjusted GFP intensity. Approximately ten larvae of each genotype were imaged in each experiment. Experiments were performed in triplicate, on three separate days. Dye filling was performed using 0.1 mg/ml DiD (Molecular Probes) as described [33].

Dauer Assays

Dauer assays were done similarly to those previously described [30]. Modified NGM plates were prepared without peptone, and Noble agar (Difco) was used. 3 ml of modified NGM was used in each dauer assay plate. Plates were seeded with 20 μl of 4% (w/v) OP50, unless otherwise noted. *E. coli OP50* was resuspended in S

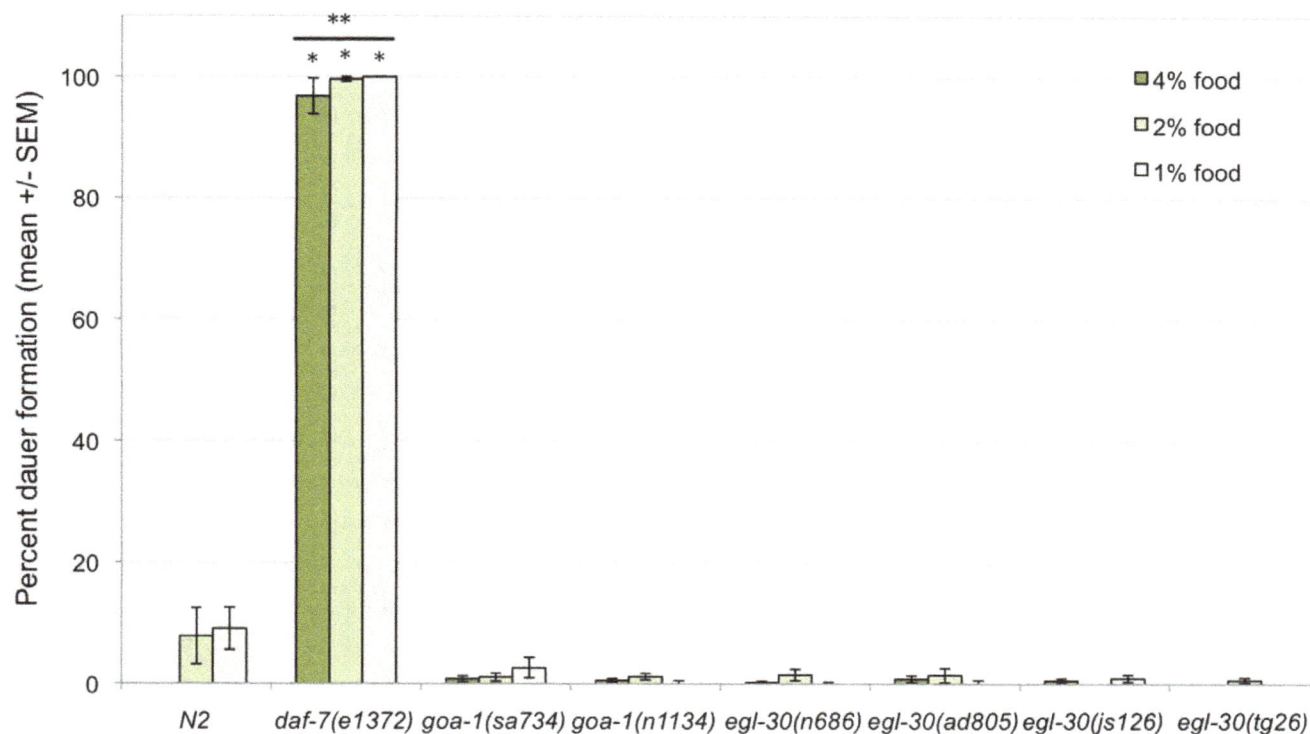

Figure 5. Neither *goa-1* nor *egl-30* are required for dauer formation in response to reduced food. Dauer formation was assayed in larvae grown at 25°C on several concentrations of *E. coli* OP50. Wild type larvae formed dauers as food concentration decreased. In at least one assay, non-dauer worms of *egl-30* (tg26gf), *egl-30* (n686lf), *egl-30* (ad805lf), or *goa-1* (n1134lf) genotype did not develop into adults when grown on 1% food. Instead, these larvae were arrested as L1s or partial dauers. Data for these strains at 1% food concentration were therefore not included in the figure. * = significant difference from wild type (N2) worms at the same food concentration (Student's t-test, $p < 0.05$). ** = significant difference between treatments in the same genotype (Student's t-test, $p < 0.05$).

Medium with 50 µg/ml streptomycin. Larvae were synchronized by first isolating eggs from bleached, gravid adults. Eggs were resuspended in S Medium. Approximately 100 eggs were pipetted onto seeded dauer plates that were incubated for 60–72 hours. Dauer worms were scored visually, and scoring was confirmed using SDS. Worms were considered dauers if they survived a several minute incubation in 1%SDS. Larvae were considered partial dauers if they were the same size and shape as dauer larvae, but exhibited pharyngeal pumping and did not survive SDS treatment.

Assays were performed with 100–200 worms for each genotype. All genotypes were tested in an assay. Assays were performed in triplicate, on three separate days.

Acknowledgments

Strains were provided by the Caenorhabditis Genetics Center, which is funded by the National Institutes of Health- National Center for Research Resources. Crosses to put G protein mutations into the *pdaf-7*::GFP background were performed while the author was a postdoctoral fellow in Michael R. Koelle's lab.

Author Contributions

Conceived and designed the experiments: EMM. Performed the experiments: EMM. Analyzed the data: EMM. Contributed reagents/materials/analysis tools: EMM. Wrote the paper: EMM.

References

1. Fielenbach N, Antebi A (2009) *C. elegans* dauer formation and the molecular basis of plasticity. Genes Dev. 22: 2149–2165.
2. Swanson MM, Riddle DL (1981) Critical periods in the development of the *Caenorhabditis elegans* dauer larvae. Dev. Biol. 84: 27–40.
3. Ren P, Lim CS, Johnsen R, Albert PS, Pilgrim D, et al. (1996) Control of *C. elegans* larval development by neuronal expression of a TGFβ homolog. Science 274: 1389–1391.
4. Schackwitz WS, Inuoe T, Thomas JH (1996) Chemosensory neurons function in parallel to mediate a pheromone response in *C. elegans*. Neuron 17: 719–728.
5. Koga M, Take-uchi M, Tameishi T, Ohshima Y (1999) Control of DAF-7 TGFβ expression and neuronal process development by a receptor tyrosine kinase KIN-8 in *Caenorhabditis elegans*. *Development* 126: 5387–98.
6. Sze JY, Victor M, Loer C, Shi Y, Ruvkun G (2000) Food and metabolic signalling defects in a *Caenorhabditis elegans* serotonin-synthesis mutant. Nature 403: 560–564.

7. Murakami M, Koga M, Ohshima Y (2001) DAF-7/TGFβ expression required for the normal larval development in *C. elegans* is controlled by a presumed guanylyl cyclase DAF-11. Mech. Dev. 109: 27–35.
8. Segalat L, Elkes DA, Kaplan JM (1995) Modulation of serotonin-controlled behaviors by Go in *C. elegans*. Science 267: 1648–1651.
9. Nurrish S, Segalat L, Kaplan JM (1999) Serotonin inhibition of synaptic transmission: Galpha(o) decreases the abundance of UNC-13 at release sites. Neuron 24: 231–242.
10. Shyn SI, Kerr R, Shafer W (2003) Serotonin and Go modulate functional states of neurons and muscles controlling *C. elegans* egg laying behavior. Curr. Biol. 13: 1910–1915.
11. Dempsey CM, Mackenzie SM, Gargus A, Blanco G, Sze JY (2005) Serotonin (5HT), fluoxetine, imipramine and dopamine target distinct 5HT receptor signaling to modulate *Caenorhabditis elegans* egg-laying behavior. Genetics 149: 1425–1436.

12. Jansen G, Thijssen KL, Werner P, van der Horst M, Hazendonk E, et al. (1999) The complete family of genes encoding G proteins in *Caenorhabditis elegans*. Nat Genet. 21: 414–419.

13. Tanis JE, Moresco JJ, Lindquist RA, Koelle MR (2008) Regulation of serotonin biosynthesis by the G proteins $G\alpha_o$ and $G\alpha_q$ controls serotonin signaling in *Caenorhabditis elegans*. Genetics 78: 157–169.

14. Albert PS, Riddle DL (1983) Developmental alterations in sensory neuroanatomy of the *Caenorhabditis elegans* dauer larvae. J Comp. Neurol. 219: 461–481.

15. Zwaal RR, Mendel JE, Sternberg PW, Plasterk RHA (1997) Two neuronal G proteins involved in chemosensation of the *Caenorhabditis elegans* dauer-inducing pheromone. Genetics 145: 715–727.

16. Gallo M, Riddle DL (2009) Effects of a Caenorhabditis elegans dauer pheromone ascaroside on physiology and signal transduction pathways. J. Chem. Ecol. 35: 272–279.

17. Burghoorn J, Dekkers MPJ, Rademakers S, de Jong T, Willemsen R, et al. (2010) Dauer pheromone and G-protein signaling modulate the coordination of intraflagellar transport kinesin motor proteins in *C. elegans*. J Cell Sci. 123: 2077–2084.

18. Lackner MR, Nurrish SJ, Kaplan JM (1999) Facilitation of synaptic transmission by EGL-30 $G\alpha_q$ and EGL-8 PLCβ: DAG binding to UNC-13 is required to stimulate acetylcholine release. Neuron 24: 335–346.

19. Miller KG, Emerson MD, Rand JB (1999) $G_o$alpha and diacylglycerol kinase negatively regulate the $G_q$alpha pathway in *C. elegans*. Neuron 24: 323–333.

20. Estevez M, Estevez AO, Cowie RH, Gardner KL (2004) The voltage-gated calcium channel UNC-2 is involved in stress-mediated regulation of tryptophan hydroxylase. J Neurochem. 88: 102–113.

21. Moussaif M, Sze JY (2009) Intraflagellar transport/Hedgehog-related signaling components couple sensory cilium morphology and serotonin biosynthesis in *Caenorhabditis elegans*. J Neurosci. 29: 4065–4075.

22. Ailion M, Thomas JH (2000) Dauer formation induced by high temperatures in *Caenorhabditis elegans*. Genetics 156: 1047–1067.

23. Ailion M, Thomas JH (2003) Isolation and characterization of high-temperature-induced dauer formation mutants in *Caenorhabditis elegans*. Genetics 165: 127–144.

24. Dong MQ, Chase D, Patikoglou GA, Koelle MR (2000) Multiple RGS proteins alter neural G protein signaling to allow C. elegans to rapidly change behavior when fed. Genes Dev. 14: 2003–2014.

25. Suo S, Kimura Y, Van Tol HHM (2006) Starvation induces cAMP response element-binding protein-dependent gene expression through octopamine–G_q Signaling in *Caenorhabditis elegans*. J Neurosci. 26: 10082–10090.

26. Hofler C, Koelle MR (2011) AGS-3 Alters Caenorhabditis elegans behavior after food deprivation via RIC-8 activation of the neural G protein $G\alpha_o$. J Neurosci. 31: 11553–11562.

27. Bastiani CA, Gharib S, Simon MI, Sternberg PW (2003) *Caenorhabditis elegans* Gαq regulates egg-laying behavior via a PLCβ- independent and serotonin-dependent signaling pathway and likely functions both in the nervous system and in muscle. Genetics 165: 1805–1822.

28. Hawasli AH, Saifree O, Liu C, Nonet ML, Crowder CM (2004) Resistance to volatile anesthetics by mutations enhancing excitatory neurotransmitter release in *Caenorhabditis elegans*. Genetics 168: 831–843.

29. Doi M, Iwasaki K (2002) Regulation of retrograde signaling at neuromuscular junctions by the novel C2 domain protein AEX-1. Neuron 33: 249–259.

30. Golden JW, Riddle DL (1984) The *Caenorhabditis elegans* dauer larva: developmental effects of pheromone, food, and temperature. Dev. Biol. 102: 368–378.

31. Cassada RC, Russell RL (1975) The Dauerlarva, a post-embryonic developmental variant of the nematode *Caenorhabditis elegans*. Dev. Biol. 46: 326–342.

32. Brenner S (1974) The genetics of *Caenorhabditis elegans*. Genetics 77: 71–94.

33. Perkins LA, Hedgecock EM, Thomson JN, Culotti JG (1986) Mutant sensory cilia in the nematode *Caenorhabditis elegans*. Dev. Biol. 177: 456–487.

22

Fluctuating Helical Asymmetry and Morphology of Snails (Gastropoda) in Divergent Microhabitats at 'Evolution Canyons I and II,' Israel

Shmuel Raz[1,2]*, Nathan P. Schwartz[3], Hendrik K. Mienis[4], Eviatar Nevo[1], John H. Graham[3]

1 Department of Evolutionary and Environmental Biology, Institute of Evolution, University of Haifa, Haifa, Israel, **2** Rowland Institute at Harvard, Harvard University, Cambridge, Massachusetts, United States of America, **3** Department of Biology, Berry College, Mount Berry, Georgia, United States of America, **4** The Steinhardt National Collections of Natural History, Tel Aviv University, Tel Aviv, Israel

Abstract

Background: Developmental instability of shelled gastropods is measured as deviations from a perfect equiangular (logarithmic) spiral. We studied six species of gastropods at 'Evolution Canyons I and II' in Carmel and the Galilee Mountains, Israel, respectively. The xeric, south-facing, 'African' slopes and the mesic, north-facing, 'European' slopes have dramatically different microclimates and plant communities. Moreover, 'Evolution Canyon II' receives more rainfall than 'Evolution Canyon I.'

Methodology/Principal Findings: We examined fluctuating asymmetry, rate of whorl expansion, shell height, and number of rotations of the body suture in six species of terrestrial snails from the two 'Evolution Canyons.' The xeric 'African' slope should be more stressful to land snails than the 'European' slope, and 'Evolution Canyon I' should be more stressful than 'Evolution Canyon II.' Only *Eopolita protensa jebusitica* showed marginally significant differences in fluctuating helical asymmetry between the two slopes. Contrary to expectations, asymmetry was marginally greater on the 'European' slope. Shells of *Levantina spiriplana caesareana* at 'Evolution Canyon I,' were smaller and more asymmetric than those at 'Evolution Canyon II.' Moreover, shell height and number of rotations of the suture were greater on the north-facing slopes of both canyons.

Conclusions/Significance: Our data is consistent with a trade-off between drought resistance and thermoregulation in snails; *Levantina* was significantly smaller on the 'African' slope, for increasing surface area and thermoregulation, while *Eopolita* was larger on the 'African' slope, for reducing water evaporation. In addition, 'Evolution Canyon I' was more stressful than Evolution Canyon II' for *Levantina*.

Editor: Shree Ram Singh, National Cancer Institute, United States of America

Funding: There are no sources of funding to this study.

Competing Interests: The authors have declared that no competing interests exist.

* E-mail: razshmu@gmail.com

Introduction

Fluctuating asymmetry, a measure of developmental instability [1], is usually estimated from bilaterally symmetrical traits. Many organisms, however, have other kinds of symmetry (i.e., translatory, radial, dihedral, or helical symmetries). Gastropods, for example, have helical symmetry, which can be the basis for fluctuating helical asymmetry. Previously, Graham, Freeman & Emlen [2] studied deviations from a perfect equiangular (logarithmic) spiral in three populations of the terrestrial snail *Cepaea nemoralis* (Helicidae: Gastropoda) in the Ukraine. Others have studied shell deformities in snails [3] and other mollusks [4,5,6]. Here we study growth, shell morphology, and fluctuating helical asymmetry of six species of terrestrial pulmonate and prosobranch snails from the opposing slopes of 'Evolution Canyon I,' Lower Nahal Oren, Mount Carmel (EC I) and 'Evolution Canyon II,' Lower Nahal Keziv, Western Upper Galilee (EC II), in Israel.

The 'Evolution Canyon' microsites are model systems for the study of adaptation and speciation. The opposite slopes of these canyons, the abiotically stressed south-facing, 'African' slopes and the moderate, north-facing, 'European' slopes, diverge biotically and abiotically, providing an opportunity to study developmental instability in a natural experiment. Hundreds of studies have been conducted here in the last 20 years [7,8,9,10,11,12].

Four 'Evolution Canyon' microsites are distributed across Israel: EC I in the mountains of Carmel, EC II in Galilee, EC III in the Negev, and EC IV in the Golan [10,11]. Most of the studies on these canyons were conducted at 'Evolution Canyons I and II' (Figure 1). They have demonstrated that the 'African' slope is more stressful for many mesic organisms (reviewed in [7,8,9,10,11,12]). The microclimatic differences produce strong differentiation of local biodiversity at all developmental levels (base sequences, genes, genomes, populations, species, ecosystems, and biota). The interslope differences at the molecular level (greater mutation frequency and recombination rate on the 'African' slope, in different taxa) are accompanied by interslope differences in species richness and abundance (reviewed in [7,8,9,10,11,12]).

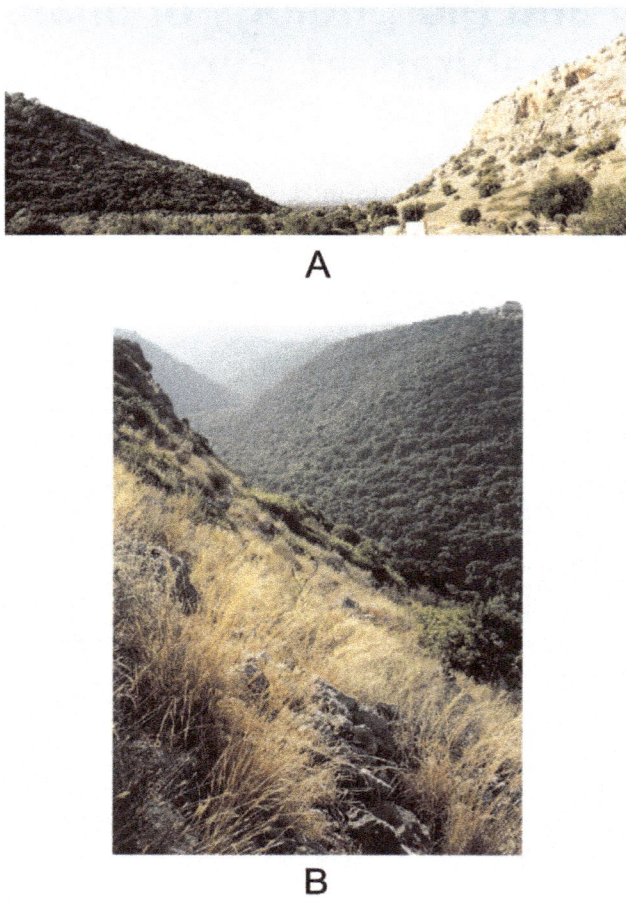

Figure 1. The opposing slopes of the 'Evolution Canyons.' The xeric 'African' slopes are on the right, and the mesic 'European' slopes are on the left. A. 'Evolution Canyon I,' Lower Nahal Oren, Mount Carmel, Israel. B. 'Evolution Canyon II,' Lower Nahal Keziv, western Upper Galilee, Israel.

In 11 of 14 model species at 'Evolution Canyon I,' Nevo and colleagues [7,8,9,10,11,12] found significantly greater genetic polymorphism on the 'African' slope than on the 'European' slope. They also found adaptive changes in other genetic characteristics. Populations of several model species on the more stressful 'African' slope had greater rates of mutation, gene conversion, recombination, and DNA repair, as well as greater genome size, more SSRs, SNPs, retrotransposons, transposons, candidate gene diversity, and genome-wide gene expression and regulation.

The coprophilous fungus *Sordaria fimicola*, for example, has heritable mutation rates 3-fold higher on the 'African' slope. *Drosophila melanogaster* has male recombination rates 4-fold higher on the 'African' slope. The filamentous cyanobacterium, *Nostoc linckia*, has higher haplotype diversity of clock genes *KaiABC* on the 'African' slope, and wild barley, *Hordeum spontaneum*, shows genetic divergence between the opposing slopes.

In addition to genetic divergence, species richness and abundance differ between the slopes. Pavlíček *et al.* [13], for example, showed that different taxonomic groups of terrestrial animals, such as scorpions, reptiles, butterflies (Rhopalocera), darkling beetles (Tenebrionidae), skin beetles (Dermestidae), and grasshoppers (Orthoptera), are more abundant on the 'African' slope than on the 'European' slope. The opposite trend occurs with springtails (Collembola), soil microfungi, basidiomycetous

fungi (Basidiomycetes), mosses (Bryophyta), and trees and shrubs. These taxa have greater species richness on the 'European' slope than on the 'African' slope.

These results [13] demonstrate that species richness and abundance vary along a climatic transect of only a few hundred meters (the geology is identical on both slopes), revealing ecological (climatic) selection strong enough to override the mixing effects of migration and stochasticity. The ecological selection caused by higher insolation on the 'African' slope leads to greater ecological heterogeneity on that slope, as well as a savanna ecosystem that accommodates more species of heat-dependent taxa.

In a previous study at 'Evolution Canyon I,' Raz *et al.* [14] studied leaf asymmetry of twelve species of vascular plants growing on the opposing slopes. Two of the species had more asymmetrical leaves on the 'African' slope, while one species had more asymmetrical leaves on the 'European' slope. Overall, the differences in fluctuating asymmetry between the slopes were negatively correlated with the differences in relative abundance. Species displayed greater fluctuating asymmetry on the slope where they were less abundant, and hence more stressed.

In the current study we explore the fluctuating helical asymmetry of six land-snail species from the opposing slopes of 'Evolution Canyons I and II.' Because land snails are susceptible to desiccation, the xeric 'African' slope should be more stressful. But unlike plants, land-snails can hide under stones and in cracks. And like plants, they can also become inactive during the dry season. In Israel's Mediterranean region, snails are typically active during rainy days from November to April [15]. Such behavior reduces temperature stress and water loss [16]. Morphology, physiology, and life history can also influence resistance to desiccation [17]. Consequently, some species of land snails can live abundantly in deserts.

Previous research on snails at 'Evolution Canyon I' shows that most snails are larger on the 'European' slope, but more abundant on the 'African' slope [13,15]. Rainfall is roughly the same on both slopes, but temperatures on the 'African' slope may be more amenable for growth during the rainy winter, when the snails are active. Broza & Nevo [15] suggested that the size differences between the two slopes might be due to *r*- and *k*-selection; snails on the 'African' slope put more energy into reproduction, while those on the 'European' slope put more energy into competitive ability. Size differences could also reflect Bergmann's ecogeographic rule extended to invertebrates: smaller body size supporting thermal tolerance on the warmer slope.

These results suggest that the interslope differences in insolation, temperature, and humidity at 'Evolution Canyon' differentially influence growth, morphology, and developmental instability of snails [18]. Hence, snails should be larger and more symmetrical on the 'European' slope than on the 'African' slope of 'Evolution Canyon.' We recognize, however, that the cool and humid 'European' slope could be stressful to land-snails adapted to more xeric and warm climatic conditions. Moreover, one expects this stress to influence snails mostly during the November-to-April rainy and cold season. Species intolerant of prolonged summer drought and heat during the May-to-October period of aestivation should have slower growth and be more developmentally unstable on the 'African' slope, while those intolerant of shade and lower winter temperatures should have slower growth and be more developmentally unstable on the 'European' slope.

Materials and Methods

Site descriptions

'Evolution Canyon I' (EC I) (Figure 1) is located at Lower Nahal Oren (32°42′51.09″N; 34°58′26.81″E), a deeply incised valley running from Mount Carmel, Israel, westwards into the Mediterranean Sea. The opposite slopes share identical geological history (Plio-Pleistocene canyon, presumably 3–5 million years old [7]), geology, soils (terra rossa on Upper Cenomanian limestone), and regional climate, although they differ in topography (dip in opposite directions; the 'African' slope dips 35°; the 'European' slope dips 25°) and aspect. Interslope distance is 100 m at the valley bottom and 400 m at the top; 'African' and 'European' slopes are 120 m and 180 m long, respectively (Figure 1). Rainfall at 'Evolution Canyon I' is 600 mm per year. The percentage of plant cover varies from 35% on the 'African' slope to 150% on the 'European' slope [19]. Life-form analysis clearly illustrates the dramatic interslope differences between the hot, xeric, Mediterranean savannoid formation of *Ceratonia siliqua–Pistacia lentiscus* on the 'African' slope and the dense maquis of *Quercus calliprinos–Pistacia palaestina* on the 'European' slope [19].

'Evolution Canyon II' (EC II) is located 38 km northeast of 'Evolution Canyon I' at Lower Nahal Keziv, western Upper Galilee (33°02'34.86"N, 35°11'05.74"E). Like 'Evolution Canyon I,' 'Evolution Canyon II' has a south-facing 'African' slope and a north-facing 'European' slope that incline 20–40° and 30–40°, respectively. The canyon is narrower and steeper than that at 'Evolution Canyon I' (50 m at the bottom and 350 m at the top). It is also further inland from the Mediterranean Sea, and more sheltered, than 'Evolution Canyon I.' The underlying rocks are upper Cenomanian limestone, with colluvial and alluvial soils at the bottom and terra rossa on the slopes. Rainfall at 'Evolution Canyon II' is 700 mm per year, which is 17% greater than that at 'Evolution Canyon I.' The plant communities also vary between the slopes. The number of vascular plant species on the 'African' slope (205 species) is substantially greater than on the 'European' slope (54 species). The percentage of plant cover varies from 70% on the 'African' slope to 100% on the 'European' slope [20]. The 'African' slope changes from *Calicotome villosa* and *Salvia fruticosa* garrigue at the bottom to a dry, Mediterranean, savannoid, open Park Forest of *C. siliqua – P. lentiscus* association at the top. The 'European' slope is covered by a dense forest of *Acer obtusifolium* and *Laurus nobilis*, which is very different from the 'European' slope of 'Evolution Canyon I,' and represents a Mediterranean maquis forest.

Sampling

We collected shells of six species of shelled gastropods (both juveniles and adults) from north- and south-facing slopes of 'Evolution Canyon I' and 'Evolution Canyon II' (Table 1). The collections were approved by the Israeli Nature and Park Authority [Permit 2010/38005 and 2010/38006 for Oren Canyon ('Evolution Canyon I') and Keziv Canyon ('Evolution Canyon II'), respectively], so all necessary permits were obtained for the described field studies.

We sampled four land-snail species from the opposing slopes of 'Evolution Canyon I,' the pulmonate snails *Buliminus labrosus labrosus*, *Monacha syriaca*, *Xeropicta vestalis joppensis*, and *Levantina spiriplana caesareana*. Two of these species, *L. s. caesareana* and *B. l. labrosus*, were also sampled at 'Evolution Canyon II.' In addition, we sampled *Pomatias olivieri*, a prosobranch snail, and *Eopolita p. jebusitica*, a pulmonate snail, only at 'Evolution Canyon II.' The taxonomy of *Buliminus*, *Monacha*, *Xeropicta*, *Pomatias*, and *Eopolita* follows Heller [21], while *Levantina* follows Pfeiffer [22] and Forcart (unpublished work).

Measurements

We scanned each snail twice on a flatbed scanner, at a resolution of 600 dpi. To support a snail for the scan, we pressed it into a cubic block of clay so that the columella was either parallel or perpendicular to the scan surface, depending upon the species. For replicate scans, and to estimate the measurement error associated with each scan, we repositioned each snail in the clay, from scratch. We also made three replicate sets of measurements per scan, using SigmaScan Pro: Image Analysis Version 5.0.0. Consequently, there were six replicate measurements made on each snail. The main measurement was the radius from apex to curve (the suture) for every 180° of clockwise rotation. All scans were done by Shmuel Raz. All measurements on the images were made by a single observer (Nathan Schwartz).

Different species of snails required different approaches. Those having a relatively depressed, flat shell (*Monacha*, *Xeropicta*, *Eopolita*, and *Levantina*) could be scanned such that the apex and entire spiral suture were clearly visible (apical view, with columella perpendicular to the scan surface). Measuring the radius from apex to curve of the suture was straightforward. This could not be done with snails having an oblong or globose shell (*Pomatias* and *Buliminus*). These species were scanned from the side, in apertural view (columella parallel to the scan surface). We measured the distance from apex to the nearest suture on the left side, and then from that suture to the next one, and so on, repeating the process on the right side.

Fluctuating helical asymmetry

Helical symmetry involves rotation, along with translation along an axis of rotation. The spiral shell approximates an equiangular (logarithmic) spiral. The equation for an equiangular spiral is $r = ae^{\theta \, \cot \, \Phi}$, where r is the radius from apex to curve, a is a constant, e is the base of natural logarithms, θ is the angle made with a reference line passing through the apex, and Φ is the constant angle at which the radius vector cuts the curve. Graham et al. [2] regressed $\log_e (r+1)$ on angle θ for each individual snail and used the standard error of the estimate, divided by the mean of the dependent variable ($S_{y \cdot x}/\bar{y}$), as an estimate of individual asymmetry (Figure 2).

Measurement error (s^2_{me}) inflates estimates of fluctuating asymmetry. It also creates problems when the researcher later corrects for size scaling [1,23,24]. A preliminary study of *Cepaea nemoralis* (previously collected in the Ukraine) suggested that the variation among photos within snails within sites accounted for 22.5% of the variation, while variation among replicate measurements accounted for 3.5%. The remaining variation (74.0%) was among individual snails within a site.

Trait-size variation is often a problem in studies of fluctuating asymmetry. Positive size-scaling of asymmetry, for example, is largely due to multiplicative error associated with the active-tissue model of growth [1,24]. We found no evidence for positive (or negative) size scaling after averaging all of the replicate measurements and $\log_e$ transforming r. The averaging of replicates removes most of the additive measurement error and the logarithmic transform eliminates the multiplicative error associated with growth.

Size and growth

As a measure of body size, we measured the height of the shell from the apex to the closest part of the aperture and quantified the rate of expansion of the body whorl as the slope of the regression of $\log_e (r+1)$ on angle θ. The number of complete rotations of the suture around the apex is also an indicator of size. This is not equivalent to the number of whorls; the number of suture rotations always exceeds the number of whorls.

Table 1. Species of snails at 'Evolution Canyons I and II' used in this study. Diet and habitat descriptions are from Pavlíček *et al.* [13].

Family	Species	Diet	Habitat
Pomatiidae	*Pomatias olivieri* (de Charpentier, 1847)	Decaying plants	Shade
Enidae	*Buliminus labrosus labrosus* (Olivier, 1804)	Lichens, liverworts	Rocky outcrops
Oxychilidae	*Eopolita protensa jebusitica* (Roth, 1855)	Invertebrates, decaying plants	Under stones, wood, and leaves
Hygromiidae	*Monacha syriaca* (Ehrenberg, 1831)	Green plants	Garrigue, terraces
	Xeropicta vestalis joppensis (Schmidt, 1855)	Green plants	Garrigue, terraces
Helicidae	*Levantina spiriplana caesareana* (Mousson, 1854)	Lichens, liverworts	Rocky outcrops

Statistical analysis

We used SPSS's GLM Varcomp procedure to estimate the variance components associated with sites, snails within sites, scans of snails within sites, and replicate measures of scans within snails within sites.

We used one-way ANOVA to compare fluctuating asymmetry, shell height, regression coefficient, and number of rotations of the suture between 'African' and 'European' slopes. Slope is a fixed effect and snail within slope (the average of six replicate measurements) is a random effect. For *Levantina* and *Buliminus*, which were sampled at both 'Evolution Canyons I and II,' we included canyon as a fixed effect.

Results

Fluctuating helical asymmetry

Variance components associated with slope, individuals, scans, and replications were estimated for *B. l. labrosus* and *L. s. caesareana* (Table 2). The among-individual variation represents both genotypic and microenvironmental variation. Measurement error includes variation among scans and among replicate measurements. For *Levantina*, which we scanned in apical view, most of the variation was due to measurement error (49–57% of the total variation was among scans and 17–30% was among replicate measurements). For *Buliminus*, which we scanned in apertural view, measurement error was much smaller (4% of the total

variation was among scans and 1% was among replicate measurements).

Eopolita p. jebusitica at 'Evolution Canyon II' showed marginally significant differences in fluctuating asymmetry between the 'African' and 'European' slopes ($F_{1, 18} = 4.146$, $P = 0.057$, Figure 3). Shells were more asymmetric on the 'European' slope. None of the other species showed significant differences in fluctuating asymmetry between the slopes ($F_{1, 22-93} \leq 1.19$, $P \geq 0.215$).

Levantina s. caesareana had greater fluctuating asymmetry at 'Evolution Canyon I' than at 'Evolution Canyon II' ($F_{1, 145} = 36.978$, $P < 0.001$). Neither the differences between slopes ($F_{1, 145} = 0.031$, $P > 0.850$), nor the interaction of slope and canyon ($F_{1, 145} = 0.279$, $P > 0.550$) were significant. *Buliminus l. labrosus*, the only other species collected at both canyons showed no differences in fluctuating asymmetry between them ($F_{1, 158} = 0.813$, $P > 0.350$).

Shell height

Mean shell height is indicative of overall size (Figure 4). There were significant differences in shell height between *L. s. caesareana* from the two canyon sites ($F_{1, 145} = 150.414$, $P < 0.001$) and from 'African' and 'European' slopes ($F_{1, 145} = 21.117$, $P < 0.001$). There was also a significant interaction between canyon site and slope ($F_{1, 145} = 9.338$, $P < 0.005$). Shell heights were greater on the 'European' slope, though the differences were less extreme at 'Evolution Canyon II,' and shell heights were greater at 'Evolution Canyon II' than at 'Evolution Canyon I.' *Xeropicta v. joppensis*, in contrast, had greater shell height on the 'African' slope ($F_{1, 38} = 19.811$, $P < 0.001$).

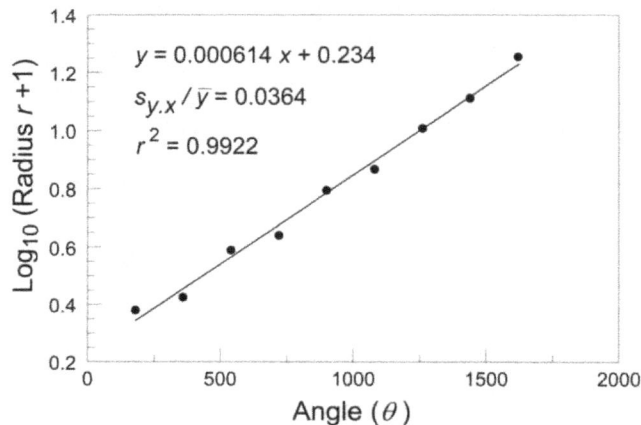

Figure 2. Linear regression of the $\log_e$ of the radius r+1 on the angle of rotation in degrees, for an individual *Buliminus l. labrosus* from the 'African' slope of 'Evolution Canyon II.'

Table 2. Variance components for shell radii: s^2_{slope} is the between slope variation, s^2_{ind} is the among-individual variation, s^2_{scan} is the among scans variation, s^2_{repl} is the variance component associated with replication, and s^2_{me} is the sum of s^2_{scan} and s^2_{repl}.

Variance component	Levantina (EC I)	Levantina (EC II)	Buliminus (EC I)
s^2_{slope}	0.00009670	0.00015303	0.00003503
s^2_{ind}	0.00063522	0.00008164	0.00304639
s^2_{scan}	0.00103574	0.00100532	0.00013466
s^2_{repl}	0.00035143	0.00052515	0.00003247
s^2_{me}	0.001387	0.001530	0.000167

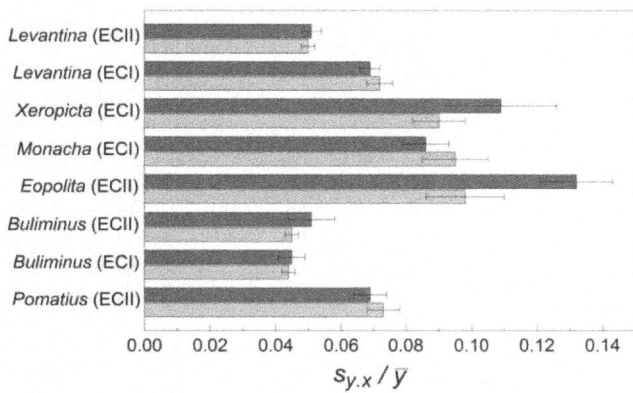

Figure 3. Mean fluctuating helical asymmetry (± standard error) of snails on 'African' and 'European' slopes. Light gray indicates the 'African' slope; dark gray indicates the 'European' slope.

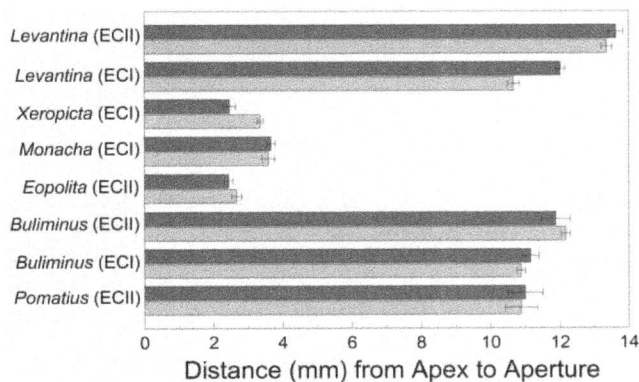

Figure 5. Mean regression coefficient (×10,000) (± standard error) on 'African' and 'European' slopes. Light gray indicates the 'African' slope; dark gray indicates the 'European' slope.

There were significant differences in shell height between *B. l. labrosus* from the two canyon sites ($F_{1, 158} = 15.916$, $P<0.001$), but not between 'African' and 'European' slopes ($F_{1, 158} = 0.001$, $P>0.950$). The interaction between canyon site and slope was also insignificant ($F_{1, 158} = 1.280$, $P>0.250$). Shell heights were greater at 'Evolution Canyon II' than at 'Evolution Canyon I.'

The shell heights of *M. syriaca*, *P. olivieri*, and *E. p. jebusitica* did not differ between the two slopes ($F_{1, 18-32} \leq 1.409$, $P \geq 0.251$).

Expansion of the body whorl

The rate of expansion of the body whorl (i.e., the slope of the regression of $\log_e r+1$ on angle θ for each individual) reflects the rate at which the spiral opens up (Figure 5). Only *B. l. labrosus* showed significant differences between the canyon sites ($F_{1, 158} = 19.560$, $P<0.001$) and marginally significant differences between the two slopes ($F_{1, 158} = 3.682$, $P = 0.057$). The interaction between site and slope was also significant ($F_{1, 158} = 20.870$, $P<0.001$). The body whorl expanded more rapidly on the 'European' slope at 'Evolution Canyon I,' but the reverse was true at 'Evolution Canyon II.' None of the other species displayed differences in the body whorl between the 'African' and 'European' slopes ($F_{1, 18-93} = 0.765-2.050$, $P \geq 0.166$).

Number of rotations of the body suture

The mean number of rotations of the body suture (Figure 6) is indicative of age and size. Populations of *B. l. labrosus* at 'Evolution Canyon II' had more suture rotations on the 'European' slope than on the 'African' slope ($F_{1, 73} = 12.374$, $P<0.001$). Populations of *L. s. caesareana* had significant differences in the number of suture rotations between the two canyon sites ($F_{1, 145} = 48.126$, $P<0.001$) and between 'African' and 'European' slopes ($F_{1, 145} = 18.731$, $P<0.001$). The interaction between canyon site and slope was insignificant ($F_{1, 145} = 0.355$, $P>0.550$). The number of rotations of the body suture was greater on the 'European' slope.

There were significant differences in the numbers of suture rotations between *Buliminus* from the two canyon sites ($F_{1, 158} = 26.637$, $P<0.001$), and from 'African' and 'European' slopes ($F_{1, 158} = 7.497$, $P<0.010$). The interaction between canyon site and slope was also significant ($F_{1, 158} = 14.704$, $P<0.001$). The number of suture rotations was greater at 'Evolution Canyon II' than at 'Evolution Canyon I,' but differences between 'African' and 'European' slopes were only evident at 'Evolution Canyon II,' where snails on the 'European' slope had more suture rotations.

In contrast to *Buliminus* and *Levantina*, *X. v. joppensis* had more suture rotations on the 'African' slope ($F_{1, 38} = 5.492$, $P<0.025$).

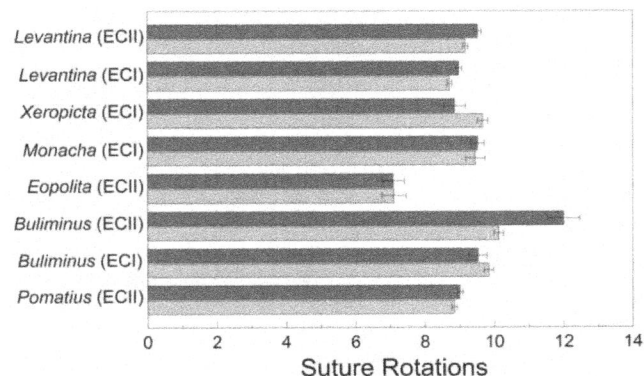

Figure 4. Mean shell height of snails (± standard error), from apex to aperture, on 'African' and 'European' slopes. Light gray indicates the 'African' slope; dark gray indicates the 'European' slope.

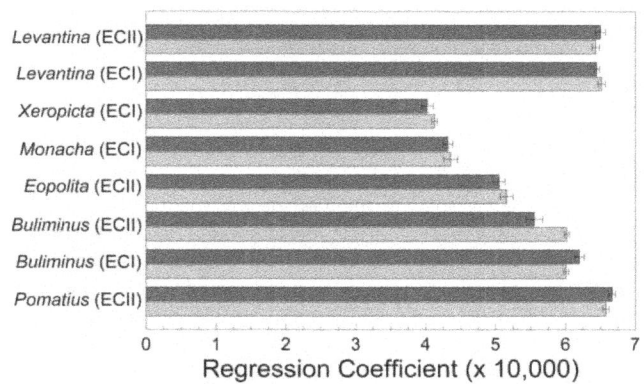

Figure 6. Mean number of rotations of the suture (± standard error) on 'African' and 'European' slopes. Light gray indicates the 'African' slope; dark gray indicates the 'European' slope.

None of the other species and populations had different numbers of suture rotations on the two slopes ($F_{1,\,18-85} = 0.002-1.833$, $P\geq0.189$).

Discussion

Gastropods are distributed from the arctic to the tropics and can be terrestrial (one-third of species) or aquatic (two-thirds of species) [25]. They are adapted to most of the habitats on Earth and in terrestrial habitats they are subjected to daily and seasonal variation in temperature and water availability. Their success in colonizing different habitats is due to physiological, behavioral, and morphological adaptations to water availability [26,27,28], as well as ionic and thermal balance [29]. The shell of a snail is constructed of calcium carbonate, but even in acidic soils one can find various species of shell-less slugs. Interestingly, land-snails also live in deserts, where they must contend with heat and aridity [30].

Species richness and abundance

Adaptation of land-snails to different regimes of heat and aridity may influence species richness and abundance of snails on the opposing slopes of 'Evolution Canyon' [13,17,27,31,32]. For example, species richness is greater on the 'European' slope of 'Evolution Canyon I,' but overall abundance is greater on the 'African' slope [13]. The greater species richness on the 'European' slope represents the addition of European species of snails and slugs at the southern limits of their adaptive range.

The greater overall abundance of snails on the 'African' slopes may reflect a better nutritional environment, less predation from small mammals [33], abiotic stress or all of these together. Early successional plants on the 'African' slope may be more palatable [34], and plant secondary compounds influence choice of food [35]. If this is true, then the snails that feed on live plants and lichens should show the greatest differences in abundance. The four species that feed on living plants or lichens (*L. s. caesareana*, *B. l. labrosus*, *M. syriaca*, *X. v. joppensis*) were rare (*M. syriaca*) or considerably more abundant on the 'African' slope, whereas the species that feed on decaying plants (*P. olivieri*, *E. p. jebusitica*) or other invertebrates (*E. p. jebusitica*) were slightly more abundant on the 'European' slope.

According to Pavlíček *et al.* [13], six of seven species more abundant on the south-facing slope fed on live plants, whereas only one of eleven species that fed exclusively on decaying plants was more abundant on the south-facing slope. Finally, the main predators, such as shrews (*Crocidura* spp. and *Suncus etruscus*), may be unwilling to venture out onto the more open south-facing slopes, which lack natural shelters [36].

Fluctuating helical asymmetry

To the best of our knowledge, the only work on fluctuating helical asymmetry of land-snails was done by Graham *et al.* [2], who studied deviations from a perfect equiangular spiral in three populations of the terrestrial land-snail *Cepaea nemoralis* in the Ukraine. The highest level of asymmetry was found in populations exposed to ammonia emissions and pesticides. Individuals in the population having the greatest helical asymmetry also showed erosion of their periostracum, which was not evident in the other two populations.

Overcrowding and nutritional deprivation can influence shell microstructure, increasing fluctuating helical asymmetry. Chunhabundit *et al.* [3], for example, raised maculated top shells, *Trochus maculatus*, a marine gastropod, under high density and inadequate nutrition. The periostracum was reduced and shell structure was dissolved in the vicinity of the shell apex. The suture lines were less smooth.

As with twelve species of vascular plants at 'Evolution Canyon I,' we cannot reject the null hypothesis of no differences in deviations from perfect symmetry between the land-snails from the opposing slopes. In the case of the vascular plants, the leaves were bilaterally symmetrical. For the snails, the shells are helically symmetrical. We suggest that land-snails are better adapted to the 'African' slope than we had anticipated. They can hide themselves in different locations, such as under stones and in cracks. Such behavior reduces their temperature and water loss, though *Xeropicta* aestivates high up on shrubs. In addition, the shells may serve as a $CaCO_3$ 'door,' separating land-snails from environmental stress.

Thermoregulation, drought resistance, and body size

There is a trade-off between drought resistance and thermoregulation in snails [32]. Thermoregulation requires water for evaporative cooling. The smaller the snail, the more effective the cooling, but this system nevertheless requires more water. Evaporative cooling is more effective for smaller snails because a smaller body size increases the surface area to volume ratio, which in turn increases both absorption and radiation of heat [32,37]. The ecological rule associated with this phenomenon is Bergmann's rule [38], which predicts larger body size of warm-blooded vertebrates in colder areas. According to Mayr [39], this is an adaptive response to environmental temperatures. Although Bergmann's rule was intended to describe body-size variation among species, it has been extended to intraspecific variation in body size [39,40] as well as to ectotherms [41], but with variable success [42,43]. This rule is exemplified by the spiny mouse, *Acomys cahirinus*, and the broad-toothed field mouse, *Apodemus mystacinus*, from the opposing slopes of 'Evolution Canyon.' Individuals from the 'African' slope are smaller [44] than those from the 'European' slope.

Is it possible that Bergmann's rule, which was meant for endothermic species, also holds for land-snails? Are the size differences of the snails between the canyons and the opposing slopes a result of microclimatic differences? We believe that, yes, the larger shells of *Levantina s. caesareana* at 'Evolution Canyon II' (i.e., Keziv Canyon) and on the 'European' slopes of both canyons do indeed represent Bergmann's rule on a microscale. Nevertheless, we collected both adults and juveniles, hence size differences may also be due to differences in age structure.

These results are supported by evidence for a correlation between habitat and body size in invertebrates. The body size of insects and spiders, for example, is smaller when the humidity is lower [45,46]. Moreover, a correlation was found between shell diameter and climate in the snail *Xerocrassa seetzenii*, from Israel [47]. The latitudinal gradient of decreasing body size, from north to south, in Israel, accompanied by decreasing rainfall, also occurs in *Levantina s. caesareana* [48].

The smaller the individual, however, the more water it loses for cooling. Hence, there is a lower-size threshold for a given individual and environment [49]. Very small animals lose almost 100% of their body mass for cooling [50]. Accordingly, the amount of water in the tissues of *X. v. joppensis* may be so small that they are selected for larger body size on the 'African' slope than on the 'European' slope of 'Evolution Canyon.'

Xeropicta lives for only one year, while *Levantina* and *Buliminus* live for several years. During years with reduced rainfall, adult *Xeropicta* and *Monacha* may be extremely small, while in a year with extremely high rainfall and numerous nights with heavy dew, they may reach very large size (Mienis, personal observations).

Morphological differences in body size between relatively large snails (*Buliminus* and *Levantina*) and somewhat smaller snails

(*Monacha*) from the opposing slopes of 'Evolution Canyon' were found in previous studies [15]. Accordingly, five-out-of-seven species were larger on the 'European' slope.

Conclusions

The differences in fluctuating helical asymmetry between 'African' and 'European' slopes were either non-existent or only marginally significant (shell asymmetry of *E. p. jebusitica* was marginally greater on the north-facing slope), hence we cannot reject the null hypothesis for no differences between the slopes. There were, however, differences between shell asymmetry of *L. s.*

caesareana from the two canyons; asymmetry was greater at the more arid Nahal Oren, 'Evolution Canyon I.'

Acknowledgments

The authors are indebted to Cathy Chamberlin-Graham for her help and comments. We thank Maayan Kotzan for helping with the field work.

Author Contributions

Conceived and designed the experiments: SR EN JG. Performed the experiments: SR NS. Analyzed the data: SR HM NS JG. Wrote the paper: SR JG. Performed the taxonomy: HM.

References

1. Graham JH, Raz S, Hel-Or H, Nevo E (2010) Fluctuating asymmetry: methods, theory, and applications. Symmetry 2: 466–540.
2. Graham JH, Freeman DC, Emlen JM (1993) Developmental stability: a sensitive indicator of populations under stress. In: Landis WG, Hughes JS, Lewis MA, editors. Environmental Toxicology and Risk Assessment. Philadelphia: American Society for Testing and Materials. pp. 136–158.
3. Chunhabundit S, Chunhabundit P, Aranyakananda P, Moree N (2001) Dietary effects on shell microstructures of cultured, maculate top shell (Trochidae: *Trochus maculatus*, Linnaeus, 1758). SPC Trochus Information Bulletin 8: 15–22.
4. Alzieu C (1991) Environmental problems caused by TBT in France: assessment, regulations, prospects. Marine Environmental Research 32: 7–17.
5. Batley GE, Fuhua C, Brockbank CI, Flegg KJ (1989) Accumulation of Tributyltin by the Sydney Rock Oyster, *Saccostrea commercialis*. Australian Journal of Marine and Freshwater Research 40: 49–54.
6. Alzieu CL, Sanjuan J, Deltreil JP, Borel M (1986) Tin contamination in Arcachon Bay: effects on oyster shell anomalies. Marine Pollution Bulletin 17: 494–498.
7. Nevo E (1995) Asian, African and European biota meet at 'Evolution Canyon' Israel: local tests of global biodiversity and genetic diversity patterns. Proceedings of the Royal Society B: Biological Sciences 262: 149–155.
8. Nevo E (1997) Evolution in action across phylogeny caused by microclimatic stresses at "Evolution Canyon". Theoretical Population Biology 52: 231–243.
9. Nevo E (2001) Evolution of genome–phenome diversity under environmental stress. Proceedings of the National Academy of Sciences 98: 6233–6240.
10. Nevo E (2006) "Evolution Canyon": a microcosm of life's evolution focusing on adaptation and speciation. Israel Journal of Ecology and Evolution 52: 485–506.
11. Nevo E (2009) Evolution in action across life at "Evolution Canyons", Israel. Trends in Evolutionary Biology 1: e3.
12. Nevo E (2011) Selection overrules gene flow at 'Evolution Canyons', Israel. In: Urbano KV, editor. Advance in Genetics Research. Hauppauge, NY Nova Science Publishers, Inc. pp. 67–89.
13. Pavlíček T, Mienis HK, Raz S, Hassid V, Rubenyan A, et al. (2008) Gastropod biodiversity at the 'Evolution Canyon' microsite, lower Nahal Oren, Mount Carmel, Israel. Biological Journal of the Linnean Society 93: 147–155.
14. Raz S, Graham JH, Hel-Or H, Pavlíček T, Nevo E (2011) Developmental instability of vascular plants in contrasting microclimates at 'Evolution Canyon'. Biological Journal of the Linnean Society 102: 786–797.
15. Broza M, Nevo E (1996) Differentiation of the snail community on the north- and south-facing slopes of lower Nahal Oren (Mount Carmel, Israel). Israel Journal of Zoology 42: 411–424.
16. Garrity SD (1984) Some adaptations of gastropods to physical stress on a tropical rocky shore. Ecology 65: 559–574.
17. Rankevich D, Lavie B, Nevo E, Belles A, Arad Z (1996) Genetic and physiological adaptations of the prosobranch landsnail *Pomatias olivieri* to microclimatic stresses on Mount Carmel, Israel. Israel Journal of Zoology 42: 425–441.
18. Pavlíček T, Sharon D, Kravchenko V, Saaroni H, Nevo E (2003) Microclimatic interslope differences underlying biodiversity contrasts in "Evolution Canyon", Mt. Carmel, Israel. Israel Journal of Earth Sciences 52: 1–9.
19. Nevo E, Fragman O, Dafni A, Beiles A (1999) Biodiversity and interslope divergence of vascular plants caused by microclimatic differences at "Evolution Canyon", Lower Nahal Oren, Mount Carmel, Israel. Israel Journal of Plant Sciences 47: 61–62.
20. Finkel M, Fragman O, Nevo E (2001) Biodiversity and interslope divergence of vascular plants caused by sharp microclimatic differences at "Evolution Canyon II", Lower Nahal Keziv, Upper Galilee, Israel. Israel Journal of Plant Sciences 49: 285–296.
21. Heller J, Arad Z, Kurts T (2009) Land Snails of the Land of Israel: Natural History and a Field Guide. Sofia, Bulgaria: Pensoft Publishers. 360 p.
22. Pfeiffer KL (1949) *Levantina spiriplana*. Archiv für Molluskenkunde 77: 1–51.
23. Cowart NM, Graham JH (1999) Within-and among-individual variation in fluctuating asymmetry of leaves in the fig (*Ficus carica* L.). International Journal of Plant Sciences 160: 116–121.
24. Graham JH, Shimizu K, Emlen JM, Freeman DC, Merkel J (2003) Growth models and the expected distribution of fluctuating asymmetry. Biological Journal of the Linnean Society 80: 57–65.
25. Ponder WF, Lindberg DR, editors (2008) Phylogeny and Evolution of the Mollusca. Berkeley: University of California Press. 488 p.
26. Arad Z (1993) Effect of desiccation on the water economy of terrestrial gastropods of different phylogenetic origins: a prosobranch (*Pomatias glaucus*) and two pulmonates (*Sphincterochila cariosa* and *Helix engaddensis*). Israel Journal of Zoology 39: 95–104.
27. Arad Z, Goldenberg S, Heller J (1989) Resistance to desiccation and distribution patterns in the land snail *Sphincterochila*. Journal of Zoology 218: 353–364.
28. Arad Z, Goldenberg S, Heller J (1992) Intraspecific variation in resistance to desiccation and climatic gradients in the distribution of the land snail *Xeropicta vestalis*. Journal of Zoology 226: 643–656.
29. Riddle WA, Russell-Hunter WD (1983) Physiological ecology of land snails and slugs. In: Russell-Hunter WD, editor. The Mollusca: Ecology. London: Academic Press. pp. 431–461.
30. Schmidt-Nielsen K, Taylor CR, Shkolnik A (1971) Desert snails: problems of heat, water and food. Journal of Experimental Biology 55: 385–398.
31. Warburg MR (1965) On the water economy of some Australian land-snails. Proceedings of the Malacological Society of London 36: 297–305.
32. Rankevich D (1997) Genetic variation and resistance to dessication in populations of landsnails on the southern and northern slopes of Nahal Oren. Haifa: Technion - Israel Institute of Technology. 143 p.
33. Yom-Tov Y (1970) The effect of predation on population densities of some desert snails. Ecology 51: 907–911.
34. Cates RG, Orians GH (1975) Sucessional status and the palatability of plants to generalized herbivores. Ecology 56: 410–418.
35. Hagele BF, Rahier M (2001) Determinants of seasonal feeding of the generalist snail *Arianta arbustorum* at six sites dominated by Senecioneae. Oecologia 128: 228–236.
36. Abramsky Z, Alfia H, Schachak M, Brand S (1990) Predation by rodents and the distribution and abundance of the snail *Trochoidea seetzenii* in the Central Negev Desert of Israel. Oikos 59: 225–234.
37. Schmidt-Nielsen K (1997) Animal physiology: adaptation and environment. Cambridge, UK: Cambridge University Press. 607 p.
38. Bergmann C (1847) Über die Verhältnisse der Wärmeökonomie der Thiere zu ihrer Grösse. Göttinger Studien 3: 595–708.
39. Mayr E (1956) Geographical character gradients and climatic adaptation. Evolution 10: 105–108.
40. James FC (1970) Geographic size variation in birds and its relationship to climate. Ecology 51: 365–390.
41. Huey RB, Stevenson RD (1979) Integrating thermal physiology and ecology of ectotherms: a discussion of approaches. American Zoologist 19: 357–366.
42. Ashton KG, Tracy MC, Queiroz A (2000) Is Bergmann's rule valid for mammals? The American Naturalist 156: 390–415.
43. Meiri S, Dayan T, Simberloff D (2004) Carnivores, biases and Bergmann's rule. Biological Journal of the Linnean Society 81: 579–588.
44. Nevo E, Filippucci GM, Pavlíček T, Gorlova O, Shenbrot G, et al. (1998) Genotypic and phenotypic divergence of rodents (*Acomys cahirinus* and *Apodemus mystacinus*) at "Evolution Canyon": micro- and macroscale parallelism. Acta Theriologica Suppl 5: 9–34.
45. Cloudsley-Thompson JL (1976) Terrestrial environments. In: Bligh J, Cloudsley-Thompson JL, MacDonald AG, editors. Environmental Physiology of Animals. Oxford, UK: Blackwell Scientific Publishers. pp. 96–103.
46. Remmert H (1981) Body size of terrestrial arthropods and biomass of their populations in relation to the abiotic parameters of their milieu. Oecologia 50: 12–13.
47. Nevo E, Bar-El C, Bar Z, Beiles A (1981) Genetic structure and climatic correlates of desert landsnails. Oecologia 48: 199–208.
48. Heller J (1979) Distribution, hybridization and variation in the Israeli landsnail *Levantina* (Pulmonata: Helicidae). Zoological Journal of the Linnean Society 67: 115–148.
49. Schmidt-Nielsen K (1964) Desert Animals: Physiological Problems of Heat and Water. Oxford, UK: Oxford University Press.
50. Prange HD (1996) Evaporative cooling in insects. Journal of Insect Physiology 42: 493–499.

23

Evolution of Body Elongation in Gymnophthalmid Lizards: Relationships with Climate

Mariana B. Grizante, Renata Brandt, Tiana Kohlsdorf*

Department of Biology, Faculdade de Filosofia Ciências e Letras de Ribeirão Preto, Universidade de São Paulo, Ribeirão Preto, São Paulo, Brazil

Abstract

The evolution of elongated body shapes in vertebrates has intrigued biologists for decades and is particularly recurrent among squamates. Several aspects might explain how the environment influences the evolution of body elongation, but climate needs to be incorporated in this scenario to evaluate how it contributes to morphological evolution. Climatic parameters include temperature and precipitation, two variables that likely influence environmental characteristics, including soil texture and substrate coverage, which may define the selective pressures acting during the evolution of morphology. Due to development of geographic information system (GIS) techniques, these variables can now be included in evolutionary biology studies and were used in the present study to test for associations between variation in body shape and climate in the tropical lizard family Gymnophthalmidae. We first investigated how the morphological traits that define body shape are correlated in these lizards and then tested for associations between a descriptor of body elongation and climate. Our analyses revealed that the evolution of body elongation in Gymnophthalmidae involved concomitant changes in different morphological traits: trunk elongation was coupled with limb shortening and a reduction in body diameter, and the gradual variation along this axis was illustrated by less-elongated morphologies exhibiting shorter trunks and longer limbs. The variation identified in Gymnophthalmidae body shape was associated with climate, with the species from more arid environments usually being more elongated. Aridity is associated with high temperatures and low precipitation, which affect additional environmental features, including the habitat structure. This feature may influence the evolution of body shape because contrasting environments likely impose distinct demands for organismal performance in several activities, such as locomotion and thermoregulation. The present study establishes a connection between morphology and a broader natural component, climate, and introduces new questions about the spatial distribution of morphological variation among squamates.

Editor: Suzannah Rutherford, Fred Hutchinson Cancer Research Center, United States of America

Funding: Funding was provided by Brazilian grants awarded to TK by FAPESP (Fundação de Amparo à Pesquisa do Estado de São Paulo; grants 2005/60140-4 and 2010/52316-3) and CNPq (Conselho Nacional de Desenvolvimento Científico e Tecnológico; grant 563232/2010-2). MBG was funded by FAPESP graduate fellowships (2007/52204-8 and 2010/00447-7), and RB is funded by a CAPES postdoctoral fellowship. The funders had no role in study design, data collection and analysis, decision to publish, or preparation of the manuscript.

Competing Interests: The authors have declared that no competing interests exist.

* E-mail: tiana@usp.br

Introduction

The general shape of a given body is recognized by its distribution along a three-dimensional Cartesian space. The morphological changes that equally increase an object in these three axes will result in a larger body with the same original shape; in contrast, when variation occurs mostly in one of these three dimensions (e.g., along the horizontal axis) the result may be a very distinct shape, such as an elongated body. Elongated animal body shapes evolve through increases in length (given, for example, by the addition of vertebrae along the trunk [1–3]), which can be either paralleled by changes in the trunk diameter [2] or coupled with decreases in the trunk height or width [4,5]. Changes leading to the evolution of elongated forms have been identified in basically all vertebrate lineages (e.g., fishes [4–6], amphibians [7,8], squamates [3,9–12] and mammals [13,14]) and often also involve a reduction or loss of the locomotor appendages [2,6,9–12,15].

The recurrent evolution of elongated body shapes in vertebrates has intrigued biologists for decades [2,7–16]. Squamata, in particular, has been used as a model system for detecting general patterns toward the evolution of serpentiform morphologies [9,16], which are characterized by elongated trunks and reduced or absent limbs [3,9,10]. There are several aspects that may explain how the evolution of body elongation relates to environmental traits. Studies on functional morphology suggest that long and thin, limbless bodies enhance the burrowing performance during subterranean locomotion such that the evolution of serpentiform squamates would be favored in fossorial lineages [2,17]. Even so, not all elongated squamates are fossorial, and there are possibly other parameters that trigger body shape diversification. For example, ecological interactions (e.g., competition and invasion of available niches) have been recently claimed to be relevant factors for the origin of elongated squamates because such interactions might explain the existence of two elongated ecomorphs in the lineage: the short-tailed fossorial species and the long-tailed surface-dwelling forms [9,10].

Regardless of the selective pressures that may be related to the evolution of elongated forms in Squamata, these conspicuous changes in body shape likely affect the interactions between the organism and its surrounding environment. The association between morphology and ecology has been identified in several squamate lineages [18–23]. Nevertheless, our current concept of the environment can be extended to encompass climate and thus improve our knowledge of the environmental effects on morphological variation. Thanks to recent geographic information system (GIS) techniques, the climate parameters obtained from specimen localities are now easily incorporated into studies in evolutionary biology [24]. Indeed, this approach has been reported in recent articles, suggesting that the general patterns of variation in some phenotypic traits of squamates are associated with climate [22,25–27]. However, it is important to emphasize that none of these studies have focused on the evolution of elongated body forms. The relationships between climate and the evolution of body shape may be predicted by the direct and indirect effects that climatic components likely have on biological traits [28]. For example, in vertebrate ectotherms, temperature and precipitation are directly related to thermoregulation patterns and rates of water loss, which may be strongly dependent on body size and form [29]. Moreover, temperature and precipitation likely determine other environmental characteristics, such as soil texture, substrate coverage and plant primary productivity (and consequent prey availability) in a given habitat [24,26,30]. Together, these environmental factors define some of the selective pressures acting during morphological evolution such that new associations between environment and morphology may be revealed when climatic variation is included in this evolutionary equation.

The associations between variation in body shape and climate are the focus of the present study. Specifically, we tested for correlations between the morphological traits that determine body shape and climatic parameters using gymnophthalmid lizards as a model system. The family Gymnophthalmidae is a good system to test for associations between morphological and climatic variations because it is composed of lineages that represent a gradient of body form ranging from lacertiform to serpentiform shapes [31]. These lizards are broadly distributed and occupy diverse habitats and, thus, are exposed to a wide range of climates [3,31–33]. Moreover, there are robust phylogenetic hypotheses available for this group [31,34], which allows the formal investigation of evolutionary associations. We first tested the hypothesis that body form has evolved to become elongated in Gymnophthalmidae, based on linear morphological traits. This hypothesis was tested using a phylogenetic Principal Component Analysis (PCA) [35], which identified clusters of species based on body shape and elongation. We then tested the hypothesis that the variation in body shape (particularly body elongation) is associated with climatic parameters in gymnophthalmid lizards. This hypothesis was tested using a phylogenetic covariance analysis between the environmental traits and the morphological component that resulted from our first analysis. These complementary approaches are innovative by adding the dimension of climate to the investigation of the evolution of body elongation in Squamata.

Results

The present study had two major goals. We first identified a composite variable clustering the linear morphological traits that likely changed during the evolution of elongated morphologies in Gymnophthalmidae. We retained only one morphological component (hereafter referred to as morphPC) on a phylogenetic PCA, which had eigenvalue equal to 6.24 and explained 78% of the morphological variation (Table 1) of the 45 gymnophthalmid species that are listed in Figure 1. MorphPC presented high positive loadings for trunk length and high negative loads for the

remaining morphological traits (head length, height and width, pelvic girdle height and width and anterior and posterior limb lengths; Table 1). Thus, in Gymnophthalmidae, increases in trunk length were simultaneously coupled with limb shortening and body narrowing, the latter represented by a decreased width and height of the head and the pelvic girdle. The use of morphPC clustered gymnophthalmid species into two groups based on the degree of body elongation: less and more elongated gymnophthalmids (Figure 2). The species classified as more elongated included lineages characterized by extreme limb reduction, such as *Bachia* (Cercosaurinae, [31]), *Scriptosaura*, *Calyptommatus* and *Nothobachia* (Gymnophthalminae, [31]), together with the pentadactyls *Anotosaura vanzolinia* (Cercosaurinae, [31]) and *Heterodactylus imbricatus* (Gymnophthalminae, [31]) and the tetradactyl *Rhachisaurus brachylepis* (Rhachisaurinae, [31]).

The dichotomy that describes gymnophthalmid species as less or more elongated was then used to test for associations between climatic parameters and morphological patterns, which was the second major goal of this study. We are reporting the results for the best fit phylogenetic linear models (lower AICc, Table 2), but results obtained for all the 31 models tested are synthesized in Table S3. Regardless of the morphological categorization, body elongation in Gymnophthalmidae increased with aridity (see Figure 2 and Table 2). Besides body elongation, the use of substrates that offer greater resistance for locomotion, by gymnophthalmid lizards, also increased with aridity (Fig. 3; slope = −0.6037, p<0.001).

Discussion

The present study investigated how the morphological traits that define body shape are correlated in the lizard family Gymnophthalmidae and tested for associations between body elongation and climate. We found that variation in the gymnophthalmid body shape involves concomitant changes in different linear morphological traits. For example, trunk elongation is coupled with limb shortening and a reduction in body diameter, and the gradual variation along this axis may be illustrated by less-elongated morphologies exhibiting shorter trunks and longer limbs. Such a morphological gradient confirms the trends previously reported in Gymnophthalmidae [16,31]. Our analyses also revealed that the variation identified in the gymnophthalmid body shape is associated with climate, with the species from more arid environments being those that are more elongated.

The general patterns of body shape in Gymnophthalmidae coincide with previously identified trends among squamates [9,16], but variation in body diameter seems peculiar in this group: more-elongated gymnophthalmids are characterized by slimmer bodies, a trend not identified in Anguidae [10] and ambiguously described in Scincidae [2,36]. It has been argued that variation in body diameter may be constrained by locomotion, as changes in this trait likely influence the effectiveness of the bending forces produced by limbless forms [36]. The evolutionary equation that explains body elongation in Gymnophthalmidae is, however, more complex and probably includes several elements in addition to biomechanics. In this sense, our study incorporates a new element into this equation: the role of climate in body shape variation.

In Gymnophthalmidae, variations in body shape are associated with aridity, a composed index that incorporates the thermal and hydric components of climate. Aridity is associated with high temperatures and low precipitation, likely affecting several additional environmental features, such as the habitat structure,

Table 1. Variable loadings resulting from a phylogenetic Principal Component Analysis (PCA) performed on the morphometric variables measured in gymnophthalmids.

Morphological variable	morphPC
Trunk length	0.84
Head length	−0.91
Head height	−0.91
Head width	−0.89
Pelvic girdle height	−0.86
Pelvic girdle width	−0.88
Anterior limb length	−0.86
Posterior limb length	−0.91
Eigenvalue/% variation explained	6.24/78%

MorphPC = morphological principal component.

which is defined by the distribution of physical elements along the three-dimensional space in which a given organism lives [37,38]. This feature may influence the evolution of body shape in gymnophthalmids because divergent environments likely impose distinct demands for organismal performance. Regarding animal locomotion, such contrasting habitats as deserts and rainforests are composed of different substrates, which may impose distinct mechanical demands for running [39]. The structural habitats used by gymnophthalmids differ in the resistance imposed for locomotion [20], and the species distributed in arid environments are those that move on substrates with greater resistance (Figure 3). Furthermore, in Gymnophthalmidae, differences in microhabitat use involve morphological specializations in the head shape because species with more compact heads are associated with microhabitats that offer greater resistance to locomotion [20]. Our data suggests that such differences also result in general changes in body shape. The association between morphology and the structural habitat is well established among squamates [18–21,23], but the incorporation of climate allows the identification of evolutionary relationships on a macro-scale that comprises all of the features influenced by climate.

In addition to the biomechanical elements considered for interpretation of the detected association between body elongation and climate, thermoregulatory components must also be contemplated, as they might also modulate the variation of body shape in Gymnophthalmidae. The preferred body temperatures (Tp) in squamates are positively related with body mass but present a negative relationship with precipitation [40]; thus, larger species in high-rainfall regions tend to have higher Tp values in comparison with small-sized species. Species characterized by small body mass have lower thermal inertia [41], and the variation in shape may contribute an additional component to this relationship because body elongation likely affects the dynamics of the thermal exchange between the organism and its surrounding environment. If the maintenance of thermal preference is under selection in Gymnophthalmidae, then less-elongated bodies may be favored in high-rainfall regions. Forthcoming information about the preferred body temperatures of species from regions with different levels of aridity may allow explicit testing of this hypothesis.

The relationship between body elongation and climate has not been explored previously, thus our study introduces new perspectives for understanding how morphological diversification occurs, with a special focus on the role of different environmental

parameters in the evolution of body form. Either directly or indirectly, climate seems to be a key element affecting the evolution of body shape in Gymnophthalmidae, without dismissing the contribution of other biological traits for morphological diversification. Although the role of some of these features may only be evaluated based on the information of behavior and natural history, which are still unavailable for most gymnophthalmids, other explanations for the evolution of body elongation in squamates can already be rejected for Gymnophthalmidae. For example, the elongation-miniaturization hypothesis suggests that the evolution of body elongation results from selective pressures for clutch-size preservation in a miniaturized body [1]; however, despite their small size and regardless of their body shape, clutch size in gymnophthalmids is fixed at two eggs [33,42,43]. A similar scenario is observed in pygopodids, which retain the same clutch size despite the evolution of elongated species from a lacertiform ancestral gecko [11]. Assembling additional ecological data will certainly contribute to the understanding of the complexity that underlies the evolution of body elongation in Gymnophthalmidae. Indeed, the present study is pivotal in the establishment of a connection between morphology and a broader natural component, climate. Moreover, the association between body elongation and aridity introduces new questions about the spatial distribution of morphological variation in squamates. For example, we detected that the more-elongated gymnophthalmids are often associated with arid environments, a pattern identified on a large geographical scale that was not accessible from focusing on microhabitat use. Many subsequent questions can be derived from this pattern, with the most general inquiry being the geographical distribution of elongated squamates along climatic gradients.

Methods

Morphometric data

The present study employed preserved specimens of 45 gymnophthalmid species (Tables S1 and S2) available at three Brazilian herpetological collections: Museu de Zoologia da Universidade de São Paulo (MZUSP), Coleção Herpetológica da Universidade de Brasília (CHUNB) and personal collection of Dr. MTU Rodrigues (University of São Paulo, specimens not yet deposited at MZUSP). The morphometric data consisted of the following nine measurements obtained using digital calipers (to the nearest 0.01 mm): snout-vent length (SVL - distance from the tip of the snout to posterior end of the cloaca); trunk length (TL - distance from the posterior end of the ventral head scales to the posterior end of the cloaca); anterior and posterior limb lengths (ALL and PLL, respectively - distances from the insertion of the fully extended limbs to the tip of the claw of the longest digit); head length (HL - distance from the tip of the snout to the posterior dorsal head scale, which was generally the interparietal); head width (HW - measurement of the widest portion of the head); head height (HH - measurement of the highest portion of the head); and pelvic girdle width and pelvic girdle height (PGW and PGH respectively - both dimensions measured anteriorly to the insertion of the hindlimbs). Only adults were measured in order to minimize eventual effects of ontogeny; however, both males and females were included due to the limitations in the number of individuals for some species. Even so, we expect that if there is any intraspecific differences regarding sexual dimorphism, it would be irrelevant when compared to the interspecific differences considered, thus having little impact on the results. The average sample size per species was 11.1 individuals; for approximately half species considered we measured 15 to 20 individuals; some rare species, however, were represented by few specimens among

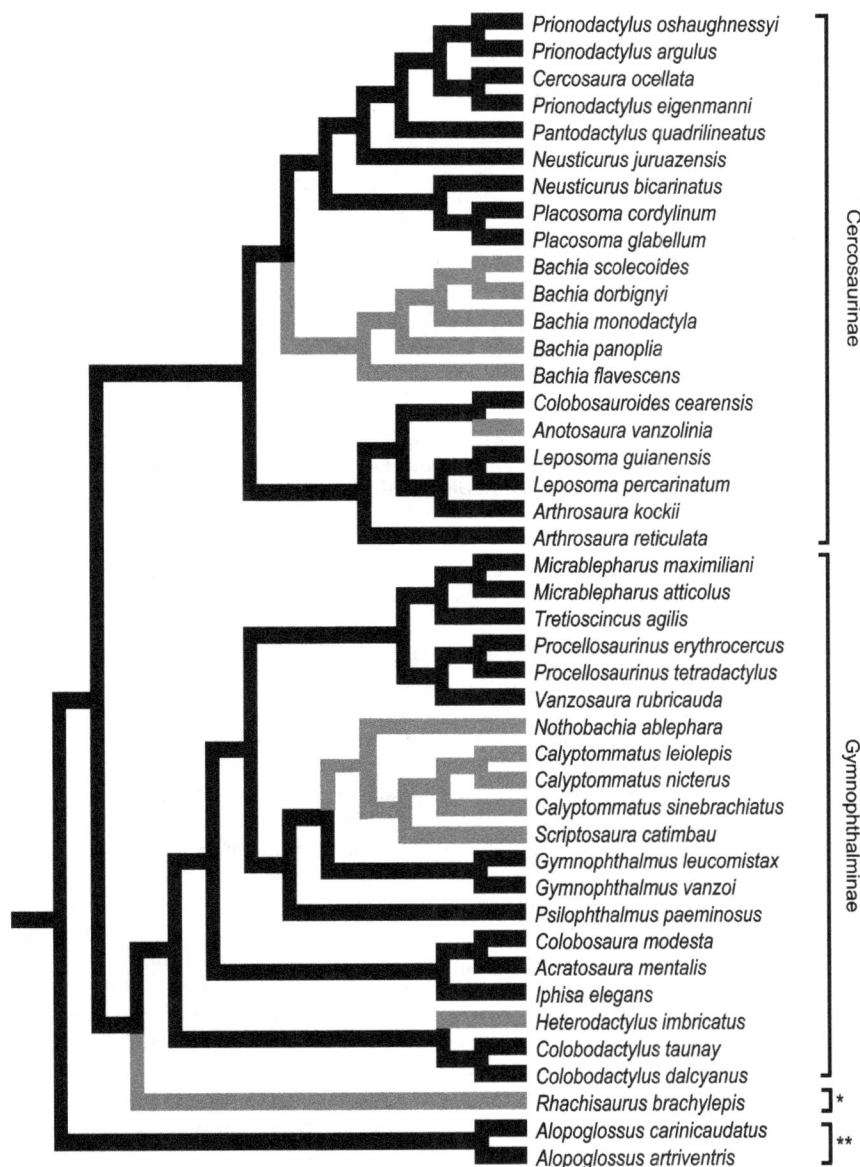

Figure 1. Topology of Gymnophthalmidae used in the phylogenetic analysis. Branch colors represent the elongation groups detected using the morphological component (morphPC): less-elongated species are shown in black and more-elongated species in gray. Asterisks represent the subfamilies Rhachisaurinae (*) and Alopoglossinae (**); taxonomy adopted follows [31].

the collections used (Table S1), which necessitated combining data from different populations for most of the species studied. The proportion of specimens available that exhibited intact tails was small, and restricting our study to individuals with well-preserved tails would have considerably decreased our sample sizes. Therefore, tail measurements were excluded from the analyses. Our study did not involve the capture or manipulation of live animals, and, therefore, there was no need of approval from the ethics committee because the measurements were obtained from fixed specimens belonging to herpetological collections.

Climatic data

The morphometric patterns detected in Gymnophthalmidae were tested for associations with climate, and the occurrence records of the specimens measured were used to extract the

climatic data. We chose the climatic variables that better represented central tendency and variation in the thermal and precipitation regimes that gymnophthalmids are exposed to, including climatic extremes. Climatic elements were extracted from Worldclim data layers [44] (available at http://www.worldclim.org, accessed 2012 Oct 18) using DIVA-GIS [45] version 7.1 and treated as single variables in linear models. These variables were Annual Mean Temperature, Mean Diurnal Range, Maximal Temperature of Warmest Month, Minimal Temperature of Coldest Month, Temperature Seasonality, Annual Precipitation, Precipitation of Wettest Quarter, Precipitation of Driest Quarter, and Precipitation Seasonality. Moreover, two additional variables were extracted from the International Water Management Institute (IWMI) World Water and Climate Atlas (available at http://www.iwmi.cgiar.org, accessed 2012 Oct 18), the Highest

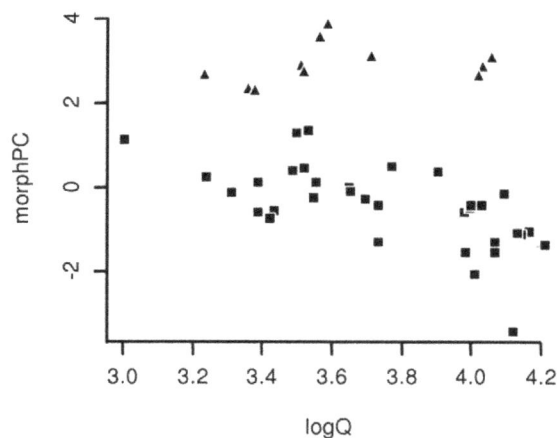

Figure 2. Relationships between aridity index (logQ) and body elongation (given by morphPC) in gymnophthalmid lizards. MorphPC = morphological principal component. Symbols represent elongation groups: triangles correspond to the more-elongated species, and squares indicate less-elongated species.

Monthly Mean Temperature (Tmax) and Lowest Monthly Mean Temperature (Tmin), which were combined with Annual Precipitation (AnnPrec) to calculate a composite variable, named index of aridity (logQ). This index was the same used by Oufiero et al. [25], according to Emberger (cited by [46]). Lower values of logQ correspond to more arid environments and it was calculated according to the following equation: $(Q) = AnnPrec/[(Tmax+Tmin)*(Tmax-Tmin)]*1000$.

The climatic data associated with those species represented by more than one population were averaged among the locations.

Statistical analyses

All of the statistical analyses were conducted using R version 2.14.1 [47] in RStudio (version 0.94.110). Some of these analyses were performed using a phylogenetic framework, which requires the use of a topology representing the relationships among the lineages. We combined two phylogenetic hypotheses available for gymnophthalmids into a single topology: Pellegrino et al. [31], for the overall relationships among the species, and Kohlsdorf & Wagner [48], for the relationships among the *Bachia* species not included in Pellegrino et al. [31]. Some relationships were modified in the topology to accommodate the specificities of our

Table 2. Best linear models testing the effects of aridity index (logQ) and elongation groups (EgroupPC) on the morphological component (morphPC).

Model	Parameter	slope	p	λ
morphPC~ log.Q+EgroupPC	log Q	−1.283	<0. 001*	0.961
	EgroupPC	2.430	<0. 001*	
morphPC~ log.Q×EgroupPC	log Q	−1.452	<0. 001*	0.945
	EgroupPC	−1.300	0.656	
	log.Q×EgroupPC	1.057	0.197	

Significant values (P<0.05) are indicated with an asterisk (*).

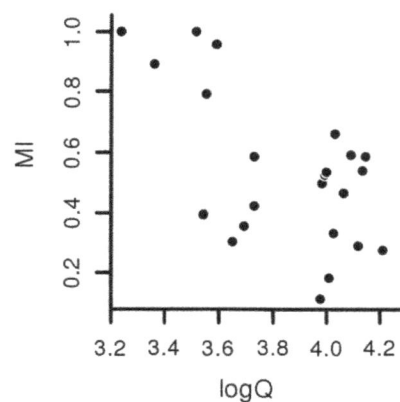

Figure 3. Linear regression between microhabitat index (MI; variable obtained from Barros et al., 2011) and aridity index (logQ) in gymnophthalmid lizards.

dataset. For example, we measured *Leposoma guianensis*, which was not included in any of the phylogenetic hypotheses available; this species was assumed to be a sister group of *L. percarinatum* (instead of *L. oswaldoi* [49]). Moreover, based on Rodrigues & Santos [50], we considered *Scriptosaura catimbau* as a sister group of *Calyptommatus* because this species was recently described and does not appear in any of the formal phylogenetic hypotheses proposed for the group.

The statistical analyses performed in a phylogenetic framework typically incorporate a topology with branch lengths proportional to the expected variance for the evolution of the analyzed traits (reviewed in [51]). Although some estimates of divergence time or genetic distance are available for specific lineages of Gymnophthalmidae [9], their use would require elimination of a significant number of species from our dataset (e.g. *Anotosaura vanzolinia*, *Bachia monodactylus*, *B. panoplia*, *B. scolecoides*, *Colobosaura mentalis*, *Colobosauroides cearensis*, *Gymnophthalmus vanzoi*, *Leposoma guianensis*, *Micrablepharus atticolus*, *Neusticurus bicarinatus*, *N. juruazensis*, *Placosoma cordylinum*, *Prionodactylus oshaughnessyi*, *Scriptosaura catimbau*). Given that the use of arbitrary branch lengths is common and well-supported in the literature using Comparative Methods [20–22,25,51–55], we maintained all species measured in our dataset and performed the statistical analyses using arbitrary branch lengths. The adequacy of four different methods for attributing branch lengths was tested following the diagnostics proposed by Garland et al. [52]: 1) All Equal One, 2) Pagel [53], 3) Grafen [54] and 4) Nee (cited in [55]). These diagnostics consist in plotting the absolute value of each standardized independent contrast versus the square root of the sum of its branch lengths, which represent its standard deviation. The method of Grafen appeared to be the most adequate, as indicated by the absence of statistically significant trends in all diagnostic plots produced using these arbitrary branch lengths. The topologies and branch lengths and the diagnostic plots of independent contrasts were built and inspected using Mesquite v2.74 [56] with the PDAP:PDTREE v1.15 [57] module for Mac OSX. The topology adopted was used to phylogenetically size-correct the morphometric variables, following Revell [58], prior to the statistical analyses.

The morphological patterns were investigated considering two main questions: 1) how the variation in linear traits defines the body shape and 2) how the body shape relates to climate. The first question was evaluated using a phylogenetic principal component analysis (PCA; [58]) implemented using the R package phytools [59]. We applied the Kaiser-Guttman criterion and retained the principal components with eigenvalue higher than 1.0 [60]. Our second question investigated how the body shape in Gymnophthalmidae relates to climate, which was tested using phylogenetic linear models. Traditionally, randomization tests of phylogenetic signals are used in such frameworks, but we chose to examine the phylogenetic regression via generalized least squares with the simultaneous estimation of the phylogenetic signal (Pagel's λ, PGLSλ; [35]). The PGLSλ model is similar to a Brownian motion model of evolution but allows for the transformation of branch lengths as it estimates the parameter λ. This parameter (λ) represents the amount of phylogenetic signal in the regression residuals: it ranges from zero to 1.0, with a value closer to zero equivalent to a star phylogeny in which no phylogenetic signal is detectable, and a value closer to 1.0 comparable to a hierarchical phylogeny defined by Brownian motion [61].

The relationship between body elongation and climate was tested using our morphological descriptor resulting from the phylogenetic PCA (morphPC), which was regressed against each of the climatic variables (all single variables extracted from Worldclim and the composite variable logQ) using PGLSλ models. Because body elongation clearly divided the species in two distinct groups (Figure 2), we introduced a new variable: the elongation group (more versus less elongated, EgroupPC). Therefore, we tested a total of 31 models relating body elongation and climate (Table S3) and used AIC as a heuristic indicator of model support based on the likelihood. The models tested followed three main categories: 1) linear models testing for the relationship between morphPC and climatic variables, 2) covariation models relating morphPC with climatic variables and EgroupPC, and 3) the same covariation models with the inclusion of the interaction effects between climatic variables and EgroupPC. We followed Burnham & Anderson [62] and also computed the AICc, which is the AIC corrected for sample size. Smaller AICcs indicate those models with the best fit, and those models within 2 units of the best model are considered to have substantial support [62].

Possible associations between logQ and soil characteristics in which gymnophthalmid species occur were also investigated by regressing a microhabitat index (MI) against logQ. The microhabitat index was calculated by Barros et al. [20] for the same gymnophthalmid species we studied. It considers the proportion of each microhabitat used by a given species together with the force of resistance to displacement that was empirically estimated for each substrate. Thus, MI is proportional to the resistance imposed for locomotion by the substrates gymnophthalmids use. It ranges from 0 to 1, and higher values of MI represent frequent use of microhabitats that offer greatest resistance to displacement. It is important to note that most of the museum specimens measured in the present study were the same used by Barros et al. [20]; even when there were inconsistencies among the populations sampled between the two studies, collection sites for each species were always located in the same Brazilian federative unit.

Supporting Information

Table S1 Morphological traits (means ± standard errors, all in mm) and scores of morphological principal component for lizards of the family Gymnophthalmidae.

Table S2 List of specimens of Gymnophthalmidae examined.

Table S3 Comparisons of linear models testing the effects of climate variables on morphological component (morphPC) with interaction of elongation groups (EgroupPC).

Acknowledgments

We acknowledge the following researchers for access to their herpetological collections: Dr. Hussam Zaher (and technical support of Carolina S Castro-Mello) at the Museu de Zoologia da Universidade de São Paulo (MZUSP), Dr. Guarino Colli at the Coleção Herpetológica da Universidade de Brasília (CHUNB), and Dr. Miguel T. Rodrigues for access to the specimens from his personal collection. We also thank the members from the Laboratory of Evolution and Eco-physiology of Tetrapods (FFCLRP-USP), particularly FC Barros, for helpful discussions regarding the early versions of this article.

Author Contributions

Conceived and designed the experiments: MBG RB TK. Performed the experiments: MBG RB. Analyzed the data: MBG RB. Contributed reagents/materials/analysis tools: RB TK. Wrote the paper: MBG RB TK.

References

1. Griffith H (1990) Miniaturization and elongation in *Eumeces* (Sauria: Scincidae). Copeia 3: 751–758.
2. Gans C (1975) Tetrapod limblessness: evolution and functional corollaries. American Zoologist 15: 455–467.
3. Presch W (1980) Evolutionary history of the South American microteiid lizards (Teiidae: Gymnophthalminae). Copeia: 36–56.
4. Ward AB, Azizi E (2004) Convergent evolution of the head retraction escape response in elongate fishes and amphibians. Zoology 107: 205–217. doi:10.1016/j.zool.2004.04.003.
5. Ward AB, Brainerd EL (2007) Evolution of axial patterning in elongate fishes. Biological Journal of the Linnean Society 90: 97–116. doi:10.1111/j.1095-8312.2007.00714.x.
6. Yamada T, Sugiyama T, Tamaki N, Kawakita A, Kato M (2009) Adaptive radiation of gobies in the interstitial habitats of gravel beaches accompanied by body elongation and excessive vertebral segmentation. BMC Evolutionary Biology 9: 145. doi:10.1186/1471-2148-9-145.
7. Parra-Olea G, Wake DB (2001) Extreme morphological and ecological homoplasy in tropical salamanders. Proceedings of the National Academy of Sciences of the United States of America 98: 7888–7891. doi:10.1073/pnas.131203598.
8. Renous S, Gasc JP (1989) Body and vertebral proportions in Gymnophiona (Amphibia): diversity of morphological types. Copeia 4: 837–847.
9. Wiens JJ, Brandley MC, Reeder TW (2006) Why does a trait evolve multiple times within a clade? Repeated evolution of snakelike body form in squamate reptiles. Evolution 60: 123–141.

10. Wiens J, Slingluff JL (2001) How lizards turn into snakes: a phylogenetic analysis of body-form evolution in anguid lizards. Evolution 55: 2303–2318.
11. Shine R (1986) Evolutionary advantages of limblessness: evidence from the pygopodid lizards. Copeia 2: 525–529.
12. Greer AE, Wadsworth L (2003) Body shape in skinks: the relationship between relative hind limb length and relative snout-vent length. Journal of Herpetology 37: 554–559.
13. Narita Y, Kuratani S (2005) Evolution of the vertebral formulae in mammals: a perspective on developmental constraints. Journal of Experimental Zoology Part B, Molecular and Developmental Evolution 304: 91–106. doi:10.1002/jez.b.21029.
14. Buchholtz EA (2007) Modular evolution of the Cetacean vertebral column. Evolution & Development 9: 278–289.
15. Lande R (1978) Evolutionary mechanisms of limb loss in tetrapods. Evolution 32: 73–92. doi:10.2307/2407411.
16. Brandley MC, Huelsenbeck JP, Wiens JJ (2008) Rates and patterns in the evolution of snake-like body form in squamate reptiles: evidence for repeated re-evolution of lost digits and long-term persistence of intermediate body forms. Evolution 62: 2042–2064. doi:10.1111/j.1558-5646.2008.00430.x.
17. Lee M (1998) Convergent evolution and character correlation in burrowing reptiles: towards a resolution of squamate relationships. Biological Journal of the Linnean Society 65: 369–453. doi:10.1006/bijl.1998.0256.
18. Warheit KI, Forman JD, Losos JB, Miles DB (2008) Morphological diversification and adaptive radiation: a comparison of two diverse lizard clades. Evolution 53: 1226–1234.

19. Irschick DJ, Vitt LJ, Zani PA, Losos JB (1997) A comparison of evolutionary radiations in mainland and Caribbean *Anolis* lizards. Ecology 78: 2191–2203.

20. Barros FC, Herrel A, Kohlsdorf T (2011) Head shape evolution in Gymnophthalmidae: does habitat use constrain the evolution of cranial design in fossorial lizards? Journal of Evolutionary Biology 24: 2423–2433. doi:10.1111/j.1420-9101.2011.02372.x.

21. Grizante MB, Navas CA, Garland T, Kohlsdorf T (2010) Morphological evolution in Tropidurinae squamates: an integrated view along a continuum of ecological settings. Journal of Evolutionary Biology 23: 98–111. doi:10.1111/j.1420-9101.2009.01868.x.

22. Brandt R, Navas CA (2011) Life-history evolution on Tropidurinae lizards: influence of lineage, body size and climate. PLoS ONE 6: e20040. doi:10.1371/journal.pone.0020040.

23. Williams EE (1983) Ecomorphs, faunas, island size, and diverse end points in island radiations of *Anolis*. In: Huey RB, Pianka ER, Schoener TW, editors. Lizard Ecology: studies of a model organism. Harvard University Press. 512 p.

24. Kozak KH, Graham CH, Wiens JJ (2008) Integrating GIS-based environmental data into evolutionary biology. Trends in Ecology & Evolution 23: 141–148. doi:10.1016/j.tree.2008.02.001.

25. Oufiero CE, Gartner GEA, Adolph SC, Garland T (2011) Latitudinal and climatic variation in body size and dorsal scale counts in *Sceloporus* lizards: a phylogenetic perspective. Evolution 65: 3590–3607. doi:10.1111/j.1558-5646.2011.01405.x.

26. Calsbeek R, Knouft JH, Smith TB (2006) Variation in scale numbers is consistent with ecologically based natural selection acting within and between lizard species. Evolutionary Ecology 20: 377–394. doi:10.1007/s10682-006-0007-y.

27. Luxbacher AM, Knouft JH (2009) Assessing concurrent patterns of environmental niche and morphological evolution among species of horned lizards (*Phrynosoma*). Journal of Evolutionary Biology 22: 1669–1678. doi:10.1111/j.1420-9101.2009.01779.x.

28. Dunson WA, Travis J (1991) The role of abiotic factors in community organization. The American Naturalist 138: 1067–1091.

29. Schmidt-Nielsen K (1989) Scaling: why is animal size is so important? New York: Cambridge University Press. 256 p.

30. Dunham AE (1978) Food availability as a proximate factor influencing individual growth rates in the iguanid lizard *Sceloporus merriami*. Ecology 59: 770–778.

31. Pellegrino K, Rodrigues MT, Yonenaga-Yassuda Y, Sites Jr JW (2001) A molecular perspective on the evolution of microteiid lizards (Squamata, Gymnophthalmidae), and a new classification for the family. Biological Journal of the Linnean Society 74: 315–338. doi:10.1006/bijl.2001.0580.

32. Rodrigues MT (1996) Lizards, snakes, and amphisbaenians from the quaternary sand dunes of the middle rio São Francisco, Bahia, Brazil. Journal of Herpetology 30: 513–523.

33. Avila-Pires TCS (1995) Lizards of Brazilian Amazonia (Reptilia: Squamata). Zoologische Verhandelingen 299: 1–706.

34. Castoe T, Doan T, Parkinson C (2004) Data partitions and complex models in Bayesian analysis: the phylogeny of gymnophthalmid lizards. Systematic Biology 53: 448–469. doi:10.1080/10635150490445797.

35. Revell LJ (2010) Phylogenetic signal and linear regression on species data. Methods in Ecology and Evolution 1: 319–329. doi:10.1111/j.2041-210X.2010.00044.x.

36. Skinner A, Lee MSY (2009) Body-form evolution in the scincid lizard clade *Lerista* and the mode of macroevolutionary transitions. Evolutionary Biology 36: 292–300. doi:10.1007/s11692-009-9064-9.

37. Moermond TC. (1979) Habitat constraints on the behavior, morphology, and community structure of *Anolis* lizards. Ecology 60: 152–164.

38. Pounds JA (1988) Ecomorphology, locomotion, and microhabitat structure: patterns in a tropical mainland *Anolis* community. Ecological Monographs 58: 299–320.

39. Lejeune TM, Willems PA, Heglund NC (1998) Mechanics and energetics of human locomotion on sand. The Journal of Experimental Biology 201: 2071–2080.

40. Clusella-trullas S, Blackburn TM, Chown SL (2011) Climatic predictors of temperature performance curve parameters in ectotherms imply complex responses to climate change. The American Naturalist 177: 738–751. doi:10.1086/660021.

41. Bartholomew GA, Tucker VA (1964) Size, body temperature, thermal conductance, oxygen consumption, and heart rate in Australian varanid lizards. Physiological Zoology 37: 341–354.

42. Sherbrooke WC. (1975) Reproductive cycle of a tropical teiid lizard, *Neusticurus ecpleopus* Cope, in Peru. Biotropica 7: 194–207.

43. Vitt LJ. (1982) Sexual dimorphism and reproduction in the microteiid lizard, *Gymnophthalmus multiscutatus*. Journal of Herpetology 16: 325–329.

44. Hijmans RJ, Cameron SE, Parra JL, Jones PG, Jarvis A (2005) Very high resolution interpolated climate surfaces for global land areas. International Journal of Climatology 25: 1965–1978. doi:10.1002/joc.1276.

45. Hijmans R, Guarino L, Cruz M (2001) Computer tools for spatial analysis of plant genetic resources data: 1. DIVA-GIS. Plant Genetic Resources 127: 15–19.

46. Tieleman BI, Williams JB, Bloomer P (2003) Adaptation of metabolism and evaporative water loss along an aridity gradient. Proceedings of the Royal Society B: Biological Sciences 270: 207–214. doi:10.1098/rspb.2002.2205.

47. R Development Core Team (2011) R: A language and environment for statistical computing. Available: http://www.r-project.org/. Accessed 2012 Oct 18.

48. Kohlsdorf T, Wagner GP (2006) Evidence for the reversibility of digit loss: a phylogenetic study of limb evolution in *Bachia* (Gymnophthalmidae: Squamata). Evolution 60: 1896–1912.

49. Pellegrino K, Rodrigues M (1999) Chromosomal evolution in the Brazilian lizards of genus *Leposoma* (Squamata, Gymnophthalmidae) from Amazon and Atlantic rain forests: banding patterns and FISH of telomeric sequences. Hereditas 21: 15–21.

50. Rodrigues M, Santos E (2008) A new genus and species of eyelid-less and limb reduced gymnophthalmid lizard from northeastern Brazil (Squamata, Gymnophthalmidae). Zootaxa 60: 50–60.

51. Garland T, Bennett AF, Rezende EL (2005) Phylogenetic approaches in comparative physiology. The Journal of Experimental Biology 208: 3015–3035. doi:10.1242/jeb.01745.

52. Garland Jr T, Harvey H, Ives R (1992) Procedures for the analysis of comparative data using phylogenetically independent contrasts. Systematic Biology 41: 18–32.

53. Pagel MD (1992) A method for the analysis of comparative data. Journal of Theoretical Biology 156: 431–442. doi:10.1016/S0022-5193(05)80637-X.

54. Grafen A (1989) The phylogenetic regression. Philosophical Transactions of the Royal Society of London Series B, Biological Sciences 326: 119–157.

55. Purvis A (1995) A composite estimate of primate phylogeny. Philosophical Transactions of the Royal Society of London Series B, Biological sciences 348: 405–421. doi:10.1098/rstb.1995.0078.

56. Maddison WP, Maddison DR (2011) Mesquite: a modular system for evolutionary analysis. Version 2.75 http://mesquiteproject.org. Accessed 2012 Oct 18.

57. Midford PE, Garland Jr T, Maddison WP (2010) PDAP:PDTREE: A translation of the PDTREE application of Garland et al.'s Phenotypic Diversity Analysis Programs. Available: http://mesquiteproject.org/pdap_mesquite/index.html. Accessed 2012 Oct 18.

58. Revell LJ (2009) Size-correction and principal components for interspecific comparative studies. Evolution 63: 3258–3268. doi:10.1111/j.1558-5646.2009.00804.x.

59. Revell LJ (2011) phytools: an R package for phylogenetic comparative biology (and other things). Methods in Ecology and Evolution 3: 217–223. doi:10.1111/j.2041-210X.2011.00169.x.

60. Jackson DA (1993) Stopping rules in principal components analysis: a comparison of heuristical and statistical approaches. Ecology 74: 2204–2214.

61. Pagel M (1999) Inferring the historical patterns of biological evolution. Nature 401: 877–884. doi:10.1038/44766.

62. Burnham KP, Anderson DR (2002) Model selection and multimodel inference: A practical information-theoretic approach. New York: Springer. 496 p.

Permissions

The contributors of this book come from diverse backgrounds, making this book a truly international effort. This book will bring forth new frontiers with its revolutionizing research information and detailed analysis of the nascent developments around the world.

We would like to thank all the contributing authors for lending their expertise to make the book truly unique. They have played a crucial role in the development of this book. Without their invaluable contributions this book wouldn't have been possible. They have made vital efforts to compile up to date information on the varied aspects of this subject to make this book a valuable addition to the collection of many professionals and students.

This book was conceptualized with the vision of imparting up-to-date information and advanced data in this field. To ensure the same, a matchless editorial board was set up. Every individual on the board went through rigorous rounds of assessment to prove their worth. After which they invested a large part of their time researching and compiling the most relevant data for our readers.

The editorial board has been involved in producing this book since its inception. They have spent rigorous hours researching and exploring the diverse topics which have resulted in the successful publishing of this book. They have passed on their knowledge of decades through this book. To expedite this challenging task, the publisher supported the team at every step. A small team of assistant editors was also appointed to further simplify the editing procedure and attain best results for the readers.

Apart from the editorial board, the designing team has also invested a significant amount of their time in understanding the subject and creating the most relevant covers. They scrutinized every image to scout for the most suitable representation of the subject and create an appropriate cover for the book.

The publishing team has been an ardent support to the editorial, designing and production team. Their endless efforts to recruit the best for this project, has resulted in the accomplishment of this book. They are a veteran in the field of academics and their pool of knowledge is as vast as their experience in printing. Their expertise and guidance has proved useful at every step. Their uncompromising quality standards have made this book an exceptional effort. Their encouragement from time to time has been an inspiration for everyone.

The publisher and the editorial board hope that this book will prove to be a valuable piece of knowledge for researchers, students, practitioners and scholars across the globe.

List of Contributors

Simone G. Shamay-Tsoory and Dorin Ahronberg-Kirschenbaum
Department of Psychology, University of Haifa, Haifa, Israel

Nirit Bauminger-Zviely
School of Education, Bar Ilan University of Haifa, Haifa, Israel

Thomas van de Kamp
ANKA/Institute for Photon Science and Synchrotron Radiation, Karlsruhe Institute of Technology (KIT), Eggenstein-Leopoldshafen, Germany,
State Museum of Natural History (SMNK), Karlsruhe, Germany

Tomy dos Santos Rolo, Patrik Vagovič and Tilo Baumbach
ANKA/Institute for Photon Science and Synchrotron Radiation, Karlsruhe Institute of Technology (KIT), Eggenstein-Leopoldshafen, Germany

Alexander Riedel
State Museum of Natural History (SMNK), Karlsruhe, Germany

Sylvain G. Razafimandimbison and Birgitta Bremer
Bergius Foundation, The Royal Swedish Academy of Sciences and Botany Department, Stockholm University, Stockholm, Sweden

Stefan Ekman
Museum of Evolution, Uppsala University, Uppsala, Sweden

Timothy D. McDowell
Department of Biological Sciences, East Tennessee State University, Johnson City, Tennessee, United States of America

Naoki Morimoto
Laboratory of Physical Anthropology, Graduate School of Science, Kyoto University, Kyoto, Japan

Marcia S. Ponce de León and Christoph P. E. Zollikofer
Anthropological Institute, University of Zurich, Zurich, Switzerland

Max Bügler
Chair of Computational Modeling and Simulation, Technische Universität München, Munich, Germany

Roei Shacham
Evolutionary and Environmental Biology, University of Haifa, Haifa, Israel

Polychronis Rempoulakis and Tamar Keasar
Biology and Environment, University of Haifa, Tivon, Israel

Frank Thuijsman
Knowledge Engineering, Maastricht University, Maastricht, The Netherlands

Helder Gomes Rodrigues, Floréal Solé, Cyril Charles and Laurent Viriot
Team ''Evo-Devo of Vertebrate Dentition'', Institut de Génomique Fonctionnelle de Lyon, UnitéMixte de Recherche 5242 Centre National de la Recherche Scientifique, Ecole Normale Supérieure de Lyon, Université Claude Bernard Lyon 1, Lyon, France

Paul Tafforeau
European Synchrotron Radiation Facility, Grenoble, France

Monique Vianey-Liaud
Laboratoire de Paléontologie, Institut des Sciences de l'Évolution de Montpellier, UnitéMixte de Recherche 5554 Centre National de la Recherche Scientifique, Université Montpellier 2, Montpellier, France

Lars Vogt, Björn Quast and Thomas Bartolomaeus
Institut für Evolutionsbiologie und Ökologie, Universität Bonn, Bonn, Germany

Peter Grobe
Forschungsmuseum Alexander Koenig Bonn, Bonn, Germany

Ana E. Escalante
Department of Ecology, Evolution and Behavior, University of Minnesota, St. Paul, Minnesota, United States of America
Departamento de Ecología de la Biodiversidad, Instituto de Ecología, Universidad Nacional Autónoma de México, Mexico City, México

Sumiko Inouye
Department of Biochemistry, Robert Wood Johnson Medical School, Piscataway, New Jersey, United States of America

Michael Travisano
Department of Ecology, Evolution and Behavior, University of Minnesota, St. Paul, Minnesota, United States of America
Biotechnology Institute, University of Minnesota, St. Paul, Minnesota, United States of America

Zhijie Jack Tseng
Department of Biological Sciences, University of Southern California, Los Angeles, California, United States of America
Department of Vertebrate Paleontology, Natural History Museum of Los Angeles County, Los Angeles, California, United States of America

Sylvain Gerber
Department of Biology and Biochemistry, University of Bath, Bath, England

Tony Gamble
Department of Genetics, Cell Biology and Development, University of Minnesota, Minneapolis, Minnesota, United States of America
Bell Museum of Natural History, University of Minnesota, St. Paul, Minnesota, United States of America

Eli Greenbaum, Todd R. JackmanAaron M. Bauer
Department of Biology, Villanova University, Villanova, Pennsylvania, United States of America

Anthony P. Russell
Department of Biological Sciences, University Department of Calgary, Calgary, Canada

Madeleine B. Chollet, Nicole Pangborn and Valerie B. DeLeon
Center for Functional Anatomy and Evolution, Johns Hopkins University School of Medicine, Baltimore, Maryland, United States of America

Kristina Aldridge
Department of Pathology and Anatomical Sciences, University of Missouri School of Medicine, Columbia, Missouri, United States of America

Seth M. Weinberg
Center for Craniofacial and Dental Genetics, University of Pittsburgh School of Dental Medicine, Pittsburgh, Pennsylvania, United States of America

Hugo J. Parker and Greg Elgar
Division of Systems Biology, Medical Research Council National Institute for Medical Research, London, United Kingdom

Tatjana Sauka-Spengler and Marianne Bronner
Division of Biology, California Institute of Technology, Pasadena, California, United States of America

Andrew Brittain, Elizabeth Stroebele and Albert Erives
Department of Biology, University of Iowa, Iowa City, Iowa, United States of America

Mathieu Molet, Vincent Maicher and Christian Peeters
Laboratoire Ecologie & Evolution – Unité Mixte de Recherche 7625, Université Pierre et Marie Curie, Paris, France
Laboratoire Ecologie & Evolution – Unité Mixte de Recherche 7625, Centre National de la Recherche Scientifique, Paris, France

Daniel de Santos-Sierra and David Papo
Center for Biomedical Technology, Universidad Politécnica de Madrid, Pozuelo de Alarcón, Madrid, Spain

Irene Sendiña-Nadal, Inmaculada Leyva and Juan A. Almendral
Center for Biomedical Technology, Universidad Politécnica de Madrid, Pozuelo de Alarcón, Madrid, Spain
Complex Systems Group, Universidad Rey Juan Carlos, Móstoles, Madrid, Spain

Sarit Anava and Amir Ayali
Department of Zoology, Tel-Aviv University, Tel Aviv, Israel

Stefano Boccaletti
Istituto dei Sistemi Complessi, Consiglio Nazionale delle Ricerche, Sesto Fiorentino, Florence, Italy,
Istituto Nazionale di Fisica Nucleare, Sesto Fiorentino, Florence, Italy

Steffen Schaper and Ard A. Louis
Rudolf Peierls Centre for Theoretical Physics, University of Oxford, Oxford, United Kingdom

Audrey Chaput-Bardy
Muséum National d'Histoire Naturelle, UMR 7205 Institut Systématique Evolution Biodiversité , Paris, France
INRA, Equipe Ecotoxicologie et Qualité des Milieux Aquatiques, UMR 985 Ecologie et Santé des Ecosystèmes, INRA-Agrocampus, Rennes, France

Simon Ducatez
Muséum National d'Histoire Naturelle, UMR 7205 Institut Systématique Evolution Biodiversité , Paris, France

Department of Biology, McGill University, Montreal, Quebec, Canada

Delphine Legrand and Michel Baguette
Station d'Ecologie Expérimentale du CNRS à Moulis, CNRS USR 2936, Moulis, France

Sigrún Reynisdóttir
Faculty of Life and Environmental Sciences, University of Iceland, Reykjavik, Iceland
Institute of Biology, University of Iceland, Reykjavik, Iceland

Arnar Palsson
Faculty of Life and Environmental Sciences, University of Iceland, Reykjavik, Iceland
Institute of Biology, University of Iceland, Reykjavik, Iceland
Biomedical Center, University of Iceland, Reykjavik, Iceland
Department of Ecology and Evolution, University of Chicago, Chicago, Illinois, United States of America

Natalia Wesolowska
Department of Ecology and Evolution, University of Chicago, Chicago, Illinois, United States of America
Cell Biology and Biophysics Unit, European Molecular Biology Laboratory (EMBL), Heidelberg, Germany

Michael Z. Ludwig and Martin Kreitman
Department of Ecology and Evolution, University of Chicago, Chicago, Illinois, United States of America

Hernán López-Fernández
Program in Ecology and Evolutionary Biology, and Department of Wildlife and Fisheries Sciences, Texas A&M University, College Station, Texas, United States of America
Department of Ecology and Evolutionary Biology, University of Toronto, Toronto, Ontario, Canada

Jessica Arbour
Department of Ecology and Evolutionary Biology, University of Toronto, Toronto, Ontario, Canada

Stuart Willis, Crystal Watkins, Rodney L. Honeycutt and Kirk O. Winemiller
Program in Ecology and Evolutionary Biology, and Department of Wildlife and Fisheries Sciences, Texas A&M University, College Station, Texas, United States of America

Edith M. Myers
Department of Biological and Allied Health Sciences, Fairleigh Dickinson University, College at Florham, Madison, New Jersey, United States of America

Shmuel Raz
Department of Evolutionary and Environmental Biology, Institute of Evolution, University of Haifa, Haifa, Israel
Rowland Institute at Harvard, Harvard University, Cambridge, Massachusetts, United States of America

Nathan P. Schwartz and John H. Graham
Department of Biology, Berry College, Mount Berry, Georgia, United States of America

Hendrik K. Mienis
The Steinhardt National Collections of Natural History, Tel Aviv University, Tel Aviv, Israel

Eviatar Nevo
Department of Evolutionary and Environmental Biology, Institute of Evolution, University of Haifa, Haifa, Israel

Mariana B. Grizante, Renata Brandt and Tiana Kohlsdorf
Department of Biology, Faculdade de Filosofia Ciências e Letras de Ribeirão Preto, Universidade de São Paulo, Ribeirão Preto, São Paulo, Brazil

Index